THE REAL GOODS

INDEPENDENT BUILDER

THE REAL GOODS INDEPENDENT LIVING BOOKS

Paul Gipe, *Wind Power for Home & Business: Renewable Energy for the 1990s and Beyond*

Michael Potts, *The Independent Home: Living Well with Power from the Sun, Wind, and Water*

Gene Logsdon, *The Contrary Farmer*

Edward Harland, *Eco-Renovation: The Ecological Home Improvement Guide*

Leandre Poisson and Gretchen Vogel Poisson, *Solar Gardening: Growing Vegetables Year-Round the American Intensive Way*

Real Goods Solar Living Sourcebook: The Complete Guide to Renewable Energy Technologies and Sustainable Living, Ninth Edition, edited by John Schaeffer

Athena Swentzell Steen, Bill Steen, and David Bainbridge, with David Eisenberg, *The Straw Bale House*

Nancy Cole and P.J. Skerrett, Union of Concerned Scientists, *Renewables are Ready: People Creating Renewable Energy Solutions*

David Easton, *The Rammed Earth House*

Sam Clark, *The Real Goods Independent Builder: Designing & Building a House Your Own Way*

Real Goods Trading Company in Ukiah, California, was founded in 1978 to make available new tools to help people live self-sufficiently and sustainably. Through seasonal catalogs, a periodical (*The Real Goods News*), a bi-annual *Solar Living Sourcebook*, as well as retail outlets, Real Goods provides a broad range of tools for independent living.

"Knowledge is our most important product" is the Real Goods motto. To further its mission, Real Goods has joined with Chelsea Green Publishing Company to co-create and co-publish the Real Goods Independent Living Book series. The titles in this series are written by pioneering individuals who have firsthand experience in using innovative technology to live lightly on the planet. Chelsea Green books are both practical and inspirational, and they enlarge our view of what is possible as we enter the next millennium.

Ian Baldwin, Jr.
President, Chelsea Green

John Schaeffer
President, Real Goods

THE REAL GOODS INDEPENDENT BUILDER

Designing & Building a House Your Own Way

SAM CLARK

CHELSEA GREEN PUBLISHING
WHITE RIVER JUNCTION, VERMONT

I've worked with and learned from many people. In the photo, I'm the kid in back, and the man sitting on the railing is my grandfather, Sam Sausser, my first and best building teacher. Much of what is presented in this book I learned working and talking with David Scheckman, David Palmer, and Barney Carlson, my partners in Iron Bridge Woodworkers.

Lou Host-Jablonski, Thorsten Horton, Jim Schley, Roger Mason, and Eli Kramer took time off to show me what owner-builders are doing in different parts of the country.

It's great working with Chelsea Green Publishing, particularly my closest collaborators, Jim Schley, Ann Aspell, Stephen Morris, and Alice Blackmer.

To Hana

Printed in the United States of America.
00 99 98 97 96 1 2 3 4 5

First printing August, 1996

About your safety: Home building is inherently dangerous. From accidents with power tools to falls from ladders, scaffolding, and roofs, builders risk serious injury and even death. I try to promote safe work habits throughout this book, but what is safe for one builder under certain circumstances may not be safe for you under different circumstances. The author and the publisher accept no liability for personal injury, property damage, or loss from actions inspired by this book. Consult the building codes and professionals when learning new processes, and please be careful.

Library of Congress Cataloging-in-Publication Data
Clark, Sam.
The real goods independent builder : designing & building a house your own way / Sam Clark.
p. cm. — (The real goods independent living book)
Includes index.
Rev. ed. of: Designing & building your own house your own way. 1978.
ISBN 0-930031-85-7 (alk. paper)
1. Dwellings—Design and construction. I. Clark, Sam. Designing & building your own house your own way. II. Title. III. Series.
TH4815.C5524 1996
690'.837—dc20 96-32231

Chelsea Green Publishing Company
Post Office Box 428
White River Junction, VT 05001

CONTENTS

Introduction 1

Part 1

NEW APPROACHES TO DESIGN

1 Ergonomics & Accessibility 13
2 Pattern Languages 26
3 Sustainable Design: A Builder's Approach 30

Part 2

DESIGN & PLANNING

4 Resources, Goals, & Priorities 41
5 Choosing a House Site 49
6 Preliminary Design Work 56
7 Making Scale Drawings & Models 72
8 Layout 78
9 Elevations, Sections, & Models 102
10 Kitchen & Bath Design 112

Part 3

BUSINESS

11 Hiring Builders & Designers 133
12 Making Good Contracts 139
13 Estimating 146

Part 4

BUILDING BASICS

14 Wood 165
15 The Structure of a House 181
16 Strength of Timbers 199
17 Heat 211
18 Heat Loss Calculations & Sun Paths 220

19 Summer Cooling & Ventilation 233
20 Plumbing & Water Systems 242

Part 5
CONSTRUCTION METHODS

21 Building in the Right Order 259
22 Post-and-Beam Construction 262
23 Foundations 272
24 Floors 303
25 Walls 319
26 Roofs 336
27 Roofing & Flashing 357
28 Windows & Shutters 375
29 Doors 393
30 Siding 400
31 Insulation & Vapor Barrier 414
32 Finish Work 420
33 Stairs 448
34 Scaffolding 462
35 Using Tools 466

APPENDICES

A. F and E Values for Timber Calculations 483
B. Joist, Rafter, Header, Sill, & Column Size Tables 490
C. Bearing Capacity of Soils & Bedrock 498
D. Frost Lines 500
E. Snow Loads 503
F. Weights of Various Materials 504
G. Nails & Screws 505
H. Sound Isolation 508

Index 513

INTRODUCTION

In 1975 Helen and Jules Rabin hired my friend David Palmer and me to work mornings with them as they built a house for themselves and two young daughters.

At crucial points we had a frustrating time finding answers to building questions that should not have been obscure, such as what size timber is needed for a floor beam, or the insulation value of some material. We had books describing how to build an absolutely conventional tract house, and others advocating various innovative building systems. But for the person interested in both designing and building, the underlying basics were hard to locate. We often had to rely on the advice of people more experienced (but not necessarily wiser) than ourselves, on intuition, or, as a last resort, on analogy: If the neighbor's barn roof is held up with 2 x 8s, and our house is a little wider, let's use 2 x 10s.

Figure I-1. Framing Helen and Jules Rabin's house, 1974. Design, Helen Rabin and John Mallery; built by Helen and Jules Rabin, with Sam Clark and David Palmer. (Photo: Jules Rabin.)

I spent much of the next year assembling from many sources all the basic design and construction information I wished we'd had. I put it together for myself, and for other builders and owner-builders. That was the 1978 edition of this book.

Now twenty years later, we're getting ready to revise the Rabin's house, to accommodate such things as grandchildren, central heat, and bathrooms on the first floor. So, we thought it's also a good time for a revision of this book.

The original edition provided basic information on design and construction, plus an introduction to some non-standard methods of building. Though written for owner-builders, the first edition of this book has been an valuable tool for builders, particularly the chapters and appendices that described how to size timbers, spec native lumber, estimate bearing capacity of soils, and do heat loss calculations.

This new edition, like the old one, provides more basic design tools in one place, in understandable form, than any other book.

Figures I-2, I-3. Rabin house, 1975 and now. (Inset photo: Thorsten Horton.)

A new version makes sense because there has been constant innovation and change in building technology and building thinking over the last twenty years. Almost every building process has been examined and redesigned by someone, who is now marketing a product that the designer or builder must evaluate or ignore.

This flood of change has affected different parts of the book in different ways.

Of course, readers will find many small updates and revisions reflecting modern practice. For instance, lumber doesn't cost 12½ cents per foot anymore, and walls aren't made of 2 x 4s. There are many small changes in detailing and materials too—gadgets and products that make building a little better or a little easier.

More important to me, there are also new sections in this edition that present ways of thinking about design that were not in circulation in 1978. Ergonomics and accessible design provide strategies and techniques for designing your house according to how people do things most comfortably and efficiently, in other words, in tune with the human body.

Christopher Alexander's Pattern Language approach is another powerful tool. Alexander and his collaborators have spent many years studying forms, details, and ways of thinking about traditional building in many cultures. They've identified which "patterns" make houses warm, human, and alive. I've found the pattern language idea extremely useful, and I think others will also. There is a brief introduction included here.

Today there is an increased emphasis on bringing ecological concerns to building. Sustainable design emphasizes the ways building design burdens the environment before, during, and after construction. This sustainable approach searches out techniques that minimize building waste, promote health, and are energy efficient. There is a chapter here on this subject, from a practical builder's point of view.

The first edition of this book had only a couple of paragraphs on kitchen design, but this version has a full chapter. I've probably spent half of my time since 1978 learning about kitchen design and construction. Since most people are by now familiar with the orthodox work-triangle approach to kitchen design (which is pretty limited and misleading), I've emphasized ideas that go against the grain or supplement this simplistic approach.

New Opportunities and Obstacles

Building your own home is an ancient tradition, a wonderful adventure, and with good planning, it's a practical and economic choice. It's probably the best way for families to establish a home—at least in rural areas—without taking on a huge mortgage. But there are new challenges to overcome compared to even twenty years ago. Land costs more, and good building sites are harder to find.

Figure I-5. *Alcoves: Pattern #179 from* A Pattern Language, *by Christopher Alexander, et al.*

Figure I-4. *Ergonomic features benefit all users.*

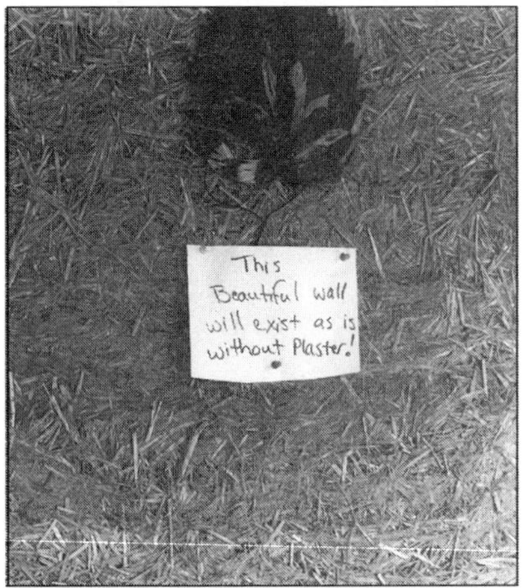

Figures I-6, I-7. Addition to Lou Host-Jablonski's house — timber frame with straw-clay wall finished in different ways, including straw-clay plaster.

Materials and wages are more expensive. There is much more regulation, more required licensing.

In the 1970s we actually built houses for as little as five or ten dollars per square foot. That is unthinkable now.

The cast of characters has changed. Not long ago, there were architects to design things, contractors who mostly followed stock plans or architect's plans, and owner-builders who worked independently as best they could. Now there are many new kinds of building professionals to hire, or not hire. Some builders specialize in certain types of building, such as timber-framing or solar. There is a small army of specialists, particularly in energy-related areas, to consult if you need to. Even more significant, there is the return of the generalist in the person of the designer-builder or design/build firm. Such people were the exception twenty-five years ago, but now they are almost the rule in small-scale custom home construction.

There are owner-builder schools where you can learn design and construction skills. And there is a far deeper library of informative building books.

In short, costs are higher, but there are more resources.

There are probably fewer people today then there were twenty years ago who are in a position to take on the entire project themselves. Many owner-builders will now be working with designers, builders, and sometimes consultants who can help them plan or construct certain aspects of the house.

Understanding Innovation

Change and innovation have become the constants in our building environment. Managing them, or at least coping with them (rather than being overwhelmed), is a critical skill. Unfortunately, it's a skill that many owner-builders, architects, and builders don't really have. Though building innovation has been lightning fast, change in good build ing practice has been slower. The underlying principles don't change as quickly as the products do. In our small building company, we've made a few major adjustments and many small ones, yet in many ways we build much as we did in the 1970s. Some revolutionary designs and products that swept through the industry have been swept right out again a few years later—usually after many people have tried out the new approaches with mixed results.

At any given moment, the person with the checkbook, the client, gets to decide which building system to choose. But in the long term, the viability of a building system is gauged by a much more authoritative judge: Reality—what works and what doesn't. Sometimes a new idea works, builders like it, and they incorporate the new method or product. But just as often, a new idea doesn't quite work, and the quiet voice of tradition—the regular way—gets the last word.

When I started building, I had a real faith in innovative methods. Like many owner-builders and self-taught builders who didn't come up through the trades, I didn't think much of the regular, suburban-looking houses I saw around me. Illogically, I made the jump to the idea that I could come up with something better. I looked to architecture for inspiration, and to the stream of building books put out by professionals, and by other self-taught builders and owner-builders. Two books were particularly popular, *Your Engineered House,* and *The Owner-Built Home,* both of which held standard construction in contempt and promoted completely non-standard building methods. *The Whole Earth Catalog* also presented hundreds of alternatives.

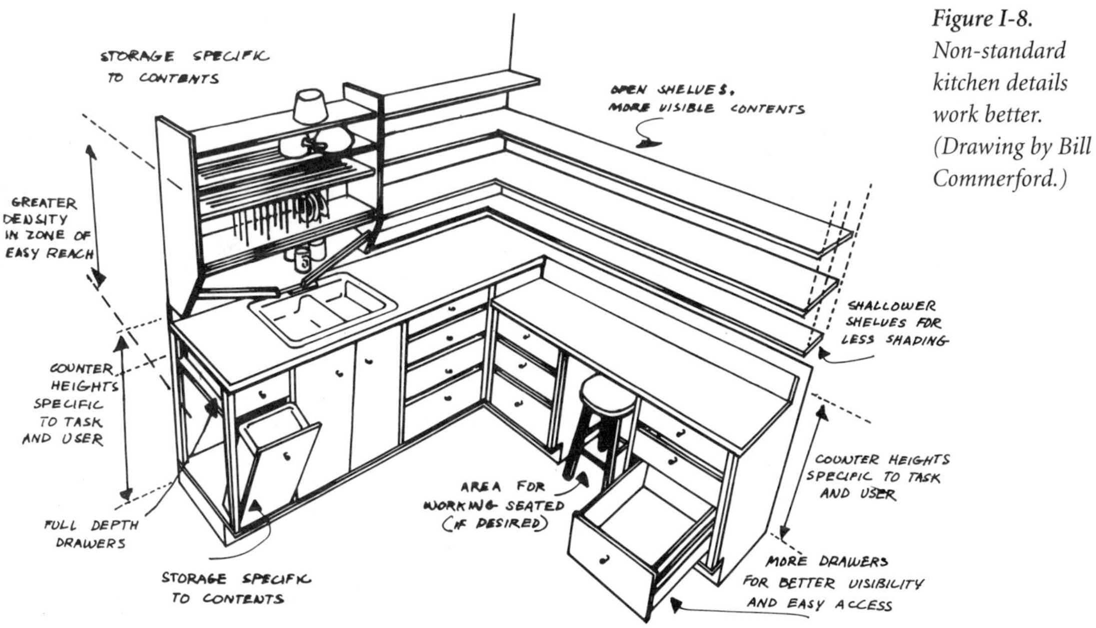

Figure I-8. Non-standard kitchen details work better. (Drawing by Bill Commerford.)

Encouraged by such books, I and many of the other builders around me tried new approaches and materials every time out. We built domes, A-frames of recycled timbers, yurts, log houses, and stone houses like Helen and Scott Nearing's. We roofed with tin, boards, shingles, and sod. We heated with all sorts of solar systems, most of which we fabricated ourselves. We built with native materials, particularly stone and locally produced lumber, and used recycled windows and other materials. We tried everything. This was the height of the owner-builder movement.

Starting perhaps in the late 1970s, reality—what works and what doesn't—made some harsh judgments. The traditional approach to building proved to be right more often than not.

Some clever building systems simply failed to work right. Geodesic domes were hard to live in, more expensive to build than promised, and prone to leaks. Other systems, like the Nearing's stone building system, worked, but added months of extra labor to the process of building a home, and saved only a little cash. When owner-builders turned professional, as many did, some of these once-appealing systems were abandoned because they didn't cost out for our customers and there were too many "callbacks."

Figure I-9. The Big House: low-pitch shed roof, post foundation, recycled windows, post-and-beam frame, a sod roof. Some parts worked better than others. (Photo: Thorsten Horton.)

Many seemingly clever or economical systems failed more slowly. Board siding was drafty and subject to rot. Flat-pitched shed roofs were subject to leaks and ice-dams, at least in cold climates. Recycled windows rotted out, and didn't keep out the wind.

Column foundations are a good example of a system that looks good, but causes trouble later. Here in New England, the conventional wisdom held that a full concrete foundation was essential. Rex Roberts and other writers argued that the basement was expensive and superfluous. Like many people, I built numerous buildings on wood or concrete posts with insulated floors. This approach did save initial costs, and buildings went up fast. The foundation system was kind to the site. But over time it became clear that houses on posts had problems. The floors were cold. Pipes tended to freeze where they came into the house. Sometimes the posts themselves were subject to frost heaves or settling. Animals got under the house, pulled down the fiberglass insulation, and sometimes made nasty smells. These problems are expensive to fix.

Early solar systems were also disaster-prone. Most of these systems were homemade, and put relatively vulnerable materials like glass, plastic, and plumbing outside the protection of the house or roof overhangs. They were subject to direct assault by sun, wind, water, and cold temperatures. In Vermont, I'd say very few of the earlier sunspaces, solar greenhouses, or active solar systems really worked well. Many were roofed over, insulated under, or "decomissioned."

What I learned from such experiences is that traditional approaches to building were often smarter than they appeared. What seems ordinary at first may prove elegant and wise after a few years. Innovation is important, but even new methods that seem well engineered and brilliant may have hidden costs, maintenance problems, or unintended consequences. I think many builders came to similar conclusions.

Today, traditional building methods, and the people who were practicing them all along, have much more respect. At the same time, standard building has been changed. We have a basic vocabulary of new, good ideas that has become accepted practice. We site houses toward the sun and insulate them better. Many new products and methods have been incorporated into routine practice. Lumberyards sell them, tradespeople can install them.

We know more. Builders have greater experience with more ways of doing things, know more about design, and have more resources for evaluating ideas. I had no way of knowing that Rex Roberts' ideas didn't all work. Today there is not only twenty years of experience to go on, but I have my fingers on the calculations for timber spans, for air movement, for insulation values, and I can see for myself why a design or building approach will or will not work. I can more easily test other schemes to see if they'll work. When I'm not sure, I know informed people I can consult.

In short, the basic level of building has risen.

Making Decisions About Technologies

Among the new ideas you'll encounter, which will work? Which will be worth the added costs? Which are right for you?

I think maintaining your perspective as you consider such innovations, and learning to make decisions that you and your family can live with, is one of the trickiest aspects of being your own designer or builder, or even of working effectively with pros that you hire to guide you. I've adopted a few approaches that I hope will help.

When I have a definite view on some building technique, I'll give it, and my reasons why, but I'll also readily acknowledge when somebody else may hold an opposing view. If I think house wraps are a waste of money—which I do—I'll give my reasons. Where I can, I'll give other sources to consult if you are unconvinced.

I also provide planning routines (including the strategic and critical questions to ask) to use when you are choosing among competing building systems.

Perhaps most important, I'd like to share an attitude, which is very simple but took me some years to work my way to. Long ago, the father of owner-builder books, Henry David Thoreau, built a cabin, and like many owner-builders, he wrote his book about it. And in this book, *Walden,* the little cabin isn't just a house, it's a moral statement, it's the "right way," which of course implies that the conventional way, and maybe every other way, is wrong or inferior. Ever since, building writers have been making big claims for their building systems—and the obvious inference is that the garden-variety way most buildings get built isn't quite up to snuff, morally speaking.

While there are ethical choices in all large undertakings, I think people do better if they don't think of their choice of a building system so much as a moral issue, or their house design as a measure of their moral worth. It will cloud your judgment. Also, I really doubt you will discover or invent a building system that is a lot more efficient or otherwise more virtuous than what your neighbor is building right now using a traditional approach.

But do you need to have this justification? I don't think so. Build your house the way you want to build it, because you want to build it that way, because it really is "your own way," and because you're willing to take responsibility for the outcome. Build with timbers because you want to look at the structure as you contemplate your home, and remember the pleasure of cutting the joints to fit just so. Build solar because you'll enjoy all the planning now and the tinkering later, and because it will give you pleasure every time you don't buy a cord of wood or a gallon of fossil fuel.

But don't claim your house is too much better than the more ordinary one going up down the road; you may regret it later.

Using This Book

There is a spectrum of homebuilders.

Many readers will be very involved in planning their house building process, but will depend on professional designers and builders to do most of the work. Part I, New Approaches to Design, introduces innovative ideas in design that are not always part of the everyday vocabulary of most building professionals.

Part 2 will help you to purchase the right piece of land, do preliminary design work, and collaborate fully with the people you hire. For those who will be working with professionals, hiring the right people in the right way is perhaps even more important to the success of your project than the design itself. Part 3, Business, will help you figure out who to hire, how to make successful agreements with them, and how to keep your partnerships with them working over the long time-span of your project. This section of the book should also be quite useful to contractors.

Other readers will be doing most of the design as well as large amounts of the actual construction themselves. For those of you who are truly "owner-builders," building a home is a learning opportunity and a means of expression. Part 4, Building Basics, goes over many of the underlying principles and concepts that apply to all good building, no matter what building system is being used. Part 5, Construction Methods, describes basic construction techniques, but also contains further design information. For example, chapter 28, Windows and Shutters, in addition to explaining how to install windows, discusses which types of windows work best in what application, how to use recycled windows, and how to make your own windows.

My suggestion for owner-builders is to use Part 1 through Part 4 to develop your basic, spatial design. Then study the chapters in Part 5 as you design the construction details for each section of the house.

Part 5 doesn't attempt to explain every building technique. I focus on relatively simple, economical, and easy-to-understand methods which in my experience have worked well for owner-builders, and which owner-builders will enjoy. Although some of these methods are slightly different than those used by professional carpenters, I've also included many of the tricks and techniques professionals use to avoid mistakes, to make corrections when needed, and to save time. I've also concentrated on methods owner-builders often favor (recycling salvaged windows, or using native lumber, for example), but which many standard carpentry books do not cover. There are also brief summaries of a range of other economical building options that some readers will want to pursue.

For another group of readers, the process of building a home will be a major life work. If you are one of these people, you want to develop or invent new systems, build the walls with straw bales, heat the house with the sun alone, power it with the wind, build it entirely with recycled materials, or otherwise push the limits of building practice — to become part of the cutting edge of change. If this is your perspective, you may not find comprehensive and detailed instructions here for your particular building system. But you can use Part 4, which explains many of the fundamentals that apply to all types of systems. If you are even further out on the cutting edge, this book will help you do your basic planning better, and keep overall control of your project as you develop approaches that go way beyond those covered here.

This book is full of ideas that excited me at one time or another. Ideas inspire people to do months or even years of extra work, to build a house that functions well but also expresses a point of view and certain values.

The ideas that make up a good design enable us to do better work, and to understand more about what makes a home more livable and enjoyable for people in complex, inconspicuous ways. Technologies always seem important, but it is ideas that get buildings built.

PART I

NEW APPROACHES TO DESIGN

1

ERGONOMICS & ACCESSIBILITY

Figure 1-1. A Windsor chair, an evolved design that works perfectly.

CONSTRUCTION SYSTEMS HAVE CHANGED DRASTICALLY over the last twenty years. But none of them have had as much impact on my thinking as the innovative design ideas that have come to the fore in the same period. One of these is ergonomics, which is also sometimes known as "human factors engineering," "user-needs design," or "motion study." The different versions of this discipline have slightly different slants on the subject. But the basic idea is to carefully study how people do things as a key part of the design process.

Accessible design simply extends ergonomic thinking to include more people, and the entire lifespan.

Ergonomic analysis will influence the kitchen, storage, and work areas in big ways, and other aspects of the house in more subtle ways. The benefits of this approach are a more convenient, efficient, safer house, that works well for more people.

Given enough years, trial and error and the marketplace will produce convenient and comfortable products. The Windsor chair, and the typical, wooden-handled claw hammer are examples of forms that have evolved over centuries and are really perfect. Nothing works better, for the intended purpose, because many generations of makers have fiddled with the design.

But if you have a new purpose—an unbreakable hammer or an office chair for computer operators, for example—and you don't have two hundred years to experiment, ergonomics provides an alternative means of perfecting your design.

Perhaps the first level of ergonomic analysis is simply to measure people, to define their critical dimensions. Scientists have measured thousands of people to arrive at average dimensions that work well for design. First come static measurements, known as link measurements, that describe all the lengths of and distances between different key parts of the body, such as the height of the eyes, arm length,

length of thigh, the typical width of a person's shoulders, and so on. All of these dimensions are in turn related to gender, height, and age. In each case, there is a different set of dimensions for the seated position, or for people who use wheelchairs.

The next level of ergonomic analysis defines movement limits, such as how high, low, or forward a person can reach, and how far the various joints can rotate. For example, people can comfortably rotate their heads left or right about 45 degrees, and about 30 degrees up or down.

Even this simple information is very useful. For example, knowing eye heights seated or standing can help you position windows, and knowing how high a five-foot-three-inch-tall person can reach upward can help you design shelves above a counter.

Such information becomes more useful when researchers study some of the more common problems which arise repeatedly in designing houses, by working with and observing people doing things. For example, five-foot-tall people work best at a counter height of 32 inches. Their eye levels will be about 56 inches. They'll be able to reach about 67 inches up comfortably, which means the top shelf should be lower than that, particularly if heavy things are on it. Working seated, they will need 15 inches of leg room, plus 4 inches more for the feet, and when seated their eye height will be about 27 inches above a typical compressed seat height of 15 inches. A six-footer, by contrast, could reach to 80 inches, might prefer a counter height of over 39 inches, and when seated would need more like 19 inches of kneeroom. A wheelchair user would have yet a different set of dimensions.

This kind of information has four effects on how I think about the dimensions of things. First, since people vary so much in size, it makes sense to look at the particular dimensions of the people I'm designing for. A shorter family's kitchen might have lower counter surfaces than a taller family's.

Second, since people's sizes vary so much, sometimes it's best to provide a variety of options. In figure 1-2, there are about five different counter heights available to make work convenient for everybody in the family, including a wheelchair user. Making things adjustable is another approach, most common in office furniture and drafting tables, but applicable to many other situations, such as shelving.

Thirdly, when you have to use an average or standard dimension, the conventions now used by the building industry may not be the best. The ergonomic chart shows a few examples of common practice compared ergonomic standards that may work for a broad variety of people.

Fourth, it helps simply to be aware of the various ergonomic categories.

Ergonomic chart of body dimensions, reach ranges, and counter heights

HEIGHT (IN INCHES)	EYE LEVEL	HIGH REACH	COUNTER HEIGHT	LOW REACH	TABLE HEIGHT
60	56	67	32	—	25
66	62	74	36	—	28
70	66	78	38	—	30
72	67	80	39	—	31
WHEELCHAIR USERS					
63	46	55	32	17	31
66	48	57	32.5	15	31
68	49	59	33	13	31
73	51	68	34	10	31

Standard dimensions vs. Ergonomic dimensions (inches)

	STANDARD	ERGONOMIC
Table Height	30	29
Kitchen Counter Height	36	varies
First Upper Shelf	55	48–50
Top Shelf	75+	70
Outlet Height	12–18	24
Switch Height	48	44
Door Width	30–32	34–36

While we all experience them, we don't always think of reach ranges, lines of sight, or high grasp, nor ask such questions as how many degrees it is comfortable to pivot your head when looking up or down. Simply identifying such categories helps focus my attention on aspects of design that might otherwise slip through the cracks until someone complains.

The ORZ

For various reasons, we usually design for mid-height, active, young, healthy adults. But for little or no extra cost, it's possible to design things for everybody. A good example is Margaret Wylde's concept of the Optimum Reach Zone, or ORZ. This is an area 20 to 44 inches from the floor, within which almost anyone

Figure 1-2. Five different counter heights accomodate different users. Birnbaum/Seeger house. Design, Sam Clark; builder, Peter Peltz Construction, Woodbury, Vermont.

Figure 1-3. Cabinet to the right of stove comes out to create knee space, and the cabinets to the far right can be adjusted by lowering the kickspace. Weiler house. Design, Sam Clark; builder, Robert Sparrow Building, Wellfleet, Massachusetts.

can reach and do things. In ergonomic design, most commonly used items in kitchen cabinets, and things like light switches and other controls, would be located in this area.

Using Ergonomics to Make Work Easier

Researchers have extended this kind of thinking by finding ways to measure how varying designs influence the effort expended while doing things. Early researchers put heart rate or oxygen-use monitors on research subjects to compare work settings. In this way, they discovered things that are obvious in a sense, but still important to remember. It's easier to work if things are available without walking too far; it takes a lot of energy to reach way down, or kneel down to get something; it's much harder to reach to the top of one's reach range than to the middle.

Other researchers have relied on how people report their experiences, and by asking people to compare different appliances, control systems and devices, or product designs.

Motion Study

I think the most powerful and interesting motion study is one of the earliest—namely, Frank Gilbreth's motion study. Gilbreth, an industrial engineer and former builder who is the unacknowledged founder of ergonomics, and who pioneered most of the ideas still used today, devised a categorization of motions known as the therblig system. He found that almost any kind of activity, particularly work activities, could be broken down into seventeen basic motions.

1. Search (manual or visual)
2. Select (can overlap with search)
3. Transport Empty (moving a hand or other carrier toward a desired object)
4. Grasp
5. Transport Loaded (moving an object with hand, or by dragging, rolling, etc.)
6. Hold
7. Release Hold
8. Position (orient an object in preparation for the next step. Gilbreth notes that it is possible to position during Transport Loaded.)
9. Pre-position (position for later use. For example, put a pen in a holder to eliminate later need to orient it.)
10. Inspect
11. Assemble

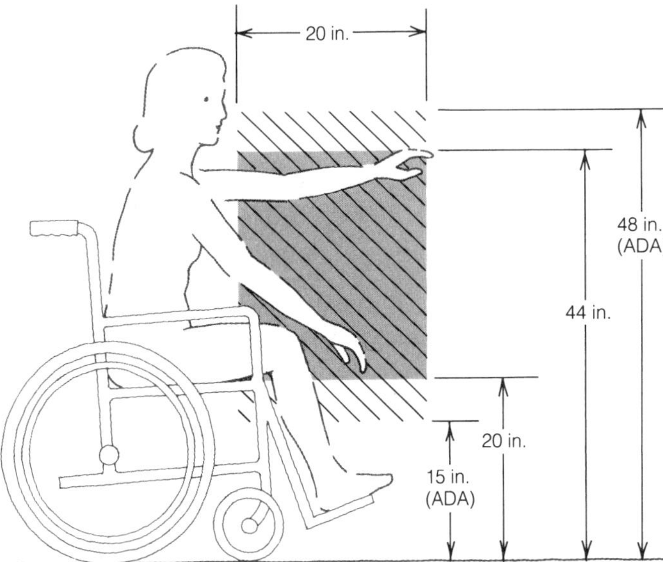

Figure 1-4. Optimum Reach zone. (From Building for a Lifetime, *Taunton Press 1994. Used by permission.)*

Figure 1-5. Knife slot pre-positions knife, so it can be retrieved using the grip required for use.

12. Use
14. Unavoidable Delay
15. Avoidable Delay
16. Plan
17. Rest to overcome fatigue

This approach is like music notation; it's a simple way to describe any sort of activity. It aids the observer, whether a university researcher or a kitchen designer, to picture accurately all of the steps or efforts involved in doing something. By comparing methods, layouts, or details, you can reduce the number of therbligs, and make the job easier.

Gilbreth developed a whole system of charting and photographic recording to document the therbligs used in all sorts of work that builders, surgeons, home-makers, and others do. But without going to that much trouble, it's possible to use this approach to look at simpler problems. For example, I used to design kitchens with cabinet doors, behind which were pull-out shelves. After I studied Gilbreth's ideas, I quickly realized that I could save my customers some therbligs (and some money) by building big drawers in the first place, and eliminating the operation of opening the door from the inventory of steps. I particularly like the idea of "pre-position," which suggests that you should store something in a position in which you can grab it in the same place you need it using the same grip needed for use.

Accessible Design

Accessible design is just a slight extension of ergonomic thinking to include everybody, in all phases of life. It factors in people with a variety of disabilities, including those that affect almost everyone who lives to become old.

In the few years in which I have been involved with accessible design, I have come to believe that almost any project, including most new homes, can be designed to be accessible (or designed to be made more accessible later), and that doing so need not be expensive or obtrusive in any way. Doing this means anyone can visit the house in comfort, including your aging parents who don't like to climb steps. It means you can stay in your house as you age or as your health changes. But it also makes a better house for all, now. The kitchen will work much better for all users. Storage will be more efficient and capacious. It will be easier to move things around in the house, and move about it yourself. The house will seem more spacious.

This chapter provides the quickest of summaries, but other information is included in sidebars throughout the book, and in references listed below.

Wheelchair accessibility and mobility: The design conventions that we think of as "wheelchair accessibility" are a good place to start thinking about access and mobility, because these conventions really are designed to make life easier for people with all sorts of mobility impairments.

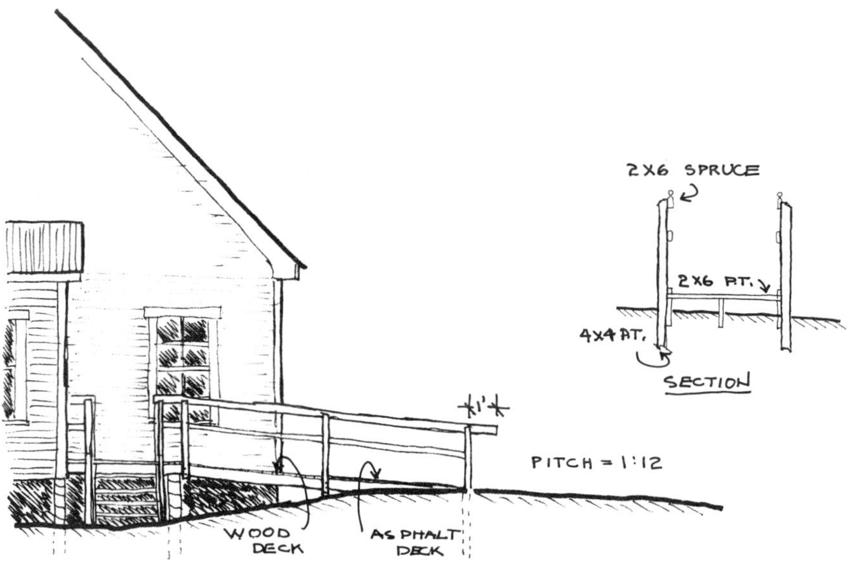

Figure 1-6. Key features of ramps.

Figure 1-7. *This is a fully accessible house, yet no details call attention to that. Costs were similar to a comparable inaccessible house. Seeger/Birnbaum house. Design, Sam Clark; builder, Peter Peltz Construction, Woodbury, Vermont.*

Figure 1-8. *No-step entry: An accessible entry can be accomplished without ramps, with careful grading leading to a covered porch whose floor is at the same level as the floor inside. Friends' Meeting House, Plainfield, Vermont. Design by Alan Walker, Jonathan Rose, David Scheckman, and Sam Clark; built by Iron Bridge Woodworkers and many volunteers.*

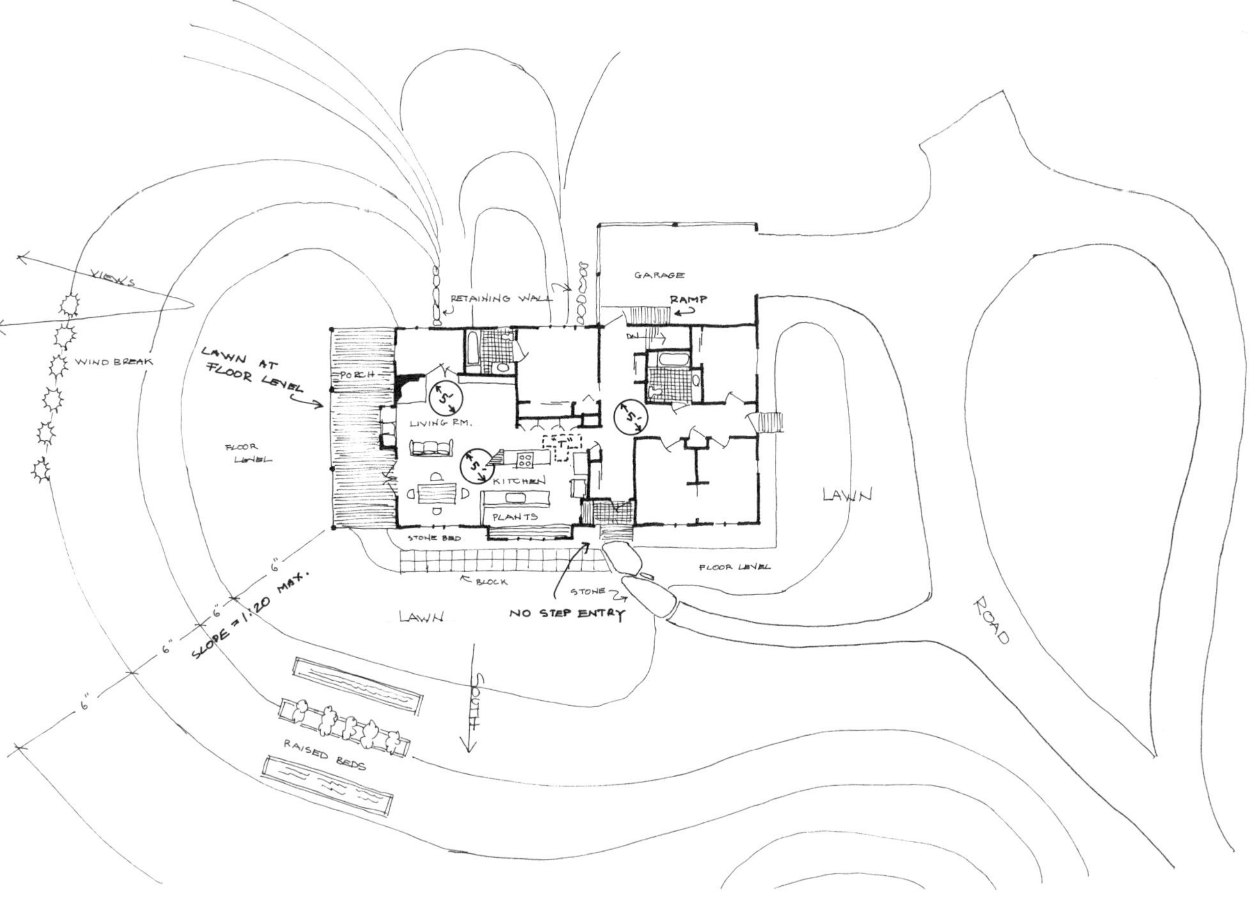

Figure 1-9. Features of an accessible layout: gentle grading, no-step entry, turning circles, big halls and doors, bathroom and bedroom on main floor.

Perhaps the biggest obstacles to mobility are steps and steep slopes. A few stairs are an obstacle for many elderly people, and even one step is an almost absolute obstacle for wheelchair users. For many people, a slope of 1:20 is fairly comfortable to negotiate. (A 1:20 slope rises one vertical inch for every twenty inches of length or "run.") Paths pitched 1:20 or flatter are referred to as "berms," and don't need railings. Steeper slopes, 1:20 to 1:12, are referred to as "ramps," are more difficult to use, and have to be designed with care, including specially dimensioned railings. I think the best ergonomic design would have gentle paths in the berm category, and no steps at all. This is referred to as a "no-step entry."

Circulation: Ideally, the whole house would be on one floor. When it isn't, there would be a bathroom on the first floor, and at least potentially a bedroom.

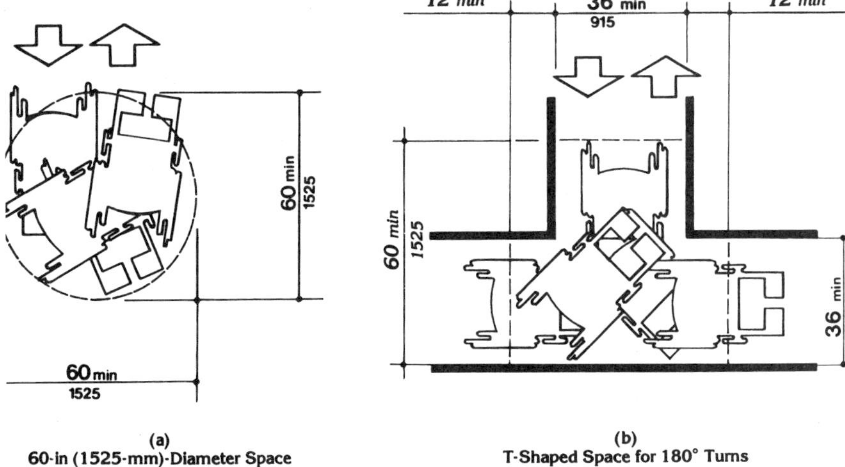

Figure 1-10. Wheelchair turning spaces. (From Uniform Federal Accessibility Standard.*)*

Inside the home, doors would be wide—usually two feet ten inches or three feet, and halls and passages would be 36 inches, or a minimum of 32 inches. Periodically there would be spots where a wheelchair user would be able to change direction. Accessibility codes call for this to be an open area five feet in diameter, which can extend into the kneespace under a sink or counter up to 17 inches. A space for a T-turn also works.

Doors and doorways: Simple doorways can be obstacles. In accessible design, the "net clear opening" has to be 32 inches. Since an open door itself uses some of the doorway space, doors should be 36 inches wide or at least 34 inches. Many people with arthritis or other impairments can't comfortably operate a door knob. Lever handles work better for everybody. The door swing shouldn't be too much of an obstruction. In tight quarters, such as a bathroom, the door should swing out, unless it's quite a big bathroom. Sometimes, with a small bath and a narrow hallway, a pocket door is the best solution.

Inside the house, omit thresholds. Exterior doors can have very low profile, ½ inch thresholds. Although standard thresholds often stick up as much as 2 inches, most doors are available with low profile aluminum thresholds. It is also possible to reduce the height of thresholds by installing the door down on the framing instead of on top of the subfloor (figure 1-11).

On the "pull" side of the door, there needs to be a space at least 18 inches wide beside the door, so a person using a wheelchair or walker can position themselves next to the door, swing the door open past them, and then go through (figure 1-10).

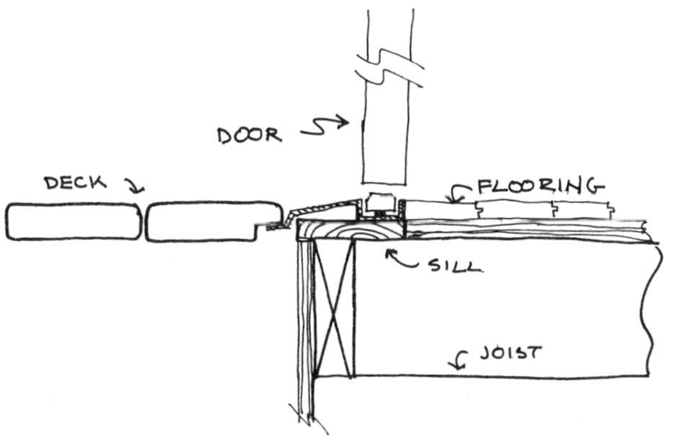

Figure 1-11. Low-profile thresholds for exterior doors.

The kitchen: An accessible kitchen will provide kneeroom under the sink, a seated work area or two, lots of drawers, lowered overhead cabinets, and some maneuvering room. These and other key features are discussed in detail in the kitchen chapter.

The bathroom: In accessible houses, the bathroom should be larger than usual, with sufficient maneuvering room inside for all functions. It should have grab bars where needed. A good scheme is to provide extra-wide blocking in the walls wherever grab bars might be desired in the future. There needs to be kneespace under the sink.

Wheelchair users need a way to "transfer" from the chair to the toilet or bathing setup. For a toilet, this usually means providing a space beside the toilet to park the chair, with grab bars as shown in figure 1-13. The seat may be raised to about 19 inches to match the seat-height of the wheelchair. For a tub, there will often be a transfer seat, as shown, plus grab bars, and the tub may be raised three to four inches to align with the wheelchair seat.

A shower would be oversize—at least 36 inches square on the inside, and preferably bigger. It might have a fold-down seat, a hand-held shower with lever control, and grab bars. Most important, the shower's floor would be flush or nearly flush with the bathroom floor, with no curbing—another no-step entry.

Some of these provisions might not meet the needs of current users. For example, many people wouldn't want a tall toilet seat. But if you lay out the bathroom for accessibility, provide good support for grab bars everywhere, and equip it for the current users, changes will be simple if needed later.

Controls: Ergonomic design places a lot of emphasis on "controls," which means handles, knobs, and dials, keypads, faucets, and similar devices. In accessible

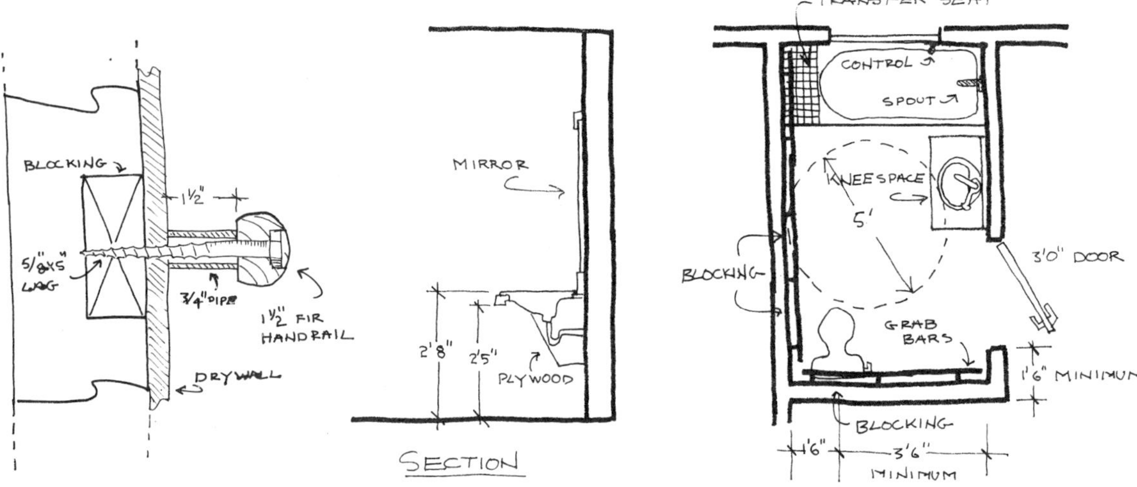

Figure 1-12. Accessible bathroom: typical features.

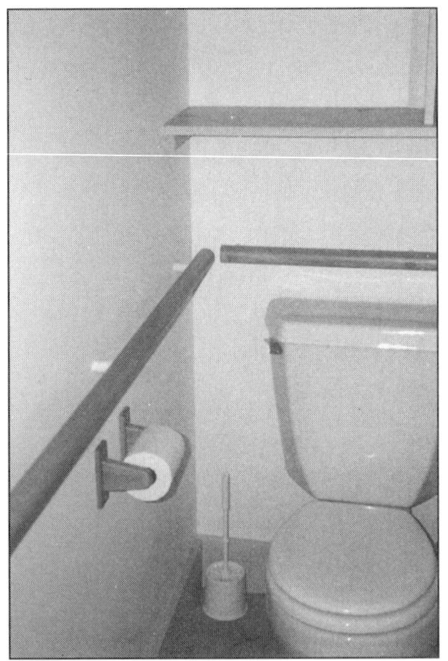

Figure 1-13. Aids such as grab bars don't have to look hospital-like. Here a strong wooden rail meets specs for a grab bar, and works great for towels, too.

design, they are even more important. As mentioned earlier, doors should have lever handles. Faucets should be lever-operated: they work better for everybody. D-pulls work well on cabinets; they are easy to grasp compared to knobs. Knobs on appliances have to be easy to see and grasp, even for those with limited hand strength or limited reach. The knobs on a stove, for example, would be near the front, big, easy to grasp, contrasting in color and texture, and they would provide some feedback such as clicks, lights or stops which tell the user when the device is on and off, and what the setting is.

GETTING MORE INFORMATION

This discussion just scratches the surface. I think it's just as worthwhile to read up on this topic as on solar or other subjects. At the same time, much of ergonomics is obvious if you take the time to imagine all sorts of people using the house you are designing, and you allow your imagination to linger over seemingly small matters: how to cook a fish, the height of a window sill, how a bag of groceries gets brought into the house.

Often the best sources of information on accessibility will be the people who will use a kitchen or a building. This is what The Adaptive Environments Center in Boston (one of the pioneers in this field) calls "User Needs Design." Most people don't realize that the dimensions and configurations of things can be so varied. If the folks who will use any space become acquainted with some of these parameters, terms, and options, they become the best source of information on what will work best for them. The more actively they are engaged with the issue, the less you will have to rely on standard dimensions or reference materials.

Design review: Although you can do a lot of this planning yourself, as with your overall layout and design, a review by an accessibility specialist makes sense. Someone who has poured over codes and books will very quickly be able to find places where your scheme can be improved, and where small oversights undermine the basic ergonomic integrity of your design.

Resources on Ergonomics

— *Humanscale 1,2,3,* by Niels Diffrient, Alvin R. Tilley, and Joan C. Bardagjy. M.I.T. Press, Cambridge, Massachusetts. This book is the handiest compilation of ergonomic information I know. For seating, counters, storage, and other design problems, *Humanscale 1,2,3,* offers useful data in a very simple format for all sorts of people.

— *Management in the Home,* by Lillian Gilbreth and Orpha Mae Thomas. Dodd Mead, New York, N.Y. 1959.

— *Building for A Lifetime,* by Margaret Wylde, Adrian Baron-Robbins, and Sam Clark. Taunton Press, Newtown, Connecticut. 1994.

— *Accessible Housing Design File,* by Ron Mace, et al. Barrier Free Environments, published by Van Nostrand, Reinhold, New York, N.Y. 1991.

2

PATTERN LANGUAGES

"A pattern language gives each person who uses it the power to create an infinite variety of new and unique buildings, just as his ordinary language gives him the power to create an infinite variety of sentences."

—from *A Timeless Way of Building*

THE PATTERN LANGUAGE IDEA IS CONTAINED PRIMARILY in two massive books entitled *The Timeless Way of Building* and *A Pattern Language.* The first is authored by Christopher Alexander, and the second by him and his associates, Sara Ishikawa, Murray Silverstein, Max Jacobson, Ingrid Fiksdahl-King, and Shlomo Angel, from The Center for Environmental Structure in Berkeley, California.

Reading these books—indeed skimming them, perhaps ten years ago—had an immediate effect on the way I designed spaces. They are full of ideas to use to make a better house. They are expensive hardbacks, though worth owning, expecially *A Pattern Language.* My initial suggestion would be to get this book from your library and read through it early in your design process.

The question the book poses is simple. How is it that traditional cultures in many times and places have built buildings and villages that are beautiful, alive, whole, comfortable, and enriching to be in, but so often we can't do so, in spite of the far greater resources, technical knowledge, and professional education we bring to the task? What have villagers in Mexico, or rural Japan, or early New England known that we don't? Why is so much modern building so unrewarding to be in, yet so expensive?

Alexander believes that all great building comes not from the conscious and self-conscious planning and drawing that we do, that we call design, but from an organic process that he calls the "timeless way of building." As he writes in his book of that name (p. 161):

> *Imagine, by contrast, a system of simple rules, not complicated, patiently applied, until they gradually form a thing. The thing may be formed gradually and built all at once, or built gradually over time but it is formed, essentially, by a process no more complicated than the process by which the Samoans shape their canoe.*

Here there is no mastery of unnameable creative processes: only the patience of a craftsman, chipping away slowly; . . . the simple mastery of the steps in the process, and in the definition of these steps.

The steps are what Alexander and his associates call patterns. The patterns can be seen as imperatives, principles, or rules, each addressing a particular problem. A good example of a pattern is "alcoves," a pattern that particularly applies to common spaces. Alexander presents the problem this way:

No homogeneous room, of homogeneous height, can serve a group of people well. To give a group a chance to be together, as a group, a room must also give them the chance to be alone, in ones and twos in the same space.

The authors then provide a lengthy discussion of this design problem, but I could easily recognize the reality from my own experience. A boxy room creates one way of interacting: we face one another in the "conversation group." A more complex form more closely suits the way people interact. When we have a group at our house, I read my paper by myself, chat with a friend, then work my way into the center of action and conversation to engage with the whole group. I then retreat again as the evening unfolds.

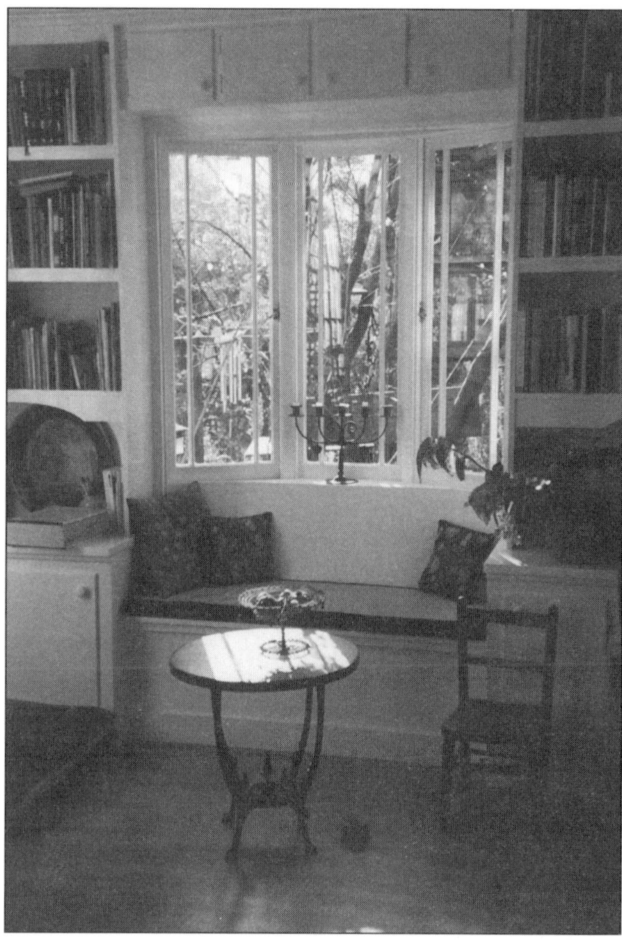

Figure 2-1. *The Tunis-McCue house, a project of Alexander's Center for Environmental Structure. This interior view expresses several patterns, including #77, House for a Couple, #128, Indoor Sunlight, #159, Light on Two Sides of Every Room, and #180, Window Place. Designed and built by The Center for Environmental Structure, Berkeley, California. Christopher Alexander, architect, Randall Schmidt, project manager. (Photo: Randall Schmidt.)*

The pattern that addresses this problem reads:

"Make small places at the edge of any common room, usually no more than six feet wide and three to six feet deep, and possibly much smaller. These alcoves should be large enough for two people to sit, chat, or play, and sometimes large enough to contain a desk or table."

Now, I'd always liked alcoves, and felt very comfortable and at home in rooms with them. What Alexander does is notice this clearly; he thinks about how alcoves function, why they matter, and how they should be included in a house.

Alexander and his colleagues identify a total of 253 patterns. Some are considered universal to all good building, but others are specific to a particular time, place, or community. The patterns that best apply to each group or people would be the basis of that group's pattern language. These compilations are not just a list of nice features. Like spoken languages, they have a characteristic grammar and logic. And like a spoken language, a pattern language has to be learned. Once learned, it can be easily spoken by all in the community.

The patterns are organized into a hierarchy, beginning on a large scale, and working down to building details. The large-scale patterns are regional ("The distribution of towns," "place of country streets"). Next is the village level ("mosaic of work," "web of public transportation"), which describes features of community. "Quiet backs," "small public squares," "pools and streams" are other examples from the village level. Following this are details of a good house ("indoor sunlight," "short passages," "Zen view," "tapestry of light and dark").

Though there is a progression in *A Pattern Language*'s organization, it isn't strictly linear; each pattern is linked to and overlaps with many others. For example, at the end of the section on alcoves is a paragraph that connects the alcove pattern to others, such as "ceiling height variety," "half-open wall," and "the shape of indoor space."

All of the material from these books is couched in a complex bed of theory; these are long books, which aim high. The authors aspire to replace how we design and build houses and cities with their method. Many of the ideas are on target, but there is a lot that I don't agree with, particularly the sections on construction techniques. Sometimes I find the visionary tone to be too much. There is certainly plenty in these books that goes against what I am arguing for, here.

But Alexander is a gifted observer of what works in and around buildings, of what features make a house work better for people, and what features are in sync with the psychology and physiology of how people thrive, or don't thrive. Though I think his ideas have to be met with the same critical thinking and skepticism that should be applied to any building book, the patterns themselves will be extremely valuable to anyone who designs or builds a house.

Figure 2-2. Entrance transition. Pat Kane's house. Designed and built by Pat Kane.

Figure 2-3. Outdoor room, and positive outdoor space. Helen Garvy's house. Design, Helen Garvy: builder, Roger L. Mason Construction.

Figure 2-4. Entrance transition, outdoor room. Mason/Wicht/Jones country cabin. Originally designed and built around 1900 as a fruit packing shed by Clayton S. Jones , it served as a temporary living quarters following the 1906 San Francisco Earth Quake. Later converted into a country cabin.

3

SUSTAINABLE DESIGN: A BUILDER'S APPROACH

THE IDEA OF SUSTAINABLE BUILDING IS SIMPLE. Building, as practiced today, accounts for a large proportion of the energy use, resource depletion, and pollution in our world. It's not just the materials that end up in the building. The energy consumed in their manufacture, transport, and installation must be factored in. Of course, the fuels used to heat and operate the house over time also play a large part in its overall ecological impact.

Sustainable building minimizes these effects through energy-efficient building design, recycling, use of local or native materials, and other methods. A related goal is to create a "healthy house," which means a house whose interior space contains clean, non-toxic air. This is usually attained through the use of natural or unprocessed building materials.

An ideal sustainable building is one that sits lightly on the land, requires little fuel to heat, operates on little or no outside electricity, and is healthy to live in. Some of the materials will have come right from the site, such as stone for the walls, or from nearby sources. Some of its fabric (doors, windows, or lumber) will have come from other buildings. Other parts of the house, such as engineered timbers, will have been processed or reprocessed from waste materials that are too low a grade to use in dimension lumber.

There are a great many very different kinds of building that could be referred to as "sustainable," from very low-tech or primitive dwellings to complex houses that contain highly engineered systems. What unites them is the concern for environmental issues.

I can think of three categories of sustainable building practice.

STANDARD SUSTAINABILITY

In the last few years, many features of sustainable building have been incorporated into routine building practice. As technologies improve and ideas spread, this list will expand.

We now site buildings better, insulate them better, and make them tighter. People understand R values (R value is resistance to heat flow) and other ideas basic to energy efficiency. Modern windows and improved heating equipment make it possible to reduce fuel use from higher levels of the past. At one time, there were no really efficient wood stoves. Now there are many, and they're getting better. In fact, all sorts of new heating equipment has become available, including more efficient conventional systems, direct-vent heaters of several types, high-efficiency kerosene heaters, in-floor systems, and so on.

Right now, more natural, less toxic finishes, fabrics, carpets, and many other products are being developed and distributed more widely.

Photovoltaics is a good example of a technology that has come a long way. The technology has been around a long time, but has been very expensive. In years past, you had to be a pioneer to power your house this way. Today, the modules cost less and work better, as does the gear that goes with them. There are organizations that specialize in the technology and the information to go with it, and there are more people available who are competent to install and maintain the components. There are also new energy-efficient products for the home, such as compact fluorescent light bulbs, which make it possible to live normally with less electricity, so that people have less to sacrifice if they decide to use photovoltaics.

As a result, what was once a marginal technology, useful only to a few innovators, is now an option worth considering for anyone, but particularly for people whose building site is remote, where bringing in regular power lines would be costly.

For a good idea to be part of "standard sustainability," it's not enough for the technology to be possible. Delivery and support systems have to be there too, including easy-to-find distribution, and local tradespeople who can install and service products.

Native lumber is another example of a typical sustainable building product. In our area, we've always had native lumber available. Now the quality and availability are improving, and native lumber is becoming a more integral part of the "standard" repertoire of available materials. Yet we don't have a fully developed system for recycling building materials such as lumber, windows, or doors. As a

Figures 3-1, 3-2. The Clearing. The house has photovoltaic power, with a generator for backup. Main house built by Iron Bridge Woodworkers; office built by Roger Hassol.

result, it can cost more to recycle a fine old wood door than to buy a new, high performance, insulated metal entrance door.

Overall, it's now possible to build "sustainable design" using what is routinely available. This is particularly true compared to what was available five, ten, or twenty years ago. There is no reason not to, and most people, given the chance, will do it.

INNOVATIVE SUSTAINABILITY

Few products are perfect. There are trade-offs. To improve your building in one area often requires a compromise somewhere else.

Pressure-treated wood allows you to build light, insulated foundations with low impact on the land (when compared to concrete foundations), but the preservatives used are chemically suspect, irritating to work with, and possibly toxic. Using wood from local forests saves transportion cost and material expense, but locally milled wood is wetter and heavier, and usually less uniform.

Many of the standard materials we use to build tight, fully insulated houses, such as plastic films and foam insulations, are highly processed and "embody" a lot of energy in their manufacture. In short, to build a sustainable house in the regular market, flexibility is necessary, as well as principles.

Figure 3-3. This house is off-the-grid. Most of the lumber came from the site, milled with a portable sawmill. Design, Rebecca Bailey; built by Rebecca Bailey and Jim Schley with family and friends.

There are people who are particularly interested in sustainable design, and want a higher level of performance from their houses than the standard. No compromises! For them, a highly sustainable design is a primary goal, perhaps even a major reason for building in the first place.

There are also people with remote sites, where it's hard to bring in power, or even difficult to truck in building materials. There are others whose sites offer special resources, such as a stand of timber that can be cut into lumber with a portable sawmill, or a good stream for hydro power.

These people can become innovators, either by choice or necessity. They might go "off-the-grid." They might build a sod roof, a fully solar design. They might build entirely with recycled materials. They might build their walls with stone, earth, old tires, firewood, or straw bales. Whatever their approach, they will need to be able to commit extra time to the project, because it takes longer to build on the cutting edge.

While this book doesn't focus on these approaches, it does emphasize the planning processes that can help any building project be successful, and that may be even more crucial when some aspects of the design are uncertain.

MUNDANE SUSTAINABLE DESIGN

From a builder's point of view, there are many building and design practices that aren't particularly seen as "sustainable" by writers and theorists, but are funda-

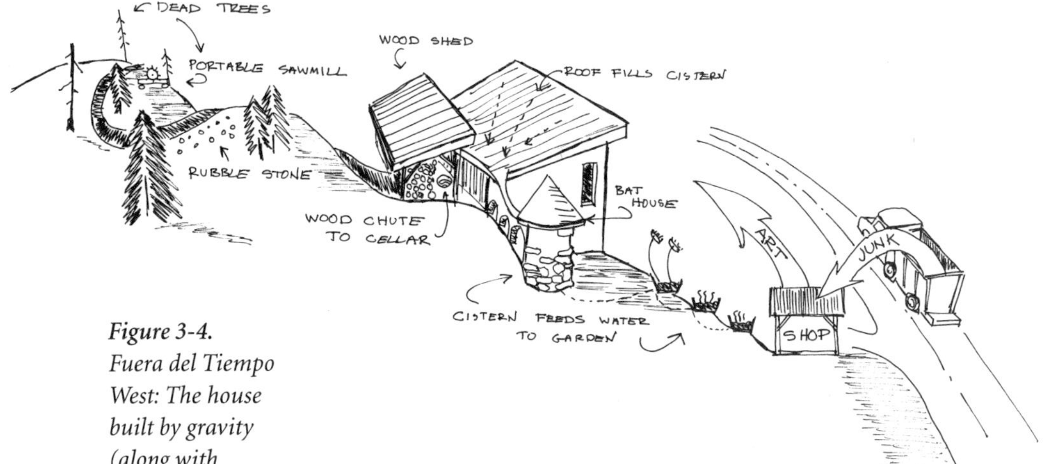

Figure 3-4. *Fuera del Tiempo West: The house built by gravity (along with sixteen years of hard work). Much of the structure is stone rubble and dead trees gathered from the hillside above, plus an amazing collection of recycled stuff. Built by Dolly and Eli D. Kramer. (Photo: Eli Kramer.)*

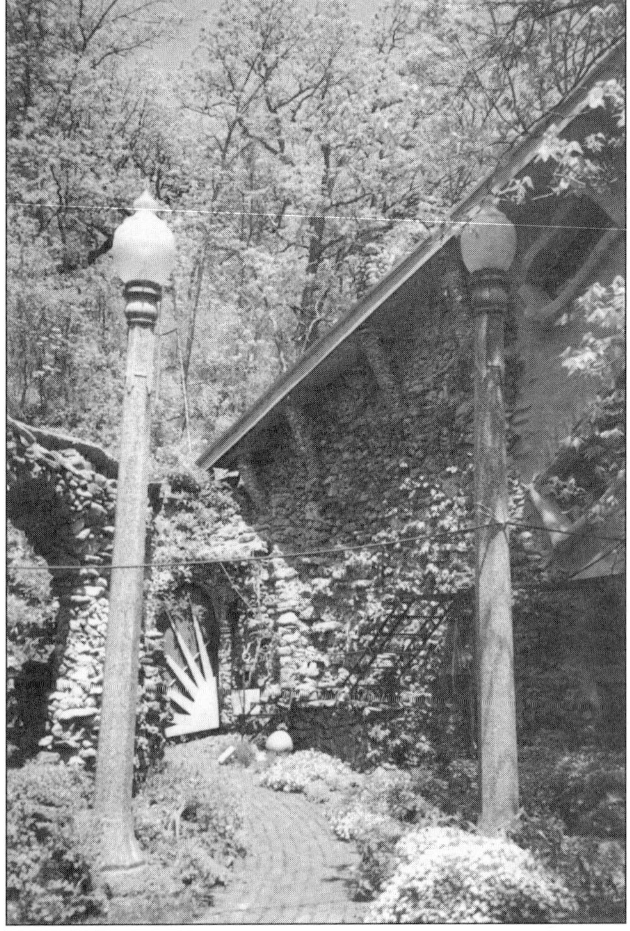

mental to ecologically sound building. These things make up what I call "mundane sustainability."

Durability: Obviously, it's wasteful if buildings have to be rebuilt or replaced; a family should be able to happily stay in a homestead over many, many years. Therefore, a house should be durable, easy to repair, and easy to change as needs change.

Ordinariness: An ordinary house has intrinsically sustainable features. A "regular" building can be permanent partly because any carpenter and most of your neighbors know how to build it and fix it, and because the systems in it are thoroughly tested. If you did a survey of all the owner-built or architect-designed buildings in your county, it would be the most innovative ones that have been rebuilt, torn down, or required the most costly repairs.

For this reason, in my work I tend to let new methods become "standard" before trying them out on my customers. I'll try a new heating system on a job when my plumber has become familiar and comfortable with it.

Standard stick-built building systems are particularly durable and adaptable. I repeat this point here because so many writers, most recently Stewart Brand, have argued to the contrary with no basis in fact. No system is as easy to build, fix, and change as the ordinary stick-built system. Unless it burns down, a house built this way will last forever if the roof is maintained and the basement ventilated. It is every bit as strong as a timber building, and can be changed more easily.

Accessibility: I would argue that a sustainable design should be accessible. It should be designed so that anybody can live and function in it over a lifetime. When people age or become disabled in any way, they often have to leave their homes, or spend very large sums on awkward retrofits. If a house is accessible in the first place, people can live in it for a lifetime. They are less likely to be injured doing ordinary things. If a house is accessible, it won't have to be replaced when people's lives change. It sustains the people.

Small: The most obvious way to conserve land, energy, and building resources now, and fuel in the future, is to build smaller houses.

A small house uses less land, less heat, and less materials. It causes less pollution at every stage. Its economies are more than proportional to size, because a small house needs not just fewer timbers, but smaller ones, since the spans are shorter. This is the point most often ignored by people who talk sustainability.

Figures 3-4, 3-5. Pat Kane's house: A small house that feels big. Designed and Built by Pat Kane.

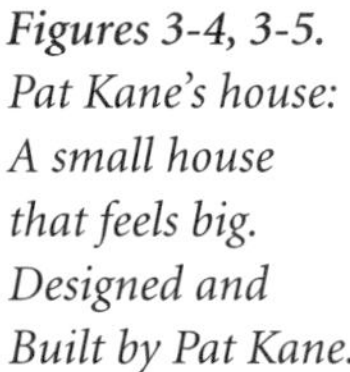

Cheap: On weekdays, we sometimes work for people building large houses—houses that are large enough to house two families, and that cost as much as $50,000 per person housed. One or two weekends each year, we work at Habitat for Humanity projects, building small, tight houses costing $10,000 per person housed, or less.

One could argue that a sustainable house should be inexpensive. The dollars that go into a house have a long history, and embody a great deal of energy and resources, just like a 2 x 4 or a can of varnish.

I wouldn't expect many people, particularly those with lots of money, to agree with me, and my livelihood depends on building expensive houses part of the time. But I think that someone who claims an interest in sustainable building should sort through this issue.

O.V.E.: Many builders and designers overbuild on principle: if a 2 x 8 will do, use a 2 x 10. I recently saw a small addition with 2 x 6-inch wall studs every 16 inches, window headers on non-bearing walls, and ridge beams as well as collar ties. This building had about fifty percent more lumber in it than needed for it to last forever. It was beefier, but not better.

Years ago I collected some pamphlets put out by the lumber industry advocating Optimum Value Engineering, or O.V.E. The idea of O.V.E. was to eliminate superfluous materials by understanding how structures really work. The house corner posts, which usually have extra studs, were simplified. Thc 2 x 6 wall studs were spaced 24 inches on center, more than strong enough, leaving more room for insulation. Headers in non-bearing walls and other superfluous framing were eliminated. The cantilever principle was used where it could save lumber. Plywood, which adds enormous strength to a wall, was factored into timber sizing, often allowing some timbers to be downsized or even eliminated.

I don't know if the O.V.E. term is still current, but the idea is. Careful engineering can reduce the amount of fabric in a house significantly, and of course, save time and money.

CoHousing, clustering, villages, renovation: There are several issues that go far beyond the scope of this book, but which are important to allude to, since sustainability is always taken as an ethical as well as a practical issue.

The best conservation of building resources has to come not from the independent homestead, but from communities. Grouping houses into CoHousing communities, clusters, villages, condominiums, or other combinations offers ways to conserve all resources. It offers opportunities for sharing gardens, trac-

Figure 3-6. The first house built by Central Vermone Habitat for Humanity.

tors, shops, forests, wells, and the labor of haying the back field, or processing meat from a cow, to list just a few of the projects the community I live in shares.

In the same vein, the best way to build sustainably is not to build a new house at all, but to renovate, to recycle an existing house.

Sources on Sustainable Design

—*The Independent Home,* by Michael Potts. Chelsea Green Publishing Company, White River Junction, Vermont. 1993.

—*Living the Good Life,* by Scott and Helen Nearing. Schocken Books, New York, N.Y. 1970.

—*Solar Living Sourcebook,* by John Schaeffer and the Real Goods Staff. Chelsea Green Publishing Company, White River Junction, Vermont. 1996.

—*Your Natural Home,* by Janet Marinelli and Paul Bierman-Lytle. Little Brown and Company, Boston, Massachusetts. 1995.

—*Guide to Resource Efficient Building Elements,* by Steve Loken. Center for Resourceful Building Technology, P.O. Box 100, Missoula, Montana 59806.

—*Architecture Without Architects,* by Bernard Rudofsky, Doubleday, Inc., Garden City, N.Y. 1964.

PART 2

DESIGNING & PLANNING

4

RESOURCES, GOALS, & PRIORITIES

THE FIRST EDITION OF THIS BOOK PLUNGED RIGHT INTO the subject of design. But I've often puzzled about why building projects frequently spin out of control. One reason may be that people aren't clear about their goals and resources, and don't evaluate whether their building plan is consistent with them as the project unfolds.

People bring a web of positive goals to their project. One or another goal can always be used to justify an addition or change to the plan. But with a successful plan, the totality of these hundreds of decisions conforms to the structure of the goals. My aim in this chapter is to show how goals might be formulated and which questions are important to ask in the process. I'll also point out some common pitfalls to watch out for.

YOUR RESOURCES

Money

Sometimes people hesitate to specify how much money they can really spend. I advocate making a real money budget, and referring to it often. If you hire a builder or designer, he or she needs to be disciplined by knowing the limits. More important, you need to remind yourself.

The process of budgeting exerts a positive force. Many owner-built or architect-built houses express wonderful ideas. The budget encourages everyone on the project to refine all these ideas into an integrated, tight, complete design, a better design.

Time

At times in my life, I've taken off months at a time for big building projects, or worked every evening and weekend, week after week. I wouldn't want to do that now.

Chapter 13 provides a way to roughly estimate how many hours it may take to build your house yourself. If you work with a builder, he or she can help you figure out

Hours it takes to do various tasks

Build a house	3 hours/square foot
Design a house	80 to 300 hours
Clean up building site	5 hours/week
Hang new window	1 hour
Hang recycled window	4 hours
Trim out a door	2 hours
Install one sheet of plywood	½ hour
Install 100 square feet of shingles	8 hours

roughly how long it might take to complete those projects you take on. Just the designing, weekend cleanups, the painting, and some of the trim work can add up to many hundreds of hours. Form an estimate of how much time you can put in, in hours or days, given your family, work, and vacation situation. Then try to match your building plan to it.

The time spent on the house is at the expense of rest, leisure, and family life. Most people love to work on their building projects. Up to a point. Beyond that point, twenty-five or fifty percent more work may be worthwhile in a good cause. But it really isn't good if the work is really double or triple, months or years beyond your emotional or physical resources.

The Hassle Budget

Taking the major role in your house-building project can be very satisfying. There is the freedom to do things just as you wish. There is the opportunity for innovation and experimentation, and a testing of your limits. When you over-extend your limits, the freedom and responsibility get transformed into stress. The adventure becomes hassle.

For many people, there is ample adventure simply in working with a paid designer and builder to build a relatively conventional house, pick colors, buy furniture, and find the money to finance the project. Others really want to be the principal designer and builder, and welcome the uncertainty and risk involved in those roles. They see the process as an opportunity to experiment with ideas and systems. Most people are in between.

Your "hassle budget" is related to your time budget, but is not the same. It has more to do with your tolerance for risk and uncertainty. You could put in a huge amount of time on a relatively straightforward project, and the uncertainty level would be low. You could invest an equal amount of time on a more innovative or experimental design, where the carpenter or architect are as much in the dark as you are, and the uncertainty would be greater, the cost of making mistakes steeper.

Though this is an aspect of building that is impossible to quantify, I think it is important to meditate on what you and your family can tolerate. In general, a larger, more complex house, built with more esoteric materials and innovative systems, raises the potential for both adventure and hassle.

Support

The final resource you should try to assess is the support of people with expertise. This resource, of course, usually costs money, and is also closely related to the "hassle budget," because experienced people can share the responsibility and uncertainty of decisions.

Your choices here are many. You can hire out everything, or do everything. Usually the middle ranges make the most sense, so you end up doing some aspects of the project—maybe the design and carpentry—and "subbing out" (subcontracting) the rest.

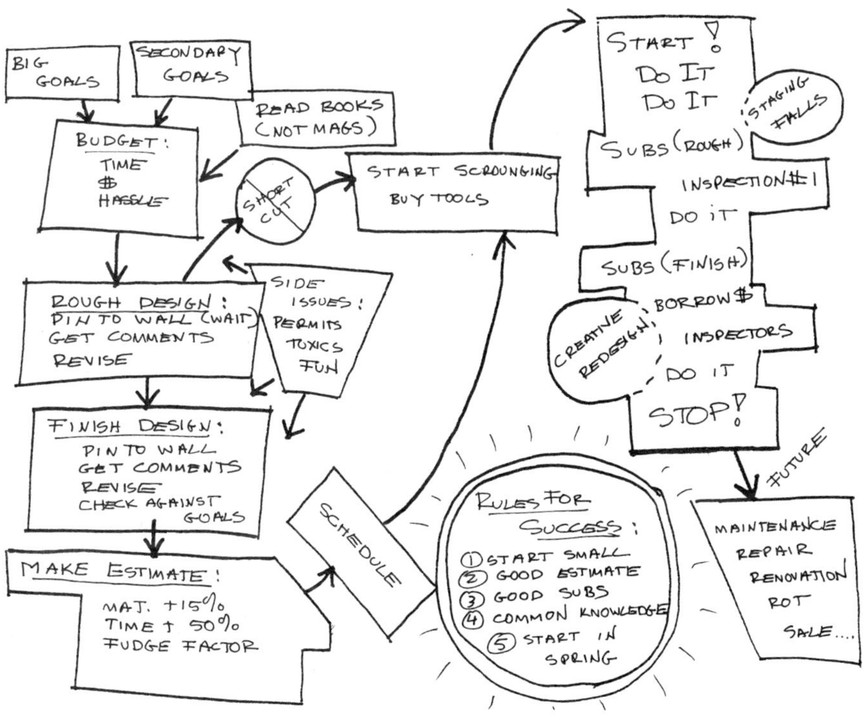

Figure 4-1. Universal flow chart.

Your level of participation may involve working alongside a professional crew (I often think that is the ideal scheme), or doing most of the work yourself. You can always have experts lined up to consult with at crucial moments.

Before designing in earnest, I think it makes sense to estimate or eyeball all these resources—money, time, your tolerance for "hassle," and your desire for the support of experts—until you have made a portrait of your project that can really work, and that defines a venture that you will truly enjoy. This is your total resource budget. Then, as your design evolves, frequently use this budget as a benchmark. Keep your design within its limits, to keep the project an adventure rather than a stressful hassle.

DESIGN GOALS

Givens

Your design goals include ideas about the size of the house, building systems, its look or style, and special ideas about space use: how you want to use the house. A few of these goals, I'd consider givens; they always should apply. The house should be:

Energy-efficient
Well built
Low-maintenance
Well sited

I'll urge the reader to add two other givens:
ergonomic
sustainability

Construction Systems as Goals

People with an interest in certain building technologies often don't merely want to build a house to meet family needs. They want a certain kind of house, a unique house, outside of building convention. Such building technologies may include heavy timber frames, solar homes, sustainable design, or log homes, which are very popular in New England.

Building Styles

Other popular ideas have more to do with certain styles, such as the Craftsman style, or traditional forms like Capes or Colonial houses, or add-on features like a greenhouse, root cellar, screened porch, or cathedral ceiling.

Size

Finally, people have definite ideas about the size of house they want, and about the types of rooms they need. Of course, the building industry is also always promoting larger space, and different kinds of spaces come and go in popularity. At one time "family rooms" were emphasized in suburban houses. Today the building industry is pushing more and bigger bathrooms, home offices, master bedroom suites, and exercise rooms.

I'll continue to make the case for small houses often in this book, and not just because they are cheaper. A well-designed small house can be as good to live in as a larger house that isn't as clearly worked out.

The Asterisk

I think your budget and other resources, and the "givens" I've listed, should be clung to tenaciously. But the other goals, like size, style, and construction system, should have an asterisk next to them, meaning: "If they survive careful scrutiny."

When people first talk to me about helping them build a house, I often hear, "I want a fully solar, super-insulated, Craftsman-style home, with hand-laid stone walls, greenhouse, two-car garage, ping-pong room, a place for huge family reunions, etc. . . ."

But when we analyze who will live in the house, what they need, and what they will actually do there, the list changes. Some of it, maybe the stone walls, gets cut for cost reasons, but the list can also change with better understanding of actual life needs. The ideal image in the future homeowner's mind may be of a greenhouse, but the need may really be some plant shelves in a sunny window. The fantasy may be of huge family reunions, but the truth may be those reunions never happen. On reflection, you might realize that however much you loved to play ping-pong twenty years ago, you don't now, and anyway your kids are moving away in six months. On the other hand, there may be other needs or functions that are important day-to-day, but haven't found their way into the house fantasy yet, like a nice warm home office, or more storage.

The next chapter, Preliminary Design Work, provides ways for looking at your building site and at your life needs and activities, in a concrete and objective way. The chapter also provides a way to look at rough building costs you might be facing. The key to all of this is to describe the house by analysis.

During this process some of the goals you have stated will get deleted, because they don't fit your site, your life, or your budget. Other things will be added, and greatly enrich your design.

Figure 4-2. *Habitat for Humanity: A crew of amateurs put up the frame of a stick-built house in a weekend. Central Vermont Habitat for Humanity. (Photo: Esther Farnsworth.)*

THE SEMI-GREAT HOUSE: MY PHILOSOPHY

As I begin to imagine building a house for myself or for a customer, my starting point is a respect for the way most American houses have been built since about 1860, which is often called stick-building, or light timber framing. Framing the house is fast because the elements are small, light, easy to cut, and easy to nail together. Plus, the material is highly adaptable; you can build a simple cottage or the most complex Victorian using basically the same method. It's easy to add or remove a wall or window, or put on an addition, because the frame is strong everywhere; it doesn't depend on a few points of strength. Stick-building is a people's technology. Many people who aren't builders by trade know how to frame, and with a little effort, most anyone can learn.

But what about all the wonderful alternative technologies, innovative systems, and new ideas? A lot of the best "cutting edge" ideas get absorbed rather quickly into routine building practice. The rest of the alternatives, from the ancient arts of timber framing or slate roofing to the most recent wall systems, become for me alternatives for special situations.

Sometimes the alternative approach solves a specific design goal: an old window looks better; the use of timber framing solves a structural problem or creates a desired look. Sometimes special site problems or resources call for special approaches. Just as often, the owner is just plain interested in or enamored of a particular approach, or wants to learn more about it by building a house. For example, some people have simply decided they want to build a straw-bale house. A strong interest is a valid reason in itself, even if there are potential practical drawbacks.

If I'm about to try an alternative approach, I ask some questions. What does it cost in time and money? Do the benefits balance the costs? Does it make intuitive sense? Are the necessary supplies available, and do local people know how to install or maintain them?

As a builder working with a budget, I have to be sure the system costs out on every level. But as an owner-builder, I can take a chance, as long as I'm not kidding myself about the costs, short- and long-term.

The Issue of Degree

Owner-builders are often drawn to complete building systems, a totally timbered house or a fully solar house, for example. My philosophy is to use these ideas or technologies in a less absolute way. I'd like the house to be energy efficient, but I don't mind burning some wood to heat it. I don't want unnecessary toxic burdens in my house (I'm careful about buying new carpets), but I'm not really convinced that a little varnish will kill me. We insulate our buildings carefully and thoroughly, but don't spend thousands of dollars extra to set records for low heating bills. We use recycled materials when they are good materials.

While I'd like to build the perfect house, it make more sense to me to design and build the pretty-efficient, largely non-toxic, mildly recycled, partially timbered, semi-great house. It will not only cost less, but wear better than a more innovative or radical design, or a more literal interpretation of a design philosophy.

Design vs. Technology

Owner-builders often put tremendous thought into the technical side of building, but the rest of the design isn't always as fully worked out. As a result, the layouts might not work quite right, the kitchens might not function very well, the windows might be the wrong size, or the house may be a bit out of proportion.

The more I build, the more interested I become in non-technical aspects of design. I spend a lot more time than I used to on window placements, proportions, ceiling heights, roof pitches, site plans, and other details that seem small, but have a big effect.

Mastery of these details is difficult. One of my favorite activities, maybe it's a

Figure 4-3. *Prettiest entry in Scotland.*

Figure 4-4. *The Cottage. The Sheldons house lives again — with a poor man's cathedral ceiling. The Cottage, Goddard College. Addition built by Iron Bridge Woodworkers.*

compulsion, is to look at every house I pass and collect details and images, as a birdwatcher collects new bird species. Until I've seen every house, inside and out, my collection will be incomplete. Of course, some of these houses are in books, too.

At times I've been particularly taken with Shaker or Japanese house design. Lately I've been thinking more about Adirondack buildings, Carl Larsson's house in Sweden, and Craftsman-style details. Much of what I design borrows, though perhaps "copies" is a better word, from these models.

The design methods I've added to this book—particularly ergonomics and Alexander's *Pattern Language* approach—are the formal ideas that have helped me most. No reader's time will be wasted by giving these ideas ample consideration.

5

CHOOSING A HOUSE SITE

"On no account place buildings in the places which are most beautiful. In fact, do the opposite. Consider the site and its buildings as a single living eco-system. Leave those areas that are the most precious, beautiful, comfortable, and healthy as they are, and build new structures in those parts of the site which are least pleasant now."

—*from* A Pattern Language

YOUR SITE IS NOT ONLY THE LOCATION OF YOUR HOUSE, but the source of what makes it livable: water, sunlight, air, wood for heat and lumber, soil for a garden, and other necessities of life. A house is really an outgrowth of the land.

A good site will be sunny. It will have good water, preferably nearby and uphill from the house. Electric lines will be close enough so that power can be brought in cheaply. The road will be well maintained and nearby, but not so close as to destroy privacy. Perhaps there will be a south-facing slope for a garden and a good wood lot. Such a site helps you build.

The things that are provided free by nature or circumstance on a good site are expensive when you have to bring them in. It can cost thousands of dollars, for example, to bring water from a remote source or to dig a well. It can cost hundreds of dollars a year extra to heat a house if there is little sunlight. Beyond cost, a house on a poor site will be less of a pleasure to live in. For these reasons, choosing a house site is the most critical decision in the building process. The fact that land has become more expensive, and more closely regulated, makes this even more important than it used to be.

EVALUATING POTENTIAL SITES

Though no site is perfect, you can do two things to make sure the site you choose will be suitable. First, compare potential sites systematically, using the procedures given below. Second, when you have found one you like, have someone who knows land and building look at it with you. This is one place where an expert opinion is worth the investment.

Your evaluation and an expert's take time. And sometimes you must make a decision to buy quickly. If this happens, you can usually secure the land for a few days for a small, nonreturnable down payment (often called "earnest money") while you inspect the land more thoroughly. You risk losing a few hundred dollars, but you avoid the possibility of losing thousands on an attractive but faulty piece of land.

Sunlight: When you are looking at building sites, take a compass. Sunlight comes to you across the southern sky. In winter the sun makes a brief, low trip across the sky from southeast to southwest. In summer it makes a slow, high trip starting in the northeast and ending in the northwest. For a house to be light, it must have good southern exposure and some east and west exposure. The southern horizon should not be obstructed too much by woods or nearby hills. A south-facing hill or a level site is ideal from this standpoint. A north-facing hill is poor.

Not everybody will find a perfectly oriented site, and many otherwise excellent sites will have only fair sun exposure. But if sunlight is a priority for you, study the sun orientation of each site carefully. Try to imagine how a house could be built to take advantage of what sunlight there is.

Water: Availability of water is crucial. In some areas you can get your water without drilling an expensive well, by using a stream or spring. A spring is a place where water surfaces from underground, rather than a place that collects surface run-off. If your land has such a source, you must find out if it can be developed into a good water supply. Taste the water. Look around for obvious sources of contamination. Manure, fertilizer, domestic sewage, road salt, animal droppings, and industrial wastes can all contaminate a water supply. Find out where to have water tested, and have a sample from the spring checked. If tests prove that the water is contaminated, find out if the contamination can be eliminated. If your land has a stream, evaluate it as you would a spring. Remember that there are relatively few places left where surface water is reliably potable.

Next, is there enough water? Sometimes your future water supply will be a wet spot in the ground when you first see it. Dig a hole, two feet square or so, down past the water level. Let the spring fill the hole you've dug with water. Note the level. Take a bucket and quickly remove about ten gallons from the spring. See how long it takes the spring to refill. If it takes one minute, that's ten gallons per minute, a large amount of water. If it takes ten minutes, that's only one gallon per minute. Such a small amount will supply your needs only if you can collect it all. Four gallons or more per minute is adequate for most households.

Is the source permanent? Many streams and springs vanish in dry months. The only way to check this is to ask someone familiar with local conditions. With persistence, you can probably find someone intimately familiar with the particular piece of land you are looking at—someone who has farmed the land, hunted on it, or who does excavation work in the neighborhood.

If you know there is enough good water, next figure out how to bring it to the house. The distance between the source and the house, the depth of the frost line, the relative height of the house and of the source, and the availability of electricity for a pump will all affect the design of a system for a particular water source. Chapter 20 describes different systems for different situations.

In general, find out how water systems work in your area. What works in Vermont won't necessarily work in California. Study the systems of local farms, because farmers often must construct permanent and reliable systems with limited funds. Check local regulations and codes.

If no spring or stream can be developed, you must have a well drilled. If the drilling must go through bedrock, it costs about $8/foot. If your are drilling through soil, clay or sand, a pipe called a "casing" is also needed, which costs another $8/foot. The pump and pressure tank add another $1,500.

When drilling a well, you always gamble on the depth at which water will be found. This can be anywhere from thirty to five hundred feet. Ask around the neighborhood to get an idea of the depths of typical wells. Find out what drillers charge in your area, and estimate the potential cost.

Electricity: To find the cost of electric power, measure the approximate distance from the power pole nearest to your building site. Note the number of the pole (usually marked on the pole) and then talk to the power company. If the line is to go through the woods above ground, you will have to cut a path for it thirty or more feet wide to keep trees from falling on the line in storms and to facilitate maintenance. While you are at the power company office, ask about a temporary electrical hookup to use while building. Find out when it can be put in, what you must do first, and what it will cost. If your temporary or permanent power must come across a neighbor's land, get a legal right of way before you are committed to that building site. Do not rely on verbal understandings.

It used to be impractical to make your own electricity. Today photovoltaic ("PV") electricity, though expensive, is an option to consider for a desirable site that is far from "the grid." PV technology has improved greatly in recent years. If it costs $30,000 to bring in power, a $4,000 to $10,000 PV system may look attractive, particularly if the land is affordable compared to other locations. The PV system won't provide the limitless power that conventional hookups provide, but many families are now living off the power grid comfortably by designing their homes and electrical systems to minimize power use.

Phone Service: Bringing a phone to your house is not always routine and inexpensive, even if a power pole is nearby. Always check with your phone company to find out what it will take to get a phone line put in.

Wind: Wind has a huge impact. Ideally, a site will afford you some protection from winter winds, while leaving you exposed to cooling summer breezes. Find out as much as you can about wind speed and direction on your site. Usually the prevailing wind will change with the seasons. Call a local radio station or airport to find out typical wind directions in winter and summer, and consult the neighbors. On site, put threads or light cloth flags on stakes in one or more spots to make direct observations. Often a site can be improved by cutting trees that obstruct the summer breeze or by planting trees to moderate the winter wind.

Access: If the land you are considering is surrounded by other people's property, make sure you have a legal right of way to it that will meet your needs. If you are contemplating building far from an access road, figure out what it will cost to put in your driveway.

Ask neighbors if the public road that leads to your land is passable year-round. If it turns to mud in spring, or is not plowed in winter, see if the town will maintain and plow it once you move in. In some cases towns must upgrade back roads when people build on them.

Zoning: Most communities have laws that regulate building. In my town, you must build seventy-five feet from the road and one hundred feet from the nearest watercourse. In our zone you must have a lot of five acres, while in other zones you can build on one acre. You can't build at all in the Forestry Zone. In the village itself there are rules that specify how a house must be built, for example, what kind of plumbing systems can be used. Get copies of zoning laws and building codes from the town clerk. Check the zoning map yourself. Somebody might try to sell you forestry land as a building site or a one-acre lot in a five-acre zone. Also, find out what permits are required, how long they take to process, and what they cost.

Drainage: Good drainage is important if you want your house to be economical to build and maintain. A well-drained site is one that (1) is not subject to flooding in extreme conditions, (2) will not be soggy during the normally wet parts of the year, and (3) has soil that is porous enough to install a well-functioning septic system, if you plan to build one or are required to by local law.

Look at the overall topography at the site. A plain by a river (flood plain) may be subject to floods. If the house site is a valley that collects surface run-off, you may need special drainage to prevent the site from being very wet during the wet season. A depressed area may collect water like a pond and stay wet for months of the year. A flat area, ridge, or gentle grade is most likely to be free of these problems.

If the topography seems suitable, look at the surface characteristics of the land. If there is a lot of surface erosion, the soil may be too unstable to support the kind of vegetation you want around your house. Do not build in a swampy area. It can breed bugs, cause your house to settle unevenly, and flood your cellar. In swampy areas the land acts as a basin to trap water, usually because the ground just below the surface contains springs or consists of clay or rock, which will not absorb water. Some areas will be dry part of the year, but very wet during the rainy season or during the spring thaw. The presence of cedar trees, ferns, or other swamp vegetation may indicate water problems at other times of the year. This does not mean that any wet place on your land is bad. But the house itself should be on a dry spot.

Most communities, even in rural areas, require septic systems, which work only in soil that is reasonably well drained. Generally, communities set standards for septic systems and require percolation (perc) tests, which measure the ability of the soil to absorb water. It may be impossible to build on a site if the perc rate is too low. Therefore, if the general topography of a site looks good, your next step is to investigate whether a septic system is feasible by digging some holes in the ground with a posthole digger. These holes will enable you to conduct a makeshift unofficial perc test, and also investigate the suitability of the soil for building. Before you do the digging, however, find out what the local regulations are concerning septic systems and what kinds of foundations are permitted. The answers will help you interpret what you find when you dig.

First, imagine where a septic system might be located. This should be an area near the house, at or below it in elevation. The area should be one hundred feet away from your water supply and not uphill from it. Local codes will probably require the system to be one hundred feet from any stream to avoid pollution. The septic system area should be five hundred to one thousand square feet in size, depending on the perc rate and the house size. The area should be fairly level, since a septic system will cost more and work poorly on a steep slope.

When you find a likely spot, dig several holes, four to twelve inches in diameter, two or three feet deep. Dig one of the holes as deep as you can go with the posthole digger. Dig another extra-deep hole in the middle of the spot where you are thinking of building.

The soil that comes out of the holes you dig will tell you what the layers of soil in the ground are like. Feel the soil every few inches with your hand. If you find that it is fine, sticky when wet, and feels like putty, you have found clay, or soil with a high percentage of clay. This can cause several problems. First, clays expand when wet, which can cause the soil to move and basement foundations to crack. Second, clay layers are often impervious to water, so the ground may stay wet and your septic system may not work well. Third, on a sloping site, a clay layer four or more feet down can cause slippage, which occurs when the entire soil surface above the clay slides downhill more or less in one piece.

If you hit bedrock (rather than boulders) you may have other problems. A rock layer can interfere with a septic system. In some towns, you may be required to have six or more feet of soil above the rock to provide four feet of soil below the septic system trenches. Rock is impractical to remove for a basement foundation, though a house without a basement can be easily built on rock. When the rock is four feet or more down on a sloping site, there can sometimes be slippage just as there can be over a clay layer.

You also have problems if you hit wet soil, or if the holes fill up with water. Unless the wetness is caused by recent rains, it may be that the ground drains very poorly or that you have a high water table. (The water table is the level at which the soil remains saturated with water.) Sometimes local regulations require that the water table be four feet below the septic system (or about seven feet below the surface) in the wet season.

A high water table in any case can cause basement foundations to flood and any type of foundation to settle unevenly or heave up from frost action. If a hole only three or four feet deep hits water you are probably on a bad place to build.

If you find soil below the topsoil that has some proportion of sand or gravel in it, and you do not find clay or water, you probably have a soil that can support the weight of the building, and that will be generally well drained.

Continue with your perc test if you haven't hit rock or water by filling the bottom of each hole with about two inches of gravel. Pour at least twelve inches of water into each hole. Add as much water as necessary to maintain the water level for four hours, or preferably overnight, particularly in a dry period. The idea is to wet the ground thoroughly, as it would be in a wet season, and then see how well the soil absorbs water. When the soil is saturated, adjust the water level in each hole to six inches above the gravel. Every thirty minutes, measure how much the water level has dropped and fill the hole up to six inches again with water. Do this for four hours. Use the last measurement to calculate the perc rate. With several holes, average the results. Perc rate is expressed in inches per hour. Thus, if the water sinks one inch in a half-hour, that is two inches per hour. One inch per hour is the minimum rate. If the rate is faster, a septic system will cost less and work better.

The United States Soil Conservation Service, or the local Agricultural Extension office may have conducted detailed soil surveys of the area. They may have soil maps of your land.

Ultimately, you will need an official percolation test to be allowed to build. If you think you want to buy a piece of land, have your offer to buy contigent on a successful perc test. Some land sellers will have done this already to make their land more saleable. This perc test is done by a back-hoe operator, working with an engineer. The test pits will be placed where a septic system might go, relative to your prospective house site or sites. The engineering of the septic system in turn will depend on the results of this test. If the soils drain well, a septic system can be economical, but if it drains poorly, it may be necessary to build a "mound system." This basically means building an area of ground that drains well to house your system. Since a mound system is expensive, you may decide to offer less money for the land.

Other resources: Big softwood trees are good to build with. Big hardwood trees are good for firewood, shade, and perhaps for sale. Old stone walls can be made into fireplaces. Is there a place to build a pond, and water to fill it? And, of course, what does the land look like? A spot may be beautiful, but will it be beautiful after you've built a house on it? Putting a big man-made object in the middle of a beautiful place will often ruin it. Visualize how your house will fit the landscape. Perhaps there will be a dip, corner, or other place in the landscape where a house will fit without undermining the natural form of the land.

How much land? When looking at a piece of land, don't just think of how big it is or how much it costs per acre. The question should be, how many usable acres are there for the money? How well will the spot suit your specific needs? A hundred acres can cramp you if the only usable place is one acre along the road. Five acres, well placed, can provide privacy and enough land for a small farm. A big forest may have been stripped of all its good wood ten years ago; three acres of good trees may contain more of actual value. A hundred acres of inaccessible backwoods might give less enjoyment than five or ten of varied land. Remember that you don't have to own land to enjoy seeing it or walking through it.

6

PRELIMINARY DESIGN WORK

HAVING FOUND SOME LAND, HOW DO YOU DESIGN A house for it? A common practice is to take an idea of a house seen in a book, in one's travels, or in one's mind's eye, and just put that house on the chosen site. I advocate something different: that you take the information you have about your land, your needs, and your budget, and develop a house design from that. Chapters 2 through 5 present a specific set of steps for this process. If you don't follow every step to the letter, be sure to consider the questions raised.

MAKING SITE PLANS

A site plan is a map that shows the characteristics of the land and helps you locate your house and important features such as well, septic system, roads, gardens, and outbuildings. This can be as simple as a sketch showing prominent features, as in figure 6-1.

The land may have been mapped by an engineer or surveyor. If not, you can make your own map. This is a great way to learn about your land, and the map you produce will be very useful as you design a building. You can also make a surprisingly precise map using an orienteering compass and some sort of leveling device, such as a transit or line level.

Basically, your orienteering compass is a compass with a rotating base. (Read the little booklet that comes with it for more detailed instructions.) Start by setting up some sort of drawing table at the site at a good starting point, perhaps the very spot you are considering building on. Set up the table with a nice piece of 18 x 24-inch drawing paper. Draw a vertical line that will represent magnetic north. You'll avoid confusion if you use your new compass to set your paper so that magnetic north on your map aligns with real magnetic north.

Standing near your drafting table, line up the arrow

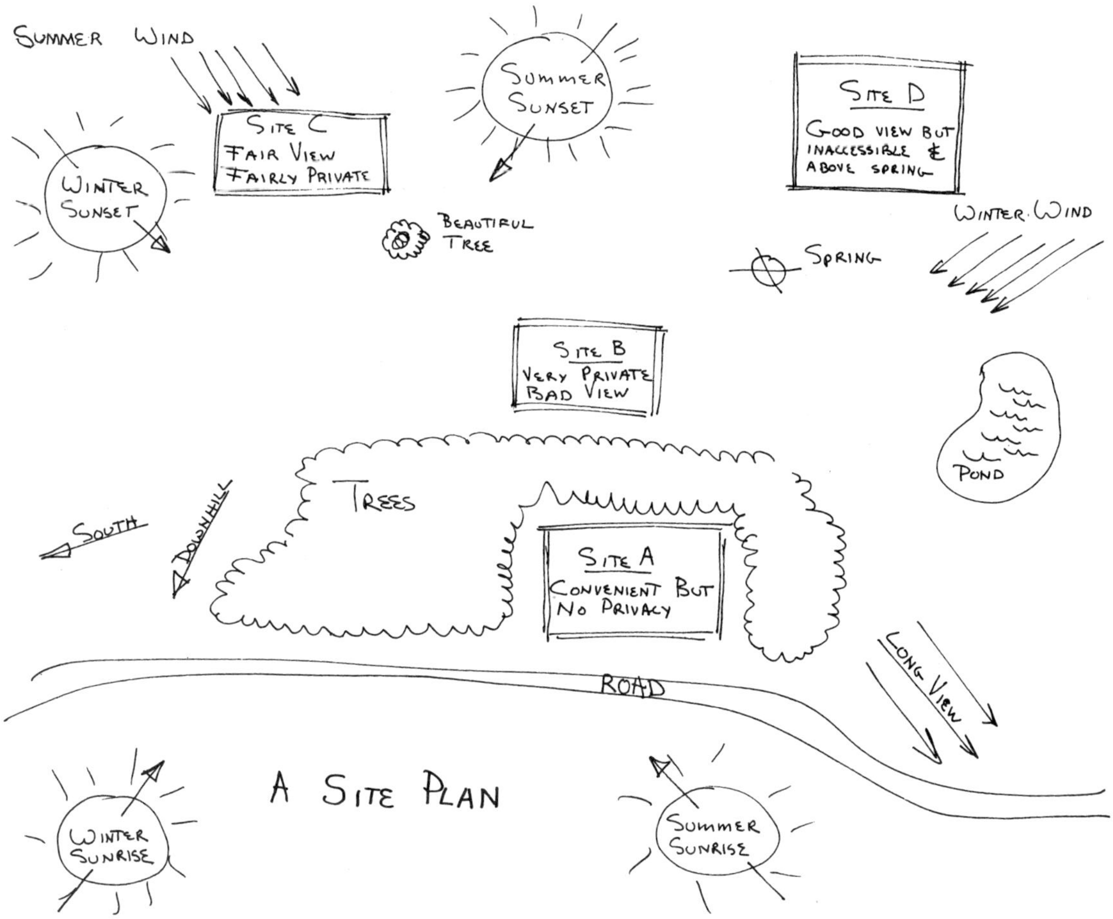

SUMMER WIND
SUMMER SUNSET
SITE C
FAIR VIEW
FAIRLY PRIVATE
SITE D
GOOD VIEW BUT
INACCESSIBLE &
ABOVE SPRING
WINTER SUNSET
BEAUTIFUL TREE
SPRING
WINTER WIND
SITE B
VERY PRIVATE
BAD VIEW
POND
TREES
SOUTH
DOWNHILL
SITE A
CONVENIENT BUT
NO PRIVACY
ROAD
LONG VIEW
A SITE PLAN
WINTER SUNRISE
SUMMER SUNRISE

Figure 6-1. Site Plan.

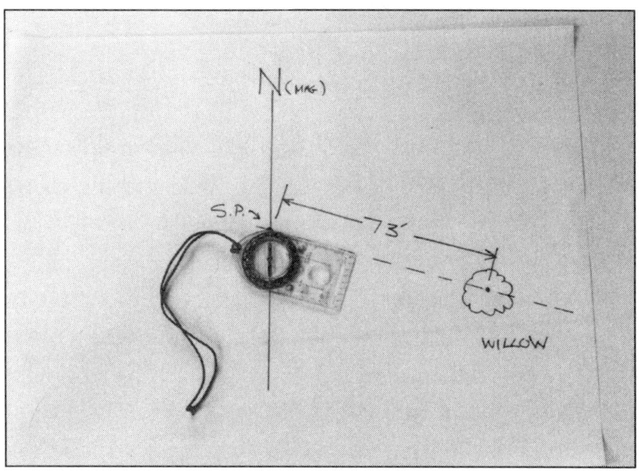

Figure 6-2. *An orienteering compass can be used to map your land. Align magnetic north on your map with magnetic north on the compass.*

with the North "N," and turn the base until its edge points toward an object you want on the map, say a willow tree, as shown in figure 6-2. It may help to use a yardstick, held against the compass's rotating base, to sight along.

Place the compass on the drawing, with the edge of the rotating base intersecting the starting point, and the dial pointing north. Draw a line along the edge of the compass's base. The willow will be on that line.

Use a fifty-foot tape to measure the distance from the starting point to the willow, then mark the willow on the drawing, in scale. In my example, the scale is ⅛ inch = 1 ft.

To figure the difference in elevation from the starting point to the willow, you need to establish a level line that goes through both points. The easiest way might be to borrow a transit or builder's level (figure 6-3), but you can also use a little level that hangs from a string, called a line level, or a little portable hand level that you sight through like a telescope.

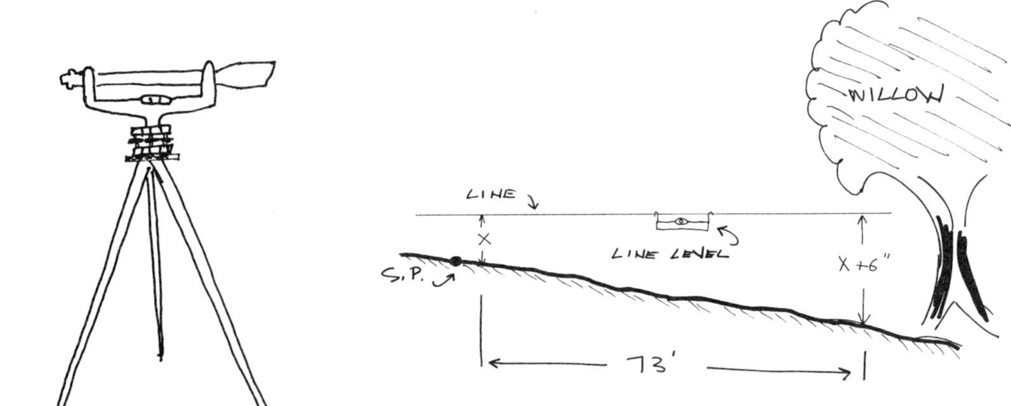

Figure 6-3. Builder's level.

Figure 6-4. Establishing the grade: the willow is 6 inches lower.

Whichever tool you use, establish your level line, then find the difference between that line and the two points. In this case, the ground is sloping off toward the willow about six inches, and I mark that on my map as S-6 inches.

In the same way, you can mark any number of key points on the site, giving angle, distance, and difference in elevation. If need be, you can set up a new starting point for convenience, but keep calculating the elevations in relationship to the original spot.

If you take a lot of readings, you can connect all the spots at the same height to establish contour lines. I'm not sure that is necessary.

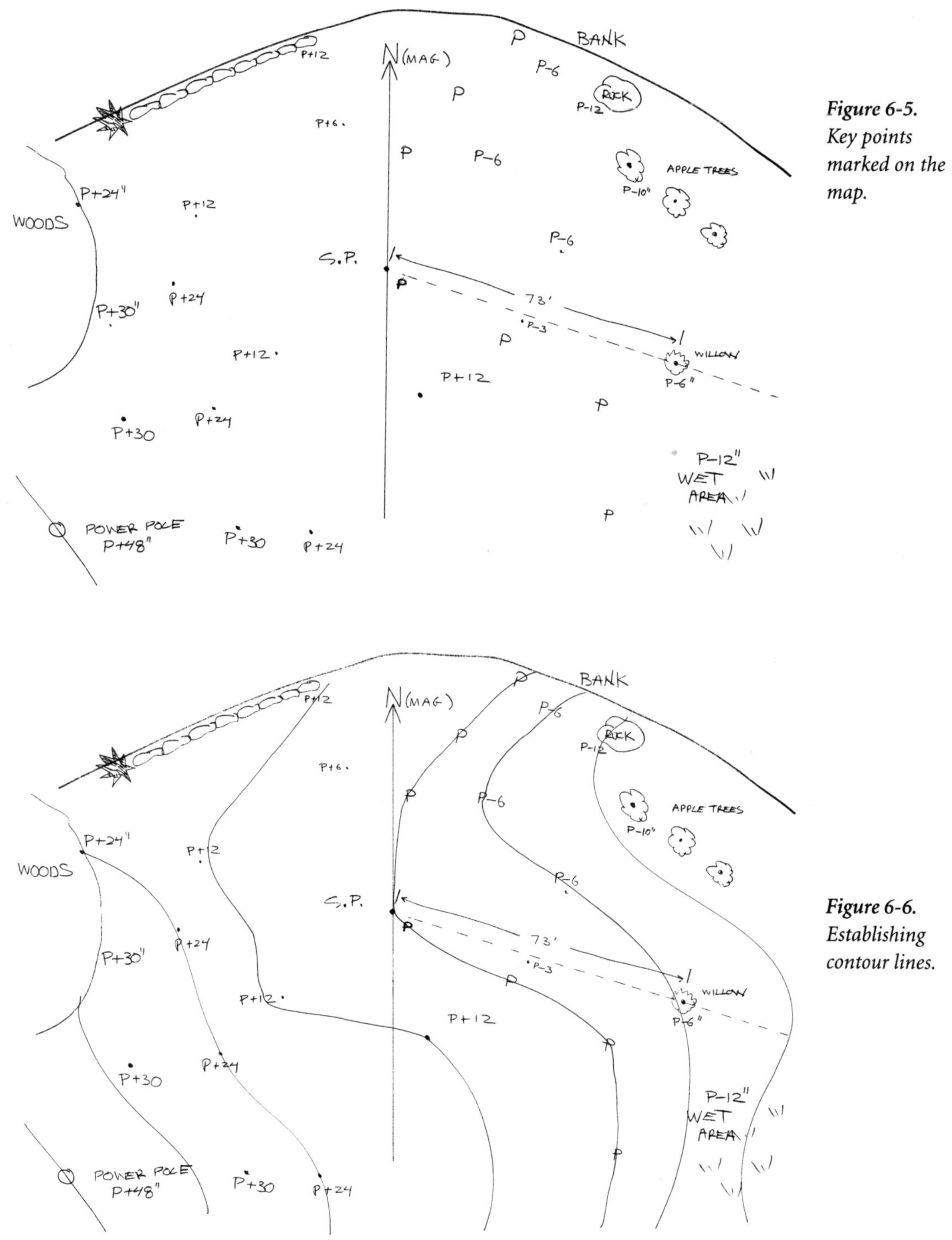

Figure 6-5. *Key points marked on the map.*

Figure 6-6. *Establishing contour lines.*

On your map, note existing features such as roadways, fence lines, wetlands or waterways, bedrock, significant soil conditions, trees or woods, views, power lines, and so on.

By referring to page 230–232, you can add in true north and some basic solar information, such as where the sun rises and sets at different seasons.

Most zoning codes will require you to build a certain distance from your boundary. Draw these "setbacks" on the site plan.

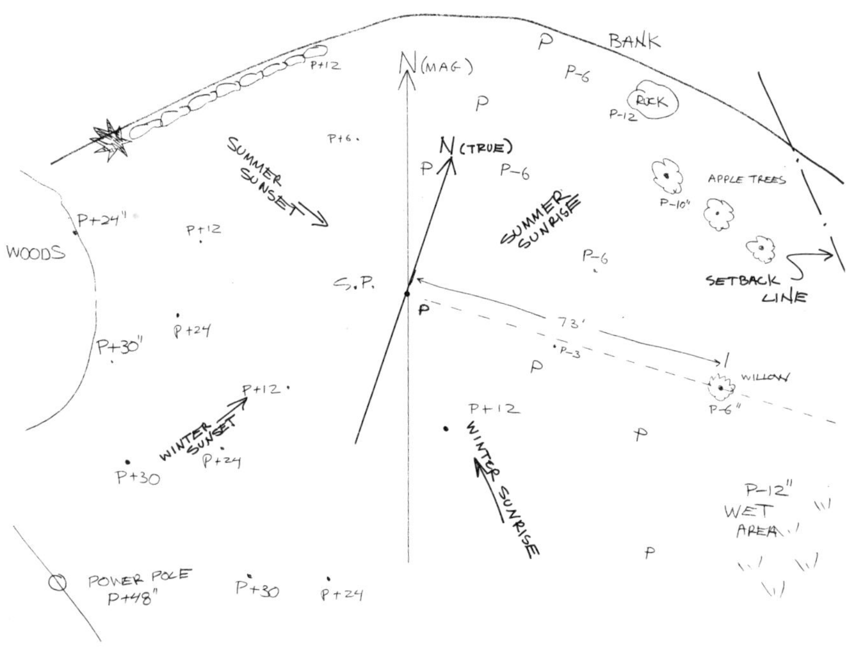

Figure 6-7. Complete site plan.

CHOOSING WHERE TO BUILD

Use tracings or multiple copies of your map to test out a variety of possible house sites. When you build a house, you are usually adding many things to a site plan, including:

Roads
Paths
Gardens
Outbuildings
Power lines, buried or on poles
Water lines
Septic tank and leach field

As you study your maps, indicate where these items might be located. From a rigorously practical point of view, you might pick the spot for your house close to the road (screened by trees), yet not too far from the power pole, where a drive could be constructed without steep grades. Another viable house location might have a better view, more privacy, or work out better for gardening or other special uses.

The cost-intensive items to remember (and keep simple) are the water system, distance to power lines, and length of the driveway.

In terms of the energy efficiency of your site, compare solar gain and wind exposure. In terms of aesthetics, compare long, medium and short views, keep in mind how they will change with the house in place and with the seasons.

Probably more long-term trouble with and damage to houses comes from wet soil conditions than from any other cause. Foundations move around, basements are wet, and sills rot. You may have learned a lot about the soils when perc tests were done. Often the back-hoe driver who does this work can tell you a lot about the soil he or she brings out of the ground, and by seeing whether the holes themselves are wet. Look for well-drained soils, and avoid boggy areas. Even if a spot seems dry now, wetland plants or trees (ferns or cedars, for example) will be a sign that the spot is boggy part of the year. Such spots might make a better place for a future pond than for a house.

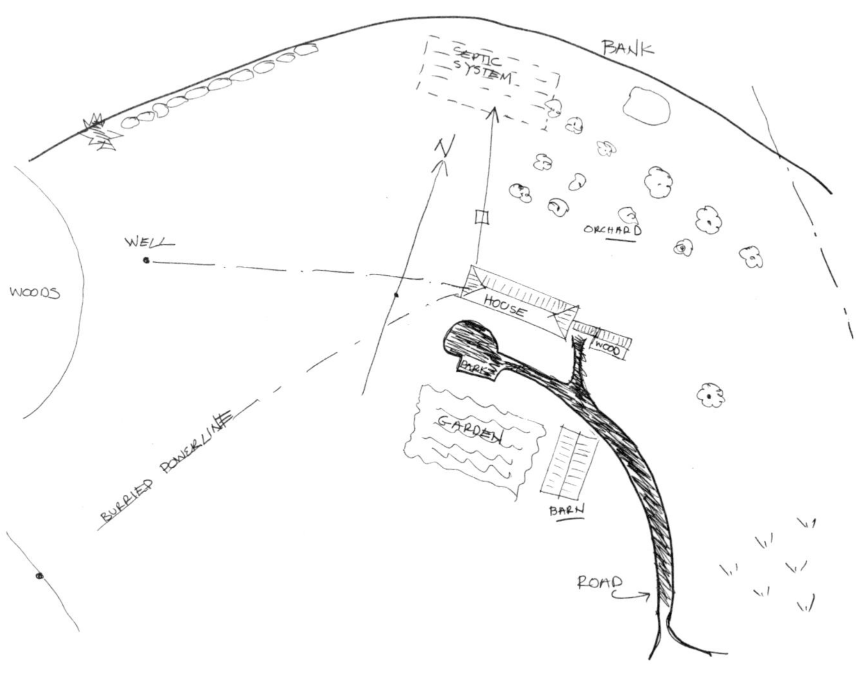

Figure 6-8.
Site development: one possibility.

The shape of the land is important to drainage, too. Pick an area that doesn't form a collecting point for a lot of uphill land, yet does provide an escape path for water. On the other hand, though it's best to work with the lay of the land as much as possible, a bulldozer and a backhoe can rectify nature's oversights amazingly quickly. Not everyone will have the luxury of choosing a perfectly arranged site.

LISTING GOALS AND ACTIVITIES

Our goal is to work toward a description of the new house through understanding how you'll use it, rather than jumping to a design based on tastes, fashions, or what your friends say you need.

Start by listing some of the general features or qualities you want, such as:

Lots of privacy
Sun
One (or two) story(ies)
Good views
High (or low) ceilings
Fireplace

Next list functions. This should generally not be a list of rooms but of activities and functions:

Cooking for big groups
Kids sleeping
Adults sleeping
Guests sleeping
Eating
Sitting
Studying
Reading
Home workshop
Storing wood
Greenhouse
Washing
Sauna
Sewing
Sheltering a car
Fixing a car
Cooking
Storing food
Root cellar
Canning
Playing games
Roughhousing
Playing loud music
Being alone
Greeting guests
Avoiding guests
Storing infrequently used possessions
Computer

Include everything. This is not the time to be realistic. Activities can be eliminated later if necessary. It might be a good idea to underline or emphasize those activities that are central.

GROUPING ACTIVITIES

Write each activity on small, separate pieces of paper. Shuffle them around in different patterns to see which can be combined. Which functions can occupy the same or nearby space? Which must be isolated? Perhaps you will discover odd combinations that make more sense for you than the usual living room, dining room, kitchen, bedroom, bath arrangement. Perhaps you will want to do your visiting mostly in the kitchen and will not need a separate living room at all. If you need both very quiet and very noisy places, you might be better off with two separate structures. You may want to save money by combining functions:

Eat-sit-visit
Workshop-noisy
games-woodshed-roughhouse
Greenhouse-bathroom

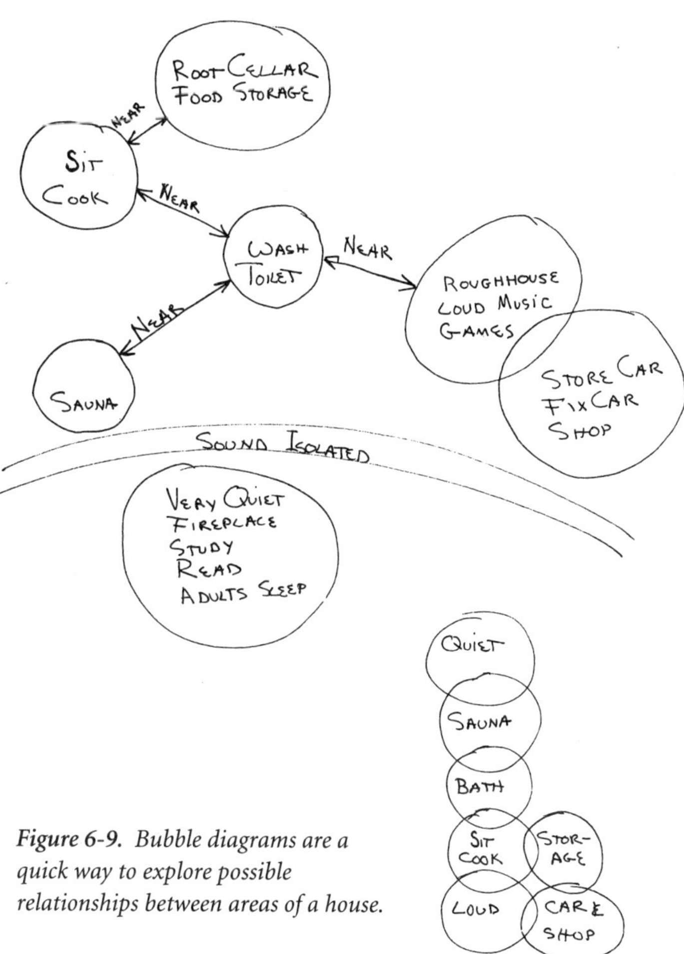

Figure 6-9. Bubble diagrams are a quick way to explore possible relationships between areas of a house.

Many people do not realize how many choices they have. I am right now designing a house with a family. We're talking about a nucleus of one huge room for eating, visiting, and cooking, with a connected open loft for sleeping. This will be an insulated "warm zone," a concentrated house. In warm weather, when people stretch out and many visitors come, the house will also stretch. Huge barn doors will open up the sides of the house, which will then include large screened-in decks to expand the public area. There will also be non-winterized, easy-to-build, cheap sleeping shelters slightly removed from the center of things so that visitors can come in warm weather without destroying everybody's privacy. What do you really need?

MAKING SCHEMATIC FLOOR PLANS: ZONING

The next step is to make rough schematic floor plans using the ideas you have developed.

The goal here is not only to identify what you really need, but to be creating a sense of how the house may be zoned, what activities belong together, which have to be separated if they aren't to interefere with one another, and how areas or rooms in the house can most appropriately connect to the virtues and peculiarities of your building site. You may be able to make the following drawings on tracing paper, placed over your site plan. You could also show the information from your site plan around the perimeter of the paper. This should include sun angles, the road, views, neighbors, wind directions, nearest power pole, slopes, compass points, and other important information.

Using this as a reference, make a series of rough floor plans looking at different design questions. These are not decisions or plans, just ways to explore potentialities and define problems. You may find large sheets of tracing paper helpful in this process.

Sunlight: Sketch how you want your life on the land to relate to the sun. Do you want the sun to rise into the kitchen, or the bedroom, or set where you can see it while you eat dinner? Arrange areas in the house accordingly. For maximum sun, a house should generally face south (figure 6-10).

Privacy and proximity: Often the easiest way to create privacy is to separate areas horizontally (figure 6-11). Ceilings don't isolate sound very well. Some areas might serve well as transition areas between quiet and loud places. Sometimes two buildings is the answer. Privacy may also mean separation from the road, or from neighbors, depending on how much privacy you want (figure 6-12).

A house should be designed so that people's activities don't conflict, but you also want to have some things close together.

If you have young children, they will need play areas, both indoors and out, located where adults can supervise. If play spaces are remote, younger kids won't use them; they'll take over the living room instead. Older kids, particularly teens, need separation. They may be happiest with a space that feels almost like a separate house.

Maybe you want to hear music while cooking, or see who is coming from the kitchen window.

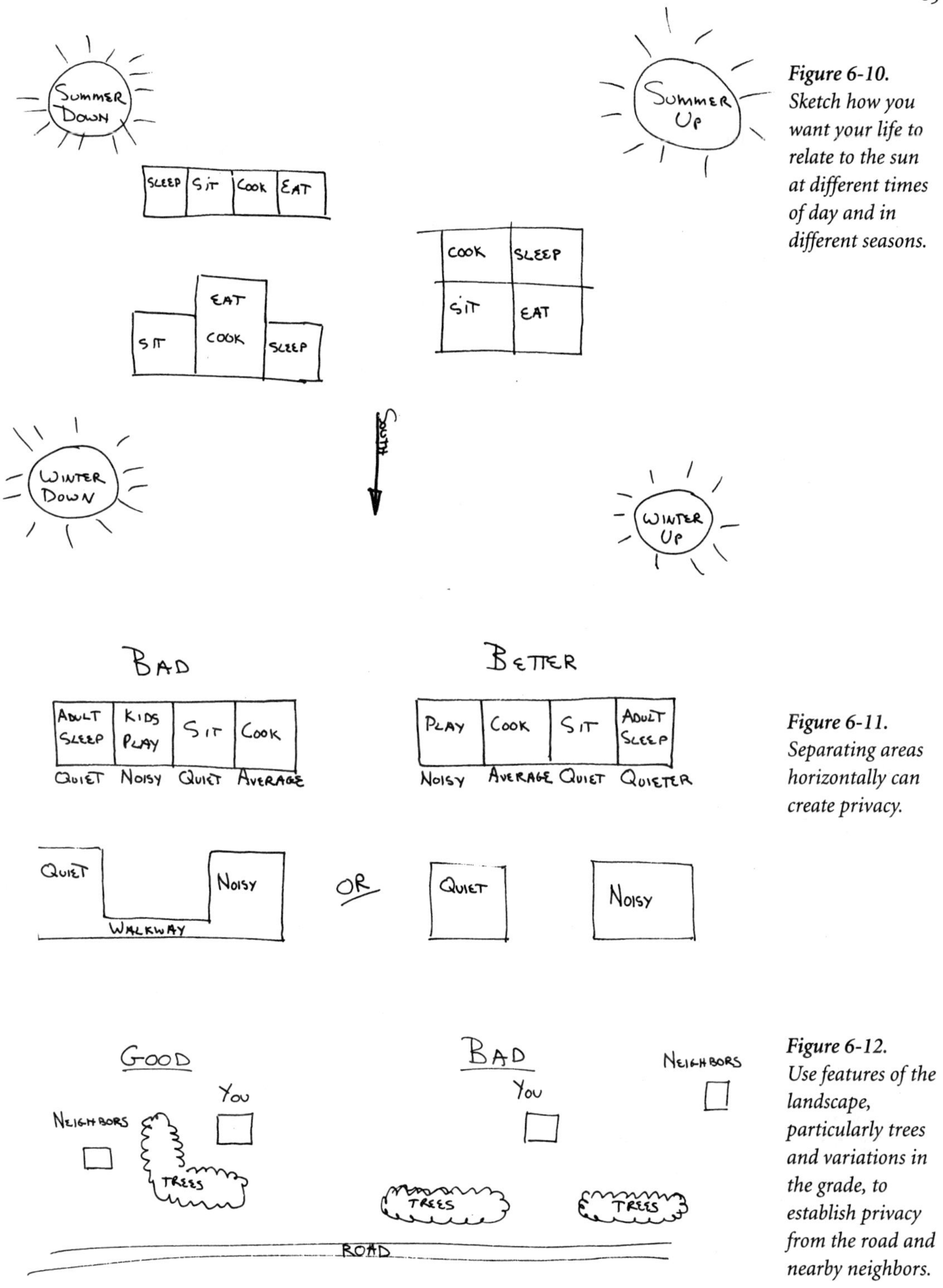

Figure 6-10. *Sketch how you want your life to relate to the sun at different times of day and in different seasons.*

Figure 6-11. *Separating areas horizontally can create privacy.*

Figure 6-12. *Use features of the landscape, particularly trees and variations in the grade, to establish privacy from the road and nearby neighbors.*

Views: "View" can include more than just the sight of distant hills. There are medium-range, short-range, or immediate views that can be just as nice. What would be most interesting to see from where? A particular rock or tree may be the perfect view from a certain bench or stairway. It is better to have a variety of different views rather than to have all windows, floor to ceiling, facing Mount Rushmore (figure 6-13).

Number of stories: The issue of whether to build a one- or two-story house is somewhat complicated. A two-story house may be more compact. It may pack more living space on a smaller foundation, and use less material in construction. A two-story house goes up quickly. It's easier to heat with woodstoves or space heaters, because heat rises more easily than it flows laterally. It can have less surface area than a one-story building of comparable size, so it has somewhat less heat loss.

Some sites suggest a two-story form. A site that slopes to the south, for example, suggests a house dug into the hill for shelter, as in figure 6-14. A second

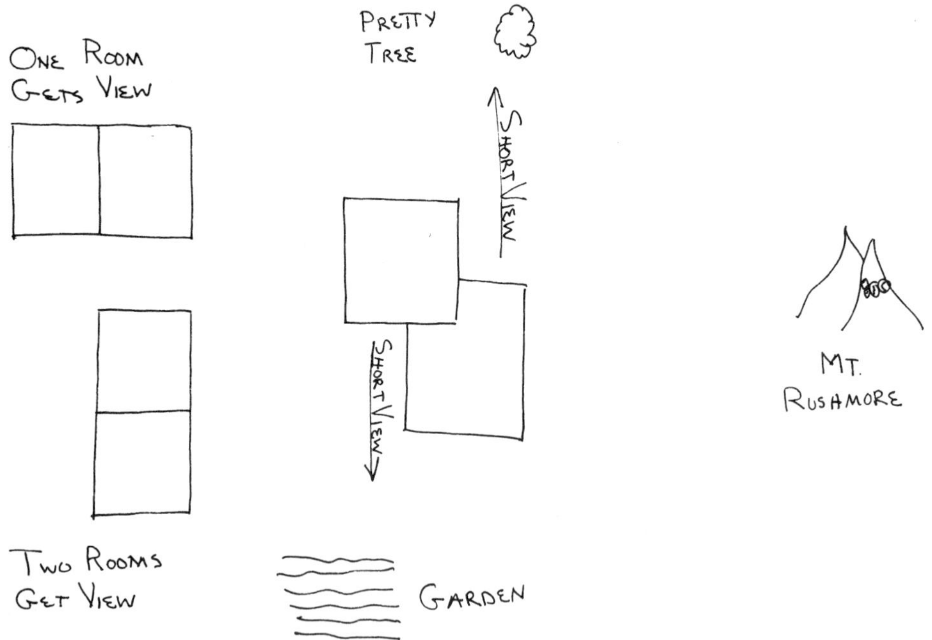

Figure 6-13.
Think about the short and medium views as well as the long, dramatic vistas.

story can get you up in the air to see a fine view or catch a breeze. A hilly site may more easily accommodate a tall house, because of the limited amount of flat land. A two-story house might suit your architectural context. In Vermont, for example, two-story houses with gable roofs look appropriate.

A one-story house is more spread out, seems much bigger, and affords more privacy. It's easy and safe to build; you don't have to lift materials high up, or work from high stagings.

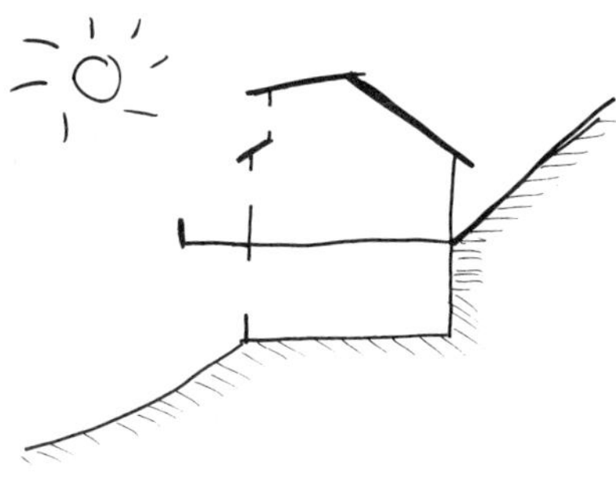

Figure 6-14. A slope may suggest digging into the hillside.

It's more accessible and ergonomic than a two-story house. That's a big point with me.

While a single woodstove may heat a two-story house better, modern insulation and modern central heating systems work as well in a one-story house as in a two-story house. Even though the single story is less compact, the difference isn't great in terms of heat loss.

While a single story seems more expensive to build, the difference is relatively small (I figure about three to five percent), because there are savings to offset the extra foundation

Figure 6-15. Wood heat flows up better than sideways.

work. For example, there's no need for stairs, or for the space stairs occupy. The single story affords more protection in some ways, particularly on an exposed site. A tall house can catch the wind on a cold, windy site, while a low house can deflect it.

Frank Lloyd Wright introduced the "prairie house" idea—the prototype of the ranch house—because he thought a low, horizontal house made sense in a low, flat, horizontal landscape; it's less obtrusive, and blends with the landscape.

If you are thinking of two stories, your drawings will now come in pairs, one for each floor. There is some more discussion of this question in a later section on house forms and technologies (see chapter 21).

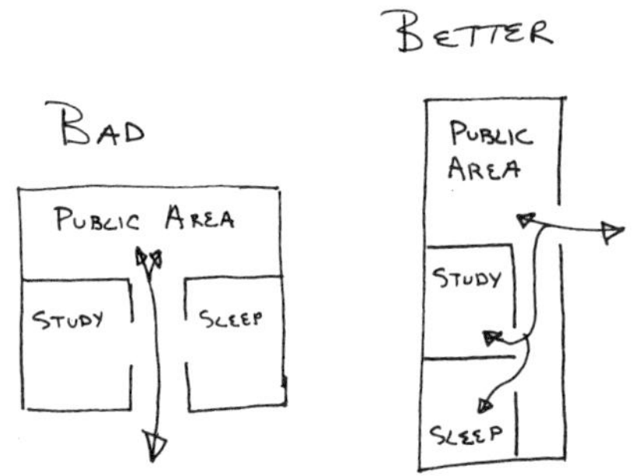

Figure 6-16. Route traffic carefully.

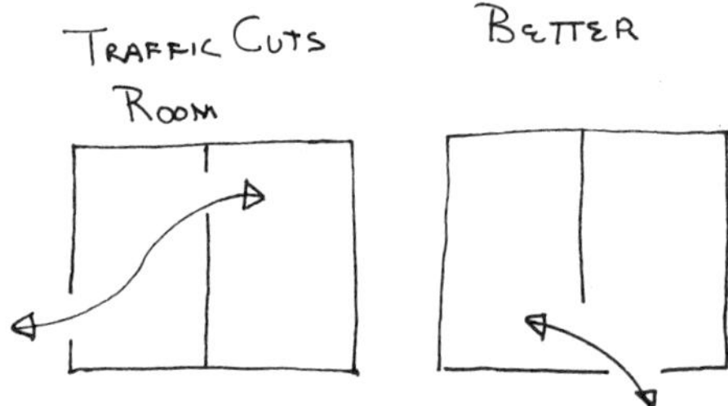

Figure 6-17. Traffic lanes may cut a space in half.

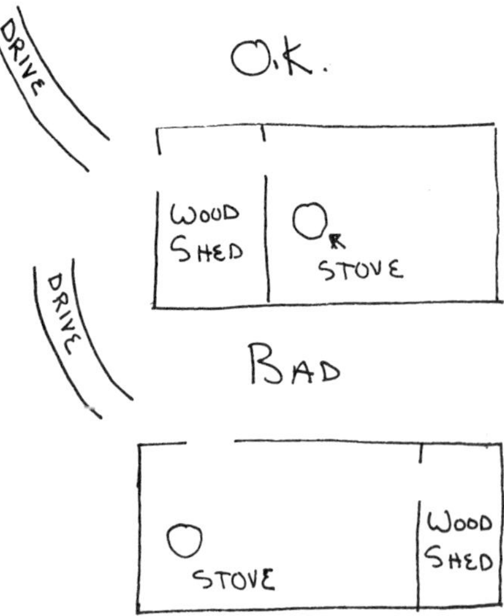

Figure 6-18. Plan outside pathways as carefully as inside halls.

Traffic patterns (circulation): Architects call the patterns of movement through space "circulation." You want your house to be designed so that circulation will be efficient, economical of space, and nondisruptive. Make sure that the private parts of the house, such as bedrooms, do not function as hallways to public parts of the house. In fact, it is best if the primary traffic does not even pass near the private areas (figure 6-16).

Public areas will (and should) inevitably serve as circulation paths to some extent, but don't arrange your house so that the traffic cuts a specific area in half (figure 6-17). It is all right for the traffic to flow through the kitchen, but it should be routed between the cooking and eating areas, rather than right through the middle of one or the other, which would drastically reduce the usable space. Often it is better to create a traffic lane across the corner of a room, or along one wall.

Think about traffic outside too, so you don't carry groceries or firewood halfway around the house before bringing them in (figure 6-18).

Often a certain part of the house will be given over to circulation. This may be a long hall along one side of the house, a central hallway with the rooms radiating from it, or a hall that winds its way deftly through the house. Just as often it will be simply an area or track defined not by walls, but by doorway locations, furniture position, and the general layout of the house.

A hall or other circulation area can be developed and interesting as well as functional. A hall can be a greenhouse, or have a nice bench to sit and chat on. A stairway can have an odd window to look out of, or be a way to communicate with people, perhaps leading past a sewing area, play yard, or garden (figure 6-19).

Don't let your circulation plan develop by chance, particularly if your house is big or if many people will use it. Sketch out possible plans, but don't be afraid to rearrange layouts, door locations, and other elements until you find an arrangement that seems to work. Some small houses are made even smaller because an unnecessarily large amount of space is given over to traffic movements.

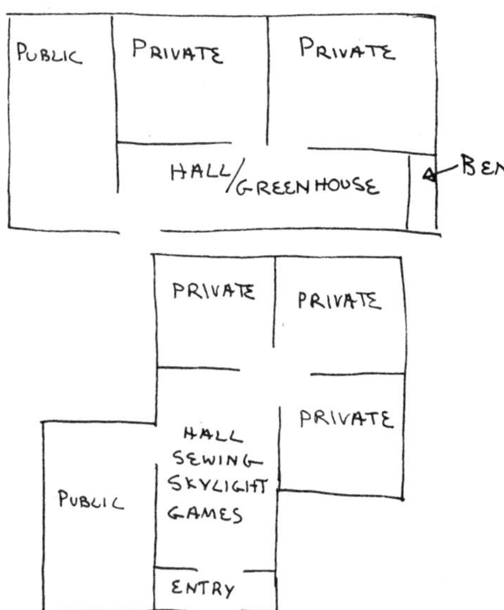

Figure 6-19. Make circulation areas into special places.

MAKING SCHEMATIC ELEVATIONS AND SECTIONS

The sketches you have made so far are in "plan view," the view of your site from directly above. Now make schematic elevation and cross-section drawings, which look at the house from the side. Such sketches can begin with the slope of the land. Then geographical features can be added (figure 6-20).

Within this framework ask some of the same questions you looked at in plan view. How will different house shapes use the sun?

Are there shade trees you can make use of? Are there leafy trees that conveniently lose their leaves in the winter to let in the sun, and grow new ones when you need shade?

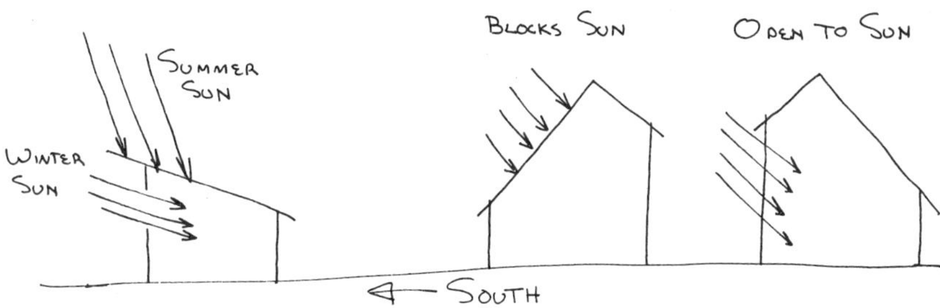

Figure 6-20. Schematic elevations help show how a house will relate to slope, sun, and other features.

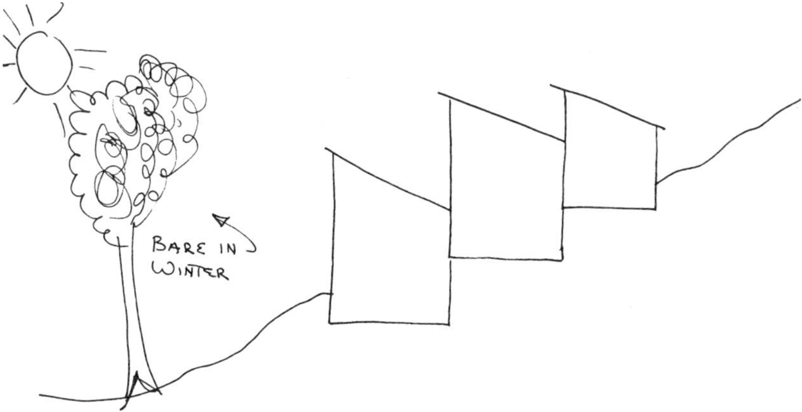

Think about the general way the house relates to the land. Use the features of the land. Try to bring your house design into harmony with it.

With these sketches you should be arriving at ideas about the overall shape of your house. It doesn't matter if there are problems. As more thinking and observation of the site go on, some more or less unified idea will suggest itself.

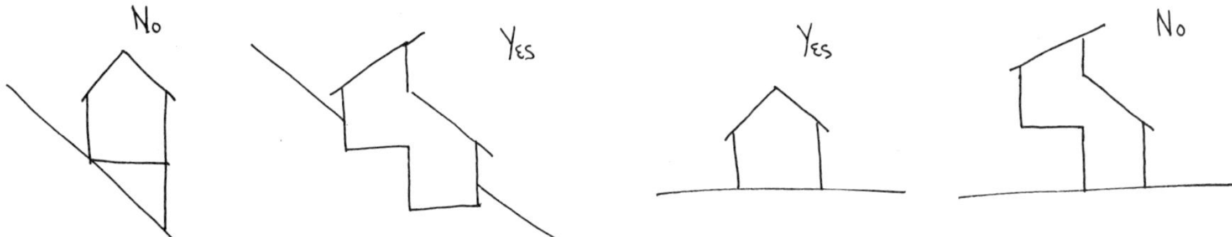

Figure 6-21. The house form should be in harmony with the land form.

CHOICE OF BUILDING SYSTEM

This book focuses on light timber framing (stick-building). It's a great system for owner-builders, and probably the most cost-effective, easy to learn, and flexible system overall. It has also been improved in recent years, as a result of the innovations owner-builders and others have introduced.

But many readers will be considering other building systems that seem more adventurous or appealing in one way or another. Right now, many people are building timber frame houses, resigning from the power grid, or considering straw bale or other sustainable systems of building that appeal to people's values, interests, or a sense of what a home should be. Plus, there are always owner-builders interested in stone houses, pole construction, or log homes.

If you are considering one of these alternatives, find out who has built with these systems in your area and visit them. See the house, find out what it cost, and how much work it took to create. Find out how the system works, and where it needs to be improved. A person who has had to experiment to get it right, may be able to help you avoid similar mistakes.

MAKING A PRELIMINARY ESTIMATE

You now know a lot about what you want in a house. But you do not know how much you have the capacity to build. This is a good time to determine the scale

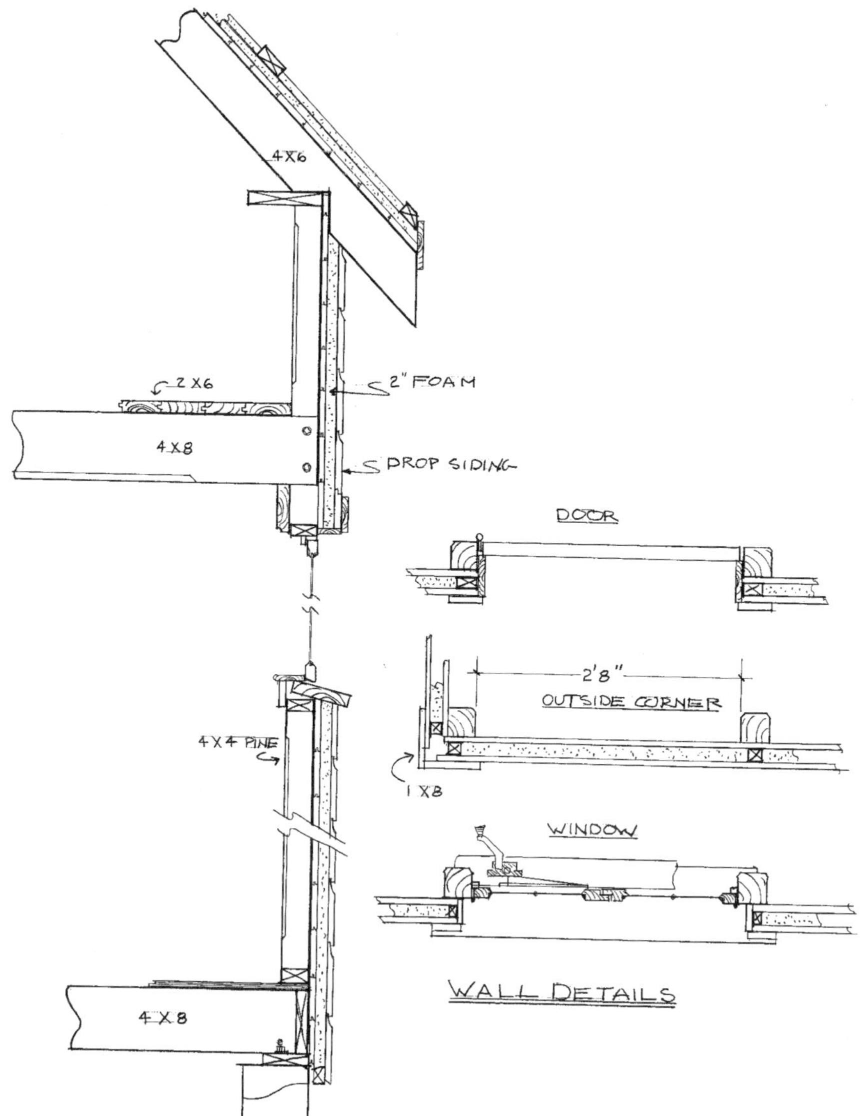

Figure 6-22. Alternative building systems require extra planning.

of your project. If you can afford only three rooms, you should know that so you won't waste time with unrealistic plans you can't execute. Estimating is not easy. In fact, it's harder than building. Things take longer than you think they will, and cost more. The rule is: intentionally overestimate the time and money for any task. Leave yourself room. If you have time and money left over, you can always start something new.

Later, you or your builder will make a painstaking estimate of your costs. In the meantime, use chapter 13 to make a preliminary estimate.

7

MAKING SCALE DRAWINGS & MODELS

It's possible to build a house with few if any drawings. I designed my first cabin on the back of an envelope, and I've designed walls by holding up used windows till they look right, then built the walls around them. But designing your house will be easier and more thorough if you learn to make and use scale floor plans, elevations, cross sections, and details.

Making drawings is not just a way to put ideas on paper. It is a way to develop ideas and make them work for you. Basically you make a scale drawing, such as a floor plan, and then systematically ask questions about it. Is the sun orientation good? Is the circulation efficient? Does the layout give enough privacy? Then you revise the drawing, changing dimensions, moving rooms around, moving doors and windows. Revise the second version the same way. Continue until you have solved as many of the problems as possible. When you have finished, your space will be efficient, privacy and communication will be good, views will be visible from the right place, light will come in where it is needed, plumbing will be compact, and so on. If your house has one or two rooms, the drawing process may be quick and easy. But if your house is larger, making drawings will be a major project. This labor may seem excessive when your real interest is not drawing but building, yet every hour you spend drawing will save you five hours of building. You can solve problems in advance on paper and avoid big and costly mistakes later.

The drawings you make may be influenced by what information the local building department requires to issue a building permit. Typically, you will have to provide a site plan (see chapter 6), a foundation plan (see chapter 23), dimensioned floor plans, and at least one cross section which shows basic construction features. You may also need more construction details, an electrical plan, or elevations. Some jurisdictions may require professional engineering or architectural drawings. Your drawings should always show very clearly any unusual or innovative building systems or details you intend to use. That way, the building department can discuss with you before an inspection what they will accept and what they will not.

The goal is to produce drawings that are relatively easy to make, change, and copy, and which communicate clearly to others. The key is to draw in scale, in pencil, on tracing velum, a heavy tracing paper that is rugged enough for repeated erasure. It can be photocopied, but can also be copied more cheaply and accurately with blueprint machines. Blueprint machines can be found in architects' offices and stores that do copying for designers and architects.

Though amateurs often make drawings with graph paper, using drafting equipment is really easier. You will need a drawing board (a flush door or a piece of smooth plywood will do), a T-square, and a 30-60-90 triangle. An adjustable 45-degree triangle is useful for making and measuring angles.

The velum size I use is 18 x 24 inches. You can usually get a couple of views of a typical house on one sheet at a scale of ¼ inch = 1 foot. For some houses, legal size 8½ x 14-inch paper will do fine. You need a variety of pencils, so that you can make thin delicate lines for dimensions and details, and thicker, blacker lines for major walls or outlines. You also need masking or drafting tape.

The critical tool is an "architect's scale," which is a special ruler that expresses feet and inches in a variety of scales. They are usually awkward to read, because two scales are often superimposed, with one starting from each end of the scale. You have to figure out which set of numbers to go by.

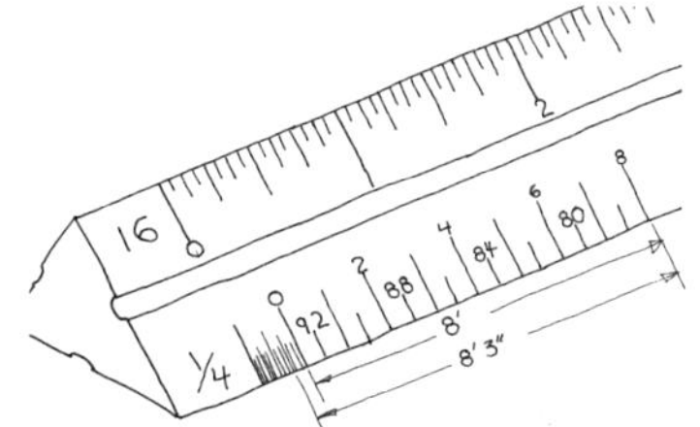

Figure 7-1. *Architect's scale.*

It's also handy to purchase one of those green templates that have cutouts of toilets, sinks, door swings, and other essentials. Get one in ¼ inch = 1 foot scale. A drafting brush is essential, for wiping away the mess from erasures before you smudge the drawing. If you want your lettering to be nice and clear, you might also want

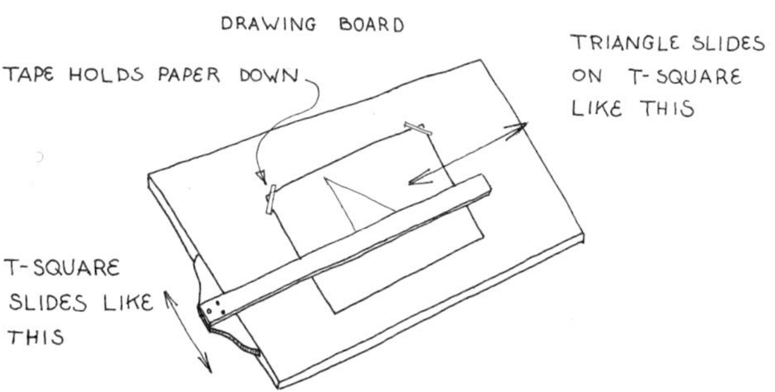

Figure 7-2. *Simple drawing board.*

Figure 7-3. Lay out light outlines

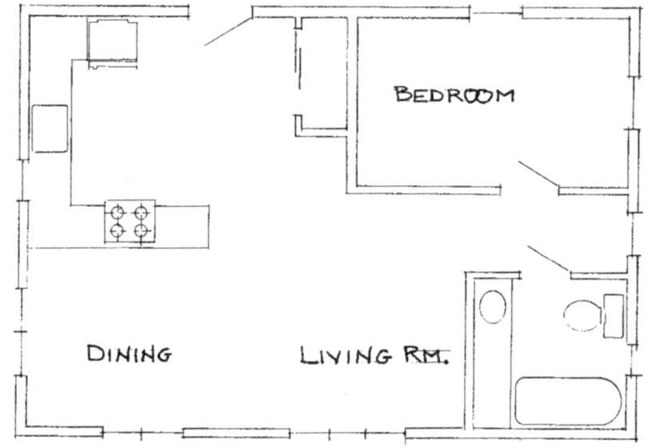

Figure 7-4. Add details

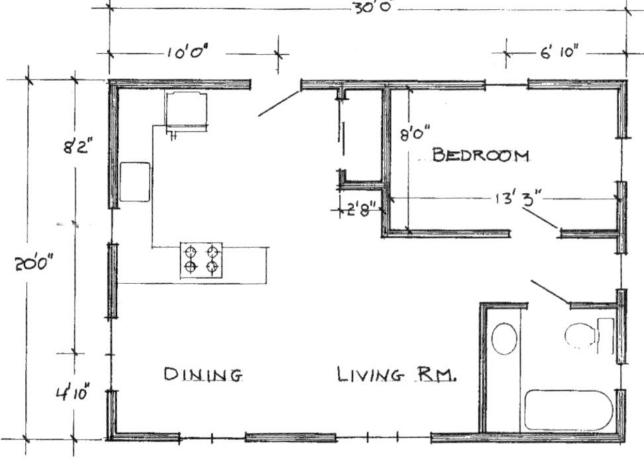

Figure 7-5. Darken and emphasize for clarity.

to purchase an Ames Lettering Guide, which allows you to make lots of uniform horizontal lines.

Figure 7-2 (page 73) shows how these tools are set up. The T-square can be set on the left or right side of the table. In either case, it slides up and down against the edge of the drawing board and is your straightedge for making horizontal lines. Vertical lines are drawn by sliding the triangle along the T-square.

First, think for a minute about where the plan should go on your sheet of paper. In figure 7-3, I've put it near a corner, so I have room for other drawings on the same sheet.

Lay out the outside dimensions of your building very lightly. Use a double line, representing the thickness of the wall. I usually make my exterior walls ⅛ inch thick, representing 6 inches. Then add window locations, doors, partitions, and so on. At this point, you can erase the double line at the doorways to make the drawing neater. You are still making careful, delicate lines.

Once the plan takes shape, you can beef up the wall outlines with a dark pencil. A normal #2 or #2 soft pencil will do it. This makes it easier to read the drawing. It also helps to really blacken in the walls. I do this with quickly drawn thin lines. They make the walls bolder, yet are still easy to erase as I revise the design.

You can then add labels and dimensions. Make your dimensions

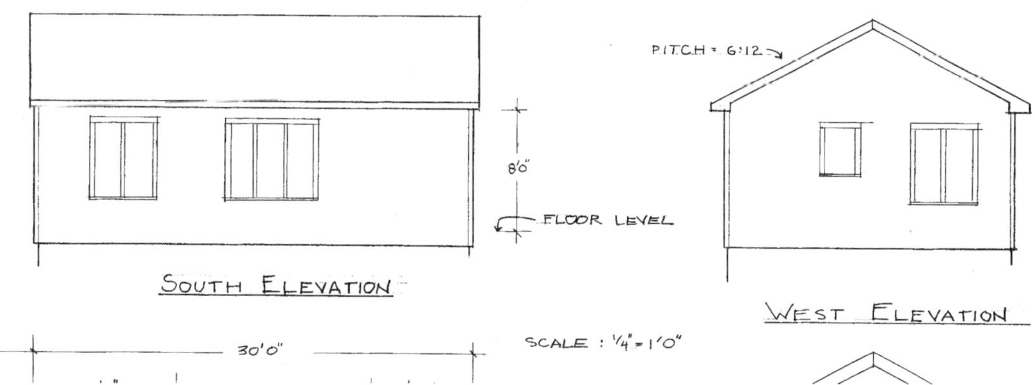

Figure 7-6. Key features of the elevation are projected up from the plan view, or drawn with a copy of the plan under your tracing paper, to show these elements in a new view without redrawing everything.

unambiguous. Normally dimensions designate framing surfaces, not finish surfaces, and indicate the center of window and door openings. It's important to use consistent dimensioning. Otherwise it is easy, for example, to confuse 36 inches with 3 feet 6 inches. Lettering will be neater if you make very light, parallel, uniform guide lines.

You can make an elevation two ways. The first method is to draw the elevation above a plan, and simply project the locations of elements up from the plan. Start by drawing horizontal lines representing floors and ceilings, then run lines up to locate windows, doors, and partitions.

The west elevation was drawn using a second method. I made a copy of the floor plan, and placed it under the sheet. I projected the floor and ceiling lines over from the south elevation, and located the positions of walls and windows by looking through at the copy underneath.

In this case, it was easiest to find the peak of the roof on the west view, and then project that leftward to complete the roof on the south elevation.

A second-story plan could also be produced by projecting from or tracing over an existing drawing that carries most of the needed information.

A cross section, or "section," is often generated over an elevation, since many of the locations will align. It's usually simplest to trace over a copy of the relevant elevation, which shows the critical shapes. The section really shows how the house is built. It often contains details the elevations don't show. This is the drawing the framers build from. The example shows some of the typical information included in a section. For convenience, I made this drawing by sliding a copy of the west elevation under the sheet and drawing over it.

You might also add detail drawings that spell out tricky areas or details that

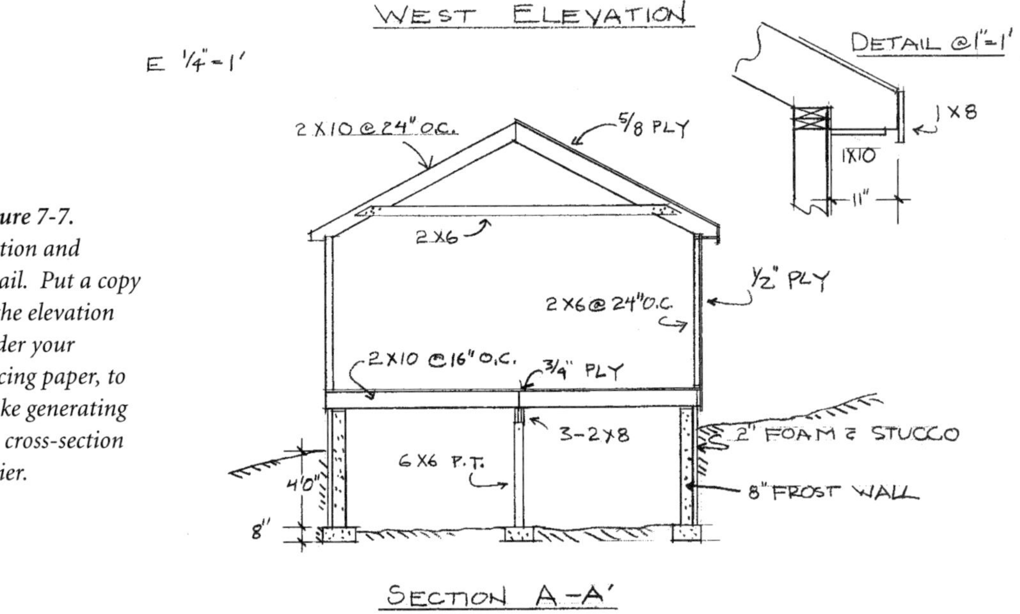

Figure 7-7. Section and detail. Put a copy of the elevation under your tracing paper, to make generating the cross-section easier.

are particularly important. Sometimes it helps to show them in a larger scale. I usually draw the kitchen in 1 inch = 1 foot, because it's more accurate and pictorial. A construction detail might be done in 1 inch = 1 foot scale. Drawing the end of a rafter in that scale, for example, allows you to read dimensions off it using your architect's scale.

The details and sections shown here are just samples. It's possible to put a lot more information on a drawing, and designers sometimes produce many pages of drawings. You might also make a foundation plan, or diagrams that illustrate how you will frame a particular part of the building.

MODELS

When I'm not sure I like the looks of my design, particularly if I'm not sure I've got the right roof pitch, I make a model. This is easier than it sounds.

First I go down to the blueprint or photocopy place, and get copies of the plans and elevations. Then I buy some ⅛-inch or ¼-inch foam-core. This is a light foamboard often used for models, which you can cut with any type of razor knife. You could also use cardboard. I also get some spray adhesive, the type sold for mounting photos. First glue the plan and elevations to some foam-core with the spray mount. Then cut out each elevation with a utility knife. If you can miter the edges where the walls will meet, you'll get a neater model. Glue or pin the walls onto the floor plan as in figure 7-8. A hot-glue gun is the easiest way to assemble the model.

Using this technique, make a roof, and put it on top of the model. Usually I glue a little triangle on the inside, to hold the roof at the right angle, so I can remove and replace it.

You can carry this further, if you want, by cutting out the windows, and adding chimneys or other features. With the roof on, you can look at the model from angles that represent how you would see it on site. You can simulate sun angles at different times of day with a flashlight (see page 230).

Drawings are most deceptive when it comes to roofs, particularly hip roofs. A model allows you to try different roofs on your design to see which looks best. You can also play with the roof's position to see what different roof overhangs may look like.

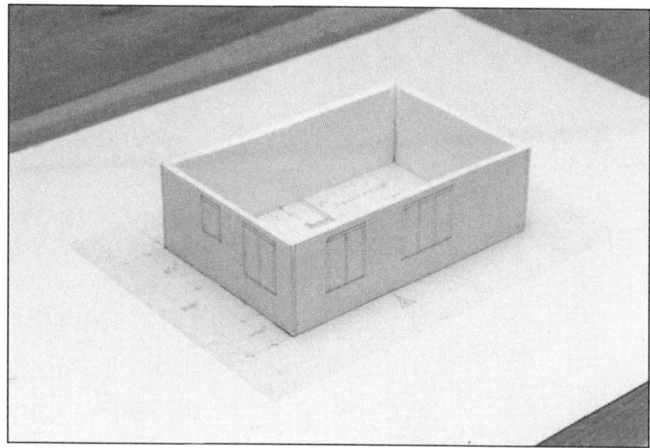

Figure 7-8. *Glue copies of drawings to foam-core or cardboard to make scale models.*

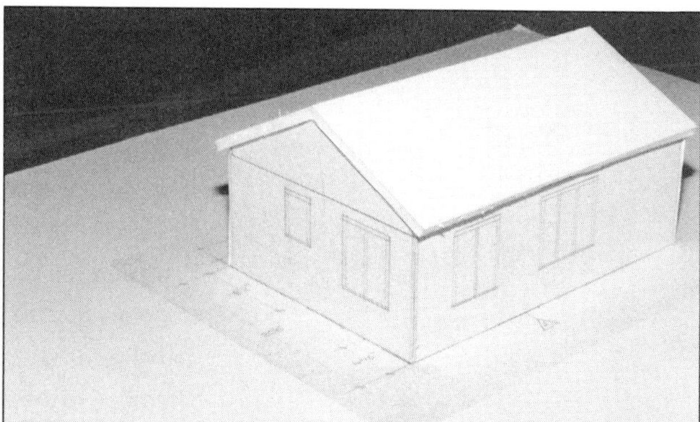

Figure 7-9. *Model with gable roof as drawn.*

Figure 7-10. *Model with hip roof.*

8

LAYOUT

To me, the layout, which designers call the plan, is the heart of a house design. It should largely dictate the structural system, the heating design, and other features.

THE IMPORTANCE OF THE PLAN

If the layout is the form, then the circulation—the halls, passages, paths—is the skeleton of the plan, the armature upon which it is built. If this layout is organized, logical, and economical, and the circulation compact and well defined, you are on your way to a good house design.

Layout is often the weak point in owner-built houses. Sometimes the structure or heating system of an owner-built house will be beautifully thought out, but the layout will not be fully realized. The house will not be as good as it could have been.

Layout seems obvious and intuitive, so people wing it. But there's a lot to learn about laying out a house. I think the plan should receive the same effort, study, and research as other aspects of the design. This is one area where an expert opinion is valuable. A skilled designer will always be able to help you improve your design through small changes in the layout.

For me the process of creating layouts proceeds in two phases. The first phase is rigorous or disciplined: crunching the spaces and large forms in the house into a sensible form that reflects your needs. Once the spaces have been subdued, the second phase starts. This is the more imaginative phase, where you engage with each corner of the plan, adjusting things this way and that, moving windows, adding detail, enhancing each part of the plan a little, as you uncover the best each part of the house has to offer.

FEATURES OF A GOOD PLAN

Relates to Site

Chapter 5 discusses siting a house, putting it in the right place. In addition, the plan of the house should tie into the *details* of the site and become integrated with it. As the floor plan grows, keep in mind sun, wind, access and topography, existing plants, and locations for gardens and other activities.

Solar Orientation

Solar orientation is crucial. The classic solar house has a long axis east-west, with a generous roof overhang to screen summer sun and most of the big windows on the sunny sides. But the main mass of the building should face within 30 degrees of south if possible, and the sun paths at different seasons should be thought through as each space is developed.

Zoning a Plan by Activities

Unless a house is quite small, it is divided into zones, areas that have unique characteristics and should be separated to some degree. In figure 8-1, there is a public area in the middle, a guest wing on one side, and, on the opposite side, an owner's area. Even with several guests, the owners can have plenty of privacy, and so can the guests. These zones are distinct places. The owner's space, for example, has its own entrance, a little deck, a study, and different views from the other zones.

There are many other possible ways to group activities, depending on who uses the house, what they do, and the levels of privacy or community that allow people to function best. Chapter 6 describes exercises to identify these for yourself. The floor plan should enhance these distinctions.

Planning for kids: If there will be children about, provide the spaces they will need. These needs vary with childrens' age. Younger children need to be supervised, of course, and usually want to be near the adults. People often locate outdoor play areas where adults inside can keep an eye on them. Inside, little kids will take over a living room if there is not another play space they like. It often works to provide an L or alcove off the kitchen or living space as a play area.

Older kids may want to be much more separate. It might be better to give the big bedrooms to older kids, rather than to the parents who usually claim the "master bedroom." Kids will also use family room spaces, attic or basement spaces, sun porches, or even garages.

Teenagers often need even more separation, ideally a room or two far enough

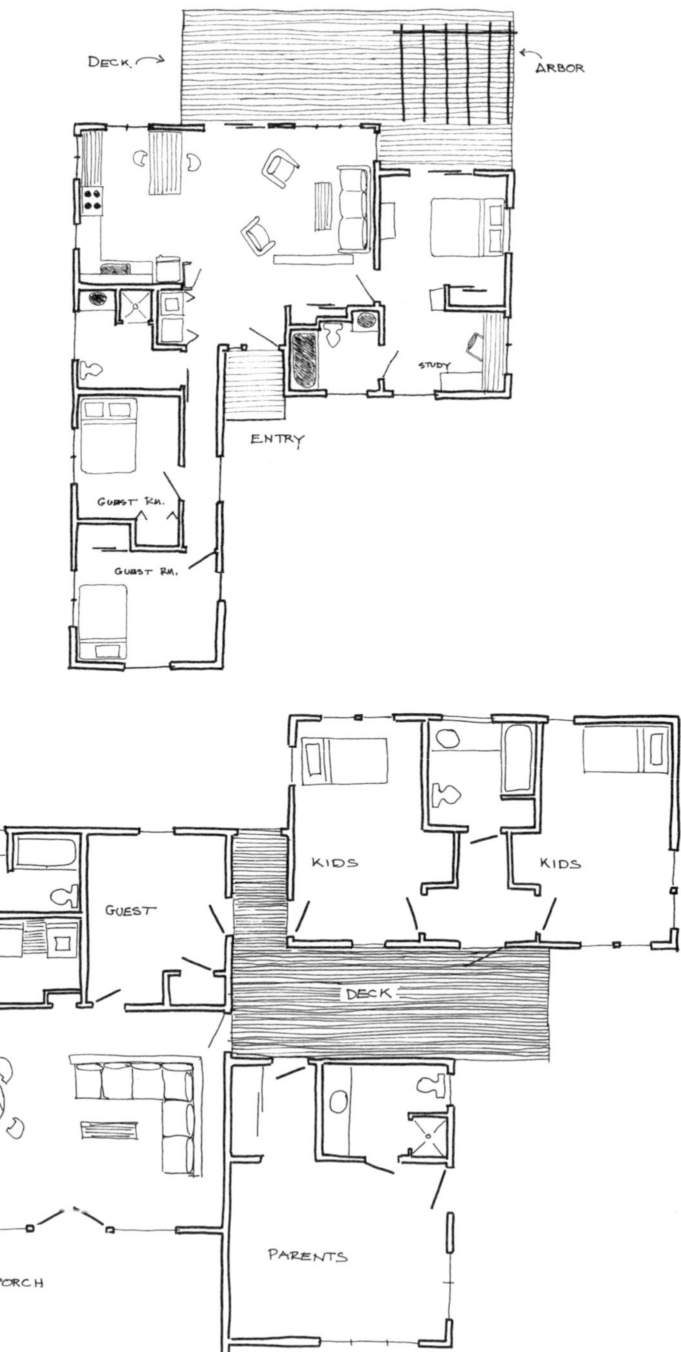

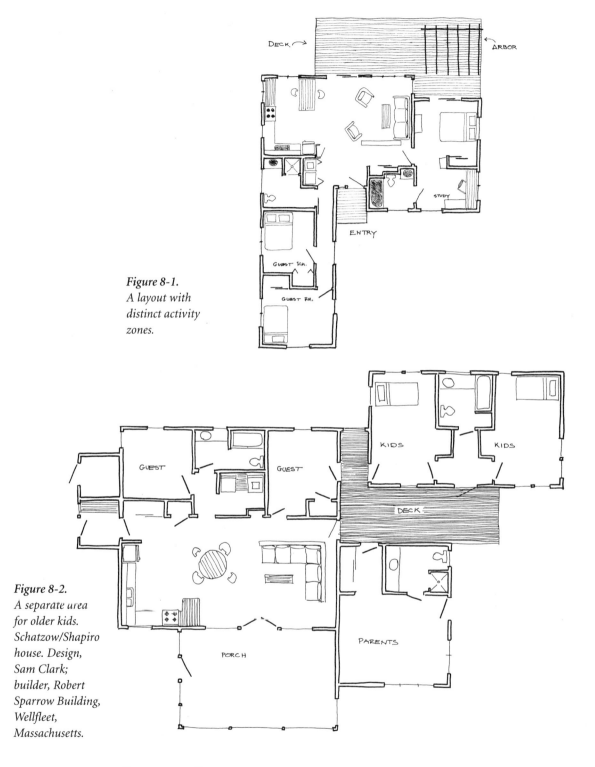

Figure 8-1. *A layout with distinct activity zones.*

Figure 8-2. *A separate area for older kids. Schatzow/Shapiro house. Design, Sam Clark; builder, Robert Sparrow Building, Wellfleet, Massachusetts.*

from the center of the house so that they have a sense of turf and privacy, plus good sound separation. Figure 8-2 is a good example of a teen space that is separate but still part of the home.

Tight Circulation

Circulation is the structure of pathways, halls, stairs, and passages. In a bad plan, the circulation uses up a lot of space. Rooms that should be private function instead as hallways. Pathways cut rooms in half. As a result, a big house can function like a much smaller one, and a small house becomes cramped.

In a good plan, circulation is direct and compact. Hallways are short, and passages skirt the edges of rooms. The total space allotted to circulation is minimal.

In a good layout, you often enter an area between zones via a distinct entry. In figure 8-3, you come in to a hall, go left to the owner's suite, right to the public area, or directly up the stairs to the bedrooms. People can come and go without making a ruckus. Very little space has been set aside for circulation, but it gets a lot done. It is interesting to note that this house has the same zones with much the same features as the one in figure 8-1, but in a two-story, Cape-style plan. Functionally, it's the same design.

This is just one example of a good scheme. I know several houses where the kitchen serves as the greeting area

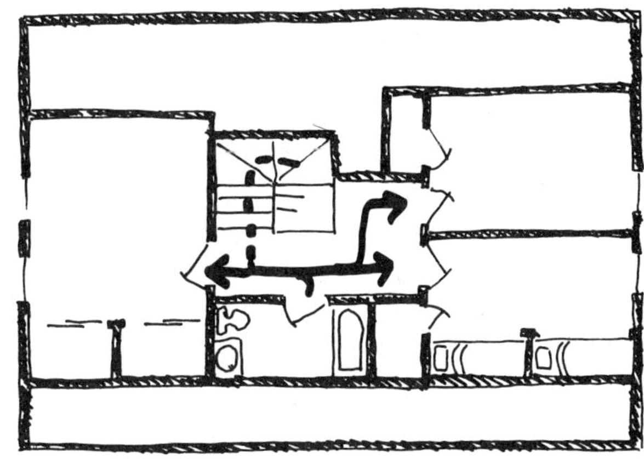

SECOND FLOOR

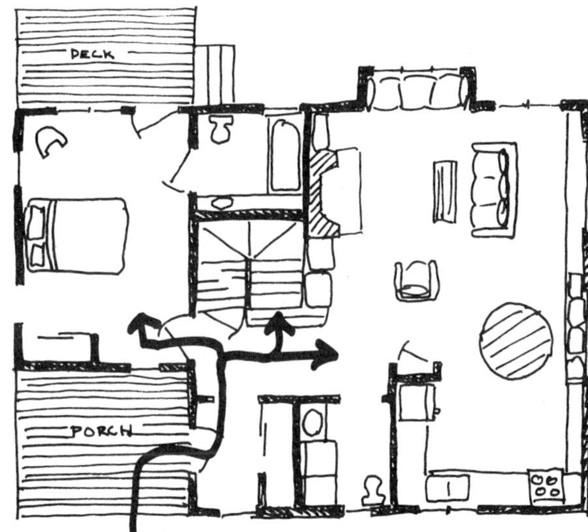

FIRST FLOOR

Figure 8-3. *Well-organized, compact, direct circulation conserves space. Wasow house. Designed by Mona Wasow, Lori Barg, Sam Clark.*

and the main social space; it is the ceremonial center of the home. The point is to figure our what suits your activities and life best.

Life goes on both indoors and outside. In a good layout, the spaces

Figure 8-4. *Kitchen as entry and ceremonial center. Rabin house. (Photo: Jules Rabin.)*

Figure 8-5. *Entrance transition (Pattern #112): Make a transition between the street and front door; bring a path through this transition area, and mark it with a change of light, change of sound, change of texture. Helen Garvy house. Design, Helen Garvy; builder, Roger Mason Construction.*

around the house are implicitly incorporated into the plan; the site is designed as a whole. Outdoor spaces include gardens, patios or decks, paths, outbuildings, and the spaces between these things. Transitional spaces include decks, porches, stairs, entries, and mud rooms. Conceptually, there are both indoor and outdoor "rooms," which together form the plan.

The traditional New England farmhouse with attached barns, though it seems rambling and informal, is an excellent example of this kind of careful, integrated planning (figure 8-6). Facing the road is the formal part of the house, with its own front yard and flower garden. To the side, facing the sun, is the kitchen garden, near the kitchen, then the back woodshed, milkhouse, cow barn, and so on. The dooryard, where so many different kinds of activities take place, faces the sun, and the house screens out the winter northerly winds.

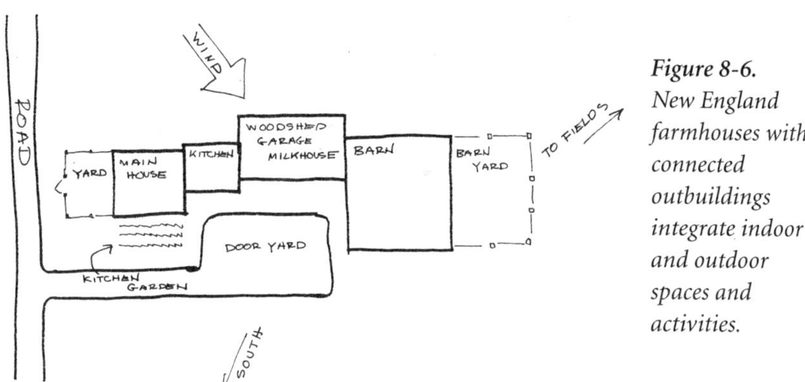

Figure 8-6. New England farmhouses with connected outbuildings integrate indoor and outdoor spaces and activities.

Ergonomics and Access

Any new house can be laid out for accessibility, now or in the future, with little added cost. Houses in figures 8-1 and 8-3 don't look accessible in any way, but have the turning circles, wide hallways and doors, and the generous bathrooms needed to make them fairly accessible now, and totally accessible after only minimal changes. With a good site plan and careful grading, there can be a no-step entry, and gentle paths that anyone can use, also without much extra expense (see chapter 1).

Simplicity

Many houses have many good ideas, but they're piled up, disorganized. In a good layout, there will always be an underlying form or order that ties the good ideas together.

This order can be of many types. It could be structural, like the colonial houses with their standardized, rationalized construction system, generating a few common house types. It could be dimensional. Traditional Japanese houses, for example, are modular; their plans are based on a standard unit, the tatami mat. Many architects have worked with other modules, such as a 2 x 2-foot or 4 x 8-foot grid.

The order could be a traditional house form, such as a Cape, if its constraints serve your purposes and your site, or the long bungalow form common in California.

Often the circulation imposes the order. In figure 8-1, everything derives from the simple L plan, which creates zones in the house and also defines the courtyard.

An orderly design is easier to build, because it's simple. It is easier for the builders to keep track of what they are doing; they can more easily imagine the entire project. An orderly design uses materials more logically, so there will be less waste.

An underlying concept or organization makes a house more satisfying to experience. The house becomes a composition, like a piece of music, with an underlying structure, themes, and variations.

MAKING YOUR PLAN

A First Scale Plan

In a minute there is time
For decisions and revisions that a minute will reverse
—T. S. Eliot

I think T. S. Eliot may have been working on his house plans when he wrote these lines. In a sense, you develop a house plan by making a bad plan, then revising it repeatedly until it's good, or until you can't stand revising it anymore.

In chapters 4, 5, and 6 you developed some house plan basics: size; a list of spaces and how they should be zoned; time, money, and hassle budgets; maybe a sense of a possible house form; and perhaps some notions of desired materials and building systems. You have a site plan of some sort. Chapter 7 shows the drawing method and the ¼-inch scale format to use.

With this as a starting point, make a very rough scale plan. As my example, I'm going to use a house on Cape Cod I designed with and for Peter and Kathleen Weiler. Our starting point was a flat, wooded, three-quarter-acre site on a quiet road. It was to be a weekend house, someday a retirement house, a house for lots of guests, including kids and grandchildren. I was pushing for a one-story plan, which would be accessible and more spread out than a two-story house of the same size.

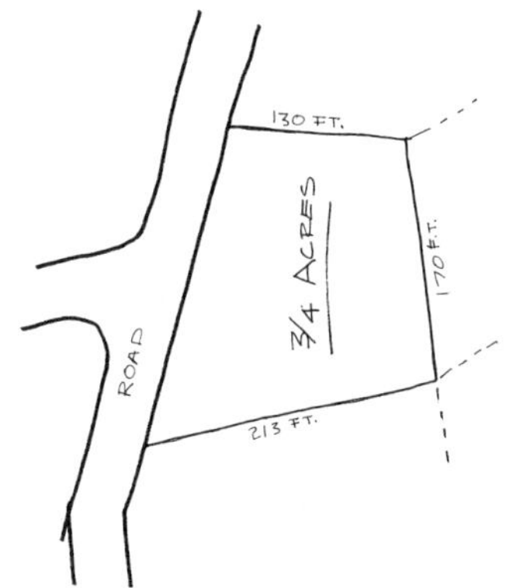

Figure 8-7. The site.

Our inspiration was Wright's Jacobs House. This Usonian house had a bedroom wing and a public wing, with kitchen and services in the middle. Designed for a corner lot, its L shape screened out the street to create a private world within the L. But where the Jacob's house (built in 1939) was over 2,000 square feet, our budget called for about 1,200 square feet. Figure 8-8 shows our first idea.

Review

At every stage, as you study your scheme, refer back to your own goals, and think about the criteria listed earlier under "Features of a Good Plan." In this case there were two glaring problems: the wing thwarted the goal of good south exposure—the passive solar goal. It blocked the sun from the public rooms, particularly in the morning. Second, the room-to-room zoning was wrong. The bedrooms were all jammed together.

Devise Alternative Schemes

Professional designers usually develop a variety of preliminary, sketchy designs. Even if you think you know the basic direction you wish to go in, these designs are a way to understand design problems, a playful way to generate options that might not surface otherwise. For this Cape design, we tried many alternative schemes, but ended up with the basic idea in figure 8-9. This plan keeps the L shape, and the idea of entering at the center, but it divides the bedrooms into owner's bedroom and guest area, and it flips the L over. The big public space now gets the sun, and there's better privacy. The guest bedroom wing has less sun, perhaps, and more exposure to the road, but the guests won't complain. The house still faces the yard. The L plan creates a little entrance courtyard at the road.

In keeping with Wright's idea of using the house to screen the road out, the house would have very little glass toward the road. The views would be to the woods, and the main mass of the house would be oriented south, to the back.

Filling in Some Details

This looks pretty good, but does it work with some of the basic features added? Specifically, what about:

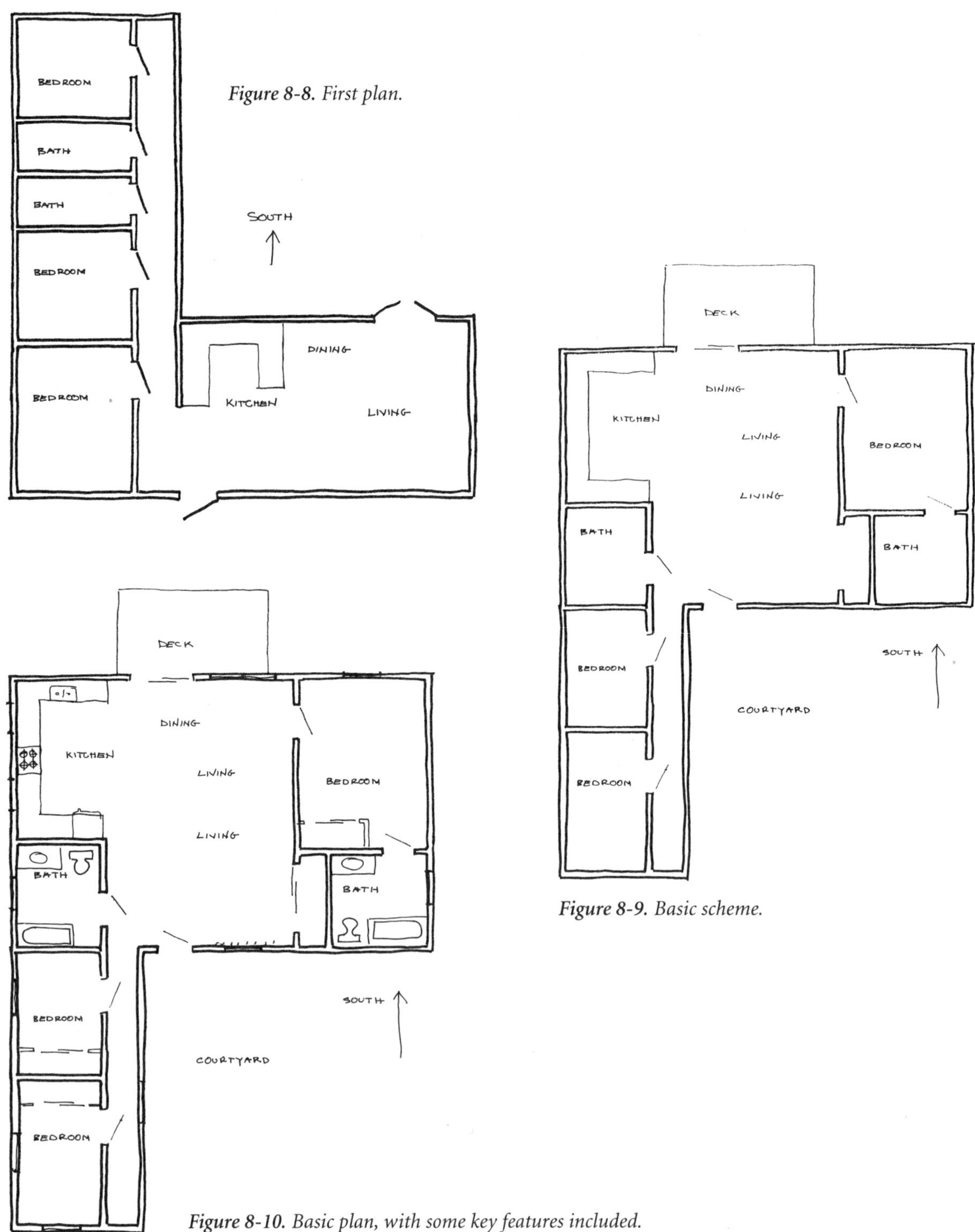

Figure 8-8. *First plan.*

Figure 8-9. *Basic scheme.*

Figure 8-10. *Basic plan, with some key features included.*

Kitchen layout
Closets
Doors and door swings
Entry
Rough window placements
Halls, as needed
Rough bathroom layout
Laundry
Fireplace or woodstove (if any)

Though the detailed kitchen design may come later, make sure there is enough space at this stage in the design for a good functional layout. This series of layouts shows the elements I look for in a functional but simple kitchen: key appliances, two or more good working counters, plus a pantry in some form (see chapter 10).

Hmm. Not bad. I don't see a reasonable spot for laundry equipment, though it might be shoe-horned into a bathroom. A study area was an important goal; it's not clear where it would go here. The bedrooms are just too small. The hallway is too long. The pathway to the main bedroom (to the right) cuts the big room in half.

ADDING IN THE FURNITURE

Figures 8-11 and 8-12 show the next version of the design.

This version shortens the hall by enlarging the end bedroom. The bath to the right, or west, side is moved over to create a study. Most important, the wing and kitchen are widened by 2 feet.

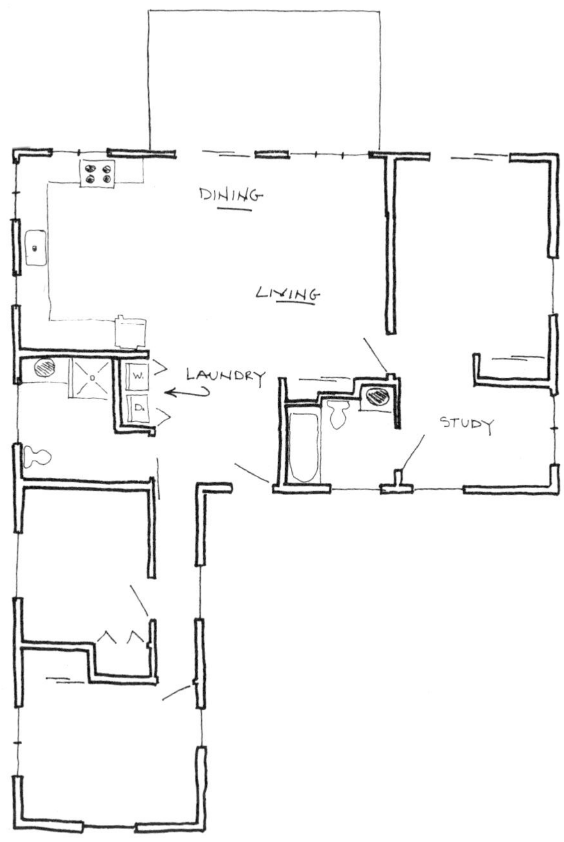

Figure 8-11. Revised standard version.

Unfortunately, a plan that looks great at this point may not actually work when furniture is added. It's not too soon to put the furniture on the plan. For adding furniture, little cutouts (pages 100–101) work well, or the furniture can be traced right on the plan.

Checking Each Area of the Plan

As I review the house plan with furnishings, I check to make sure each sub-area works.

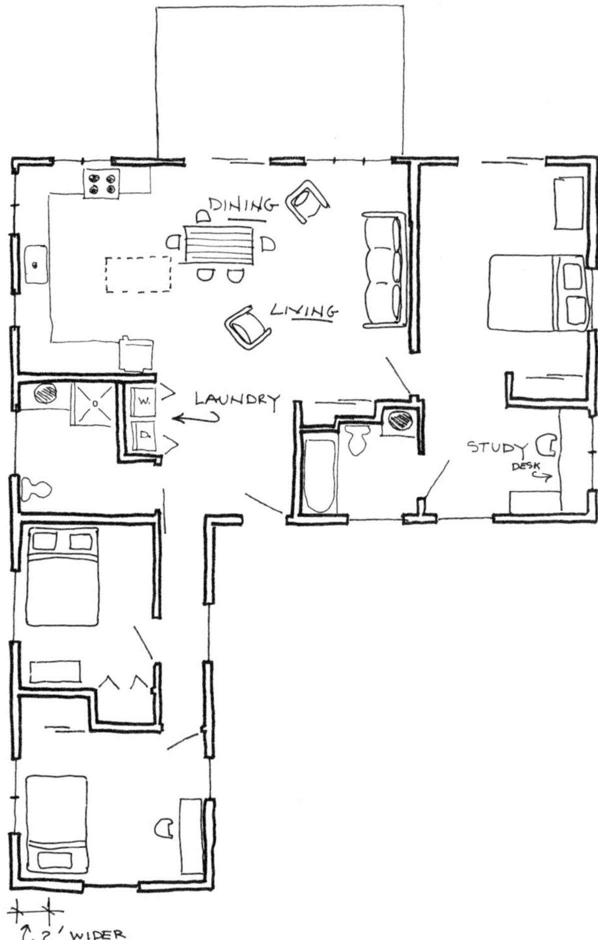

Figure 8-12. Revised standard version with furniture.

Dining area: The cutouts on page100 show what size table is needed to seat two, four, six, and eight or more people. The table should be big enough to seat the household plus a couple of guests without hassle. The space around it has to accommodate the chairs, plus additional space to push the chairs back as people seat themselves or get up. Also, even if there isn't a separate "dining room," the dining area should still be a distinct space with its own boundaries and features.

In figure 8-12, the table is big enough, but the space behind some of the chairs is cramped; you would bang into other furniture as you seated yourself. The chairs would also be in the way when they weren't tucked under the table. The big problem, though, is that there is no distinct dining space. The table is jumbled in with the living room and traffic lanes. There's little sense of place. It would be disconcerting and unrestful to sit here.

The dotted line shows an alternative location for the table, which would solve some of these problems, yet obstruct the kitchen area.

Living areas, conversation groups, and psychological space: A successful living room or living area should also be a distinct space. Even if it is just a zone of a large great room, the living area's boundaries should be obvious, even if it is just defined by furniture placements. Basic traffic should go around, not through it. There should be enough comfortable seats, usually arranged in an L or circle so people can comfortably visit and talk. Coffee tables or end tables of some sort are needed as a place for snacks, cups, and reading materials.

These configurations are referred to as "conversation groups." Where a big house may have several of these areas indoors, many houses like this have only one. A conversation group ideally should provide for very different ways of interacting. An intense, intimate conversation might call for seats that faced each other with only two to four feet of separation. A more casual discussion might work best with a little more distance, say six or eight feet, between facing seats. When people sit and read quietly, a little more separation might help, and you might not want to be directly facing someone else. Today, many families will also often sit theater-fashion, watching TV.

Without consciously thinking about it, people should be able to choose how they will interact with one another simply by which seat they select to sit in, and how they position themselves in the seat. A well-designed conversation group will try to accommodate all these kinds of interactions gracefully by providing a variety of options.

In figure 8-12, there is room for about five people to sit, and there is some variety in the seating. A person in the chair at the top of the page could have an intense discussion with someone on the nearest end of the couch, while another person reads the paper in the other chair. Sitting at the other end of the couch would create a little more psychological space. So far, there is no place to put a book or a snack or a TV. The space as a whole is indistinct, and some traffic crosses it. More revision is needed.

Bedrooms: Bedrooms require room for the bed or beds (see page 101 for sizes), storage for clothing that hangs, and room for piled or folded clothes. It's important to be able to comfortably get into and out of a bed, and to make it up conveniently. For a double or queen-size bed, this may mean access from both sides. Other bedroom furnishings could include a work table or desk, a dressing table, and one or more chairs. In the Cape design example, the right-hand bedroom, which is the main bedroom or owners' bedroom, has room to move around, but at this point in the design process the clothing storage is limited. The middle bedroom is tight for a double bed. The third bedroom is a little more spacious.

Studies and study areas: Designing a study can be as involving and complex as designing a kitchen, particularly if there is a lot of equipment, or a great many books. My own study, which is about 9 x 11 feet, has an area for the computer and phone gear, a very large drawing board area, plus a copier and blueprint machine. However, without this specialized equipment, a simple L such as is shown in figure 8-12 can comfortably house a desk-type work area plus computer gear.

Egress and safety: Ideally, there would be two safe means of escape from every part of your home, so that no matter where a fire started, you could escape in the other direction without having to risk injury. Building codes often require two exit doors from the house, and they have to be "remote," so that a fire near one exit doesn't block the other. This is a very safe system for single-story houses, or houses dug into a hill. Both floors can have a doorway leading directly to safe ground outside. It gets more complicated, though, with two-story houses, where the second story isn't connected to the ground.

Most codes don't require a separate stairwell and doorway out of a second floor of a single-family house, even though that would be the safest provision, because providing this level of egress would be expensive. Instead, codes require "egress windows" for all sleeping rooms. These are windows large enough to climb out of, and for a fire department to access from outside in a rescue (see page 378 for dimensions).

It's important to consult local codes on this subject, and conform to them. But I'm not always happy with the level of safety this achieves. I'm not sure a large window is sufficient egress for an upstairs bedroom. I'd prefer to see a second stairway from the upstairs to the outside, covered if possible, particularly in snow country. In old farm houses, a second stairwell will sometimes lead out through a barn or garage; that is a good solution. Many houses have a second staircase leading from an upstairs porch or deck.

The egress window is a safe exit from a ground-floor bedroom, as long as the occupant is fit enough to use it safely. For an elderly or disabled person, a door can be substituted for one of the windows. But upstairs, the egress window leaves me worried. The usual solution is to provide a chain ladder that can be hung from the window in case of fire. But I can readily imagine someone falling from such a contraption. When there is no better solution, one idea is to build a small deck outside one or more of the upstairs bedrooms, with a simple ladder to the ground. Even with no emergency, it is difficult to step from a window sill to a ladder. The deck provides a safer transition to the ladder. Even so, the ladder will not be safe for everyone.

Looking for the Voids

Usually at this stage in the design, some parts of a layout seem very cramped, and it is a temptation to immediately enlarge the house's footprint. But at this stage there will also be voids—spaces that aren't being used at all, or aren't being used well. There may be circulation pathways or halls which could be incorporated into living spaces, or put to better use.

Tour the plan to find these areas. In this case, the middle of the kitchen, and the space inside the front door, are large vacant areas that don't contribute that much to the design. Often it is possible to incorporate these voids into adjacent areas that are currently cramped.

Revision

The revision proposed in figure 8-13 addresses these issues and makes other changes that improve and vary the design without departing from the simple scheme.

The key is to go to a more compact (and more efficient) L kitchen layout in place of the U kitchen in figure 8-12. This opens up the whole scheme in the middle.

The dining table can now move out of the middle of the room and out of the traffic lanes. It's in the kitchen, yet it doesn't interfere with kitchen activities. The living room becomes more distinct. A long bookcase helps define its boundaries, and serves as a place to set things. With the dining table moved, there is enough space to devise other furniture placements, a second couch, for example. Figure 8-2 shows a layout where built-in benches have been used to create a wider variety of places to sit in a house of similar size.

A small covered entry is added to the plan. Similarly, a 4 x 12-foot area initially included in the main bedroom has been turned into a covered deck. This provides a private outdoor space for the bedroom, and a west window for the living room.

Though I've represented this design version as one step removed from the previous one, in reality there were probably five or six revisions in between. Also, it might seem from this presentation that a single designer makes a plan, reviews it, revises it, and reviews it again until it works. It is actually a collaborative and messy process. I make a drawing, leave it around, make a few changes, and send a copy to the Weilers. They think about it for a few days, we talk on the phone,

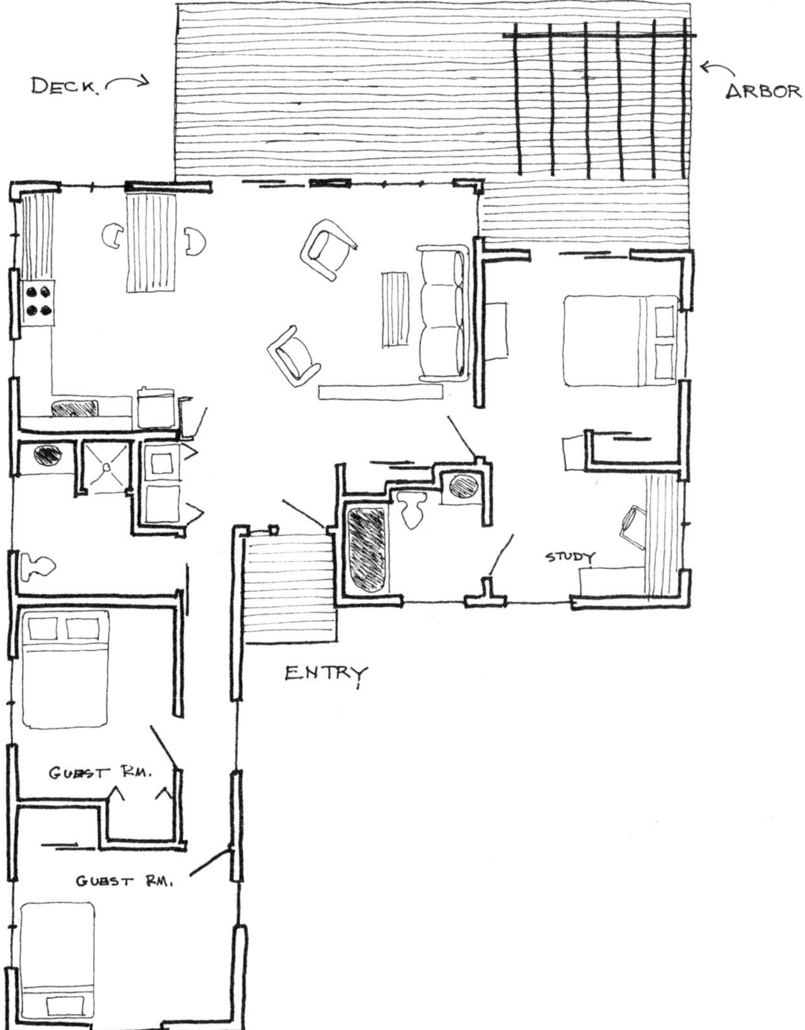

Figure 8-13. *Enhanced revision.*

their friends look it over, and they send it back for another version. The drawing sits on many desks, many heads are scratched in puzzlement—what's wrong here?—and many small inspirations find their way into the plan over a fairly extended period.

The builder—in this case Bob Sparrow of Wellfleet, Massachusetts—should be involved as early as possible, reviewing the design, and suggesting alternatives and details that take full advantage of local resources. Other improvements may arise during construction. Even later, good ideas become additions and renovations.

Key Points of Small House Layout

Many readers will be building relatively small homes, perhaps quite a bit smaller than the one in this example. It is worth calling attention to those design devices that allow small layouts to work well, and make them seem bigger than they are.

Common areas: First, if the common area feels big, the house feels big. Therefore, the basic strategy in small house layout will be to make the public spaces as big as possible at the expense of bedrooms and studies. Use every device that makes the public room or rooms seem bigger than they are. Provide plenty of windows with good views and long lines of sight. A nice high ceiling with its own dormer, clearstory, or skylight will expand a basically modest space. In the Weiler house, the dining space, kitchen, and living room area are all small, almost minimal. Joined together, they make a grand "great hall."

Nooks: Sometimes a nook or bay can be used to make a cramped "conversation group" area spacious. The photos in figures 8-14 and 8-15 show simple built-in seating added to the living room. Without the seats, the sitting area was very cramped. With the seats, the sitting area functions as it would in a bigger space.

Figures 8-14, 8-15. *A bay with a built-in couch makes the house seem much bigger with little expense. And built-in benches create nice seats, with storage underneath. Design by Mona Wasow, Lori Barg, and Sam Clark; built by Lori Barg, Patti Garbeck, Curtis Karr, and Denise Wands.*

Dining area: In the dining area, experiment with different table shapes. Some spaces will work better with a round table, while others will work best with a square or rectangular table. When space is really limited, a built-in breakfast nook or booth shape is probably the most compact, because it eliminates the space usually required for moving the chairs in and out.

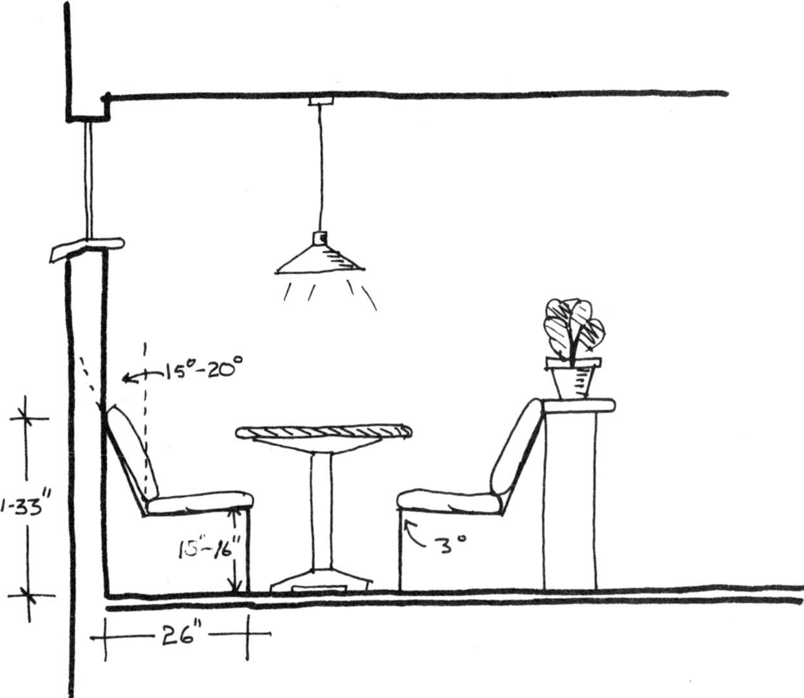

Figure 8-16.
A breakfast nook or booth is probably the most compact seating scheme.

Built-ins: The bay and the breakfast nook are examples of built-ins, a standard device for conserving space. The house in figures 8-14 and 8-15 has built-in seating along the west wall, and on either side of the fireplace. To the left of the fireplace, the bench is really an extention of the stairs, and the steps become a kind of seating, too. Built-ins are usually more compact than free-standing furniture. In this plan, there would not have been room for regular furniture in these four parts of the plan. While using very little space, these built-ins provide additional places to sit, each with a little different feeling. They can also provide storage under the seat.

Compared to stock furnishings, built-ins provide maximum use of available space at low cost. In the Weiler house, the study has a long, built-in desk that is simply a flush door supported on cleats nailed to the wall. It uses every inch of the available space, and cost perhaps $75 installed. It has a bigger work surface than any bought desk could have, and the fact that it is so simple contributes to a feeling of uncluttered spaciousness in the room. The room feels bigger than it would with free-standing furniture. Built in tables, bookcases, storage units, and beds (including loft beds) often will often work this way.

Kitchen: In the kitchen, the L kitchen layout we've looked at here will often make a big difference. In place of cramped kitchen and dining areas, you can have a spacious kitchen-dining space. To make the kitchen space seem large and more open, use fewer, lighter overhead cabinets, tucked away (figure 8-17). Big windows bring in outside space.

There are three other devices I've used dozens of times when designing kitchens for tight spaces. First, leave out the dishwasher, which uses up prime kitchen real estate. Second, use a 25-inch sink instead of the 33-inch double. As I mention elsewhere, a large European-style Franke single sink has almost as much volume as a conventional double sink, yet leaves more room for counters. Third, recess storage in the walls, if necessary, to create the volume of pantry storage that can't be found in the kitchen area itself. There is a good example of this idea in figure 10-19.

Figure 8-17. *A small kitchen that feels big. Weiler house. Design, Sam Clark; builder, Robert Sparrow Building, Wellfleet, Massachusetts.*

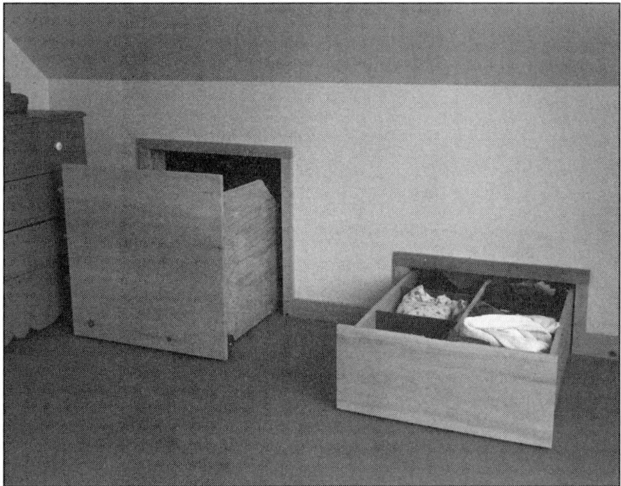

Figure 8-18. Well engineered storage, with lots of drawers, is a key to designing small houses.

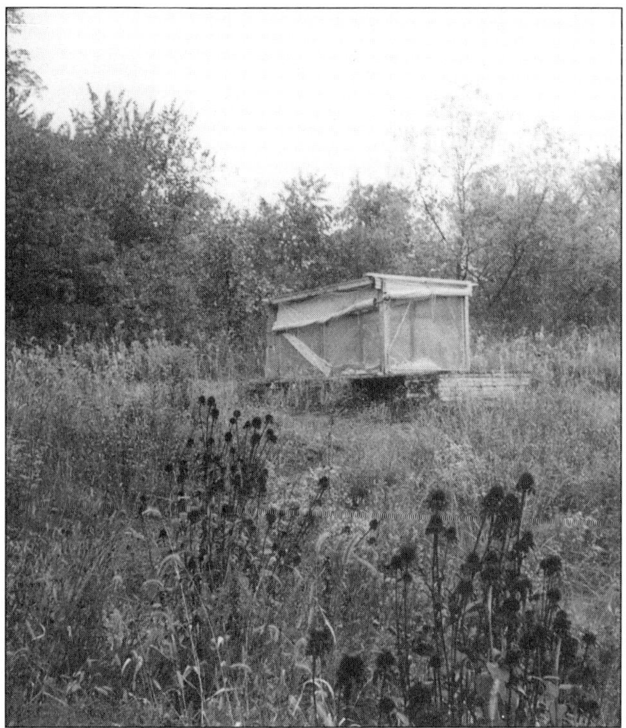

Figure 8-19. Pat Kane's sleeping shelter.

Bedrooms: In bedrooms, clever storage is the key. By building big drawers under the bed, dressers can be eliminated. Closets can be designed with a double row of hangers, or with drawers or shelves below hanging clothes. If uncluttered and well-lit, a tiny bedroom can be very pleasant and cozy.

Careful storage in the rest of the house is equally important. I find that drawers, if designed with their contents in mind, often hold the most stuff in the least space.

Compact objects: Carefully examine the scale of every object that comes into the house. Buy compact TVs, sound systems, computers, and other equipment. Sofas, kitchen chairs, and other furniture all come in very functional but diminutive sizes that take up less floor space and less visual space. These things make the house seem much bigger by comparison.

These simple devices can make a small, well-planned house seem as comfortable and spacious as a larger home.

Sleeping porches: The small house I describe here works well, but has one liability: it can't take a crowd of guests for very long. There is no privacy if a second batch of guests or grandchildren are camping out in the living room. Since guests most often come during good weather, it works very well to build cheap, uninsulated sleeping shelters for warm weather use. This can be simply a good tent platform and tent, with the platform extended to create a little porch or deck. I've spent many summer nights in this kind of accommodation.

In warm weather, the main problem is just to keep the rain off and the bugs out. We often have built sleeping shelters which consist of a platform the size of a small bedroom, say 8 x 8 feet or so, a frame of scrap lumber, a roof of recycled tin sheets, and walls consisting of old storm windows, screening, and other surplus building materials. Visitors can comfortably spend a week or a summer living in these shelter annexes.

A more permanent version would add simple, uninsulated, screened porches, positioned for private use of guests rather than public use as porches. These spaces cost at most half as much as heated spaces. Provided with bed, small table and chair, and some light, they make a great guest space, and can double as a summer study or play area. If simple windows can cover the screens, such porches can be comfortable in spring or into the late fall.

PUTTING THE PLAN ON SITE

Next, combine the floor plan and site plan. We had a tiny site plan from the town. At our neighborhood photocopy place, we blew this up to a scale of 1 inch = 50 feet. (A scale of 1 inch = 32 feet would also have been handy).

We added important items to the site plan, such as location of neighbors' houses, and zoning constraints; the house has to be 25 feet from the road, and 25 feet from its own lot lines.

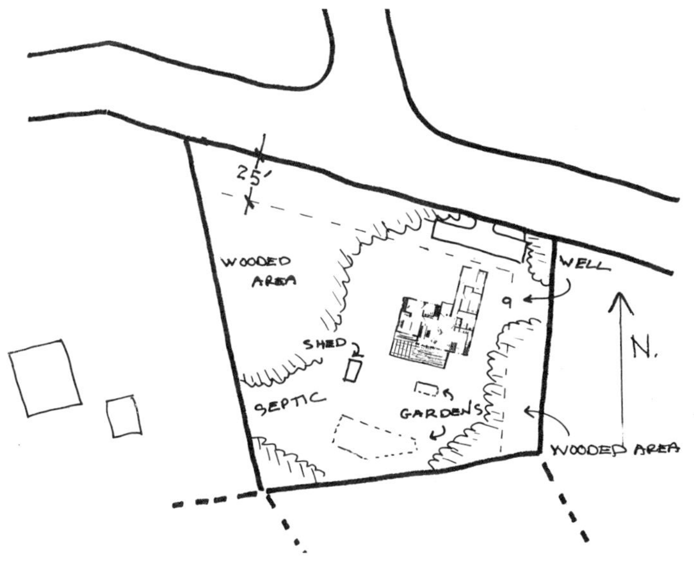

Figure 8-20. The house positioned on the site, with gardens, septic, tool shed, and so forth.

Figure 8-21.
The entrance.

We then reduced our house plan to the same scale, and made a cutout, which we could move around the site to find an optimum location.

You can also go to the site with your compass, some stakes, a long tape measure, string, and a stepladder. Stake out the house location you are considering, and ask the same questions you asked when you made drawings on paper. This is more objective than any review done on paper.

Outside, there was an excellent view of the neighbors' new garage, breaking down any illusions of being out in the country. Our review of the site plan, on paper and at the site itself, included repositioning the house, adding potential locations for well, septic, and parking, and thinking about views and privacy.

To begin with, the site was entirely wooded. The idea was to cut enough trees to let in the sun, allow construction, and provide outdoor spaces, while maintaining the woodsy flavor of the site, as we screened out the neighbors' houses. To maximize the yard in back, we put the house as close to the road as zoning would allow. Then we added garden and yard sites.

The septic area was positioned beyond the gardens, to give more open area in the deepest, southerly part of the lot to admit light, without cutting trees too near the edge of the lot. The well was located near the front (it has to be away from the septic) where the huge well-drilling rig could get in without taking out any trees. Our goal was to preserve a band of trees along the road except where the drive entered.

A tool shed was placed where it would block out the view of the neighbors' garage and also help create a more defined, enclosed yard space.

Review: Looking at the Plan as a Whole

It is possible now to look at the whole site as a floor plan. There are now outdoor rooms or spaces such as the decks, gardens, and yard. There is a whole-lot circulation plan. There is an entry sequence that begins outside at the parking, enters through the little court, outside porch, and inside hall, then divides quickly to various zones.

This plan can be analyzed much as we analyzed interior floor plans for economy, orientation, zoning, and circulation.

The next step is to look at the house in three dimensions.

Books on Small House Design

—*Small Houses for the Next Century,* by Duo Dickinson. McGraw Hill Inc., N.Y., 1995.

—*New Compact House Designs,* by Don Metz. Storey Communications, Pownal, Vermont, 1991.

—*Tiny Houses,* by Lester Walker. Overlook Press, Woodstock, N.Y., 1987.

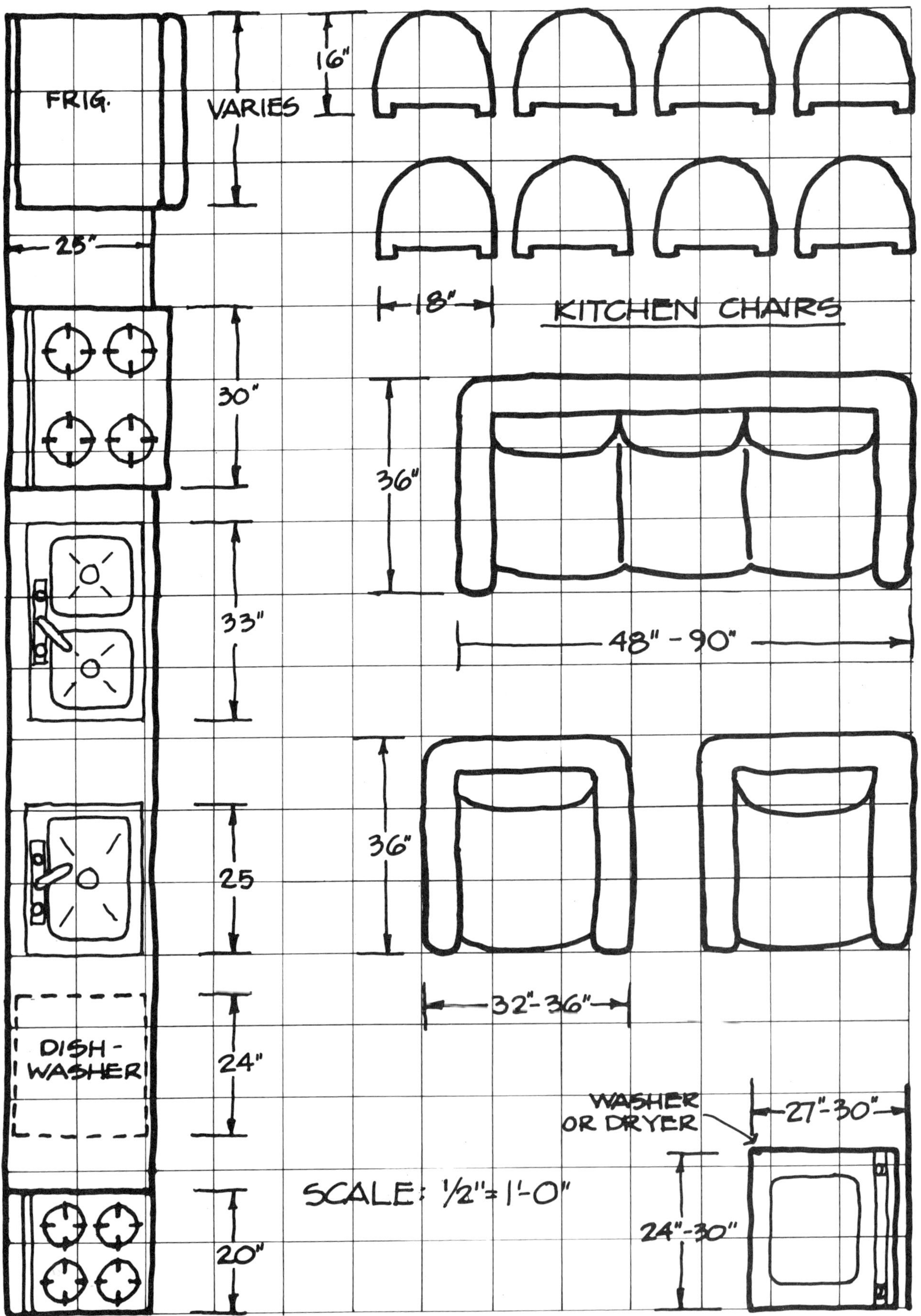
FRIG.
VARIES
16"
25"
18"
KITCHEN CHAIRS
30"
36"
33"
48" - 90"
36"
25
32"-36"
DISH-
WASHER
24"
WASHER
OR DRYER
27"-30"
SCALE: 1/2"=1'-0"
24"-30"
20"

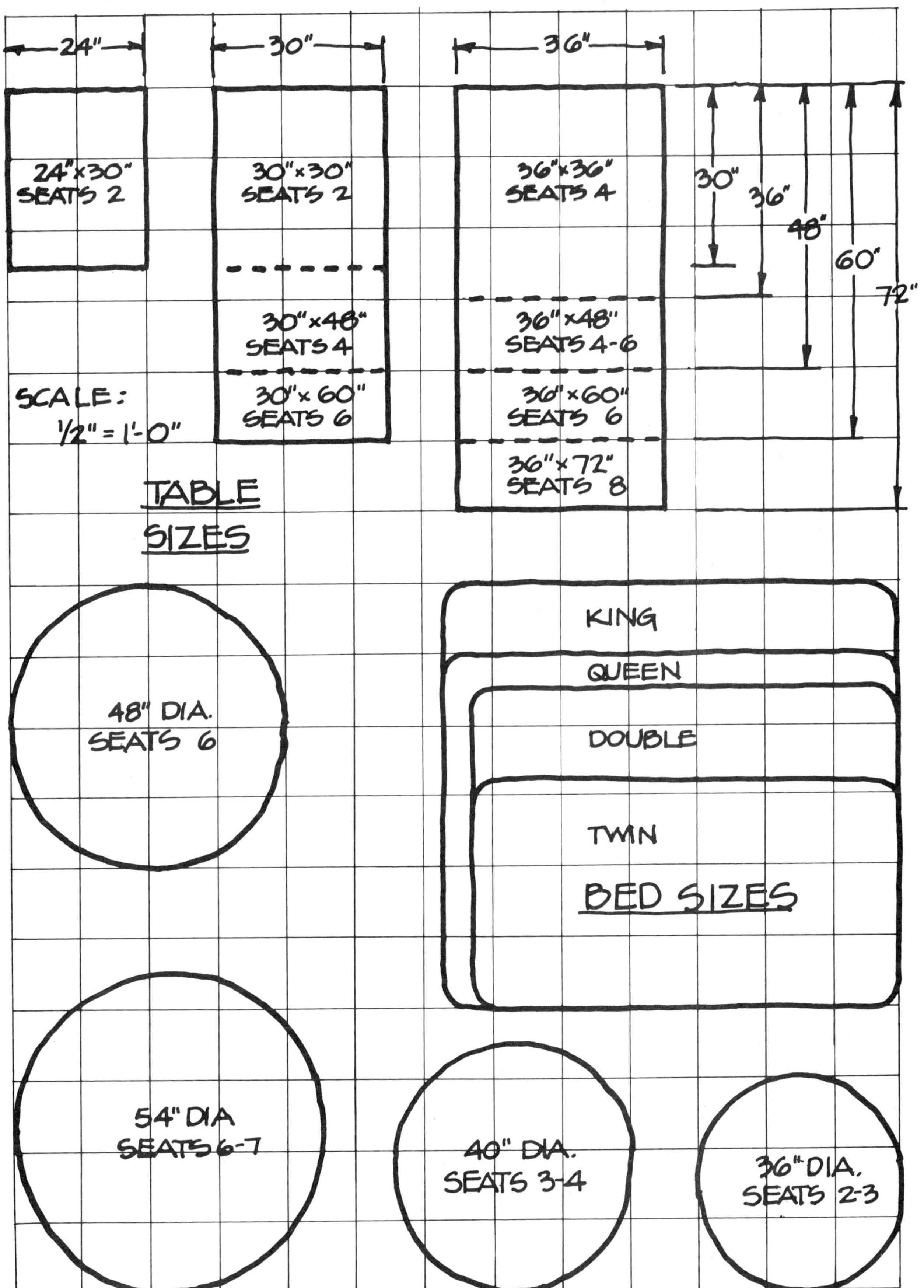

24"
30"
36"
24"x30"
SEATS 2
30"x30"
SEATS 2
36"x36"
SEATS 4
30"
36"
48"
60"
72"
30"x48"
SEATS 4
36"x48"
SEATS 4-6
SCALE:
1/2" = 1'-0"
30"x60"
SEATS 6
36"x60"
SEATS 6
36"x72"
SEATS 8
TABLE
SIZES
KING
QUEEN
48" DIA.
SEATS 6
DOUBLE
TWIN
BED SIZES
54" DIA
SEATS 6-7
40" DIA.
SEATS 3-4
36" DIA.
SEATS 2-3

9

ELEVATIONS, SECTIONS, & MODELS

An elevation drawing shows a vertical or side view of the house from a point of view either inside or outside. A cross section, or "section," shows a plane sliced through the house horizontally, vertically, or diagonally. Your next step is to make elevations and sections of your house. These help you design your roof, locate windows, and study how well different parts of the house will relate. At this time you try out your ideas about trim details. These drawings will be developed and revised much like floor plans until you get the house to look and function the way you want. Chapter 7 shows how to make these drawings.

MAKING TRIAL ELEVATIONS

Make a set of four exterior elevations. If your house is oriented with the compass points, these will be labeled north, south, east, and west. Later you will make detailed elevations and sections to study particular design problems that come up, such as stair locations. The same hypothetical house I have used earlier is used here as an example.

To make the south elevation, tape the plan to the drawing board with the south wall at the bottom. Put another piece of tracing paper over that. Then build up the elevation as described in chapter 7. Basically, I establish some horizontal lines to represent floor and ceiling, then get window and door locations from the plan below.

The elevation should show the foundation and indicate the lay of the land right near the house. There might be other features you could include, like a special view, or sun angles. This drawing allows you to study several basic features of your design.

If I knew where I was going with the roof design, I could now draw the roof the way I envisioned it. If I weren't sure, I could draw the scheme up to the roof line, make a half-dozen copies, and draw different roof lines on each one to see what looks best. I could also make a model and try out different roofs on it.

Each elevation of your house is a composition. It can be orderly or chaotic. It can be in balance, subtly balanced, or out of balance. It can be boring and bland, comfortably typical, interesting and pleasant, unique but irritatingly chaotic, or unique but well composed and pleasing. A few basic elements make up the composition: the windows and their placement, the panels of wall around the windows, the roof form and roof pitch, and the trim elements of windows, siding, and roof. As you begin to develop elevations, you are adjusting these elements for both functional and aesthetic reasons. Ideally, the practical and aesthetics work together.

Windows

Often the window locations will be determined as you make the floor plan, based on interior function, sun angles, and views. The exact window dimensions will also be based on the windows available. If you are using new windows, you will be using a product catalog to select sizes. Chapter 28 discusses types of windows in detail.

Figure 9-1 shows how I often position windows. People often select windows that are too small for the application. Small windows can be worse then none at all. They let you know there is light to enjoy and views to see, yet they don't deliver them. A regular window, say a living room window, should allow you to see outside views both up and down, either seated or standing.

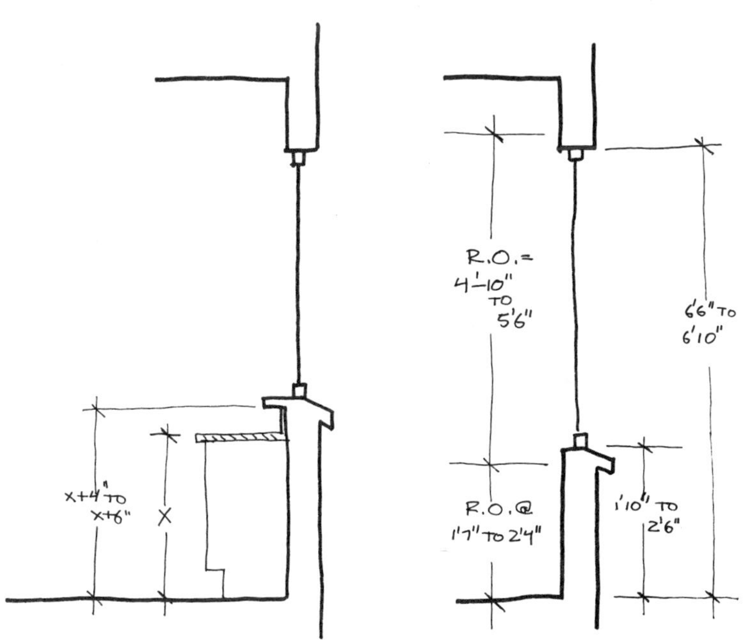

Figure 9-1. Typical window heights.

Figure 9-2. Clean detailing. Window size and location are carefully coordinated to make the window sill and the top edge of the splash coincide. Designed and built by Sam Clark and Larry Duberstein, Cambridge, Massachusetts.

Windows over desks or counters should be set fairly close to the work surface, to make it easy to see out and down.

On the other hand, some windows, such as in a bathroom, may be set as high as five feet above the floor, to admit light while preserving privacy.

Small adjustments in window size or placement may provide an opportunity to make the elevation more orderly or interesting. Sometimes this means aligning two windows that are slightly out of line. It could mean setting all the windows tops at the same elevation to create a consistent horizontal line. It could mean aligning window with some other building element, such as the edge of a porch.

Though you might specify window placement for alignment and order, you could also place them to enhance liveliness and variety.

Roof Form

With the windows tentatively located, you can try out different roofs on your house.

The roof lines in figure 9-3 differ in more than looks. They enclose different kinds and amounts of space. They heat differently, and they afford different possibilities for windows. Some roof types leave room for an attic. Some give a nice high ceiling. Some provide space for a loft. A high ceiling lets in more light,

Figure 9-3. Roof forms

but makes the space harder to heat. A steep roof sheds snow; a shallow one doesn't. A gable roof spans a longer distance than a flat or shed roof. A roof that is too low on the sunny side might block the sun. A roof that is high on the windy side may gather cold air you would rather deflect.

In general, any roof design should help you put sunlight where you want it, and support the layout inside.

Figure 9-4 shows an east elevation of our example house, with a couple of roof options. The shed roof option dramatically affects the look of the inside space.

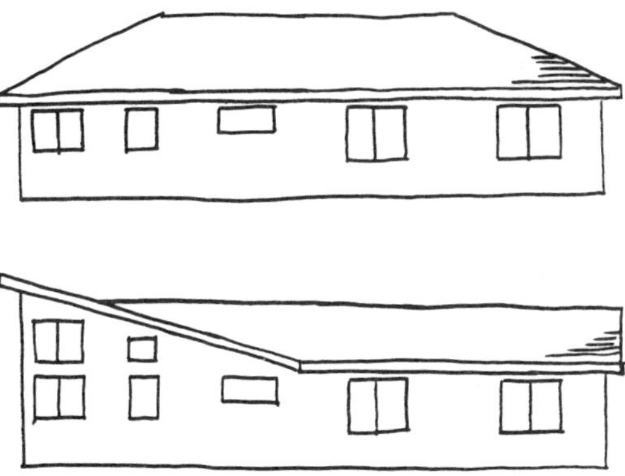

***Figure 9-4.** Sample house, east elevation, with hip roof. Same view with a shed roof.*

There are several other features of a roof besides its basic design or type, including the pitch, the overhang, the roofing material, and the trim detailing.

The roof pitch and overhang work together. A shallow pitched roof, with a small overhang, looks skimpy. Give

Figure 9-5. *Clearstory roof lets in south light. Schatzow/Shapiro house. Rebuilt by Robert Sparrow Building, Wellfleet, Massachusetts.*

Figure 9-6. *Exposed rafter ends look very different than boxed-in rafters. Seeger/Birnbaum house. Built by Peter Peltz Construction, Woodbury, Vermont.*

it a stronger overhang, and it looks modern and more sheltering. Adding to the roof pitch adds roof mass.

In the example, we had already settled on a hip roof. I drew several pitches and overhang until I thought thc roof looked sheltering but not too heavy. The overhang was simple but did provide some shading. I was juggling looks and cost. There is further discussion of ways to detail rafters and overhangs in chapter 26.

MAKING A SECTION DRAWING

As you zero in on an idea that you like, make a section to see how it would be built and what it will be like inside. Usually what you learn making the section ends up influencing the elevation. For example, when you detail the roof overhang, you might end up changing the width of the roof trim.

In the example, we wanted a big main room that was spacious, light, and a little dramatic, with plenty of south light. The section in figure 9-7 shows how these goals were accomplished within the simple hip roof scheme. The collar tie was lifted up about a foot to create a high ceiling—sort of a domed effect, but easy to build. A simple dormer with its own window lets lots of light into this dome, and makes the main room seem really big.

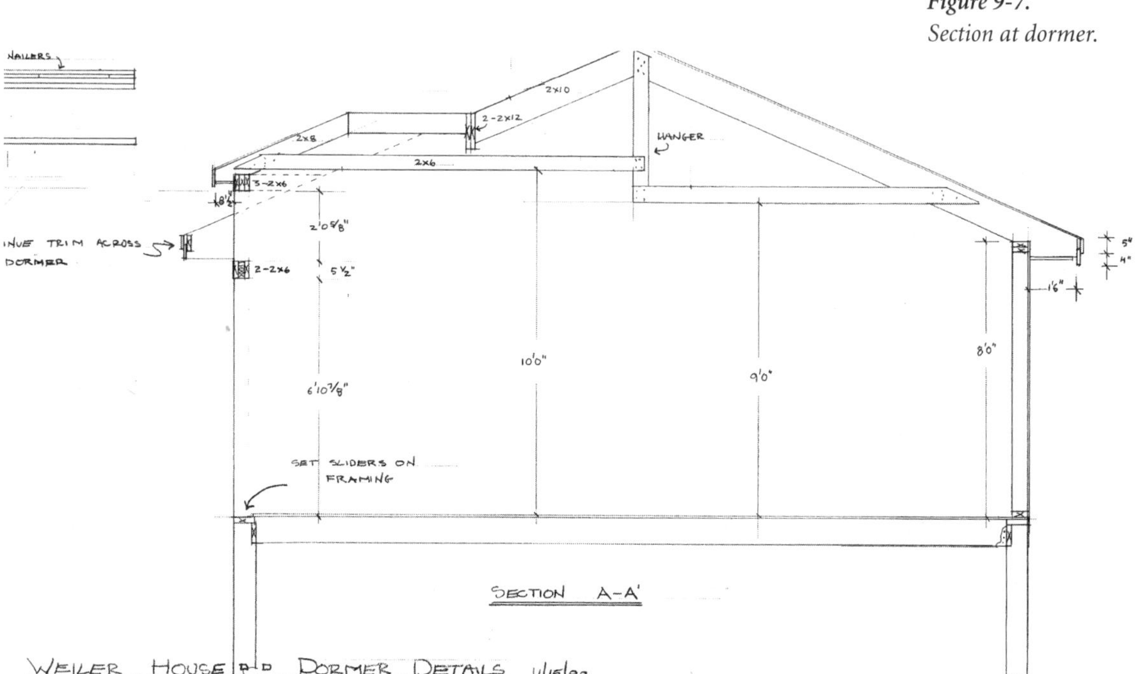

Figure 9-7.
Section at dormer.

Every element, angle, and dimension was adjusted until the design features each worked. There wasn't really a lot of space to work with.

TRIM AND SIDING DETAILS

Many houses are built without drawing the trim or siding on the elevations. I like to spend the time to draw these, because I think they have a major effect on how the house looks. Trim that's too minimal looks anemic or bland; too-heavy trim and it looks clumsy. Try out different ideas on your design.

Perhaps window trim comes first. Traditional windows usually had flat casings, 3 to 6 inches wide, with thick 2-inch sills. Modern windows often also come with a 2-inch trim, called "brickmold." Another style of windows have virtually no trim beyond a thin, half-inch flange against which the siding is butted. Functionally, they all work well. But as a traditionalist, I want a window to look like a window. I'll always order windows with flat trim when possible, and if for some reason we get the trimless style, the windows look better when trim is applied around them. This additional trim isn't absolutely necessary, but makes the house look the way it should (figures 9-8 and 9-9).

The sample house shows the kinds of design choices you can make during the elevation stage of design. The key elements here include the window sizes, the window and door trim, the corner boards, two types of siding, and the size and style of the roof overhang "fascia." I drew these many times before I got the elevation the way I wanted it.

Shingles can be put up with or without corner boards. Corner boards and window or door trim can be wide, narrow, or in-between. There can be a horizontal trim band at the bottom of the wall, often called a "barge board," or at the top. The roof fascia can be wide or narrow, simple or complex. In this case, a single wide board looked totally different than the double, lapped fascia shown. The single wide board would have been much heavier and clunkier.

These differences have very little influence on cost. Basically, the whole facade of the house will be covered in one material or another, which costs about $1/ square foot. So I can be fairly free in fiddling with these elements to create an Adirondack lodge, Craftsman cottage, a modest Cape bungalow, a quaint Cape bungalow, or other style. I spend a lot time adjusting the composition of trim elements to get the look I want.

Here I'd say I was going for a fairly understated cottagy look. But the same budget could have bought quite different-looking houses, though they would have been functionally identical.

Figure 9-8. *Changing trim details changes the style.*

Figure 9-9. *One approach to trim details.*

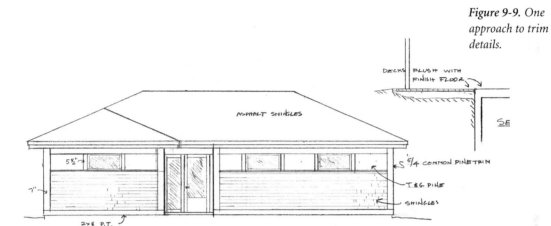

DECKS FLUSH WITH FINISH FLOOR

SE

ASPHALT SHINGLES

5½"

5/4 COMMON PINE TRIM

7"

T&G PINE

SHINGLES

2×8 P.T.

NORTH

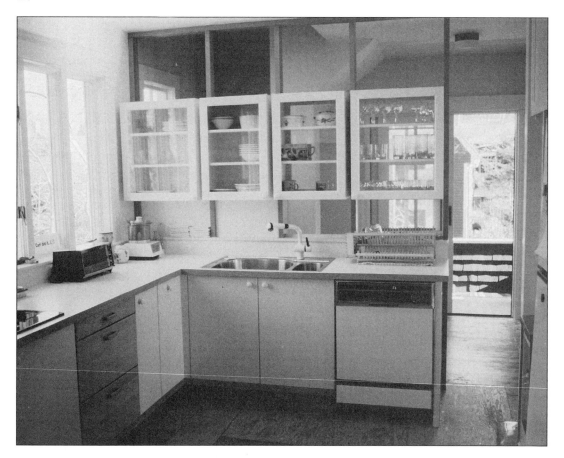

Figure 9-10. Cabinets with glass on both sides let light into kitchen. Hill/Popper house. Built by Clark and Duberstein.

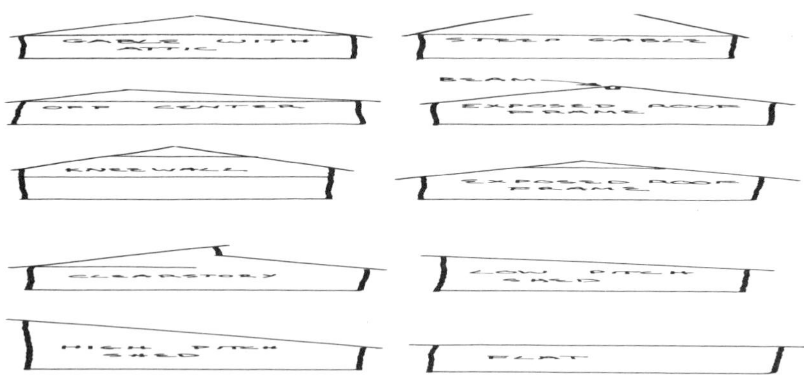

Figure 9-11. Closet layouts: among the many special drawings worth the effort to make.

ANALYZING ELEVATIONS AND SECTIONS

Light: Your elevations and sections allow you to see things that are not evident on your plans. Make sure every space gets light, preferably from more than one direction. There are several ways to admit light to the middle of a house, including skylights, internal windows, half-walls, and even mirrors. Your sections provide the best tool for figuring this out in your case.

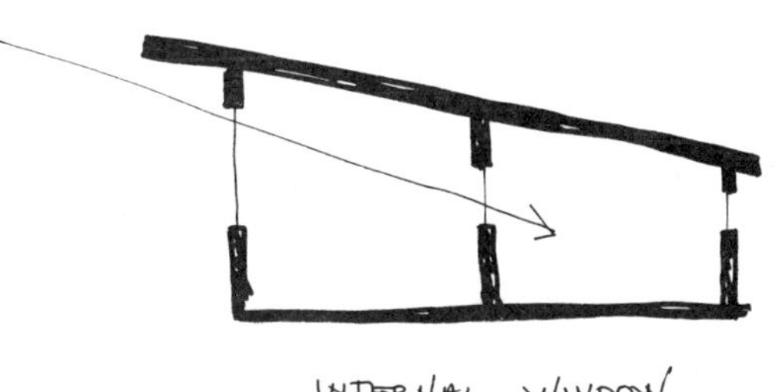

Figure 9-12. Internal window lets light into back spaces.

No-step entry: Sometimes builders don't put enough thought into the height of the house relative to the grade of the earth. My goal is generally to keep the house low to the ground to minimize or eliminate steps. At the same time, for drainage you want the earth to slope away from the foundation a bit, to carry water away. Building codes often specify that any wooden aspect of a wall start eight inches up from the ground. Figure 9-13 shows ways to keep the floor level near the ground level.

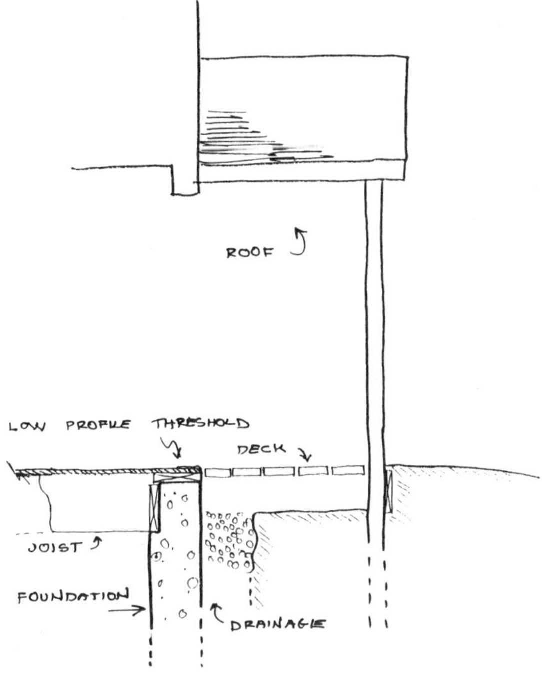

Figure 9-13. No-step entry details.

10

KITCHEN & BATH DESIGN

After 50 years of successful marketing campaigns, the standard way we as consumers visualize kitchens and their design emphasizes cabinets and appliances. The general idea put forth is, more is better. My approach, on the other hand, emphasizes creating an overall space that is pleasant and homey, and then designing simple, carefully detailed cabinetry that is efficient without dominating the space.

SPACE FIRST, CABINETS LAST

The best cooks I know often have very simple cabinetry. They know that a compact layout is much more productive to work in, if you know what you're doing, and if everything is where it should be.

This chapter outlines the steps I follow, more or less, in designing a kitchen, in a new house or a renovation.

Since commercialized images of the kitchen are so pervasive, it is good to begin with an exercise to shake off some stereotypes, and conjure up your own design goals. Think back over the nicest kitchen spaces you can remember, and recall the features or qualities which made these places feel right. What made them a pleasure to be in whatever you were doing? I don't mean cabinets or equipment, but the light, the decoration, the activities that occurred there, in short the overall form and feeling of the room.

When I do this, I think of sunny places, with lots of windows and a garden view. I think of a fireplace or an old-fashioned cook stove. I do a lot of my reading in the kitchen; I think of kitchens with a couch in them. I remember plants, even a miniature garden of herbs to pick to go directly into the pot. I remember looking out the window at the deer in the field. Of course, I also think of a big table to eat at and do my taxes on.

My ideal kitchen has a work-place feel to it, maybe a little time-worn in spots, not too shiny. It's a social space

Figure 10-1. More than cabinets. Brown/White kitchen. Built by Breslaw, Clark, and Duberstein.

Figure 10-2. A place to read in the kitchen. Hill/Popper house. Built by Clark and Duberstein.

Figure 10-3.

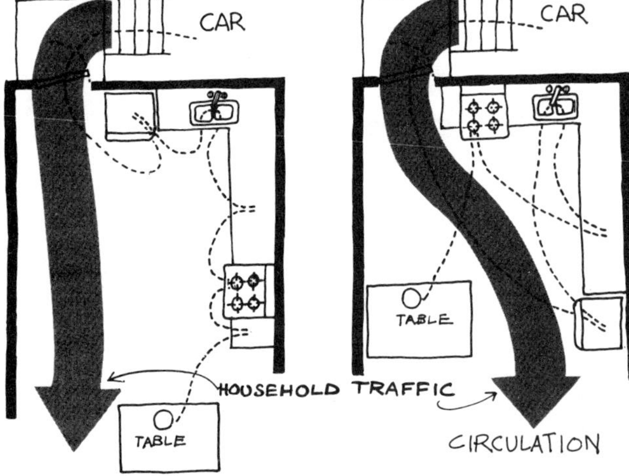

Figure 10-4. Logical kitchen circulation, compared to a less functional layout.

most of all, and an informal and busy space.

You may have a very different image. Perhaps you'll add other features and functions to it. Many kitchens I've built have a sound system, phone and message center, maybe a desk, an attached greenhouse, a homework desk, a sewing area, a laundry section. Think of the things that will get done in the kitchen, whether it is designed for them or not.

In my experience, these qualities or features count for a lot more than marble counters or wall ovens. If there are too many appliances and cabinets, there is no room for the home-like features.

Let's get down to work. Schematically plan the circulation in and around your kitchen. Circulation is the traffic pattern through the house: hallways, stairs, passages, etc. Often, the kitchen is the hub most journeys pass through. Getting this circulation right is crucial; it is the design underlying the design.

The scheme should take into account not only the kitchen work area itself but pantries, recycle area, entrances, parking and unloading packages, passages to the dining and living spaces, a route to the garden or compost, and porches or other outdoor eating areas.

It should be very easy to unload groceries from the car, or to bring food directly in from the garden. Similarly, taking out trash, compost, or recyclables should be short and simple tasks. The kitchen should lead directly to the eating places. Be sure to consider the outdoor places to sit or eat as eating places, just as you would treat the inside dining table. If possible, the kitchen work area should be along the traffic lane, but not bisected by it, although I've known successful kitchens that violate this rule.

Diagrammatically, a classic house design scheme has a front door leading to the living room and other public areas of a home, which guests perhaps use for formal events. A back door near the garage leads into the kitchen.

In Vermont, one often sees these front doors boarded over to save heat. The kitchen door is the real entrance.

Many contemporary houses acknowledge modern informality and functionality by letting the main entrance be the kitchen entrance in some sense.

Diagram these room-to-room spatial relationships before laying out your kitchen.

KITCHEN LAYOUT

The idea that underlies all kitchen layout is the "work center concept," which comes from efficiency planning in industry. The basic idea is to design any work station to reflect the work to be performed, whether it be bricklaying on a scaffold, working at a carpenter's bench, or cooking. Whatever the work at hand, the work surface should be at the height that is most convenient and causes the least strain. Everything habitually used would be stored right in the work center, in the location that worked best, and in the position that makes retrieval quick and simple. Researchers have spent a lot of time observing workers in different situations to find the best arrangements, and this observation phase is a valuable part of designing kitchen work centers. Later I'll review some of the details of good work center design.

The basic work center concept has been applied to kitchen layout in a variety of ways. The researchers who applied "motion study" to the home defined three principal work centers:

The mix center, or food preparation center, was the main work counter, with most-used ingredients, pots and pans, mixing bowls, spices, tools and utensils.

The cleanup center included the sink, cleaning supplies, a place for dirty dishes, a drainer, and probably dish storage. It might also contain things first used at the sink, such a saucepan you would fill with water to prepare cereal.

The cooking center included the stove, and storage for items such as griddles and spatulas which are first used at the stove. The cooking center would also contain a work counter, and a heat-resistant area to place hot pans.

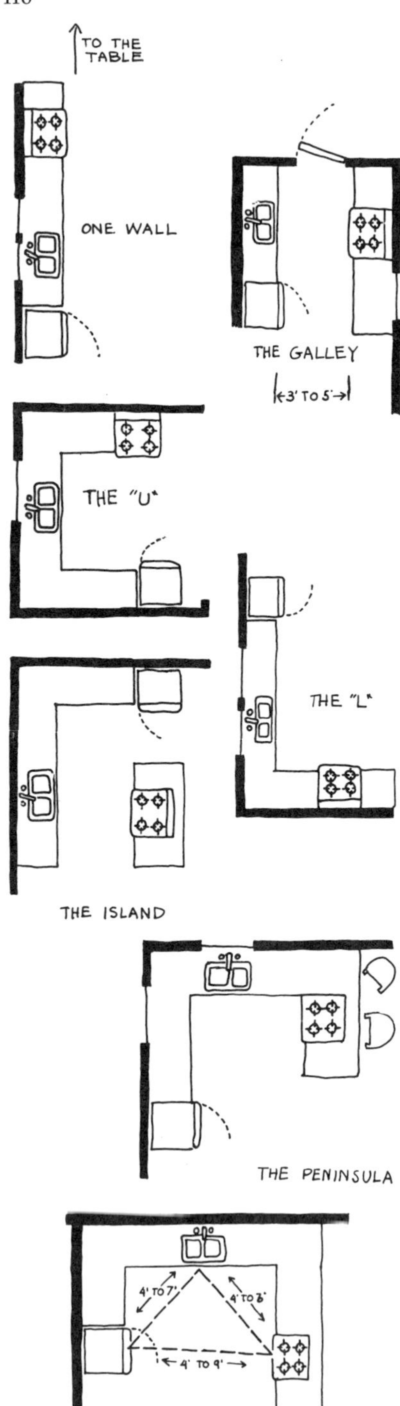

Figure 10-5. Work triangle and standard kitchen layouts.

This last function could be elaborated into a fourth center, the serving center, where serving dishes, napkins, and other items used at the table were stored. This area would be near the dining room or table.

Though the centers merge, they are also distinct. For example, it doesn't work to pile the dirty dishes on a counter designated as the "mix center."

The kitchen industry re-cast this formulation—dumbed it down—by standardizing layouts and by promoting the "work triangle."

The work triangle is a one-step test for kitchen layouts, devised for government-financed housing in the early 1950s. It's basically valid. The idea is that if the work areas and appliances are too far apart, the cook will have to do a lot of unnecessary walking back and forth. If the appliances are bunched, and not separated by counters, it's even worse, in effect splitting each work center in half. For example, the work you would usually do next to the stove, like chopping up things for a stir-fry, would have to get done across the kitchen somewhere, resulting in many extra trips and lots of wasted effort. If you follow the work triangle, there is always a decent counter next to an appliance (figure 10-5).

Food Flow

There is a logical order to the centers. The fridge and pantry are near where you bring things in to the kitchen. Next you get items out of the fridge or pantry, and often proceed to work at the sink, perhaps to add water or rinse some produce. Then you process the food at the food preparation area, cook it, and serve it. This order saves steps, and keeps work simple.

Sometimes other design considerations, including cost, make the ideal sequence unworkable. If the work centers themselves are carefully planned, the kitchen can still be very efficient. In our house, the stove was near the door, and the sink nearer the table, but it is still a great kitchen to work in.

I often see kitchens arranged backwards for no thoughtful reason. Usually this seems to be because kitchen design planning went no further than the work triangle idea. Another common reason for illogical work sequences is the conviction that the sink must be under the window. This idea dates to the period when kitchens were in dark basements with one small window, and most work was done at the sink itself. In today's sunny kitchens, with dishwashers and big counters, the window at the sink doesn't contribute much, and can work against kitchen function.

Standard Layouts and Their Advantages

The standard layouts such as the U, L, and galley are the most common permutations of the work triangle formula (figure 10-5). Probably most of the kitchens that get built, and that I have built, fall into these molds, and should. In new houses, these layouts usually make sense. In renovations, it is sometimes more difficult to devise a layout that strictly follows these patterns.

The U layout is probably the most efficient, at least for one cook at a time, because it puts the largest amount of storage and counters in the most concentrated space. The island or peninsula layouts are often more social. They face the rest of the house in some sense.

The L plan is quite efficient, and is compact in a special sense. All the other schemes require a room-sized place, whether it's a separate room or just an area within a larger space. The L doesn't require its own space. The space is simply a two-foot-wide margin along the edge of an area. This feature is particularly worth considering when you are designing for smaller houses.

This points out another advantage of the L. By minimizing the kitchen visually, without sacrificing function, you can create a big "great hall" type of space in which the kitchen equipment is downplayed.

The Power Kitchen

Sometimes I've worked in kitchens that seemed to conform to these rules, but were in fact awkward and clumsy to work in. Perhaps the area that should have been the primary work area was dark and antisocial, and people just didn't like working there. More often, the layout worked, but things were stored and placed in illogical ways.

On the other hand, I've worked in kitchens that didn't fit the standard layouts, but were very efficient and pleasant to work in.

After puzzling over this, I've come to the conclusion that there is really one feature that separates efficient kitchens from merely okay kitchens, namely a really well-thought-out primary work area that is very handy to both stove and sink. This area needs to be three or four feet wide, and either adjacent to sink and stove or one step away. It should be equipped with the items used the most. These supplies would include some pans and bowls, the favorite knife or knives, and the most-used foods (spices and grain, in my case). Another requirement is that this work area not be used for storing dirty dishes, miscellaneous small appliances, or book bags. Since trash and compost disposal is almost constant during cooking, these services have to be located in this primary area.

The work area shown in figure 10-6 works best because there is a counter between sink and stove, but there are many configurations that can work.

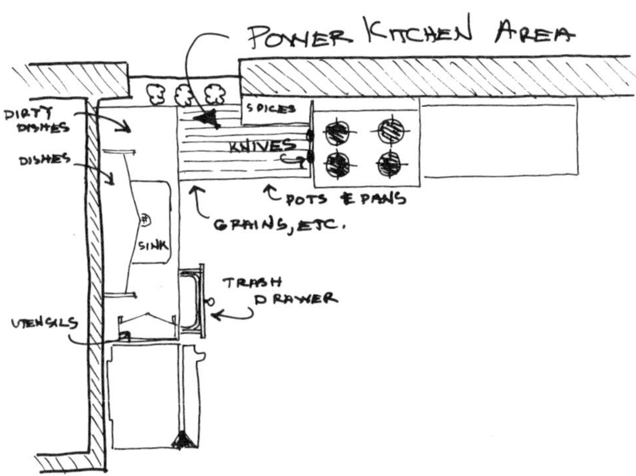

Figure 10-6. *My favorite kitchen, very compact, very easy to work in.*

If a kitchen design incorporates these simple features, a good cook can organize the room as an efficient, comfortable place to work. It doesn't matter if the fridge is located a few steps away in the pantry, or if secondary work areas are not ideally situated. These things matter less because the predominant trips when cooking are from the primary counter to stove and from counter to sink, and from counter to trash can or compost. There are fewer trips to the fridge or pantry.

Figure 10-6 illustrates this idea. To get from my forty-inch counter to the sink, it is a half-turn to the left. The stove is right there, too. One drawer below holds all the flours and supplies I use to bake. Another drawer holds my favorite pans. My favorite knife is in the slot beside the stove. The spices are in a shelf along the wall. Utensils are in a bin just by the sink. The trash is right there; I can pitch stuff in by simply reaching back. Under the sink are all the bags and paper products. The dirty dishes go in the inside corner, so they don't obstruct the path from my counter to the sink. I have almost everything I need, and still I have a window right in front of me. I know some people like a more spread-out approach, but I like to stand in one place and really cook!

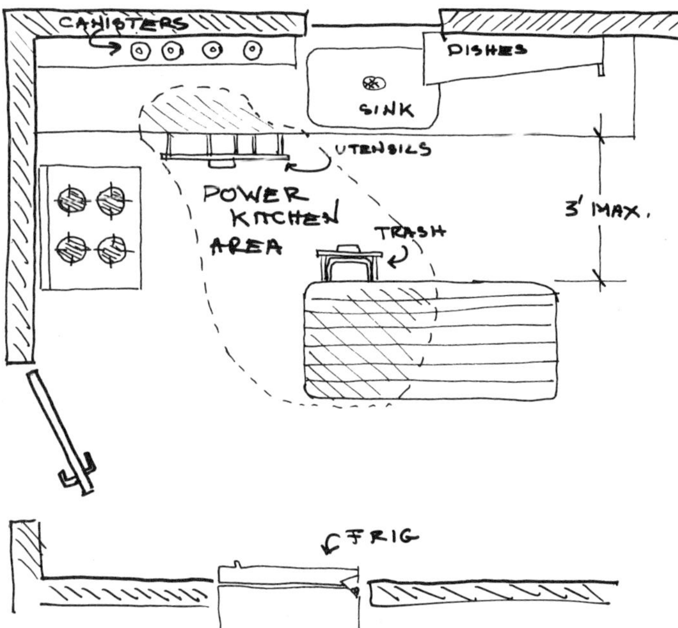

Figure 10-7. With a narrow aisle, the island is handy to both sink and stove.

When I design a kitchen anywhere, I first think about the overall layout, what I call circulation. Then I look for this "power kitchen" area. If I can locate this area on the plan, I know everything else will fall into place.

WORK CENTER DESIGN

Some Basic Principles

There are a few basic principles that apply to the design of a work center in a kitchen or elsewhere. Most of them go back to the research done by Frank Gilbreth, who founded Motion Study in the 1920s, and Lillian Gilbreth, who later applied these concepts to the home. (The novel *Cheaper by the Dozen* is about the Gilbreths.)

The first law is "storage at point of first use." Kitchen stuff is usually stored by category, for instance all the knives together. But different knives, or pots, or bowls have different uses and often go with different places in your work area. If you tend to use your paring knife at the sink, store it nearby. If you use your carving knife near the stove, store it there. The first corollary to this rule is to give the best storage spots to the things you use every day. I have one knife I use for most everything. I want it within instant reach. The infrequently used knives can go anywhere.

The second corollary is another Gilbreth term, "one-motion storage." Gilbreth broke labor down into therbligs, basic motions of work, for example searching for something, accessing it, grasping it, transporting it, using it, cleaning it, putting it away, etc. (see pages 17–18 for a complete list). Take my knife for example. If my favorite knife is in a drawer with a bunch of knives, I have to grasp the drawer handle, open the drawer, locate the the right knife, grasp it, close the drawer, then perhaps reposition it in my hand before I can use it. If the knife is in a slot right in front of me, carefully positioned at a comfortable angle to grab, I can simply grasp it with the grip needed for cutting, and start work: two motions. In thinking about any particular kitchen storage scheme, breaking it down into these units helps find the simplest way.

Drawers Below

Almost everything under counter height stores best in drawers. You can pack a drawer solid front to back, and by using good quality full-extention slides, easily pull it open to make everything easy to find and retrieve without leaning far over. By comparison, the standard door cabinet requires you to stoop down, and rummage around in a dark area where only the frontmost layer is at all easy to use. I put all kinds of things in drawers that usually go on shelves, including spices, canned goods, cereal boxes, cannisters of flour, lids, small appliances, and the good dishes.

Since drawers are expensive, I make wide ones, 24 to 40 inches wide if possible, to get the most for my money. I carefully figure out the drawer contents in advance, so the drawer is big enough, yet no deeper than necessary.

I don't build door cabinets at all, except at the sink or for very large pot storage. Trays store best in a vertically divided rack.

There are some economical ways of creating drawer type storage at lower cost, such as pullouts, wire baskets, and woven baskets.

Shelves and Racks Above

Kitchen designers usually fill every wall with overhead cabinets. My main beef with oveheads is that they crowd the space; they make the room seem smaller, and I don't like them in my face as I work.

Open shelves work best. It's easier to find and retrieve things because the door isn't in the way. Shelves need be no deeper than the particular items they hold, which often means they can be less than the 12- or 13-inch depth of standard cabinets. If shallower, they can be lower on the wall, putting things in easier reach. I often make shelves that start right at the splash as a very narrow 3- or 4-inch shelf, then gradually get wider as they go up. Of course, shelves are

Figure 10-8. Put everything possible in drawers.

Figure 10-9. Open shelves need be no deeper than the things they hold.

much cheaper than door cabinets.

Nice racks of various kinds are another alternative to the door cabinet.

Of course, people often want door cabinets anyway to cover up the mess. If so, I try to position them where they won't show so much, perhaps next to the fridge.

With lots of drawers and strategically placed shelves, you can leave some walls blank for windows, pictures, or nothing at all. This makes a kitchen much nicer than it would be with wall-to-wall cabinets. It's more of a place and less of a warehouse.

Figure 10-10. Open shelves can be low on the wall.

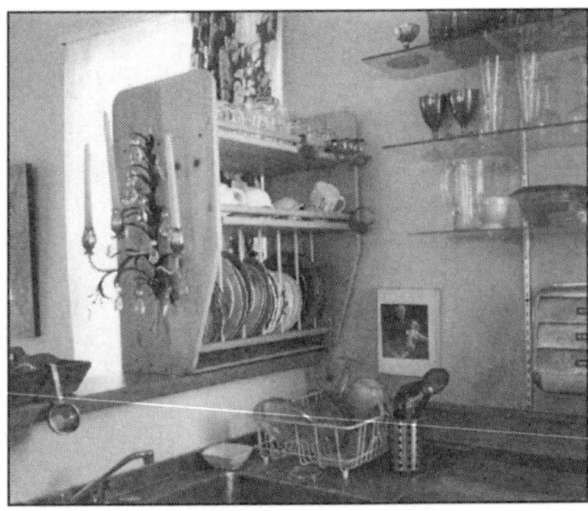

Figures 10-11, 10-12. Racks provide cheap, visible storage.

The Margin

The most reachable, convenient storage in a kitchen is the place most cabinet configurations don't use, the space at the back of the counter, below the upper shelves. This "margin" can be accessed with a very easy reach, no bending, and little searching. It is an ideal location for frequently used smaller items, or things that can be stored flat on the wall. I almost always make a little 2- to 3-inch "cap" on the splash, which is great for spices and tall bottles. This can be elaborated a bit to provide a home for canisters, little coffee makers, and similar items. A saw kerf (a blade-width through-cut) in counter or splash cap makes a safe storage area for knives.

Figure 10-13 shows some of the many ways to put this valuable margin space to work.

Varied Counter Heights

The ergonomics of counter heights is discussed in chapter 2. There are three basic reasons to vary counter heights from the standard 36 inches. First, people who aren't average height often prefer counters a little lower or higher than standard. Second, different tasks work better for some people at different heights. Cooks often prefer a low counter for kneading bread, or when mixing big batches of things. Sometimes a sink works better as high as 39 or 40 inches, because you are reaching down into the sink. Thirdly, some people, including many with disabilities, are more comfortable working seated.

In a given case, it often works to tailor counter heights to the preferences of the cooks. Another approach is to provide two or more counter heights, so that users can find the most comfortable spot for them.

The Fourth Major Appliance: The Trash Can

The trash can and the compost bin are used constantly, more frequently than sink or stove. Yet often these functions are tucked awkwardly under the sink, or hardly planned for at all. Locate them carefully. I use the compost

Figure 10-13. *There are many ways to use the margin area for convenient storage.*

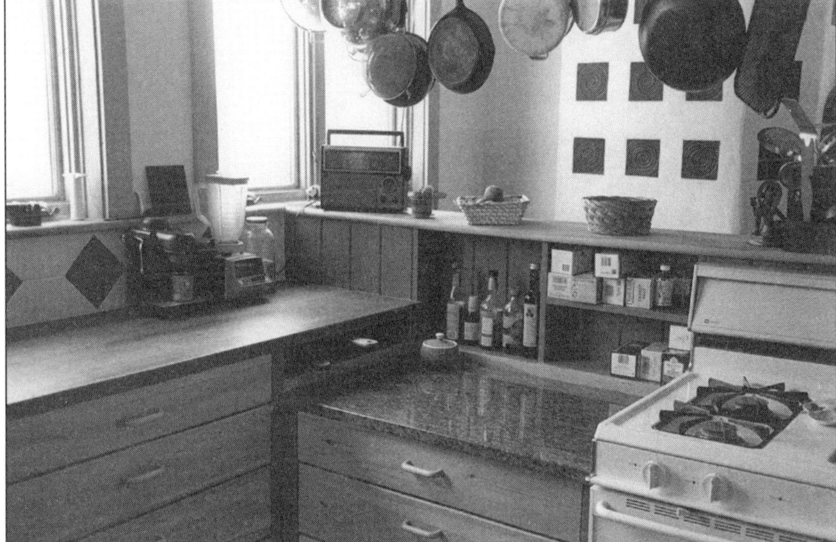

Figure 10-14. *Different size people and different tasks often call for different counter heights. Baldwin kitchen. Built by Iron Bridge Woodworkers, Plainfield, Vermont.*

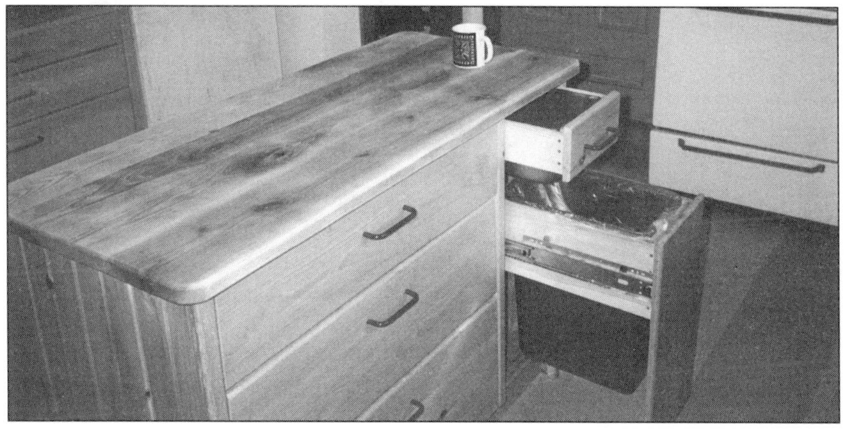

Figure 10-15. *Best setup for compost and trash.*

as I prepare virtually any dish more complicated than coffee, and I use the trash can almost as often. They're also used during cleanup, when scraping dishes. To meet these needs, I often locate both items between the mix center or "power kitchen" and the sink. I have experimented with many systems. The arrangement in figure 10-15 works best in my opinion. The top bin is a removable stainless steel liner (purchased from a restaurant supply house) resting in a bottomless drawer. The trash is a convenient-sized plastic trash can mounted the same way. I include it in every kitchen design unless I'm forcibly restrained from doing so. To be fair, though, simply leaving a space under the counter is also a good system for a trash can, and the old-fashioned triangular garbage drainer, left in the corner of the sink, works well, too.

Detailing Work Centers

The kitchen shown in figure 10-16 is a very compact, accessible kitchen. It illustrates some of the basic features worth considering as you design specific work centers.

Figure 10-16. *A simple kitchen with all the basics. Karen Fiser's house. Built by Clark and Duberstein.*

Cleanup center: At the sink, it pays to carefully envision exactly how you will clean up. Figure 10-16 shows some of the features I like. This is an accessible kitchen, but the idea can be broadly applied. The sink is at the corner, which allows the dirty dishes to get stacked at the inside corner, a section of counter that is otherwise not that useful. The trash is handy, in this case in the kneespace to the right. Everyday dishes are stored right at the sink. A dish gets washed, rinsed, and put away wet on the dish rack, which drains right into the sink. The little dishrack to the right is for utensils. This setup makes routine dishwashing very quick, with no moving about. I think this is as quick as using a dishwasher, because there is so little handling of the dishes.

Instead of a window above the sink—prime territory for storage—the window is to one side.

Mix center: The counter to the right of the sink, with the kneespace, is what we're calling the mix center. Utensils are just to the left in the little drainer. Spices and commonly used foods such as oil, sauces, and teas are just above the counter. Knives and such are in the top drawer. Bowls and pans not in the drainer are in the drawers. All it takes is a half-turn one way to the sink to get any of these things. The fridge is a couple of steps away.

Cooking center: An equally basic cooking center is to the right. It has a two-burner gas hot plate, with supplies below. A built-in tiled hot-pad provides a serving platform at the table.

This particular kitchen is extremely compact, perhaps more so than many people would like. My client here could literally stand or sit in one place and prepare, serve, eat, and clean up after a meal without having to move more than a few feet.

Accessibility

Chapter 2 has a more detailed discussion of accessibility, including accessibility in the kitchen. I try to design accessibility into every kitchen. Almost all of the basic ergonomic ideas presented here are elements of accessible kitchens. There are a few other features needed for a higher level of accessibility. There has to be maneuvering room if wheelchair users are to use the kitchen. There must be good counters, perhaps 32 to 34 inches high, with a 36-inch-wide kneespace below. The sink must also be open below if it is to be usable by disabled people.

For my own use, I would build a fully accessible kitchen, even including the open area under the sink. Sometimes I like to work seated, and I like the variety and simple look of a kitchen such as in figure 10-16. But often people don't want

Figure 10-17. Doors and kickspace lift off to make sink accessible.

or need such features now. In that case, it makes sense to provide ways to simply adapt a kitchen later, if your needs should change. Figure 10-17 shows a kitchen where cabinets can be reconfigured later to a more accessible form. The photo shows how doors at the sink can lift off to create a kneespace.

Small Kitchens

It is worth mentioning a few considerations that apply to smaller kitchens, some of which may be suggested in figure 10-16. As I've said elsewhere, consider the L layout; it often conserves space best. Putting the sink at the corner often works well, because it gets the most out of the corner.

People often insist on a full-size 33-inch double sink, but in a small kitchen, this decision can leave you without enough counter space. It is better to use a 25-inch single sink. Franke and perhaps other companies make 25-inch sinks that have much narrower margins at the side, and a bowl which extends all the way to the back. They have much more capacity than the more common type.

Standard 30-inch stoves can eat up too much counter. The easiest fix for this is to buy compact, 24-inch European cooktops or ovens, though they are very expensive. It is also possible to get economical 24-inch or even 20-inch domestic stoves.

Consider doing without a dishwasher. Apart from the cost, a dishwasher, located next to the sink for plumbing reasons, eats up two feet of cabinet space, prime storage that might be needed for most-used items of various kinds that should be centrally located. Dishwashers also create duplication of storage. The dishes stored in the dishwasher will later have to be stored somewhere else.

In general, use smaller tools, and fewer of them. A small kitchen, uncluttered, seems bigger than it is. Packed with small appliances, unnecessary tools, and rarely used supplies, it will be cramped to look at and work in.

In small kitchens, you may have to go to extremes to create the ancillary storage or pantry areas that are easy to locate in a bigger house. Useful storage can often be recessed into a wall. Shelves 3 to 6 inches wide can be used to store canned goods, brooms, cereal boxes, and many other items. Figure 10-19 gives an example.

Figure 10-18. European cooktops and ovens cost more, but use about 24 inches of space rather than 30 inches.

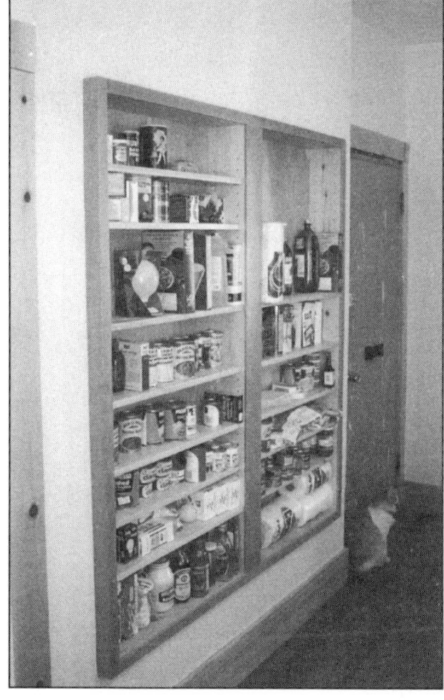

Figure 10-19. Recessed shelves.

Ventilation

A kitchen ventilator of some kind is very helpful if you cook with grease a lot, boil stocks for hours on end, or need to control cooking smells. It also helps remove carbon monoxide created by gas ranges, which can be a problem in very tight houses.

Recirculating-type range hoods, which do not vent to the outside, are basically useless. Conventional hoods work well, though most are noisy. My big objection to them, however, is that they protrude far into the room; they make the room seem smaller. I have experimented with ways to create ventilation without these problems.

You can now find high-quality, low-profile hoods that retract in various ways, and are relatively quiet. They are usually quite expensive.

A wall fan (figure 10-13) doesn't stick out like a hood, but it is not as effective as a hood either, because it doesn't gather the air before extracting it. A lot of what you are trying to vent will escape. I think this is a good solution if your need for a hood is moderate. It will work well as a way to create some air circulation in hot weather.

There are a variety of down-draft systems, which pull air down, and then duct it out. It takes a more powerful fan to pull the air down. These units are expensive. The type in which the draw fan is mounted outdoors will be quieter.

Often I've designed or built hood arrangements that attempt in various ways to look better than what is available commercially.

In figure 10-20, the overstove construction serves as plant shelf, pot rack, lighting valance, and hood. I like the looks of this, but its small size means it is not as effective as a larger unit. The hood in figure 10-18 folds forward in use to gather more air. Figure 10-21 shows the open position. This hood has a powerful fan because of the barbecue mounted below it.

Both units are ducted to the outside, where restaurant-style exhaust fans pull the air out. These can go in walls or roofs. These units are very quiet, because the fans are outside.

Figure 10-20. Hood serves many purposes.

Figure 10-21. Hood in open position.

FLEXIBLE BATHROOM LAYOUT

A lot has been written about bathroom design. Rather than review that material, I want to focus on alternative approaches to layout that may not be familiar to readers.

Bathrooms are the most expensive part of the house, per square foot, with the possible exception of a kitchen. At my grandparents' house in Ohio, there was one bathroom. Its only luxury was a clever laundry chute, built by my grandfather. To avoid the morning rush and nighttime inconvenience, each bedroom was provided with a large porcelain washbowl and pitcher of water on a nightstand or dresser, and a chamber pot.

Though this scheme seemed okay from my point of view, it definitely had disadvantages for my grandmother, who had to empty the chamber pots each morning.

Though designers usually provide convenience and privacy for families and their guests by simply building more bathrooms, there are some alternatives that may solve the problem with less cost.

These necessitate breaking a conventional assumption, which is that the toilet, sink, and bathing apparatus have to be in one room.

In England and Europe, where space is at more of a premium, it is not unusual to see the functions separated. The tub can be in one little room, a toilet in another. One or the other can have a little sink, or sinks can be provided in the bedrooms.

This scheme allows two or three people to attend to matters at the same time with reasonable privacy. If that is not enough, extra fixtures could be added, but not necessarily whole bathrooms. For example, in our house, I don't think we need two toilets, but there is often a line for the shower in the morning. My scheme is to convert a closet into a second shower.

There are lots of possible variations. Some friends of ours installed a fine old claw-foot tub right in the middle of the bedroom. One of the cleverest schemes I know of was devised by Ken and Judy Gleason. There is a small bathroom with tub and toilet, located between their bedroom and the kids' bedrooms, with doors leading both ways. The adults have a nice sink area on their side, as part of a walk-in closet, and the kids have a sink on their side.

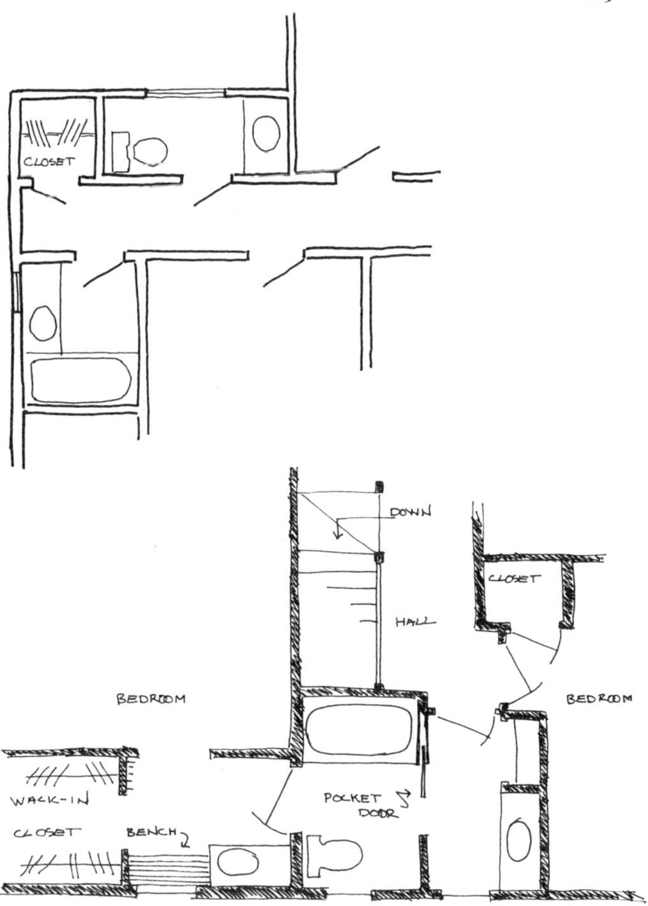

Figures 10-22, 10-23. Clever layouts that give most of the advantages of two bathrooms with little extra expense. Ken and Judy Gleason's house. Design by Parva Domus.

Accessibility

When laying out a house, my goal is always to provide a basically accessible bathroom on the first floor, whether there is any apparent need for it or not. Failing that, I try to provide a bathroom that can be converted easily as needs arise.

This approach makes the bathroom more comfortable for everyone, and more safe. It means anyone can visit the house, and attend a

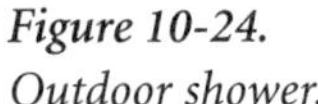

Figure 10-24. Outdoor shower.

meeting there comfortably. It also makes the house more saleable. A convertible bathroom also means that if any household member is injured or ill, or simply old, they can still use the house in comfort.

These provisions cost very little if done when the house is built, yet they are very costly to do after the fact.

The basic requirements are a wide door, a little extra space inside the bathroom, grab bars (or blocking that makes it easy to add them) and a planned "transfer"—a way for a disabled person to safely get in or out of a tub or shower. At the sink, there would be a plan that allows a kneespace to be created under the sink if needed.

There is more discussion of this topic in chapter 1. Figure 1-12 shows these basic features.

PART 3
BUSINESS

11

HIRING BUILDERS & DESIGNERS

Most people will be hiring others to do some of the work on their homes. This chapter reviews possible choices, and goes over some of the issues to think about.

THE COLLABORATIVE MODEL VS. THE MIND/BODY SPLIT

Here is a common scenario: One person designs a house, down to the littlest details. Builders compete for the job through a bidding process. During construction, the designer walks through the jobs at prearranged times, to make sure everything matches the design. Design changes require another round of documentation called "change orders."

This is the mind/body split. The designer does the thinking, the builder does the work. The designer knows everything, and the builders know nothing. I don't like this model. As a builder, I never subject myself to it, no matter how juicy the job being offered.

This kind of relationship creates many practical problems, and some of them are discussed later in the sections on contract form. But in human terms, there is an unsound premise here. It is very rare for any one person to know the perfect way to do everything. There is also a Catch 22: The designer has all the authority, but it's the builder who is held accountable.

Collaboration

Whatever the job title of the various parties, I advocate collaboration. I've met a lot of very intelligent people in the building trades, and you want the intelligence of such people working on your behalf. I always seek to establish a team approach, with owner, builder, subcontractors, designer, and any other consultants that may be needed contributing what they can.

The ideas that follow are designed to help you put together a team that can function well, keep close to the budget, and keep everybody happy during the relatively long and stressful process of building a house.

DESIGN

Being Your Own Principal Designer

The premise of this book is that people can design their own homes. Many people have been thinking about their house plan for years, and the freedom to design may be a primary motive for the project. It is one of life's great learning opportunities. Since this whole book is made up of advice on how to do this, I'll only add one point here: Always get your design checked over by people who have designed many houses. Find someone whose work you admire to look over your plans at various stages. Professional designers get their colleagues to critique their designs. It's called *review.* Another person will notice flaws, suggest new ideas, and perhaps identify problems the designer is partially aware of, but has not wanted to admit. This is even more important for the amateur designer.

If You Hire a Designer

Today a wide range of people design houses. They bring a range of skills, training, and points of view to the task.

The traditional architects are still available. Most of them make their living designing big buildings, but some are interested in houses, including modest houses, and they may have designed many of them.

Today the biggest new class of building professionals are the designer/builders, building companies that normally design the things they build. This could be a builder who also designs, an architect who also builds, or a builder and architect working together.

In general, I would urge readers who are planning to hire a designer to look into local designer/builders. A person who builds on a daily basis may have a better idea of what details work best and what the overall project will cost.

There are also architects who do small houses well, and have a builder's way of thinking about cost. An architect friend who had once been a builder told me, "I build everything I design in my mind." That's the attitude you want from the designer you hire.

Look for someone who is experienced in the kind of building you want, a designer who can collaborate with you without taking over; one who can give you a convincing answer when you ask, "How do you keep costs under control?" Go see their work; talk to previous clients. Make sure to ask how the budget held up.

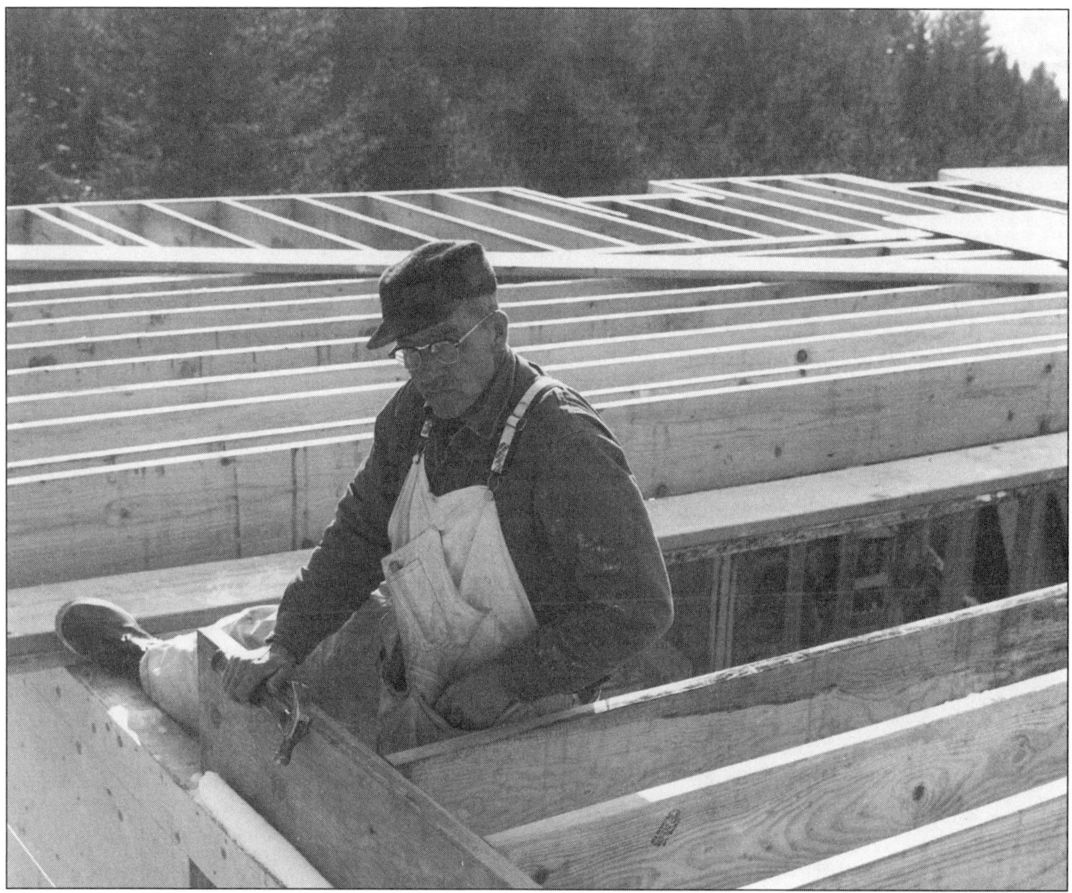

Figure 11-1. *(Photo: Thorsten Horton.)*

Design Costs

Be sure to ask potential designers what their services cost. I would stay away from "free design" offered by some builders and home centers. Such services are as much marketing as design, and in the end are not really free.

It is best to pay for design services by the hour. That way, you can obtain the help you need, but you need not pay for unnecessary work. The hourly rate doesn't necessarily measure what your design costs will be. A more experienced person may charge more, yet work more efficiently. Further, your design work costs you twice: once when you pay the designer, and again when the building gets built. A good designer might spend a bit more of your money in the design phase, but save larger sums by producing an efficient, easy-to-build design.

It also makes sense to hire only the design help you need. You may only need to contract for an occasional review of your drawings, plus a couple of on-site consultations. Perhaps you only need floor plans and an elevation or two. On the other hand, you may need a complete write-up, down to the last nail.

When I'm hired to design a project, I spend a lot of time on the layout. I provide carefully dimensioned drawings in plan and elevation. I often provide drawings suggesting ways to solve special framing problems or other details. I make a very detailed, large-scale kitchen plan. But I trust the builder to work out most of the construction details that work for him or her.

It is common practice to hire your architect or designer to supervise the contractor, if you have one, even to the point of having the architect perform an inspection before the builder is paid at each stage. On huge buildings, this is necessary. On modest houses, it is a needless expense that makes the builder's work more difficult.

The design cost is partly controlled by you, the client. If you are clear in your thinking, can state your budget openly, and are able to make decisions, your design costs can be relatively low. I recently designed a house for some friends in considerable detail for about $2,000—less than two percent of the cost—including a detailed preliminary cost analysis. If you are indecisive or unrealistic in your goals—you want everything in the design at first but don't really have the money—your design may go through many more alterations, and cost several times as much.

CONSTRUCTION

What Should You Hire Out?

Only a few people do all the design and building themselves. Even if you are the primary builder, and perform the role of general contractor, you will probably hire some help. Local codes may require licensed people for certain tasks. Septic systems usually must be put in professionally. Wiring and plumbing may have to be done by licensed people, although many jurisdictions allow owners to do their own work and then have it inspected. Excavation will almost always be hired out, unless you have the simplest of post foundations. Though people used to do their own concrete form work, I would always subcontract this out for a conventional concrete foundation.

Sub out the drywall, also called "sheetrock." Most owner-builders do this themselves, as it is a task an ordinary person can certainly take on. But the pros are fast, and hence very cheap. You'll never do as good a job. It will take you weeks to do what a pro will do in three days. If you have the money, sub this out.

What Should You Do Yourself?

I'm biased, but I'd say do the carpentry yourself; it's the most fun. Many people have successfully done their own wiring, though I'd have a pro put in the panel (breaker box). Painting, varnishing, and other finish tasks are very often done by families. If the roof form isn't too complex, you can do the roofing yourself too.

Your ability to take on tasks beyond these depends on your interests and aptitudes. If you have a strong desire to learn electrical work, plumbing, solar design, stonework, and cabinets, you can do these, too.

I think there are two, maybe three keys to success in taking on construction tasks yourself. One, read up, do your homework. There are good books on everything. Two, don't take on too much. Three, get the backup or support you need.

If you haven't done a lot of building, I like the idea of hiring a pro to work with you, at least part-time. There are so many techniques, setups, and time-saving skills. If you can find a very experienced carpenter, mason, or electrician to work with you ten or twenty hours per week, or even to advise you periodically, it will be worth it.

HIRING BUILDERS

Trust

Half the battle is picking the right builder. This person is going to be in charge of spending all of your money, and some of the bank's. After talking to former clients, and asking potential builders all sorts of hard questions, pick somebody you feel you can trust. Even if you have to wait a few months to start, or if the hourly rate seems a little high, it is better to hire someone you can trust as your guide on a journey to an unmapped country. Because that is what you are undertaking.

Here are some other suggestions: Of course, walk through projects by anyone you are considering, and talk to their clients.

Hiring a small company makes sense. A big outfit will often carry a lot of overhead you don't want to be subsidizing. You want to be paying primarily for work done, plus the necessary overhead that supports it. You don't need to be supporting a big office, lots of shiny trucks, heavy equipment, or other resources that are not needed to build a small home.

Look for someone whose tastes you share in many ways, a person you trust to make the right decisions in your absence. You don't want someone who will be obliged to call you every day to discuss details, or meet with you every morning (at your expense) to go over every little thing.

ASSEMBLING YOUR TEAM

If you are working with a design/build outfit (whatever term they use to describe themselves), the mind/body split is more or less solved. The folks who do the work will have input into the drawings, and the person who can estimate the job will be collaborating with the person making the drawings. They will already know how to work together. They will often be the same person.

If the designer and contractor are not the same person, you may have to carefully structure a productive collaboration. Having worked both sides of the aisle, I can suggest the following key elements.

The builder and designer should both be hired early. They should be asked to collaborate from start to finish.

During the design phase, the builder should be producing and updating estimates from the moment there is a design clear enough to estimate. That way the designer can scale the project to the budget, and avoid dumping untold hours into a design for a project beyond your means.

My suggested team model is essentially this: During the design phase, the builder is supporting and consulting with the designer, checking costs, and suggesting those details that he or she prefers and is set up to execute. Plumbers or electricians may also have valuable suggestions when they look at the plan.

Later, the roles reverse: During construction, the designer supports and consults with the builder on detailing and other problems that arise.

There may be other team members, such as a kitchen person, a solar or energy consultant, or someone to do an accessibility review. If so, they should be brought in early too.

The owner or owners are key team members, and they do a lot of work. Even if they do almost no building, they will be reviewing plans, picking colors, buying materials, cleaning up, and handling money. Often the owners will be coordinating the team.

12

MAKING GOOD CONTRACTS

THE DEAL YOU STRIKE WITH YOUR BUILDER HAS, OR should have, several primary purposes. The first is simply to clarify each person's obligations, the schedule, when money is to be paid, and what the basis of charges will be. The deal's second purpose is to make sure the owner gets a fair price on the work, without too many financial surprises. The third purpose is, or should be, to make sure that the builder prospers, too.

People sometimes worry that the builder will take advantage of them financially. It is actually more common for the builder to make promises that in the end don't quite work out fairly to him or her. When this happens, it is hard to concentrate fully on the important work at hand, and things can go wrong. I think it is in the owner's interest to make sure the financial part of the deal is not just the best deal for the owner, but fair all around.

The contract or agreement is also a metaphor for the relationship you are setting up with your builder. A contract can presume and promote honesty and trust. It can also foster a basic antagonism of interests between parties. As a builder, I always want the agreements I enter into to assume and promote trust, and in general to conform to my sense of self. I avoid contracts that assume a divergence of interest between me and my customers.

THE BIDDING PROCESS

Putting your house "out to bid" is one of the most common ways to choose a builder. Putting a project out to bid is usually favored by architects, and many builders are accustomed to it. In some areas of the country, almost every job is done by bid. The basic idea is that the designer or architect creates the drawings and specifications that constitute a complete description of the building. Selected builders are given the opportunity to bid on the project. The bid proposals include the exact cost of the job, and often specify when and how the project will be done. Bid proposals can also include expectations of the builder's qualifications, such as previous experience, special skills, or community reputation.

The architect and owner look at bid proposals, and pick the one that is most attractive overall—not always the lowest bid. The bid and the architect's documents become the basis for the contract or agreement. Sometimes during construction the architect supervises the builder to make sure the building conforms to the design.

Cost-Plus

There are many versions of "cost-plus," but the basic idea is that the builder is paid by the hour, plus the necessary costs of doing business. For example, a bill for a certain period of construction would include a charge per hour for each worker, including taxes and benefits. Then the materials costs get added in. A percentage is added on top of this sum to cover the business overhead, which could include office costs, insurance, equipment, etc. This percentage commonly includes a business profit for the builder.

There are many ways this total can be put together. For example, sometimes the contractor's time is billed out per hour, and sometimes included with the overhead. Sometimes the builder's discount on materials is passed on to the customer. Often this discount serves as part of the builder's profit. When looking at cost-plus builders, it is important to ask exactly how the bill is calculated.

Advantages and Disadvantages of Bid and Cost-Plus

The bid system is very clear, and protects the owner from financial uncertainty once a bid is accepted. But there are potential problems. First, since the estimating that the builder does only begins when the design is complete, it often happens that the project has to be redesigned because the original design was too ambitious.

Second, creating bids adds costs. The design has to be very detailed, so a great many drawings are required. The preparation of the bids takes many hours, which is expensive, particularly when every builder is estimating far more jobs than he or she is getting. Changes to the design are difficult, requiring in effect an additional bid or "change order." Having the architect supervise the builder is expensive. Finally, the builder's price guarantee is itself expensive, because the builder adds a percentage to the total as protection from the risk he or she is assuming. I would say that these factors can easily add five to twenty percent to the total cost of the project.

The bid system can squeeze the builder into an odd and difficult position. At the beginning, the builder has to keep the price low to get the job. But later the builder is asked to guarantee the costs resulting from factors over which he or she has only partial control: The weather, the price of lumber, the amount of time spent interacting with the architect and client, and most of all, the possible mistakes and probable uncertainties of a custom design created by someone else. While the designer's and clients' decisions actually determine the cost of the building, the builder is asked to guarantee it, and it is always a gamble.

The cost-plus approach has some advantages, too. It's flexible; changes can be discussed and acted upon with minimal negotiating. The builder doesn't have to be a gambler, and can concentrate on the task at hand.

The owner, on the other hand, takes more risk. While I don't think there is a great liklihood that the ultimate cost of the project will be higher than it would be if bid, the owner won't know the total in advance. If the design is too ambitious, it might not become apparent until it is too late to cut back.

In addition, with cost-plus the builder has no direct financial motive to work quickly and get the job done. In fact, it could be (and has been) argued that the builder benefits by dragging things out.

Finally, there is considerable room for confusion about exactly how the "cost" and "plus" are calculated (as explained above).

Alternatives to Bid and Cost-Plus

There are some systems that attempt to combine the advantages of cost-plus and bidding. I have sometimes been asked to agree to an "upset limit" plan. In this system, the builder is paid by the hour as in a cost-plus arrangement, but the client or customer is protected by an "upset limit"—a figure beyond which the price cannot climb. I would never consider agreeing to such a scheme, because it saddles me with the risk of overruns without rewarding me for underruns. I would rather bid.

A mason who did a large repair on my house in Cambridge, Massachusetts, introduced me to a "split-the-difference" contract, which I like. I hired Michael DiBlasio to tear down and redo a mammoth brick wall that had been collapsing for many years. It was a very unpredictable job. His best guess was that the work would cost $7,600 (this was about fifteen years ago). Michael would do it cost-plus, or bid it at $9,000. But he also offered a third choice, which was that we would split any cost above $8,300 or any savings below $8,300. That way, we shared the risk of the cost climbing up, and we shared the savings if things went well. As it happens, we chose to go with the bid to avoid worry. But we would have been better off with either of the other deals, because the actual cost was somewhat below the initial estimate of $7,600. Though I haven't used the "split-the-difference" approach, it's an interesting idea I would consider, particularly with respect to a straight bid.

What We Do

When I worked in Cambridge, I read about the early-20th-century builder named Frank Gilbreth, who later became famous as the developer of "motion study." Gilbreth championed something called the "fixed fee contract," which he used on all his large-scale commercial projects. The basic idea was to "make the builder's company an extention of the client's organization for the duration of the job," and to make the clients' and builders' interests run parallel.

In the fixed-fee contract, both labor and materials are billed to the client at the builder's net cost, with no markups or profits of any kind. The builder's overhead and profit are contained not in a percentage, but in a fixed management fee negotiated at the beginning of the project, and based on a percentage of the estimate. Once the figure is set, it doesn't go up if the job is increased, and it doesn't go down if there are cutbacks.

There are two principal advantages to this system. One, there's no confusion. The builder knows his or her costs are covered, and the owner knows exactly what the builder is getting out of doing the job. Secondly, both owner and builder now benefit equally from a timely and economical completion of the job. If the job drags on, it takes the builder longer to earn the fee. If the job is done quickly, the builder earns the fee sooner, and the client saves building costs. It also benefits the owner to assist the builder in all possible ways.

As soon as I read about this system, my partner Larry Duberstein and I adopted it, and were very happy with the results. It took a bit of explaining, but I never had problems with my customers using this system.

When I moved back to Vermont, my new partners had developed another good approach. This company, Iron Bridge Woodworkers, charges on a cost-plus basis. We charge an hourly rate that covers our take-home pay, unemployment insurance, tax withholding, and health plan. We add to this materials and subcontractors, at our cost. Then we add a percentage to cover our business overhead, which includes insurance, bookkeeping expenses, and our shop and equipment.

We don't take a personal or business financial risk to guarantee the total price of jobs. But we do take several steps to control the overall cost. First, we estimate carefully and, we hope, conservatively. When one partner prepares an estimate, we have another partner go over it for mistakes or over-optimism. During the job, we update the estimate several times, to see how costs compare to the estimate. If it appears that we are running over the estimate in some areas, we look for areas in the design where costs can be reduced. We might go to a softwood floor instead of hardwood, or delay finishing off an area. In other words, we can stay within the budget, but we do it by adjusting the design as needed.

How to Decide

Find a kind of deal that allows all parties to be comfortable. No contract form works best for everybody.

My personal preference is for cost-plus or modified cost-plus systems, such as the fixed fee. I'm not enough of a gambler to feel comfortable with bidding jobs. I want to be paid for my time.

But some builders prefer to bid jobs. Sometimes it's just local practice; it's what people do. But also I think many builders find that by guaranteeing a price they avoid conflicts with clients, even if they occasionally lose some money.

Similarly, some clients are more suited to one or another type of contract. Certain people really can't handle the uncertainty of cost-plus; it makes them anxious. Distrust develops, arguments break out because of the worry factor. These kinds of folks should ask for bids, even if the dollar cost may be somewhat higher. Other people don't worry quite as much, and can handle cutbacks in the design or extra costs if things take longer than expected. They may benefit from a cost-plus deal of some sort.

The nature of the project will also influence the decision. If the design is basically complete and set and the construction methods fairly standard, a bid can work. If many decisions about the design will be made during construction, or if the owners are doing some of the work themselves, a bid won't really be meaningful. If the design is unusual or innovative, a bid is also not likely to work out well, because the level of uncertainty is too high.

If you are borrowing the money from a bank, its policies may dictate the type of contract you sign. Although we have done large bank-financed projects with the cost-plus and fixed-fee contracts described above, banks may insist on fixed bids, and dictate the payment schedule or other features of the deal you construct with your builder.

THE PRIMACY OF THE ESTIMATE

While I put a lot of thought into the form of the business deals I enter into, in a sense the estimate is more important.

If the builder's estimate is basically realistic, and the changes to the design are monitored, things will work out. If the estimate is too low, there is a serious problem, no matter what form of contract is used.

Estimating is very difficult. Even a simple building is hard to estimate well. A custom design is harder. A difficult site, uncertain weather, inflation in material costs, or a little unanticipated bedrock can be hard to predict and can significantly throw costs off.

The builder has to estimate the client, in a sense, as well as the job itself. It costs much less to work for a customer who can make decisions, stick to them, and stay out of the builder's way. A customer who changes things or leaves decisions to the last minute can use up hundreds or thousands of dollars worth of extra time over the course of a project. A customer who makes decisions on time, or is happy with the decisions the builder makes on-site, saves the builder time. Of course, a customer who is willing to help out saves even more money.

In short, estimating is probably the most difficult building skill, and no estimate is guaranteed unless the design has been built before, which is rarely the case with owner-designed or custom homes.

My strategy is to estimate high, factor in the client's personality, and keep a list of items that can be eliminated or delayed if necessary, with minimal loss to the basic design. If I think the customer is certain to add $5,000 worth of extras to a job, I add it in ahead of time.

I always aim to come in under or at the estimate. I'm always happy if I'm within a few percentage points either way. I think a project that is within ten percent of its estimate has been successfully estimated, and even twenty percent is not bad when estimating projects that are complicated.

I urge my readers not to look for low estimates—in fact to be suspicious of them—and to keep the design well within your financial comfort range. A low estimate feels good at the start of the job, but can be painful at the end. A high estimate hurts at the beginning, but feels great at the end, if the project comes in on target.

LOVE YOUR BUILDER

Your builder is undertaking a preposterous task. He or she will be spending more time on your "case" than any other professional you hire in your lifetime. For the better part of a year, the design, estimation, and construction of your home will be at the forefront of your builder's mind. He or she often starts at first light and works into the evening, when the phone calls that set things up for the next day occur.

The builder is under pressure, start to finish. Whatever the contract form, he or she is responsible for your money, and most builders take this very seriously.

He or she has to keep the work flowing, keep the workers happy, and—often most difficult of all—keep you happy.

Having a house built is a very emotional, meaningful, powerful, and sometimes disorienting experience. It is an object being built, but people aren't objective about it. Building engages ones hopes and dreams, but also the deepest wells of worry, particularly because of the money involved. In my years as a builder, I have only had a handful of customers on big jobs who could keep their feelings under control through the whole project, even when things go well. A big part of my skill as a builder is being able to handle blowups.

People get upset when progress seems too slow. There may be a crew member working on your house whom you don't like that much. Maybe the estimate will go a little high, particularly if you've been kidding yourself all along about the probable costs, adding up the savings in your mind, but not the extras.

When such times occur, there is always somebody handy right there to blame: your builder. The builder is a convenient target, and to some extent, a vulnerable one: You owe him or her money.

I would urge people to pick a good builder, then trust him, her, or them. Don't blame them for the enormity of the project. After all, you chose to do it. Don't get angry because of delays or occasional errors—these are intrinsic to all work. Don't try to renegotiate the deal in mid-project. Don't resent the fact that your builder is making some money; it's probably a lot less than you make.

In short, most builders are conscientious about their clients' money; they wouldn't be in business long if they weren't. To keep their attention on the work at hand, they need your support and loyalty. Without it, they won't serve you as well.

13

ESTIMATING

To be in control of your building project, and have fun doing it, you need realistic expectations of what your house will cost and how much work it will take to make it livable. If your house will take five or ten years to build, you want to know that ahead of time. If you need to get it done this summer, you need a plan that makes this possible for you.

Estimating is more difficult than building. If you tried to estimate the time it takes to walk to town or toast a piece of pumpernickle, you'd be doing well to get it right within twenty percent. Think about what is involved in estimating a house. Many good builders never learn to precisely estimate their work, and that of their subcontractors, and many house projects are unique enough that accurate estimating is impossible.

People naturally underestimate. They are optimists. Always be skeptical and conservative about the estimates you make, or that a builder makes for you, unless he or she is known for good estimates. Even then, allow for a possible overrun of ten or fifteen percent.

Traditionally, design is done by one person and estimating by another. But in truth, they go together. Estimating is an extension of designing, and it usually results in improvements to the design. It is a creative and imaginative process.

As you begin your design, you will need a preliminary estimate to make sure the scale and nature of your project conforms to the money and time available. If you're hiring out most of the work, look at typical cost per square foot figures. For those who plan to do work themselves, I'll also provide a way to gauge the labor that might be required.

A detailed estimate should be prepared once the design is worked out. The contractor provides this estimate, but if you are your own contractor, I'll show you to how to do this yourself.

The detailed estimate does more than look at costs. It is

a way to visualize how the job will get done. As you design, you figure out you want ⅝-inch plywood on the roof. As you estimate, you figure out how to get the sheets of plywood up on the roof, and how to lay them out to minimize cutting. Some of this analysis is likely to result in adjustments to the design. It can also form the basis of your work schedule.

PRELIMINARY ESTIMATES

Square Foot Costs

In this section, we will consider the square foot cost of the structure itself, including the foundation, porches, and garage, if there is one. The septic system, a power line to the house, a water supply, and other site costs will be considered separately.

To find the total square footage to use in a square-foot-cost calculation, take the square footage of the living space, then add half of the area of a garage or enclosed porches, and add one-fourth of the area of decks.

The recent contractor-built house projects I've worked on have ranged from as little as $40 per square foot up to about $85 per square foot or more.

The $85 house is typically a three-bedroom house with a full basement, two full baths, lots of top-quality windows, nice cabinets, and solid core doors. It's complete, all new, fully painted, with nice siding. It isn't extravagant. The fixtures are well designed but inexpensive; the finishes economical. It does have some special features, like a dormer, high ceiling, or window seat which aren't strictly necessary.

The houses in figures 8-21 and 13-3 are the type of houses in this $85/square foot category.

A few strokes of the pen in the design phase could run the cost up quite a bit. Expensive cabinets, an extra bathroom, lots of ceramic tile, hardwood floors, or added complexity of shape could easily push the cost up. For example, the floorings in the $85/square foot house would be limited to about $3/square foot installed. But a hardwood floor could run $7/square foot installed, pushing the house cost almost to $90/square foot. It would be easy to spend another $3,000 on more or better kitchen cabinets or bathroom fixtures. Without really changing the underlying design, the cost could easily exceed $100/square foot with a few extras or higher quality items added on.

Obviously, cost depends somewhat on the market. The same house that costs $80 in an expensive market could cost $70 here in Vermont.

In Vermont, we can build quite a nice house in the $50 to $60 dollar range, if the design is simple, and economical materials are used. But that price is still for a complete, new house. The house in figure 3-1 is an example of such a house.

"Simple" here means using a boxy shape that takes advantage of the standard dimensions of building materials, and using such products as softwood or vinyl floor instead of hardwood, and plain doors rather than nice panel doors. Very often keeping a house simple means using part of the basement as living space, particularly when the slope of the site allows windows to be placed there. Perhaps the owner does the painting and varnishing, and some of the weekend cleanup.

Of course, it is possible to build a house in the range of $40/sqare foot, but every economy must be taken.

In the $40/square foot range, the space can be bright and well thought out, but it will be simple. The roof will be a simple gable or shed, probably covered with easy-to-install galvanized sheets. It will have only one bathroom. Windows and doors will be low end, and some of them will be recycled, or from the lumberyard bargain section. Flooring, wall surfaces, and other materials will be inexpensive. Cabinets will be minimal, and the overheads will be open shelves. Appliances will be second hand. Some things will be left for later completion by the owners, such as finish flooring or shingling. Heat may come from high efficiency space heaters, or a wood stove. Houses in figure 3-4 and 3-5 are examples of this kind of house.

None of these economies necessarily detract from the design. Some of my favorite houses have been very inexpensive, yet they look teriffic because the design is so strong.

I'd consider $40/square foot to be about the least you could spend to have a house built without doing a lot of it yourself, and it takes considerable imagination, flexibility, and skill to build a nice house for that cost.

Additional Savings for Owner-Builders

Of the total house cost, square-foot costs often break down roughly like this:

Materials: 30%
Labor: 30%
Subcontracts: 30%
Builder's overhead: 10%

There are possible savings in all of these categories, but what you save in one category, you may pay out for somewhere else. For example, the cheaper materials take longer to use; the builder's overhead, which seems superfluous, turns out to cover real tasks and real expenses.

Our somewhat hypothetical $40/square foot example provides a good way to

Figure 13-1. Simple house with wood post foundation, approximately $45 per square foot. Hendee house. Designed and built by Bruce Caswell. (Photo by Kevin Ireton, courtesy of Fine Homebuilding.*)*

put forth strategies that you as an owner-builder could use to cut costs further by doing work yourself.

Cheap lumber: In New England and other rural areas, there is often native lumber sold through small sawmills. It typically costs about one-third less than commercial lumber, but it usually takes extra work in one way or another.

If you own a stand of usable timber, in most areas you could hire a sawyer with a portable sawmill to saw up the lumber on your property, for some further savings. Again the trade-off here is increased lumber handling time.

Recycling: It is possible to build a house almost entirely from recycled materials, bought or scrounged. Building this way is an art in itself. But it's very common to use a significant amount of recycled materials on any project. Windows and doors are often recycled. Windows and doors on a house often cost $4,000 to $10,000 new, so there is a major potential for savings here. But where a new window can be installed in one or two hours, a used window is likely to require repair work, and will take longer to install.

New kitchen appliances, plumbing fixtures, and heating equipment such as

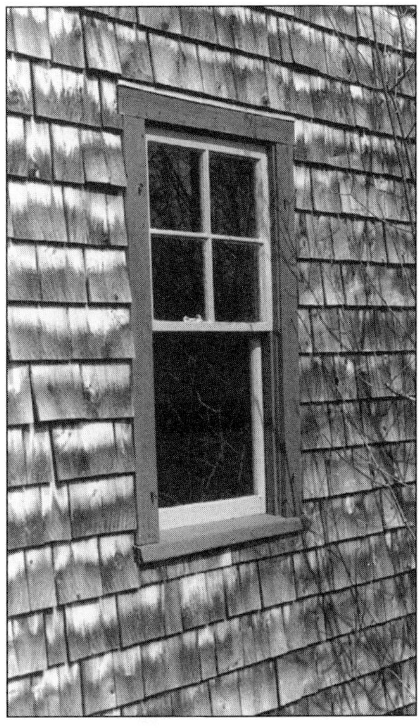

Figure 13-2. The sash might cost about $16, the wood to make a frame for them, about $20. A storm panel to double glaze the window, another $40 or $50. It might take two or three hours to build the frame, costing about $70, for a total of roughly $150. A more energy-efficient new window would cost $200 to $300.

space heaters are big expenses. All these items can be purchased used, or even scrounged from people remodeling their houses. From visiting just a few building sites, we recently picked up good windows, refrigerators, direct-vent space heaters, insulated doors, and any number of nice sinks and toilets.

In general, recycling materials often saves money, but using the recycled items always takes more time. If it's your time, the savings are real. If you are hiring people to use or repair recycled materials, some of the savings will evaporate. Look at each case carefully to see what makes sense.

Carpentry: I think doing some of your own carpentry is one of the best ways to save money. It's a major expense, it's a skill you can learn, and most important, it's fun and satisfying to do.

On a recent project in this $40/square foot category, the carpentry budget was about $14,000 for about 660 hours of work. If all of the carpentry were done by the owner, the total house cost would have been reduced from $50,000 to about $33,000, and the square foot cost to around $26.

It's important to note that an inexperienced person will usually take at least twice as long to get each job done. As I argue elsewhere, doing some of the work yourself makes the most sense when a more experienced builder can provide some guidance.

Subcontracts: The next possibility for savings (we're already down to $26/square foot with you doing the carpentry) is in aspects of the work usually done by subcontractors. Here are some examples.

The drywall, which covers half of the interior wall in our example, cost $600 for materials and $800 for installation by a professional. Owner-builders often do this themselves, but where the pro will need five days start to finish, the amateur will need a month to finish, and the results won't be great.

By doing the plumbing and wiring yourself, you will save at least $1,500. If that figure represents a total of 120 professional trade hours, you will end up working in the neighborhood of 400 hours.

By the time you reach this point in the project, the general contractor is no longer needed. Leaving out that expense saves about $4,000.

If you realized all of these savings, you'd be down to around $18.40/square foot. Getting below this cost level takes some doing.

Site Costs

You can't really use rules of thumb to approximate site costs. You need information specific to your site.

Septic system: The simplest system can cost as little as $2,000. But if your drainage is poor, you may need a "mound system," which is a man-made patch of well-drained land. If your septic isn't directly downhill from the house, you may need a pumping station. An expensive system could cost more like $8,000. If your land isn't suitable for a septic system, you may want to consider an alternative system such as a composting toilet. Usually you will have perc tests dug during the process of buying your land. Using the perc test results, the excavation contractor and engineer can give you a rough idea of what a system will cost on your site.

Electrical service from pole to house: Your electrical service from the nearest pole to your house can cost almost nothing if you are near the road. But I've met a couple of people recently who were far from the pole, and were looking at $30,000 for a power line. The local utility can quickly determine what a line extension would cost. If it seems too high, you can consider photovoltaics, which in today's market can cost between $3,500 and $10,000 for a basic system.

Water: Unless you have "town water," you will probably drill a well. A well-drilling company can quote you a per-foot price for well drilling, and tell you what typical wells in your area have cost, including the pump and other hardware required. Costs average around $4,000 for a standard 200-foot well. Chapter 20 also describes other types of water systems, such as springs.

Road work: There will usually be some costs for a driveway or roadwork. An excavation contractor can help you figure out a good plan, and give you some cost information.

Design and Permitting Costs

If you design the house yourself, make some allowance for a little professional consultation from experienced designers or architects to review your scheme. This could run a few hundred dollars depending on whether this person has to make drawings or merely review yours.

A design/build firm can design a house with you, based on your ideas, for anything from $2,000 to $5,000. They typically charge on an hourly basis. Hiring an architect is likely to cost somewhat more plus an additional fee of five to fifteen percent of the project's gross total if the architect manages the construction. In any case, the design cost depends on how realistic and decisive you are as a client. It's best to find someone in sync with your goals who can work to your budget and provide the level of design help you actually need.

As with site costs, you must first decide whom to hire before finding out what your costs may be.

Permit costs vary. In California, permitting can cost thousands of dollars. Earthquakes, endangered species, habitats, and all sorts of other factors come into play. In other rural areas, permits for a house might cost only $200 to $500. Find out the details at your town hall, and add that cost to your total.

Time Budget for Owner-Builders

If you are the primary builder, how long will it take you for the whole project? Since every house and every crew are different, it is difficult to make an accurate estimate of labor time. I have found that many owner-built houses take between two and four hours of work per square foot of living space to build.

Many owner-builder projects tend to fall in the middle at around three hours per foot. Two hours applies to a very simple, efficient design using easily installed materials like plywood or galvanized roofing. Four hours or more are needed for more elaborate shapes or other complexities. A more complete discussion of this calculation is provided below, on page 156.

Making Your Preliminary Estimate

If you are having your house built, start by using the square foot costs summarized above. Then work with your builder to create a preliminary estimate that accounts for local cost differences. Subtract the value of your labor. But remember, you may not work as efficiently as a pro.

For owner-builders, identify the type of building you are imagining in terms of complexity and size, and gauge the cost per foot as if you were having it built.

Our oversimplified cost summary is:

Carpentry materials:	30%
Carpentry labor:	30%
Subcontracts, including:	
electrical	
plumbing	
foundation	
drywall	
heat	
materials:	15%
labor:	15%
Contracting:	10%

If you will literally do everything yourself, take 45% as your dollar cost; that represents the materials. If you will have the foundation and mechanicals done professionally, your cash outlay will be 60% (this is more realistic for most people). If you will hire out some of the subcontractor work, choose a number in-between. Then add site, design, and permitting costs.

Add in your labor—in hours, days, or weeks of work—by multiplying the area in square feet of the house by three, factoring in the simplicity or complexity of your house if you want to do so.

Be sure to factor in hidden expenses like wages you will lose by taking time off to build, lodging, and additional transportation expenses.

DETAILED ESTIMATE

The preliminary estimate disciplines your design process. It tells you how much house you can reasonably expect to build with your resources. Once you've finally designed the house, you need a detailed and more accurate estimate for the project, which includes dollar costs, time estimate, and a basic schedule of the project, start to finish. All three elements are needed for a well-coordinated, smooth-running operation.

If you are hiring a builder to run the show, he or she will do this estimating. The trick here is to make sure that your builders estimate as well as they build. Talk through the estimate with them. Talk to former clients to see how accurate they were on earlier projects. If you are unsure about accuracy, add twenty percent to the total to protect yourself. Even if your builder is a crack estimator, costs can run over. Most clients I've worked with, for example, add quite a few "extras" to the project at some point, sometimes without knowing it. If you can't go one penny over the estimate, simplify the design by ten percent, or figure out a few items that can safely be delayed as the project unfolds.

Preparing Your Own Detailed Estimate

If you are your own contractor, you'll create the estimate yourself. It will be the basis of your materials purchases and the work schedule. When the estimate is done, you might get a second opinion from a full-time builder.

Here I'm going to provide two ways to generate an estimate. Both approaches require an enumeration of all materials, which of course is essential to ordering them, anyway.

The labor times can be generated step by step by gauging how long each piece of material will take to install. This is how I estimate projects I build. But it is also possible to make a more global estimate based on hours per square foot of living space, and factoring in the complexity of your design. Unless the cost doesn't really matter to you, I would suggest estimating the labor step by step, then using the global method as a double-check.

How I Estimate: Step-by-Step Method

Estimating is one of the most interesting mental aspects of my work. I'm writing a screenplay, imagining the job as a story, visualizing every step. I imagine how I will complete each step and I think of what special gear I will need. If the drawings call for diagonal boards for the wall sheathing, as I estimate I'll figure out where to get the boards, how much they will cost, how to get them to the site, where to pile them to make installation easiest, and where to toss the cutoffs. I also estimate the time I should allot to each project. Those boards will figure in the schedule at least three times: when I order them, when I stack them to dry out on site (hopefully out of the way, but handy) and again when they're installed. In this mental process I also think of other items I will need to get the job done, such as stickers to use as spacers in the woodpile, staging planks, nails, and so on. If I want to, I can then compare the costs and scheduling effects of alternatives, like plywood, which costs more but saves time.

As you visualize the steps, your material orders and your schedule begin to take shape. Your design becomes a plan for action. You use it to anticipate problems, refine your method, and fiddle with your design, so that there is less head-scratching on site and more productivity for everyone who works on the project.

My written screenplay looks like the example in the sidebar. It is a single list that describes the materials I need, what they are for, and how much they cost.

I'll also specify when I need them and how long I expect it will take to install them. Let's look at the materials first.

A complete list of materials should include all items and expenses, large and small. Make a separate list for each part of your house, based on this format. The construction chapters include tips on what to include.

Usually your drawings will provide the best tool for figuring quantities. For example, you can figure out what studs you need by counting the studs in your framing plans, then adding a few extras. You can figure your siding by measuring the wall area with your architect's scale right on your elevation drawings.

For your calculations use the units of measurement that the material is commonly sold in, as I have done in the example. For example, wood lengths come in multiples of two-foot lengths between 8 and 16 feet. Sheet goods are usually 4 x 8 feet. Show how the major timbers will be used. If some 2 x 8s are for a certain beam, it is easier to have a note about it than try to remember these distinctions in the heat of battle.

Some materials will not have found their way onto one of the materials lists you have made. Add a list of these essential extras, which may include the following items:

50 pounds 8d galvanized common nails (if your siding consists of boards)
50 pounds 16d common nails
20 pounds 20d common nails
5 pounds 16d duplex (double-headed) nails, for scaffolding
5 pounds 8d finish nails
5 pounds 6d finish nails
5 pounds 4d finish nails
Plenty of extra 2 x 4s and 2 x 6s
Shim shingles
A roll of 6-mil plastic for covering piles of materials

My list is organized chronologically. Foundation costs come first, then floor, and so on down to trim and painting. As I create the list, I'm building the house, writing my screenplay.

Estimating times: As I put down each item, or each group of items, I'm trying to figure out how long each step will take, and I put that down as hours in the appropriate column. Factoring in the work force available, I put down a date I need the material in question.

But how do you know how long it will take to do something? The more imaginatively you are able to picture the task at hand, the more accurate will your time estimate be. That's why the movie or screenplay metaphor is a good one. If I include all the steps and mis-steps, my time estimate will be somewhere near correct.

Over the years, I have periodically kept records on the job of how long things take to do. The box below gives some of the rules of thumb I use. I always adjust them for the particular job. Things go quicker on a very simple project, near the

Typical labor times (but don't blame me if they don't work for you)

Frame floor: .05 hours/square foot
Wall frame: 3 hours/linear foot
Roof frame: .06 hours/square foot
Install plywood: .5 hours/sheet
Install a window: 1-2 hours
Install prehung door: 1-2 hours
Hang door from scratch: 5 hours
Trim a door: 1-2 hours
Case a window: 3 hours
Install wood shingles: 8 hours/100 square feet
Install vertical boards: 4-5 hours/100 square feet
Install metal roofing: .03 hours/square foot

Owner-builders estimating format

NEEDED	QUANTITY	UNIT	ITEM
Oct. 12	1	Roll	Foam sill seal
Oct. 12	7	each	2 x 6 x 12 p.t.
Oct. 12	17	each	2 x 10 x 16 spruce
Oct. 12	7	each	2 x 10 x 12 spruce
Oct. 12	13	4 x 8	¾-inch T and G ply
Oct. 12	1	50# box	8d galv. box nails
Oct. 12	6	tubes	const. adhesive

Subtotals:
Plus:
"Contractor hours" (± 2 hours per day)
15% odds-and-ends factor
Add for comfort level
TOTAL:

ground. They take longer for complex projects, or projects two or three stories up. Err to the high side when you estimate. When you are done, I recommend that you add three important factors at the bottom of the page.

Adding 15 percent: Were you to simply total up materials and labor, with zero error, you would be underestimating. At one time in my work as a builder, I found myself running over estimates, even though my allowances for given tasks are accurate. I learned that there are a lot of costs that are easily ignored. On the labor side are trips to the lumberyard, unloading tools, covering the pile at night, and daily cleanup. On the materials side are staging, replacing lumber that isn't up to snuff, and damaged materials. My research came to the conclusion that these factors represented as much as fifteen percent on a renovation project, and seven to ten percent on new construction. For the first-time estimator, add fifteen percent. In the estimating format, this gets added globally at the bottom.

Contractor time: Next, add "contractor time." This is the time each day that the lead worker (probably you) spends telling the crew what to do, resolving details, talking with people, or ordering materials. With a crew of me plus three other people, I would allow about two hours per day for this. I'm lucky to get six hours each day for actual work.

USE	PRICE	TOTAL	HOURS
seal sills	$8.00	$8.00	2
sills	$7.50	$52.50	5
joists	$19.00	$323.00	8
beam	$15.00	$105.00	4
subfloor	$21.00	$273.00	7
	$43.00	$43.00	
subfloor	$2.00	$12.00	1
		$816.50	27
			4
		$122.00	4.5
		$61.50	4.5
		$1,000	40

Fudge factor and comfort level: Even though some of the costs above seem amorphous, I don't doubt they will be needed. But other things can happen too: bad weather, design changes, mistakes, missed orders, damaged materials due to bad weather, disputes about the design among interested parties, and miscalculation. For instance, you might hit bedrock when you try to dig the foundation.

The fudge factor is simply a dollar amount that subjectively accounts for these possibilities. If I'm very confident of my planning, it might be $1,000—a couple of percent. Usually it will be more—say five percent. I put a lot of weight on the personalities involved. As for customer-builders, I allow a smaller fudge factor for those who are unfussy and decisive, and who trust my judgement. Customers who are indecisive, picky, and want to decide every detail themselves (while I wait), are more likely to run up the bill one way or another.

My aim is to bump up the estimate (the estimate, not the bill) to a comfort level where I am confident enough in the total so that I can enjoy the project.

The Computer

I do my estimating on computer, using a spreadsheet program. If you are a good spreadsheet user, you may want to try it for your project. The spreadsheet allows you to quickly convert your estimate to a lumber order by sorting the file by date or project. The spreadsheet also allows you to easily update the estimate to reflect design changes or costs. And it allows you to track your progress by "zeroing out" the items completed. That is, by plugging in zeros for all items completed, the computer will retotal to indicate costs of work still to be done.

Double Check: Global Labor Estimate

For a variety of reasons, not every home builder will want to do the procedure described here. But it is also possible to do a global labor estimate, based solely on the size and complexity of the house. This approach can also provide a double-check on the step-by-step method.

As I said earlier, it seems impossible to estimate roughly how long it will take you to build a unique house. But I've found that owner-builder projects often take in the range of two hours to four hours to build per square foot of living space. To help you figure out where you fit in this range, I am going to introduce two useful measures: the typical house and the mixed crew.

By a typical house, I mean a house that is fairly simple in layout and fairly economical in range of materials, but that has some special features or complications that will take some extra time to complete. Figures 9-9 and 13-3 are examples.

Such a house might have a complex roof or an an octagonal addition. It might have a very simple plan but include materials that are labor-intensive to install, like clapboards and real plaster walls. Still, a simple house is essentially straightforward, without a lot of stonework, tons of recycled wood, or a hard-to-build design. If the form of a simple house seems complex, that complexity was achieved by combining simple boxlike shapes.

By a mixed crew, I mean one that has both experienced and inexperienced workers. A mixed crew might consist of:

One professional carpenter, plus two inexperienced people, or
One somewhat experienced but nonprofessional builder, plus one inexperienced person, or,
Two energetic and handy but inexperienced people, plus daily advice and supervision from a very experienced carpenter or builder, or,
A larger crew with the same basic proportions as those above.

These crews are typical of owner-built projects. It seems as though one is as equally productive as the other. If you have what I call a typical design and a

Figure 13-3. Design and construction by Sandy Lillidahl. (Photo: Thorsten Horton.)

mixed crew, your house will take about three hours of labor per square foot to build. You can add up the square footage of the house and multiply by three to figure out how many hours you might need.

If the house is 800 square feet, plus a deck of 300 square feet, your calculations might look like this:

Floor area	800 square feet
½ of 400-square-foot garage area	200 square feet
½ of 200-square-foot covered porch	100
¼ of 400-square-foot deck area	100 square feet
Total Area =	1,200 square feet
Times 3 hours/square foot = Total hours needed	3,600 hours
Average crew size	4 people
Average work week in hours	40 hours
Weekly output of crew	160 hours/week
Weeks needed to finish	22.5

Many additional features could make the job take longer.

Complicated design: If the design has lots of cupolas, diagonal walls, bay windows, complex roof shapes, curved windows, four stories, or similar complications, you can expect the job to take you an extra hour per foot, at least.

Stonework: Any significant amount of stonework will add to labor time. If all the walls of a house are stone, for example, the extra work could easily add two hours per square foot to your labor time.

Recycled lumber: If you are building your house with recycled lumber, you can expect to spend an extra one hour or more per square foot if you include the work of tearing down the old building.

Inaccessible site: If your site has a difficult access, or if you have no electricity for power tools, allow one extra hour or more per square foot.

High degree of finish: If you are striving for a highly finished appearance throughout, with lots of trim, fancy joints, precise finish work, etc., expect to spend an extra hour per square foot, or more.

Special building system: Straw bale, timber frame, or other special building systems are almost always going to take extra time compared to conventional building. The only system I know of that is as fast as stick building is the pole-building type of timber frame. The best way to factor in time required for methods other than stick building is to meet with people in your area who built using the method you plan to use.

There are other factors that can make your house go up *faster* than a typical house. Let's look at some.

Plywood: Plywood as subfloor, sheathing, or siding is much quicker to put up than boards. If you use plywood extensively, expect to save about one half-hour per square foot. This will be particularly true if the materials you put over the plywood go up quickly. For example, vertical boards go up faster than shingles.

Simplified design: Our typical, three hour per square foot house is not overcomplicated, but it does have some extras and complications. You can design a house with a very simple shape, a very simple roof, and none of the extras mentioned above, and save about one half-hour of work per square foot, in addition to any time you save by using plywood.

Total the factors that add to or subtract from building time, starting with the basic figure of three hours/

Figure 13-4. Complicated shapes take longer to build. Ian and Margo Baldwin's house. Built by Eric Darnell; remodelled by Blake Spencer Associates, Strafford, Vermont.

Figure 13.5. Simple shapes go up fast. Built by Central Vermont Habitat for Humanity.

square foot. If you build a stone house in an inaccessible place with a cupola, a bay window, and a curved wall, you may spend six hours/square foot or more. If you build a simple pole house and use plywood extensively, your house may take only two hours/square foot.

All these calculations assume a mixed crew. Now adjust your calculations to account for your work crew. A crew that consists entirely or mostly of experienced people could get a house built much faster than a mixed crew (perhaps in sixty-five percent to seventy-five percent of the original time estimate). On the other hand, a completely inexperienced crew, with little or no support from skilled people, will require maybe half again as much time as a mixed crew. Some inexperienced people can work as productively as a professional, if the person supervising the day's work is good at supervising, and if the inexperienced person is handy, energetic, and fairly confident. Even without supervision, some inexperienced people work very quickly because they have a knack for building or have experience in other areas of life they can apply to building. Others may work more slowly.

The Total

Your estimating total now consists of:

- Materials total, with fudge factors
- Estimates or bids from subcontractors, such as:
  - Roadwork
  - Septic system
  - Excavation
  - Foundation
  - Plumbing
  - Drywall
  - Electrical
  - Well or other water system
- Your labor time, calculated two ways (use highest figure)
- Site costs
- Design and permitting
- Costs for tools and equipment
- Living costs while building

PART 4

BUILDING BASICS

14

WOOD

MOST PEOPLE BUILD HOUSES THAT CONSIST LARGELY OF wood. But even a stone house will be partly wood. Wood is an easy and forgiving material. It is light, yet strong. It is soft enough to cut easily, but hard enough to last a long time.

There are two problems with wood. First, there are many different kinds, and not every kind of wood does every job well. Some kinds are strong, some are weak; some are hard, and some soft; some are durable and some are not; some rot quickly and some resist rot. Second, though wood is dead when you use it, it seems to have a life of its own: it shrinks and expands continually, it splits and cracks, it warps, and, if not kept dry, it rots. Designing a house consists partly of contending with the varieties of wood and the changes all wood goes through when you use it. If you are fully aware of these factors, you can build your house more easily, and it will look and function better. Furthermore, you can lower your building costs significantly by using less expensive grades of wood in most parts of your house.

Shrinkage

A tree is full of water, like any living thing. When it is sawn into lumber, the moisture slowly starts to evaporate into the surrounding air. This process is accelerated if the surrounding air is warm, if the air is relatively dry, or if the air is moving. The process is slower if the air is cold, humid, or stagnant. The drying-out process continues until the moisture content of the wood comes into equilibrium with the air. If the humidity drops, the wood starts to lose moisture. If the humidity rises, the wood starts to absorb moisture. Since the humidity of the air frequently changes, wood is often absorbing or losing moisture. The process is important to anyone building a house because when a piece of wood gives off moisture it shrinks, and when it picks up moisture it swells. Since the humidity constantly varies, the wood in your house is always slowly changing size.

Wood dries out in two phases. The first phase is the initial seasoning, in which most of the water is removed from the wood permanently. This usually takes a year at most. The second phase consists of the more minute variations that continue indefinitely as the humidity varies. Indoors, in a heated house, these second-phase variations are essentially seasonal. In the winter months, the heating system dries out the house and the wood shrinks to its minimum size. When the heat goes off in the spring, the wood gradually starts to expand, particularly if the weather is wet. Even your varnished furniture will go through these seasonal changes, only more slowly than raw wood. Outside or in unheated buildings the variations may be more frequent, since the outside atmosphere varies in temperature and humidity much more often. When it rains your siding will swell. When the sun comes out it will shrink again.

Lumber shrinks mainly across the grain, that is, in width and thickness. The lengthwise shrinkage is negligible. An 8-foot 1 x 10 board can shrink ⅛ inch in thickness and ½ inch in width, but the length will not change measurably. The amount of shrinkage is affected by how the board is positioned in the log. As figure 14-1 shows, a board sawn out tangentially to the growth rings will shrink more than one sawn out radially. A tangentially sawn board will also cup away from the center of the tree as it dries. Since most boards will inevitably be tangential to some extent, most boards can be expected to cup slightly as they dry.

Carpentry practice continually takes these facts into account. Wood is usually carefully seasoned until it is in equilibrium with the air in the place where it will be used. Since houses are drier inside than out, boards to be used inside are dried more thoroughly than boards to be used outside. Since humidity varies seasonally, it is the usual practice to dry wood to match the driest conditions that will occur. Many construction methods are specifically designed to permit wood to shrink and expand without causing cracks or other problems. For example, wood clapboards can slide by each other when they change size. Shingles are usually nailed up with spaces between them so that when they are soaked by the rain they can swell without buckling.

Structural Defects in Wood

Big, tall, straight trees tend to produce straight-grained, knot-free lumber with little tendency to warp. Lumber from big redwood or Douglas fir trees is sometimes so uniform that the lines of the grain are almost perfectly straight and evenly spaced (figure 14-2). Wood like this is strong and easy to work, but very expensive. Most reasonably priced wood is full of structural defects. These defects cannot be avoided, but their structural effect can be minimized.

Knots are the most obvious defect. Structurally, wood is a cluster of parallel fibers that we see as the grain of wood. A knot interrupts and weakens this structure. How much of a problem a knot will cause depends on its size and on where it is located. A knot on the upper, or compression, side of a beam is less harmful than a knot on the tension (underside) side, at least if the knot is tight in its hole. A knot in the center of a piece of wood usually presents no problem, because the center of a beam is not really under stress. A knot near the end of a piece weakens the piece less than a knot located midspan. When you build, put knots where they will do the least damage. Select and cut the wood so that the best pieces are used for heaviest loads and longest spans, and the worst pieces are cut up for shorter spans or nonstructural uses. Place the bigger knots on the compression (upper) side. Only the worst pieces need to be eliminated outright.

A second defect is wavy or angular grain. A piece of wood is strong as long as the grain is more or less parallel to the piece, even if it is quite wavy. But when the grain actually bisects the piece, it will be weak. When no grain lines can be followed from one end of the piece to the other, don't use the piece as a beam.

Warps do not weaken wood, but they do make it harder to use. Warping can mean cupping (which is to some extent unavoidable), bowing, or twisting (figure 14-3).

Bowing and twisting are particularly likely to occur in boards sawn from crooked trees. Such a tree is full of tremendous strains in every direction, like crooked sticks tied together in a bundle. When the tree is sawn into lumber, these tensions are released and the lumber warps. Usually this warping happens as the lumber dries. If wood is carefully dried in straight stacks (see figure 14-5), the amount of distortion can be reduced. Warped lumber is still usable up to a point. Often 1-inch boards can be simply forced into position and nailed down flat. Warped or twisted framing lumber is much harder to use, and you can usually

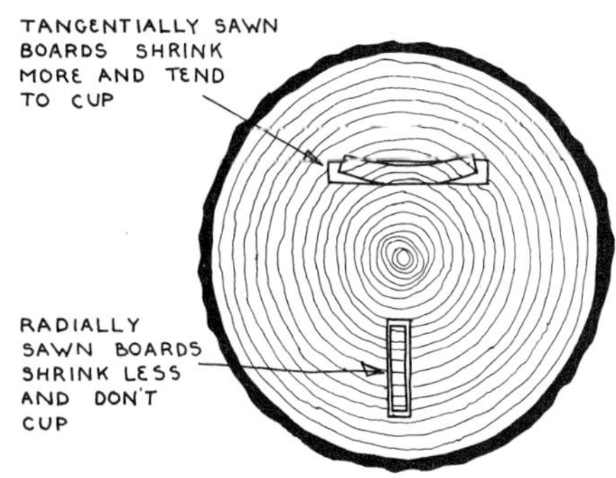

Figure 14-1. Shrinkage.

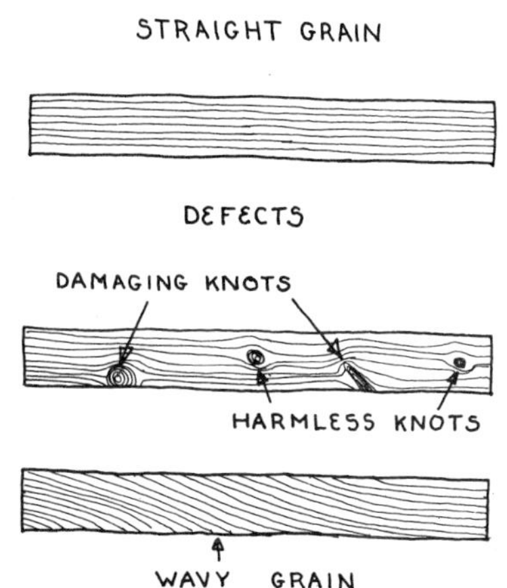

Figure 14-2. Grain patterns influence strength of a timber.

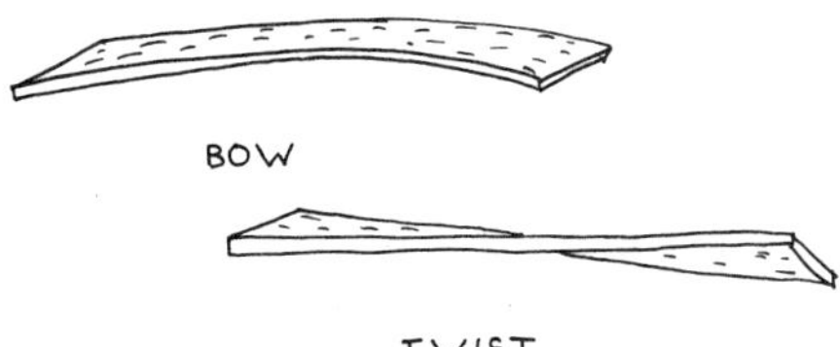

Figure 14-3. Few boards are perfectly straight.

count on some of your framing lumber being useless for this reason. When native lumber from small, relatively crooked trees is being used, the framing members are often nailed up green, before the warping occurs, and allowed to season in place. The pressure of the adjacent framing keeps them from warping. The effect of a warp can be reduced by cutting and using the piece in shorter lengths.

HOW LUMBER IS PROCESSED

Wood falls into two categories: hardwood and softwood. Hardwoods come from broad-leaved trees, that shed their leaves in winter. Softwoods are evergreens. Maple, oak, beech, and ash are hardwoods. Spruce, fir, pine, and hemlock are softwoods. While most hardwood is in fact harder than most softwood, there are some very hard softwoods and some very soft hardwoods. Butternut, a hardwood, is softer than yellow pine, a softwood. Softwoods are used for house frames, siding, flooring, paneling, trim, and in fact for every part of a house. Hardwoods are commonly used in building only for certain specialized purposes: flooring, door sills, and occasionally for paneling and trim. Hardwood is difficult or impossible to nail through without drilling a hole first, and it takes much longer to saw. It warps more as it dries. For all these reasons, softwood is the basic housebuilding material.

Sawmills saw logs into boards, planks, and timbers of various sizes. A log might be sawn entirely into 2 x 4s, or perhaps into a variety of sizes according to what the sawyer thinks is the best use of the particular log. After initial milling the lumber pieces are in the rough, which means they have the rough surface left by the sawmill blade. The lumber is approximately full size; a 2 x 4 is about 2 inches by 4 inches. The lumber is not perfectly uniform, however, since a sawmill saw is not a precision machine. A 2 x 4 may vary as much as ½ inch from the nominal size.

At this point, the lumber is green and full of moisture. Lumber is sometimes used in this form, particularly for framing and for boards that will later be covered, such as subflooring. Rough, green lumber is usually bought directly from local sawmills. But much lumber has to undergo two further steps before use: drying, to minimize further shrinkage, and planing, to increase uniformity.

Drying

There are several ways of drying lumber. The simplest is called air-drying. Sometimes it is done by local sawmills and it is the way you would dry native lumber yourself. The lumber is stacked outdoors under cover in specially spaced piles to allow air to circulate freely around each board. Air-drying will take from a few weeks to several years, depending on the type of wood, the drying conditions, and how the wood will be used. Lumber can also be kiln-dried in a huge storage room where heat and moisture are controlled to dry out the wood as fast as possible without damaging it. Much of the wood available at lumberyards has been kiln-dried.

Planing

Wood is planed or milled to uniform dimensions by feeding the pieces through huge planers. A rough 2 x 4 will be milled down to 1½ x 3½ inches. A 1 x 8 board will end up ¾ x 7¼ inches. The actual thickness will generally be at least ¼ inch less than the nominal thickness. The actual width will be ½ inch to ¾ of an inch less than the nominal width. In addition, some lumber will be milled to special shapes for flooring, molding, door frames, and other items.

PRICING

Lumber is often sold by the board foot. The easiest way to think of a board foot is as a piece of wood 12 x 12 x 1 inches or any equal volume (144 cu. in.). A 1-foot length (one running foot) of 1 x 12 contains one board foot. So does a running foot of 2 x 6 (half as wide but twice as thick) or of 3 x 4. A running foot of 2 x 4 is two-thirds of a board foot, because its cross section is only two-thirds that of a 2 x 6 (8 square inches instead of 12). One way to figure board feet in a piece is to find the board feet per running feet and then multiply by the number of running feet. Say you have an 8-foot 1 x 6. A 1 x 6 contains one-half of a board foot for each running foot; one-half times eight is four, or 4 board feet.

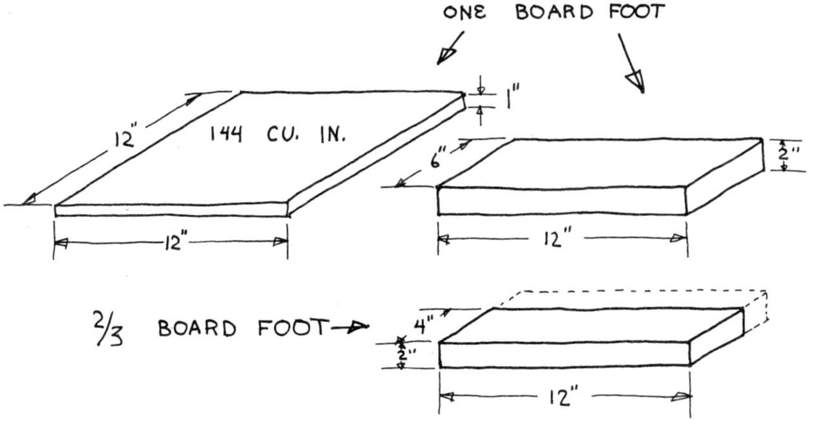

Figure 14-4.

Another method is to multiply the width of a piece in inches, times the thickness in inches, times the length in feet, and divide the result by twelve. If you have a 2 x 8 that is 10 feet long, you multiply 2 x 8 x 10 to get 160, and divide by twelve, which comes out to 13⅓ board feet.

Board measure is figured for pricing before planing. Thus a planed 1 x 12 is still one board foot per running foot, even though the piece is only ¾ x 11¼ inches after planing. Any wood under 1 inch thick is considered 1 inch thick for pricing purposes.

Large quantities of lumber are often sold by the thousand board feet. A price of "180 a thousand" means $180 per 1,000 board feet. That price would be written on a price list as $180/M and would be the equivalent of .18 per board foot.

SOURCES

In most areas there are two main sources of supply. Local sawmills sell mostly green or air-dried rough lumber from local trees. Lumberyards sell mostly planed, kiln-dried lumber imported from the large forests of the Pacific Coast or Canada. The source you choose will have a big influence on the appearance and cost of your house.

Lumberyard Wood

Yards stock different items in different parts of the country, but the following description is of typical offerings.

The 2 x 4s, 2 x 8s, and other framing lumber are usually spruce, hemlock, white fir, hem-fir, (hemlock and white fir mixed), or S-P-F, (spruce, pine, and fir mixed). This lumber is planed, dry, straight, fairly economical, and strong. These qualities vary considerably with the grade of lumber. Though yards usually refer to their framing lumber as construction grade, there are actually several construction grades of wood, of which the most common are (in descending order of quality) Select Structural, #1, #2, and #3. Select Structural, #1, and #2 are the premium grades. Number 3 is the cheap grade, which is generally inferior.

The construction grades are usually covered up by the finish materials of the house, except when used for decks or other somewhat rough work. Lumber yards may also sell Douglas fir, redwood, yellow pine, or other species in a special clear or appearance grade. This lumber is usually knot-free and is intended for exposed uses. It is much too expensive for framing but may be good for railings, window sills, window frames, or other jobs where you need very straight knot-free lumber for practical or aesthetic reasons.

Lumberyards also sell kiln-dried, planed 1-inch boards in various widths, such as 1 x 4, 1 x 6, 1 x 8, 1 x 10, and 1 x 12 inches. The most common species is pine, which is used for finish work of all kinds. This comes in a very expensive knot-free clear grade and a more knotty common or #2 grade. These grades cost about half as much as knot-free clear lumber, but they are more trouble to work with. Redwood, Douglas fir, cedar, and yellow pine boards are also often available, particularly in a clear grade.

Lumberyards also sell specially milled pieces for specific purposes. Stock for door frames, window frames, and moldings is usually made of clear, white pine. Stair treads and door sills come in oak or clear Douglas fir. Siding materials (shingles, clapboards, shiplap siding, etc.) are available in pine and in rot-resistant cedar or redwood.

Lumberyards sell a lot of plywood for sheathing, subfloors, and cabinets. Plywood usually comes in 4 x 8 sheets and in thicknesses from ¼ inch to 1 inch. It is expensive, but goes up fast if the house is framed for it. In addition, lumberyards often supply other building materials, such as sheetrock, insulation, concrete blocks, nails, and sometimes hardware and tools as well.

Today many people want to avoid using lumber from sources whose logging practices damage the land or animal habitats. I don't think there is one certain way to identify which lumbers are harvested in the way you are comfortable with. Almost any species of lumber can be harvested in a responsible and sustainable way. But you can't tell from looking at a piece of maple whether it was clearcut in Virginia or selectively cut in Vermont. But you can ask suppliers where their wood is from, and gather information on species that are under pressure in some way, such as redwood or hardwoods from the rainforest.

Sawmill Wood

You can also patronize local suppliers. Sawmills are local institutions. They sell whatever wood grows nearby. In the northeast this usually means pine, spruce, hemlock, and smaller amounts of hardwoods like oak, maple, birch, and cherry. In parts of the south, certain hardwoods, such as oak, are so plentiful that they are sold by sawmills as framing material. The quality of the wood depends on the quality of the local forests and the way the logs are treated by the mills.

The framing lumber from local mills is usually green and rough. Some mills stock a variety of sizes and lengths, but often mills custom-saw logs into the particular sizes you need, including large timbers, such as 6 x 8s. Usually timbers are sold at a standard board-foot price, irrespective of the size of the pieces. (At lumberyards, the larger timbers are usually much more expensive). Some sawmills also have planing equipment and sell framing lumber planed to uniform dimensions.

The boards that sawmills sell are often unplaned and green like the framing lumber. Often these boards are sold random-width, random-length. If you buy 1,000 board feet of boards, some will be long, some short, some wide, and some narrow. Other mills sort out the sizes you need, which is essential for some purposes and convenient for others.

In addition, some mills have a supply of last year's boards that they have air-dried. Some may also have planed tongue and groove or shiplap boards.

Sawmill vs. Lumberyard Wood

The primary advantage of sawmill lumber is that it costs up to one-third less than lumberyard wood. The primary advantages of lumberyard wood are that it is uniform and light (planed and dry) and convenient to obtain. All you do is call the lumberyard, tell the clerk what you need, and the next morning the lumber will be delivered to your site. Whether this convenience is worth the extra price depends on what the lumber is being used for.

For framing purposes, sawmill wood has several advantages. First, you may save money, especially with large-size timbers like 4 x 6s or 6 x 10s, since lumberyards charge a premium for these big pieces. Second, unplaned sawmill lumber is bigger, and therefore stronger, all other factors being equal. Third, rough framing lumber looks better exposed. Ordinary construction-grade wood from a lumberyard is not usually very presentable. Fourth, a sawmill will saw lumber to the exact sizes and lengths you want.

If you use rough lumber for framing, you'll have to put up with some disadvantages. It is much heavier than lumberyard stock because it is green. It usually has to be ordered in advance, especially when custom-sawn to your order. Finally, roughsawn lumber is not as uniform in dimension, which makes it harder to work with. You may, for example, have to trim each joist with a saw or hatchet to make the floor come out flat.

My experience is that for heavy timbers like floor joists, sills, and rafters, rough lumber is worth using, even if you do have to spend extra time. Its appearance and economy outweigh the disadvantages. But walls should be framed with planed timber, because they are finished on both sides and will be uneven if the studs are irregular. Also, windows and doors take much longer to install in an uneven wall. If a sawmill has planing equipment, this may not be a problem.

With boards, there are even stronger reasons for using native lumber. One-inch boards bought from a sawmill are considerably cheaper, about half the price of #2 pine or plywood. For the most part sawmill boards are perfectly adequate, especially if they are dried and planed. If a sawmill sells planed, dry boards, there is probably no reason to buy any boards at all from a lumberyard, except for special projects that require clear stock, like window frames.

If your local sawmill offers only rough, green boards, you can still use them "as is" on many parts of the house. Subfloors and sheathing can be made of green boards because they will be covered later. So can board-and-batten siding, since the gaps that appear between the boards are covered by the battens. But boards that will remain in view, like paneling, finish flooring, door and window frames, and trim should be dried and planed. If your sawmill can't plane the boards you should find one that can. If the boards come green, you'll have to air-dry them yourself. If the boards are to be used inside, you will need to dry them further by stacking them in a heated place for a few weeks. All this moving and stacking and drying and restacking takes time and planning. But what you get may be worth it in beauty and economy.

BUYING NATIVE LUMBER

Visiting Sawmills

Before you buy native lumber, visit several sawmills to see what is available. They are usually small, informal operations, and you can find out much of what you need to know by talking to the owners. Here are the main questions to ask:

1. What species are available? Different woods have different appearances, and some are stronger than others (see appendices A and B).
2. What can the mill supply? Will it custom-mill to your specifications? Does it sell batches or boards the same lengths and width, or do they come in random sizes? Can the mill provide extra-wide boards? Does it sell large timbers? Can you get air-dried boards?
3. When can the mill supply what you need? Can it promise you a specific date? How far in advance do you have to order?
4. What special milling operations can the mill perform? Some mills offer only rough-cut lumber. Others have planers that can plane the pieces smooth and uniform. Some do special work such as tongue and grooving or shiplapping, valuable for floors or horizontal siding. Some can mill boards down to ½-inch thickness. Some make clapboards.

You can check out a number of questions yourself if you bring a ruler and square with you to the mill. Here is what to look for:

1. Squareness. Lay a square on some of the larger timbers, 4 inches or 6 inches thick. Sometimes they will be sawn so that they look like parallelograms in cross section, which you don't want.
2. Straightness. Take a board or a 2 x 4 and sight down an edge with one eye. Do this with pieces from different piles. Take note of twisting, bowing, and cupping. Twisted pieces are hard to use. Bowed pieces are generally usable but only with extra work, and they are harder to fit tight. Cupping is not so serious and to some extent is unavoidable.
3. Strength. Look at the structural pieces. How straight and parallel to the piece is the grain? How knotty are the timbers? All framing lumber has knots, but too many, especially along the edges, can seriously weaken a beam.
4. Uniformity. Take a board or timber and measure the width at both ends to see if they're the same. Compare different boards with one another. A dimesional difference of plus or minus ⅛ inch is common on rough-sawn board widths. Plus or minus ¼ inch is about the maximum tolerance for width. For board thickness, plus or minus 1⁄16 is about the tolerable limit. If you are using native 2 x 4s unplaned, check the consistency of the dimensions, especially the 4-inch dimension. On tongue and groove or shiplap boards, check the fit of the edges with one another. Bad milling makes wood hard to use.
5. Waste. Check the ends of boards and timbers to see how much will have to be cut off. Some board sections may be only marginally usable due to knots, splitting, or sloppy sawing. Sometimes the sawyer uses a log that is a little too thin and the end of the piece won't be square.
6. Appearance. If you are planning to leave any of the wood exposed, look over its surface. Some wood may be blemished by rot or fungus, although interesting grain patterns or colorations can be a positive asset. Check if the surface of planed lumber is smooth or chipped.

Planning and Ordering

A lot of preparation is necessary to use native lumber. Start visiting sawmills as soon as you think you might be using native wood. It's worth it to buy some 1-inch boards even before you know exactly what your design will call for. Mills may have a supply of last year's boards already air-dried, or you can arrange in advance to have the mill dry boards for you.

It helps to have a schedule, so you can work the various steps of the lumber-getting process into your overall building scheme. An ideal schedule might be to visit sawmills in summer or fall, make your plans over the winter months, order in the early spring, and do your drying in the spring and early summer while working on the foundation. Don't count on your lumber being ready at a given time unless you know the mill is reliable. Ask around to get some idea. Some mills are slow but dependable. Others will promise work soon but rarely deliver on time. It is not a bad idea to fix a delivery date a few weeks in advance of the time you will actually need the wood.

When you order, give the sawmill operator a detailed list. For framing, specify the exact sizes of the pieces you need and order a few extras of each size. For boards, the order should show the total board feet needed, along with any other dimensions you can specify. Show which items you need dried, which should be planed, and how they should be planed. Finally, the list should show when you need each item. Keep an exact copy for yourself to keep track of what has been delivered and how much money you owe the mill.

DRYING NATIVE LUMBER

It is usually best not to dry framing lumber. As it dries it tends to warp and twist and become unusable. Thicker timbers, 4 inches or 6 inches thick, will almost certainly warp terribly if they are air-dried. Two-by-fours are less likely to warp, but there is still a risk. Therefore, framing lumber should be prevented from drying if possible, and used green. Stack the lumber tightly in a pile off the ground, and cover it with plastic so it will not dry out much. However, make sure the pile does not get so damp that rot sets in. Inspect the pile periodically by moving a few pieces around and looking for stains or other signs of decay and dampness. If you find any, air out the pile enough to dry the surfaces and cover it up again. If you have a stack of 2 x 4s sitting around for a long time you might try to air-dry them, using the stickered-pile method described below for boards.

Boards, or any thicker pieces that will be used for interior finish work, must be thoroughly air-dried. Wood is air-dried by being stickered in a pile. Stickers are long strips of wood about 1 inch thick used to separate lumber in a pile so that each board is surrounded by currents of air that carry moisture away. Figure 14-5 shows a properly stickered pile, which is made as follows: Find a breezy spot, since the wind will definitely accelerate drying. A drafty barn with the doors open will do, or simply a spot outside. The ground should be flat, but it need not be level. In fact, if the pile slopes, the boards will shed water better. Next, put down logs or other ground supports, 4 feet to 6 feet long, every 4 feet, to keep the pile off the ground. If you're drying hardwood, the supports should come every 2½ feet. If you take the trouble to elevate the entire pile 18 inches off the ground, the lumber will dry faster. If the support logs vary in thickness,

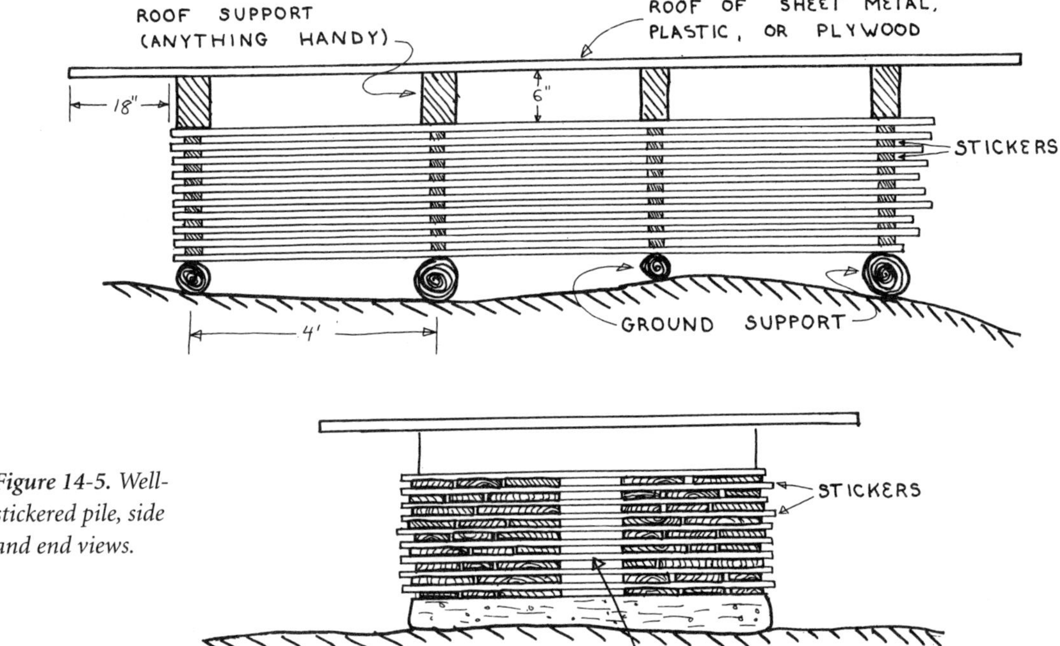

Figure 14-5. Well-stickered pile, side and end views.

the variance can be used to compensate for any unevenness in the ground. Have the tops of the supports form a flat bed for the pile. Sight along the tops to make sure.

Now put a layer of boards on the logs, spacing them at least 1 inch apart. If you have a pile containing over 500 board feet, create vertical flues in the middle of the pile as shown in figure 14-5, by leaving a wider 4-inch to 6-inch space every 15 inches or 2 feet. Next put a row of stickers on top of the boards, directly above the ground supports. Put down another row of boards, again leaving spaces between them. Put the next row of stickers directly on top of the first row, add another row of boards, and so on. The pile can be very high, but 4 feet is a good height.

Protect the pile from rain with a roof of plywood, plastic, or old tin roofing. Leave 6 inches between the roof and the top row of boards so that the flues will work. The roof should overhang 18 inches.

Figure 14-6 shows a badly stickered pile.

In poorly constructed piles, the stickers aren't lined up, some are thicker than others, some of the boards are unsupported at the ends (not always avoidable), and the log foundation is not flat. All this makes for bending and sagging in the pile. Any kind of sags will become permanent features of the boards as they dry.

Well-stickered softwood boards will usually air-dry about as much as they ever will in six to ten weeks of hot, dry summer weather. Colder, wetter weather will slow the process tremendously. You can test the dryness of the pile by including in it some short sample pieces that can be measured for shrinkage. As you build the pile, include about eight pieces 2 or 3 feet long. Draw a line across each one, and measure the width to the nearest 1/64th of an inch. Write the width next to the line. These samples should be chosen from boards cut tangentially, since boards of this kind shrink more than those cut radially. Place the sample pieces where they can be easily removed, perhaps with a crowbar. After the pile has been exposed to a month of good drying weather, remove a few of the test boards and remeasure their width. If your test board is 8 inches wide and has not shrunk at least 1/4 inch, the pile is probably not dry yet. If the test board is 6 inches wide, look for a shrinkage of at least 3/16 of an inch. You can also measure test pieces several weeks in a row. When the pieces stop shrinking, they are as dry as they'll get outdoors.

You can also make an estimate of the further shrinkage you can expect. Take two or three of your test boards

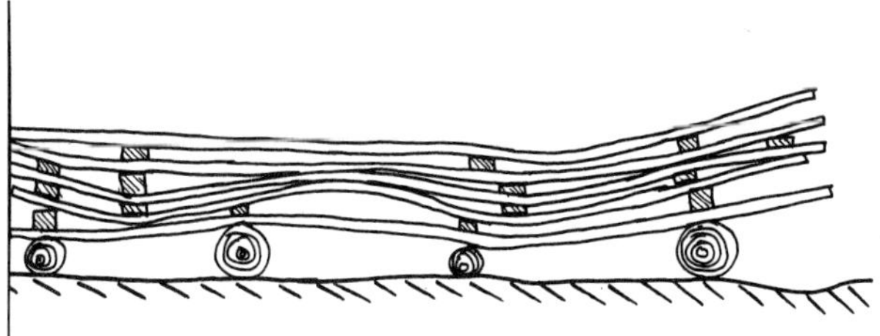

Figure 14-6. Badly stickered pile.

and cut 1-inch strips across the center of each one, as shown in figure 14-7.

Measure the width of the 1-inch strips precisely. Put the strips in a spot that has similar humidity to the part of the house where the wood will be used. For siding or exterior uses, place the strips in direct sunlight for three hot days, or until they stop shrinking. Rotate the pieces periodically, and keep them from getting covered with dew or rain, which will slow down the test. To test lumber to be used indoors, place the sample in a heated place, directly above and near a heat source. If the heat is not on yet, put the sample in the oven with the pilot on. Rotate the pieces for two days, or until shrinking stops, and take a second accurate measurement of the width. The difference in width

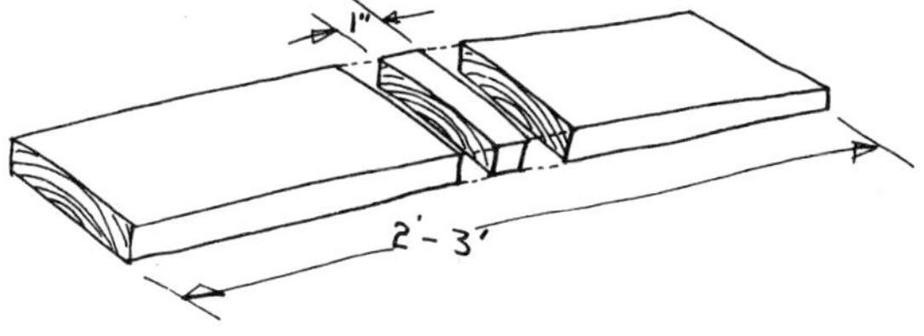

***Figure 14-7.** Making a test sample to determine possible shrinkage.*

before and after will approximately equal the further shrinkage you can expect and therefore the size of cracks you can expect to see between boards.

The following chart gives a rough idea of how much shrinkage is acceptable under various conditions. Many buildings are built with wider spaces, and time schedules will not always permit perfect drying procedures. Dry wood as thoroughly as you can without making the chore a nuisance.

	MAX. SHRINKAGE:
Board-and-batten siding	1/4 inch
Shiplap or T&G siding	1/8 inch
Vertical square-edged board siding	1/16 inch
Shiplap or tongue and groove paneling	1/16 inch
Window casings and trim	1/32 inch or 0 inch
Door and window jambs	1/32 inch
Flooring	1/32 inch or 0 inch
Ceilings	1/8 inch

Shrinkage is proportional to board width. If your boards are wider than the test pieces, the cracks between boards will be greater. If the boards are narrower, the cracks will be proportionately smaller.

No matter how long you air-dry boards outside, they will shrink further when used in a heated house. This shrinkage may be within acceptable limits for ceilings or paneling, especially if the boards are relatively narrow. But for flooring, trim, and paneling, I think an additional drying period indoors in a heated part of your new house is worth it. This extra drying is equivalent to the kiln-drying done by lumber companies and is just as effective. Sticker up the boards again inside, preferably in a loft or upstairs area that will be quite warm. The boards should probably be left this way for about six weeks. However, as before, the best method is to test 1-inch sample pieces exactly as you would with air-dried lumber. An alternate method is to use a few of the boards just as you will later use the whole pile. Pick a location near the heat, and observe the cracks after two weeks.

Boards dried this way should be used in the winter or spring, before humid summer weather causes them to swell again. If your boards must sit around in hot, humid weather, wrap the pile with plastic to minimize swelling.

The point at which you move boards indoors may also be the time to have shiplapping, tongue and grooving, or thickness planing done. Do this shaping work after the boards have been air-dried, so that once planed they won't shrink much more, but before bringing them indoors, to minimize moving piles around.

OTHER SOURCES OF LUMBER

Having Your Own Lumber Sawn

If your land has tall, straight softwood trees at least 12 inches in diameter, you can have them sawed up into lumber by a sawmill. This service will cost roughly twenty cents per board foot. However, you must include other costs in your figuring. First you must buy or borrow a chainsaw to cut down the trees. Next you need a jeep, horse, bulldozer, or farm tractor to drag the logs out of the woods. Then you must find or hire a truck or wagon to haul the logs to the mill and haul the lumber back. All this costs money and takes time. Usually it will cost enough money to make getting your own logs sawn a marginal economy at best. If your plans already include a chainsaw for wood heat, a tractor for farming, or a truck for general purposes, doing your own logging may be practical if you have the time. If you decide to do logging, you should definitely get an experienced logger to teach you. Logging is a dangerous trade, perhaps the most dangerous.

Today there are increasing numbers of sawyers who have portable sawmills—usually band-saws—that can be brought to the woods. This has two major advantages. First the bandsaw makes a very thin cut, which wastes much less wood than the traditional circular saw mill. Second, if the mill comes to the woods, some of the costly handling of logs can be avoided. If you have good trees on your property, having them sawed on-site may be a true economy, saving money without vastly adding to the work you must do.

Used Lumber

You can get used lumber in two ways. You can buy it from a salvage company for about half the cost of new wood. This wood varies greatly for appearance and usability. It may be already dry. But it often takes more work to use than new wood. You can demolish decrepit buildings yourself to recycle the wood. Sometimes people will sell cheap or give you the wood from an old barn if you promise to demolish the building cleanly and completely. Lumber from such a source is cheap and sometimes beautiful, but it takes a long time to demolish a barn. If you build entirely with used lumber, you can probably expect your labor time to be about fifty percent more than with new wood.

Before you make a deal to buy or demolish a barn, poke around extensively with an icepick or knife to make sure the wood is not rotten. Remove a few boards to make sure they do not conceal rotten timbers. After all, the building is being demolished because it is defective. Make sure you will not be building these defects into your house.

New Sources

As attitudes change and some sources of lumber get criticized, new sources will arise. You can now buy yellow pine lumber recycled from old mill buildings; you can buy lumber sawn from cyprus logs dragged from the swamps. More practically, I think local sawmills will improve their production quality, install kilns, and provide better delivery services. This has already happened in New England.

15

THE STRUCTURE OF A HOUSE

It is possible to design a house without understanding much about how structures work. Sizes of rafters and joists can be looked up in rafter and joist tables, or determined by comparison with existing buildings. But it is worthwhile to understand some of the basic concepts that apply to structures, and to know some of the math used to apply these concepts. The concepts, covered in this chapter, will help you envision your structure in a more imaginative way. The math, covered in the next chapter, will help you make the best use of the resources you have available.

TENSION AND COMPRESSION

All questions of structure reduce to one basic problem: to design a house that will withstand the forces acting upon it. These forces include the weight of the house itself, the weight of snow, the weight of the contents of the house, such as furniture and people, and the wind. In some areas, earthquakes could also be a factor.

The forces acting on a house may be differentiated according to how they act on a given part of a house. A compression force is one that tends to compress a building element or shorten its dimensions. A weight, or "load," resting on a column exerts a compression force on that column; the column is compressed between the weight and whatever the column is resting on. A tension force is a force that tends to pull apart or elongate something. A weight hanging from a wire exerts a tension force on the wire, stretching it between the weight at the bottom and its anchorage at the top (figure 15-1).

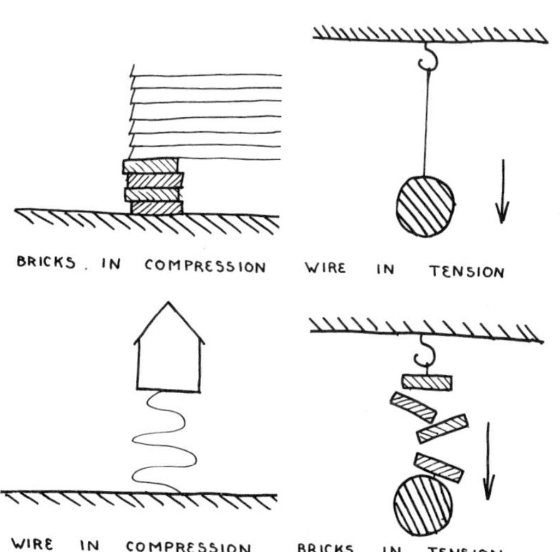

Figure 15-1.

Very few pieces of a house, however, are purely in tension or compression. Take a wooden beam with a heavy load resting on its center. It may seem to be purely in compression from the load, but a closer look reveals both kinds of forces. The top of the beam is in compression; it is being squeezed together. The bottom part is in tension; it is being stretched (figure 15-2).

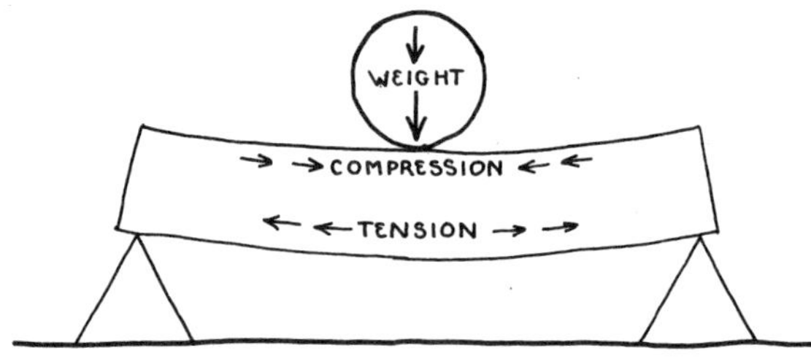

Figure 15-2.

The middle of the beam is not being strained much at all. The design of steel I-beams takes advantage of this dynamic, since they are wide at the top and bottom for strength, yet narrow in the middle (called the "neutral zone") for lightness and economy (figure 15-3).

Figure 15-3.

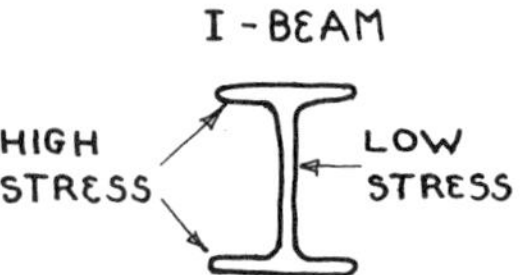

Any structural member will be subjected to both tension and compression forces whenever the loads on the member are exerted along more than one line in space, or if they tend to bend the piece. In the case of a weight on a column, there is an upward force from the base, and a downward force from the weight. These forces act in opposite directions along the same line in space. When a weight hangs from a wire, here too the forces are in the opposite direction but along the same line. With a beam supported at the ends, the downward force is applied to the middle (or all along the beam) and the upward force is applied just at the ends, in the opposite direction but along different lines in space (figure 15-4).

Different materials have different strengths under tension or compression. Masonry is very strong in compression, but not strong in tension. A pile of bricks will hold a huge weight in pure compression, but no weight at all in tension. Even when bricks are mortared together the tensile strength is not great. Concrete, which has low tensile strength, is often reinforced with steel, which has tremendous tensile strength. Bars or mesh can be cast into a concrete beam or wall where the tension loads will be greatest.

Perhaps an example will clarify this. A pile of bricks or a brick wall above ground might simply be subjected to a downward, compressive force. If that same wall is a basement wall, it will also be subjected to the sideward pressure of the earth. When forces are exerted on part of a building along different lines in space, that building element will be subjected to a combination of tension and compression forces. The wall is like a beam on its end. The outside of the wall is in compression and the inside is in tension. In masonry construc-

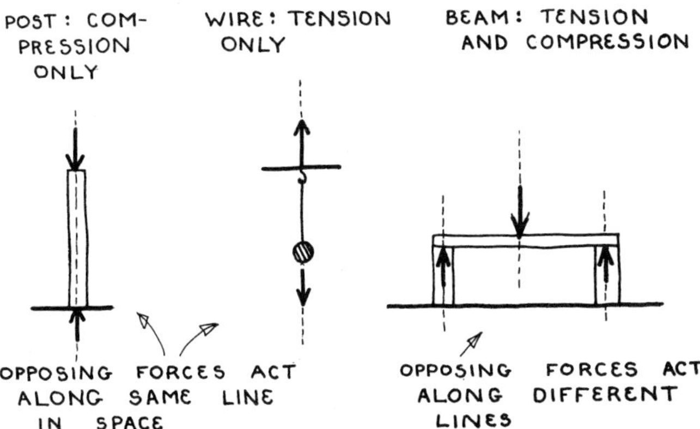

Figure 15-4.

tion, this pressure on a basement wall can be small or large, depending on soil conditions. If the pressure is small, as is typical with a well-drained soil, the pressure can be resisted simply by the weight of the wall, by mortaring the bricks together, or by making the wall thicker. If the pressure is greater, as it might be on a poorly drained site, the wall could eventually fail, as shown in figure 15-5.

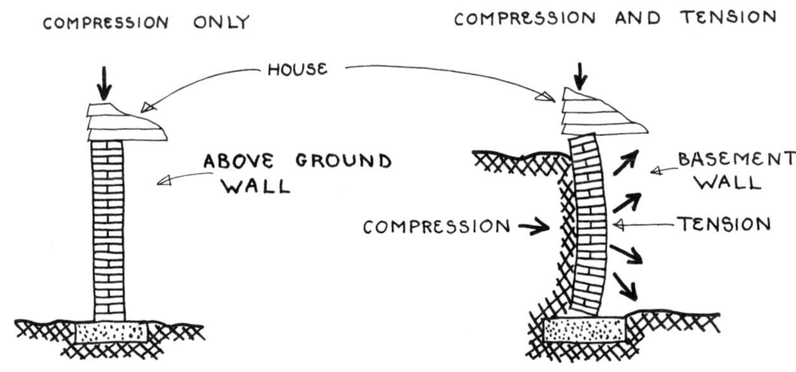

Figure 15-5.

This sideward pressure can be resisted by making the wall thicker, by using mortar to strengthen the wall, or by using concrete, which is stronger because it is monolithic. The wall can be further reinfoced by casting steel reinforcing bars into the concrete.

Wood, on the other hand, is strong in both tension and compression. A relatively slender wood column will support a huge weight in compression. A 4 x 4

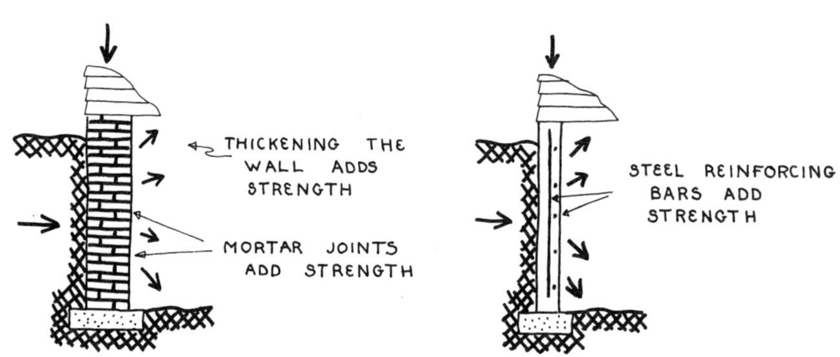

Figure 15-6. *How masonry walls resist tension forces.*

post that is four feet long, for example, would probably be able to carry most of the weight of a small house. You could hang a great weight from a ¼-inch dowel—hundreds of pounds—as long as the pulling force is straight and the joints at the end are strong. The joints would break before the wood. Wood is a good material for beams because it has both kinds of strength.

WHAT MAKES A STRUCTURE STRONG

These forces on building elements are called "loads." Dead loads are those imposed by the building's own weight. The weight of the roof, for example, is a dead load on its rafters. Live loads are other forces acting on the building, such as the weight of furniture, people on the floor, snow on the roof, and so on. When calculating the loads a structure will have to bear, it is important to take both kinds into account.

To withstand the loads on it, a house structure must meet three criteria. One, the individual members of the structure, such as beams, joists, and studs, must be strong enough. Two, the members must be attached to one another properly. The joints must be strong. Three, the lumber must be assembled so that the structure as a whole is rigid. We will be talking here mostly about wood, since the main engineering problem will be to make the wooden parts of your house strong. The masonry parts in small-house construction are usually engineered by rule of thumb unless particular conditions such as a potential for earthquakes are present.

STRENGTH OF TIMBERS

Columns or posts are structural members in pure compression. Typically they are very strong. Consider a 4 x 4, 4 feet long, made from the cheapest grade of wood. Such a post can hold approximately 9,800 pounds. If you make it 6 feet long, it can support only a little less, about 9,200 pounds. Higher than that, the capacity quickly drops off. At 8 feet high, the post can hold about 5,000 pounds, at 10 feet, about 3,000 pounds, and so on. As a column gets taller, it has an increasing tendency to bow out in the middle, as shown in figure 15-7. That is the way columns fail. One way the strength of a column can be improved is to brace the middle in both directions to prevent bending. If you have a 4 x 4 that is 8 feet long, well supported in the middle, it will hold as much as a 4 x 4 that is 4 feet long.

Figure 15-7.

Appendix B gives loading capacities for timber posts. In carpentry practice, it is rare for post size to be carefully scrutinized because posts are so strong. For example, the 2 x 6 studs in a wall are the main posts that hold up a house. Each is braced every few inches, since the inside finish and outside sheathing of the house are firmly nailed to it. The worst grade of 2 x 6, braced in this way, can support 3,000 to 5,000 pounds. Under the worst conditions, with five feet of snow on the roof, wide, heavy floors, and a party upstairs and down, each stud could be called upon to support perhaps 2,400 pounds. A more normal load would be 1,000 pounds.

Even an open, unbraced post can support a great load. Generally, a post in the middle of a room supporting one floor can be a 4 x 4 or 4 x 6, and a post supporting two floors can be a 4 x 8 or 6 x 6. If you think the load on a post is extra large, you can compute the amount of load using chapter 16 and find the column size by using appendix B. Usually a post size will be determined not so much by the load it must bear as by other considerations. For example, a post may be larger than the load demands for the sake of appearances, to provide bearing surface top and bottom, or to make the joints at the top or bottom strong.

Members in pure tension are rare in most houses. Occasionally something will be hung from a wooden frame, but usually the only real tension members will be the collar ties that hold the rafters together (figure 15-8). A tension member commonly fails at the joints at the ends. There-

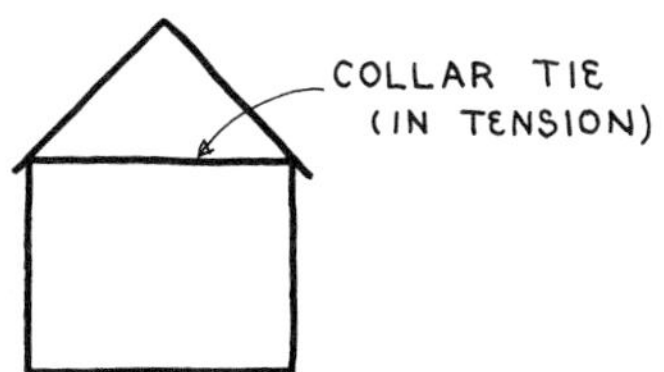

Figure 15-8. Collar ties are often the main pure tension members in a house.

fore, the size of a tension member is usually determined by the size and shape of wood needed to make a strong-enough joint. The piece will be large not because the wood is weak, but to make the joint strong. For your purposes, follow rules of thumb given in this book unless your design has some unusual tension member. If so, get some experienced advice.

The timbers that require the most engineering from you are the horizontal beams, joists, and rafters that are subject to both tension and compression loads. Beams, joists, and rafters must be strong enough to carry their live and dead loads without breaking or "deflecting" (sagging or bouncing) too much. At the same time, they should be no bigger than necessary. Oversize timbers will cost more and require more work to put up.

The load a particular beam can carry depends on the size of the beam, the grade of wood in it, the span, and the location and size of the load on the beam. Chapter 16 shows how these elements are quantified and how you can use simple formulas to compute timber sizes. Here are some of the basic principles these formulas are based on.

1. The longer a beam is, the weaker it is. A 2 x 4 that is 4 feet long will carry twice what a 2 x 4 that is 8 feet long will carry, assuming both are supported at the ends.
2. The wider a beam is, the stronger it is. The strength is directly proportional to the width. A 4 x 4 will carry twice what a 2 x 4 will carry over the same span (figure 15-9).
3. The deeper a beam is, the stronger it is by far. Mathematically, strength is proportional to the square of the depth. If you double the depth of a beam, the strength increases by four times (figure 15-10). That is why most timbers are rectangular in cross section. The rectangular shape gives the most depth, hence the most strength, for the amount of lumber used (figure 15-11).
4. The strength of a timber depends on how it is loaded. Figure 15-12 shows a 10-foot 2 x 6 under two different loading conditions. Figure 15-12A shows the most common situation, a single span loaded more or less uniformly along its length. This kind of load is referred to as a uniformly distributed load. The people, furniture, snow, and most other loadings on floors and roofs are normally considered to be uniformly distributed loads, even though at times a part of such a load might be concentrated in one place or another. Under these conditions a typical 10-foot full-size 2 x 6 can hold a safe working load of 800 pounds. Another way of saying this is that the maximum span of a typical 2 x 6, if it has a uniformly distributed load of 80 pounds per foot of beam, is 10 feet.

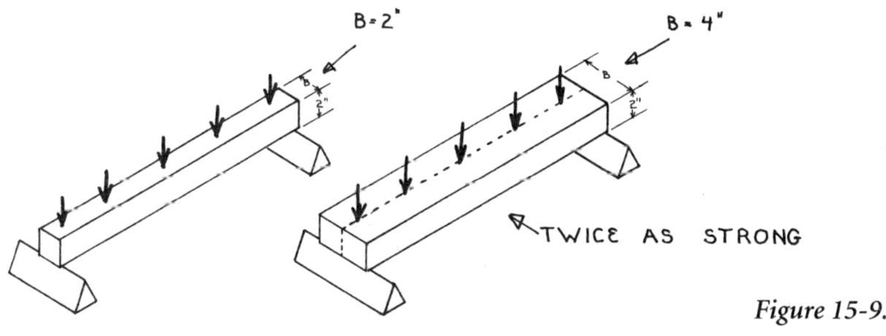

B=2"
B=4"
TWICE AS STRONG

Figure 15-9.

D=2
D=4
4 TIMES AS STRONG

Figure 15-10.

TWICE AS STRONG

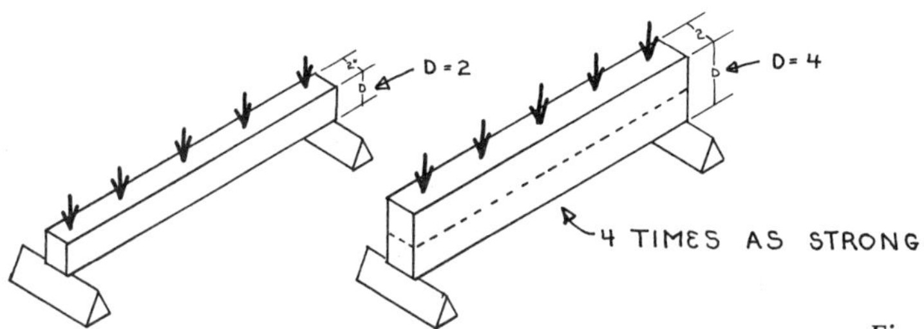

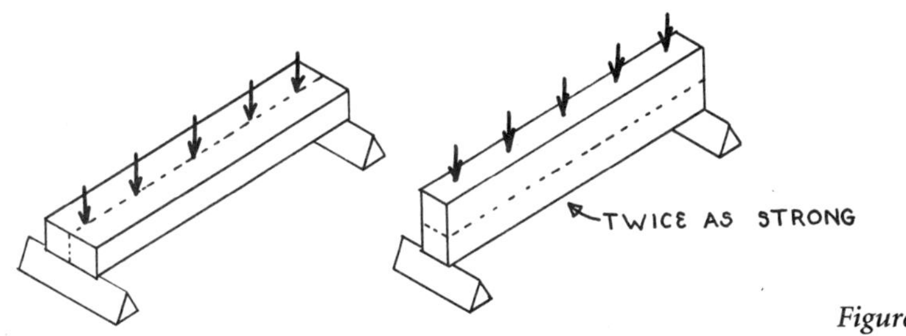

Figure 15-11.

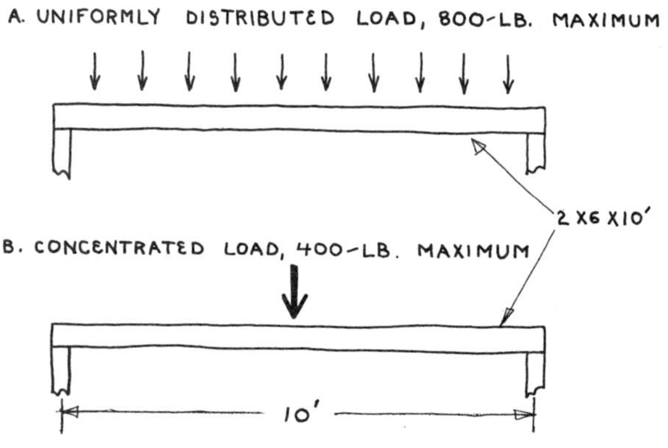

A. UNIFORMLY DISTRIBUTED LOAD, 800-LB. MAXIMUM
2 X6 X10'
B. CONCENTRATED LOAD, 400-LB. MAXIMUM
10'

Figure 15-12. Loading conditions.

A heavy weight, such as a wall, column, or a heavy machine, is called a concentrated load because its force is focused on a single area of a beam. If there is a load concentrated in midspan of the same 10-foot 2 x 6, it could carry only half as much, or 400 pounds, as shown in figure 15-12B. This does not mean that any 2 x 6 in the world will hold exactly these amounts. These figures, and the other examples mentioned below, are based on calculations for eastern spruce or a similarly strong timber. For other woods the figures might be proportionately higher or lower.

Cantilevered beams present a more complicated case. A cantilever is an overhanging end of a beam supporting a load, as in figure 15-13. The cantilevered portion of a beam is much weaker than the same portion would be if it were supported at both ends. For example, a spruce 2 x 6 spanning 10 feet can normally carry an 800-pound distributed load. But a similar 2 x 6 cantilevered out 10 feet can only support a distributed load of 200 pounds, or one fourth as much (figure 15-14A). If the load were concentrated at the end of the 10-foot cantilever, the maximum load would be only 100 pounds, or one-eighth of the regular load (figure 15-14B).

Figure 15-15 shows this same mathematical relationship in a more usable form for carpentry purposes. In figure 15-12 the maximum span was 10 feet for the 2 x 6 with a distributed load of 80 pounds per foot of beam. In figure 15-15A, the maximum cantilever is 5 feet, with the same loading of 80 pounds per foot of

Figure 15-13.

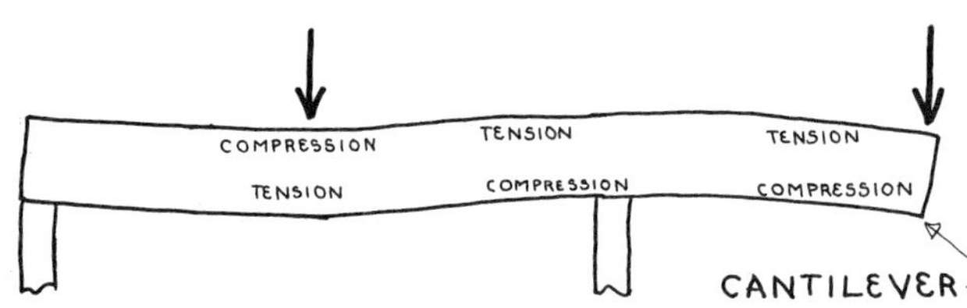

Figure 15-14.

2X6 CANTILEVERED TEN FEET

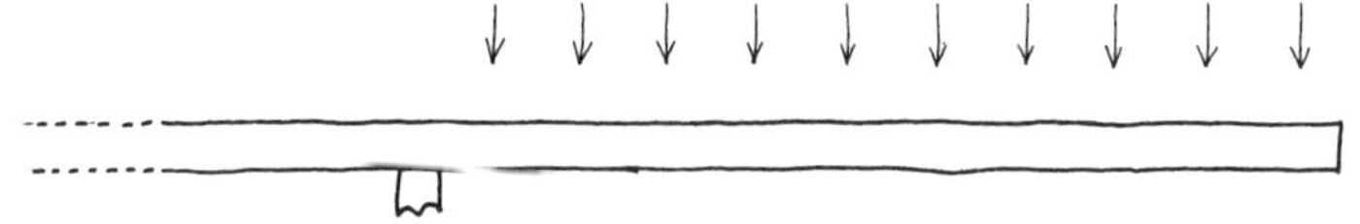

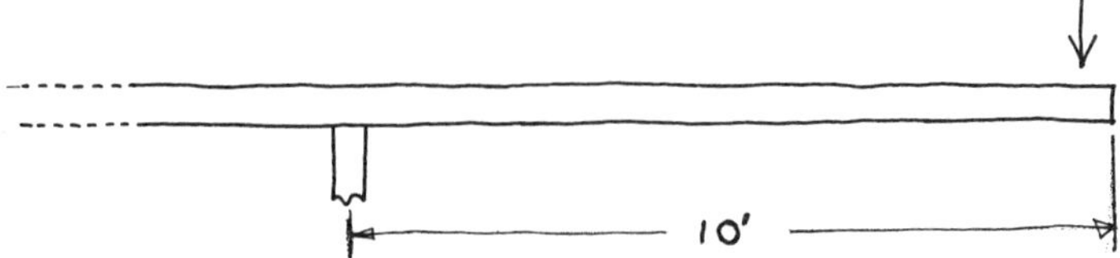

beam. In other words, a beam can cantilever half the distance it can span, if the distributed load per foot of beam is the same. Figure 15-15B shows that if this load is concentrated at the end of the cantilever, the maximum cantilever will be 2½ feet, or one-fourth of the normal span.

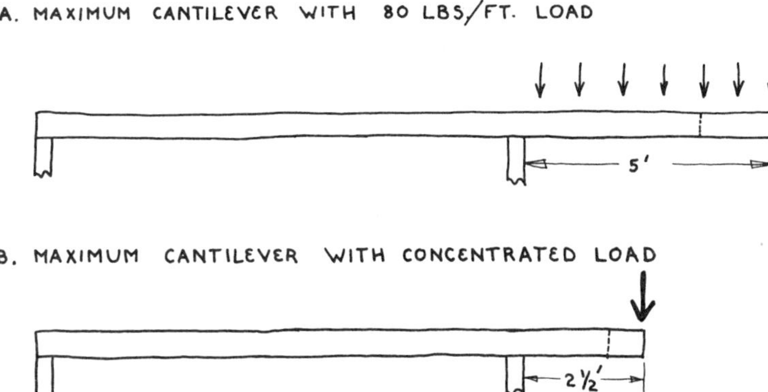

Figure 15-15.

If you actually extended cantilevers to these maximums, they would be strong but quite springy. In building terminology, they would "deflect." So in practice it is unwise to push cantilevers to the limit. A common rule of thumb for distributed cantilever loads is that you can cantilever at most one-third of the distance you could normally span between supports, assuming the same loading per foot of beam in both cases. In figure 15-15A, this means that in practice the *maximum* cantilever would be 3 feet, 4 inches (one-third of 10 feet) instead of the theoretical 5 feet, at the same loading per foot of length.

The corresponding rule for concentrated loads is that you can cantilever one-sixth the distance you could normally span, assuming the same per foot of length both cases. In figure 15-15B, the maximum cantilever would be 1 foot, 8 inches, which is half of 3 feet, 4 inches and one-sixth of 10 feet. These limits are shown in the drawing as dotted lines. Chapter 16 gives the formula for each kind of loading.

You might think that cantilevers are inefficient because they hold so much less than spans supported on both ends. But in certain situations cantilevers are actually more efficient.

Suppose you were building a deck 10 feet wide and the load was going to be, as before, eighty pounds per foot of beam. If you support the deck at the ends of the 10-foot span, you will have to use 2 x 6s. If you move the supports in one foot at each end, the span is now only 8 feet, so the joists can be 2 x 5s. If you push the cantilever as far as it will go, you can reduce the span to 6 feet, 8 inches and use 2 x 4s.

Figure 15-16.

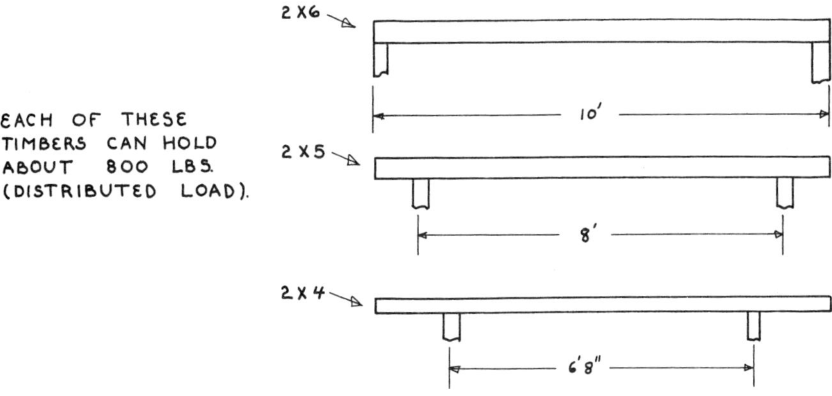

In addition to enabling you to reduce spans, the cantilever actually strengthens the beam inside the cantilever. You can see why in figure 15-13. The force down on the end of the cantilever provides leverage to lift the load between supports, making the noncantilevered section stronger and stiffer. This means that the 2 x 4 in figure 15-16 is actually somewhat stronger than the 2 x 6 above it.

This effect is put to use when a beam is continuous over a series of supports, such as in figure 15-17. The downward force on each section of beam helps lift the adjacent sections. Thus, the middle section of the 2 x 6 can now support 1,200 pounds, which is fifty percent more than it could support with the same span without the cantilever effect.

Figure 15-17.

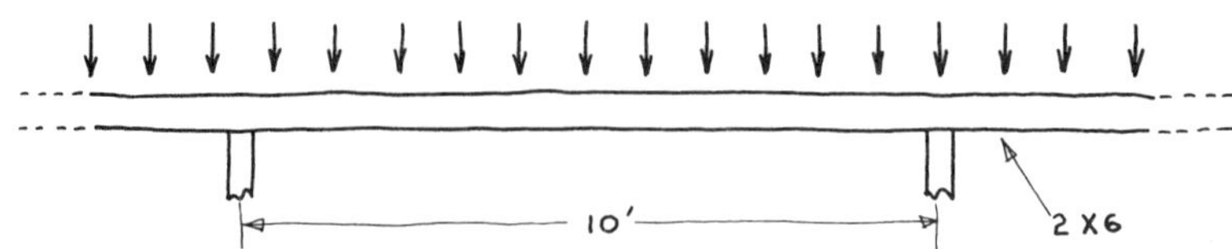

CONNECTIONS

Once you're sure the pieces of lumber in your house are strongly enough, you must make sure they are strongly joined. The joints as well as the lumber must carry the load. The kind of joint needed will be determined by the forces acting upon it.

To understand joints, you should understand bearing. A bearing surface is a surface where two members of a structure meet and through which a load is transmitted from one to the other. Any joint is, among other things, a combination of bearing surfaces.

The simplest kind of joint is one piece sitting on top of another, bearing down on it through gravity. An example is the stones in an old foundation. Each stone bears on the ones below to form a joint. A joint consisting of nothing but a common bearing surface will work if two conditions are met. First, the joint must be in compression. The forces or loads on the joint must be in opposite directions along the same lines in space, and there must be no great forces tending to push the two parts of the joint away from each other or make them slide against each other. Second, the bearing surface must have adequate surface area. In a stone foundation, if the bearing surface between the wooden house and the top of the wall is too small, the wood fibers will crush. If the bearing surfaces between the stones are too small, the stones will break. And if the bearing of the stones on the earth is too small, the house will settle (figures 15-18 and 15-19).

Of course, most joints are subjected to forces in several directions at once, so a mere piling will not suffice. Even a post supporting a second-floor beam could be accidentally subjected to a large sideways load. Therefore, most, if not all, the joints in a house need strong fastening as well as large-enough bearing surfaces.

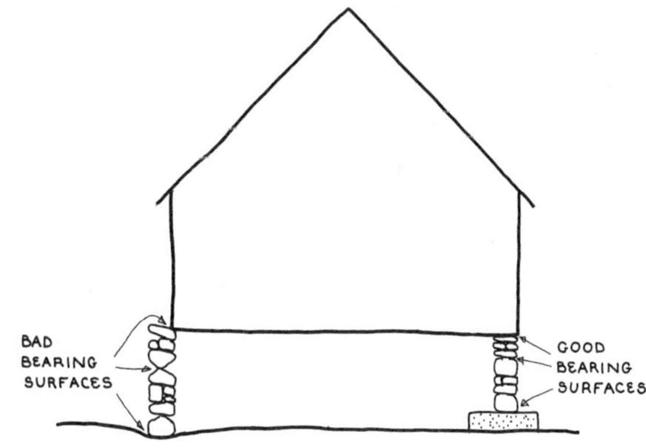

Figure 15-18.

Figure 15-19.

Some of the common fastening devices used in building are nails, screws, bolts, glue, dowels, mortise and tenon, and pegs (figure 15-20).

Before nails were cheap and plentiful, carpenters had to fasten joints using wood alone. They did this using notches, dovetails, dowels, and similar techniques that depended on making complicated cuts in wood (figure 15-21).

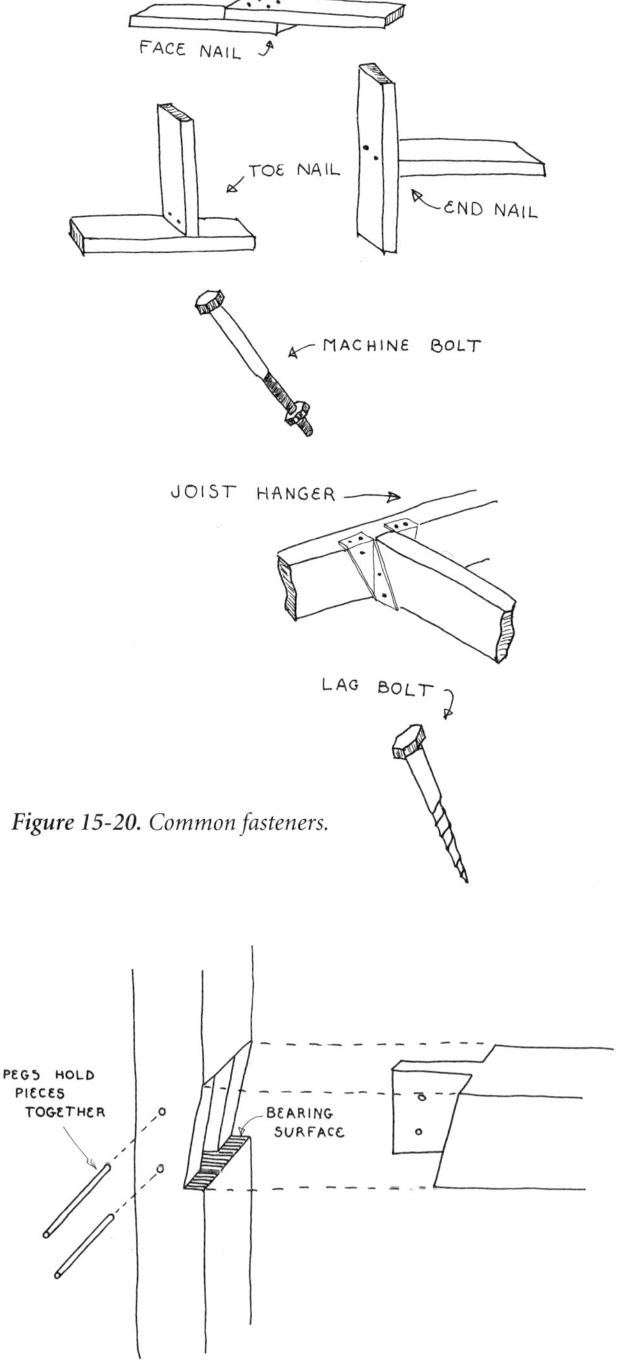

Figure 15-20. Common fasteners.

Figure 15-21. No hardware in traditional timber frames.

But in a modern wood-frame house, most joints are held together with nails. The nails perform the same functions dowels or dovetails used to: They counteract forces that might pull the joint apart. Different nailing methods are used depending on the kind of joint and what the joint is doing. The most common joint is a simple butt-nailed joint (figure 15-22).

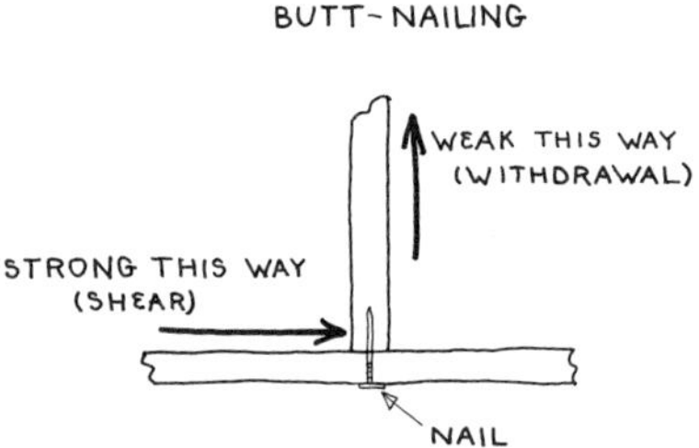

Figure 15-22.

The problem here is that such a joint is only really strong against some forces. If the top piece in the figure is forced upwards, parallel to the nail, the joint could come apart fairly easily, because all that is holding it is the friction between the nail shank and the wood. The joint is much stronger against sideways forces, perpendicular to the nail. For the joint to break this way either the wood fibers must be destroyed, or the nail must shear off at the mating surfaces between the two pieces. Technically speaking, butt-nailed joints are much stronger in shear than in withdrawal.

Many of the joints in a house, however, are perfectly strong even though they are only butt-nailed, either because the nail joint is in shear or because the withdrawal load is fairly small. Siding and trim for example, are only held to the wall by butt-nailing, but there is very little force pulling them off, so there is usually no problem.

There are many ways to improve on a simple butt-nail joint. First, you can use a kind of nail that has more friction with the wood. Ring nails, threaded nails, and cement-coated nails (which are coated with glue) are often used when more strength is needed. Flooring is often laid down with threaded nails. Second, you can use screws or lag bolts (figure 15-20), which hold even better. Third, you can toenail, as shown in figure 15-23.

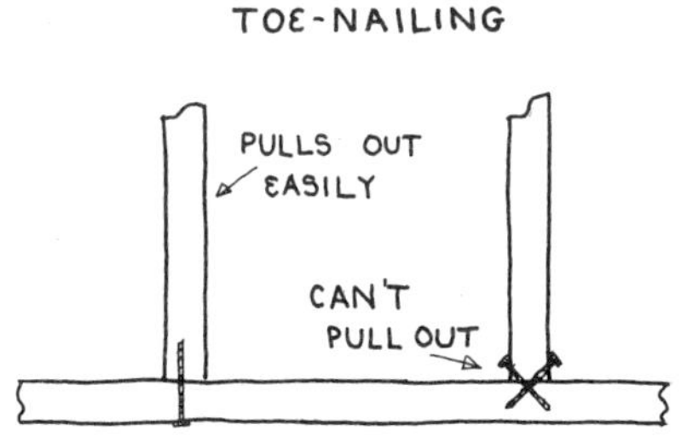

Figure 15-23.

Toenailing is stronger than simple butt-nailing because some of the nails will be in shear to some extent against a force in any direction. Any nail joint can be similarly improved simply by varying the nail angle. Fourth, you can strengthen a joint with a gusset. A gusset is a piece of wood nailed across two others to reinforce the joint between them. In figure 15-24, two butt-nailed 2 x 4s are strengthened with a gusset. In this joint, a force in any direction will be resisted by some nails in shear, either the nails between the 2 x 4s or the nails between the gusset and the 2 x 4s.

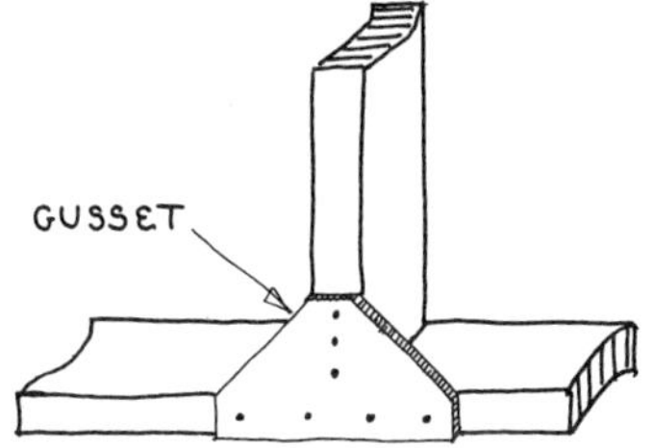

Figure 15-24.

Any piece of wood properly nailed across a wood joint between two others acts as a gusset. Most of the joints in the frame of a house are in effect gusseted by the sheathing and subflooring. Even if each individual nail joint is weak in some directions, the combination is strong.

Strength is especially important in the joints at the ends of a structural member in tension, such as the collar tie that holds a pair of rafters together and keeps the roof from spreading at the eaves. Suppose you end-nailed the collar tie to the rafter as in figure 15-25. The forces on the joint could easily pull the nails out. The joint is weak in the one way it must be strong. It would help a lot to toenail the joint (figure 15-26), but the nails could still pull out.

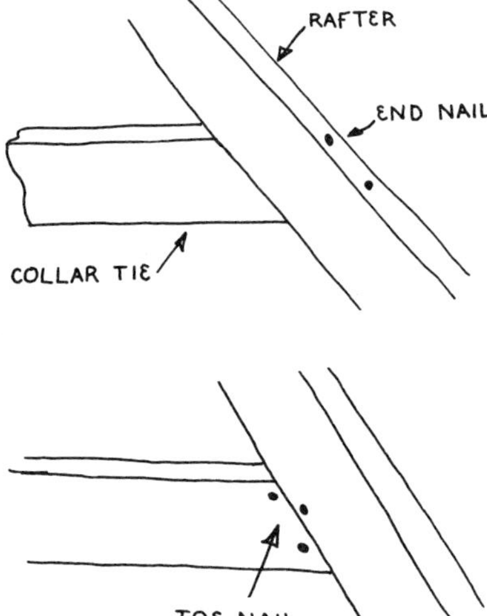

Figure 15-25.

Figure 15-26.

A better solution would be to lap the collar tie onto the rafter, as in figure 15-27, because the nails will then be in shear. Notice that the actual bearing surface in all of these joints is not where the wood pieces touch each other, but where the nails meet the wood. The nails transmit the load. The wooden pieces are not being pressed together by any outside force, which means that the amount of bearing surface and the strength of the joint will depend upon the number and size of the nails. In tension joints it is always important to use many nails and to use the biggest nails you can that won't split the wood.

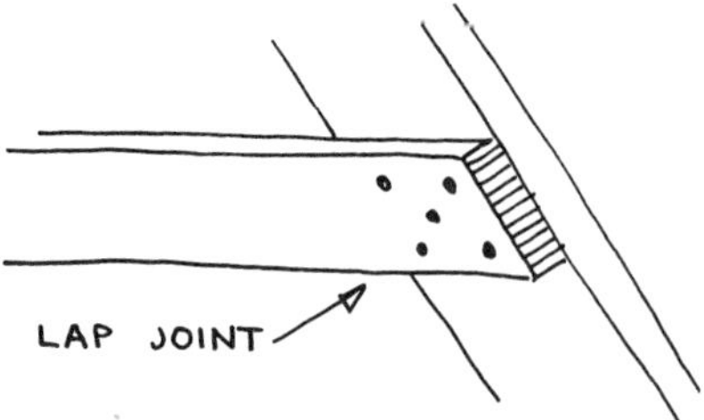

Figure 15-27.

Figure 15-28 shows how the same joint would be done with a gusset. This is just as good as the lap joint, particularly if the gusset is made of plywood, which won't split from nailing. One advantage of using a gusset is that you can make it very big to accommodate lots of nails. A gusset joint can be made twice as strong by using two gussets, one on either side of the joint, as shown in figure 15-29.

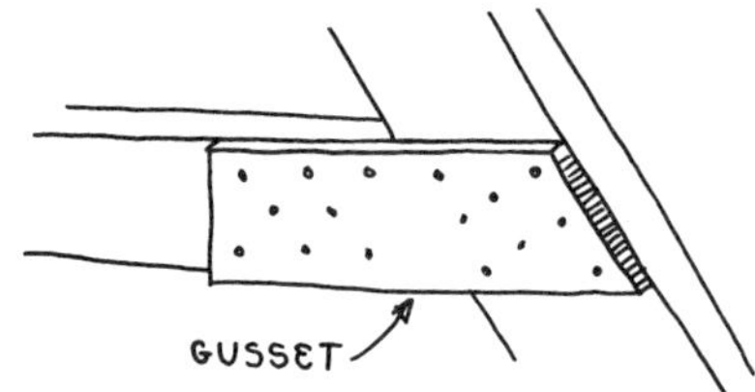

Figure 15-28.

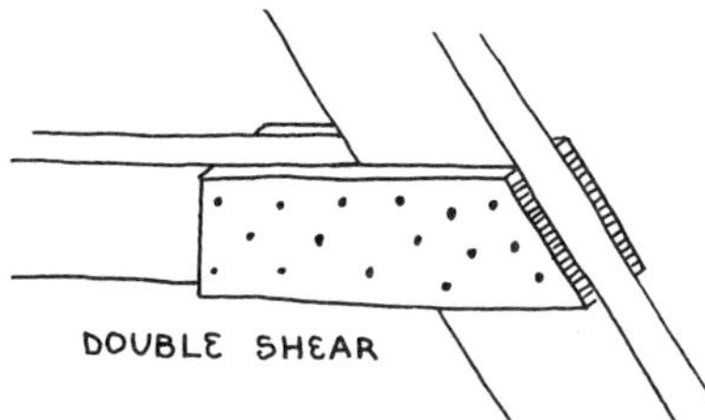

Figure 15-29.

This is called a joint in double shear. It is twice as strong, not only because there are more fasteners, but because the pull is symmetrical and there is less likelihood of the pieces turning slightly, which would allow the nails to withdraw slightly.

Rigidity. Strong pieces strongly jointed do not necessarily make a strong structure. The forces acting on a house, if they do not break the pieces of the house, or the joints tying them together, can still distort the shape of the house. That is why you must design your house to be rigid.

A house is basically a box or a combination of boxes. Yet a box shape is not intrinsically rigid. It can fold up like a cardboard carton. The simple box that is your house can be distorted by a sideward force, like wind or movement in the house, or by the weight of its own roof (figure 15-30).

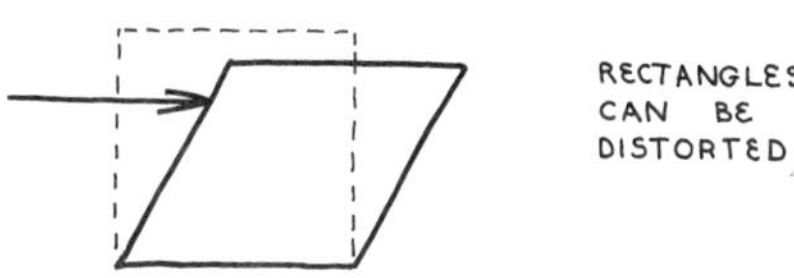

Figure 15-30.

This problem is solved with triangulation, often called diagonal bracing. Unlike a rectangle, a triangle will hold its shape against a force in any direction as long as the joints don't break apart (figure 15-31).

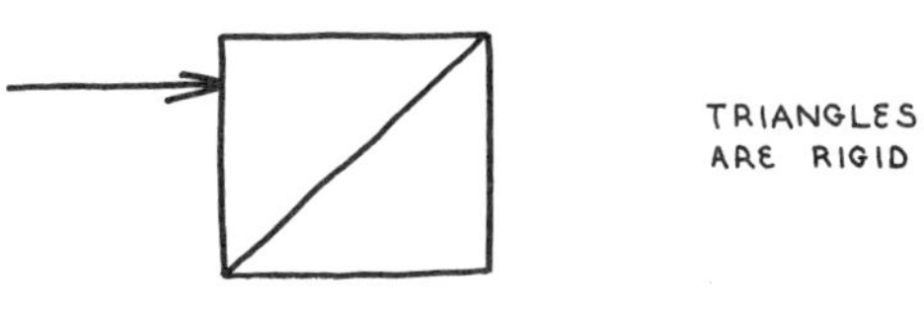

Figure 15-31.

The way to make your house rigid is to introduce triangles into its basically rectangular shape. One way to do this is with diagonal bracing in the frame (figure 15-32). Another way is to put your sheathing boards on diagonally (figure 15-33).

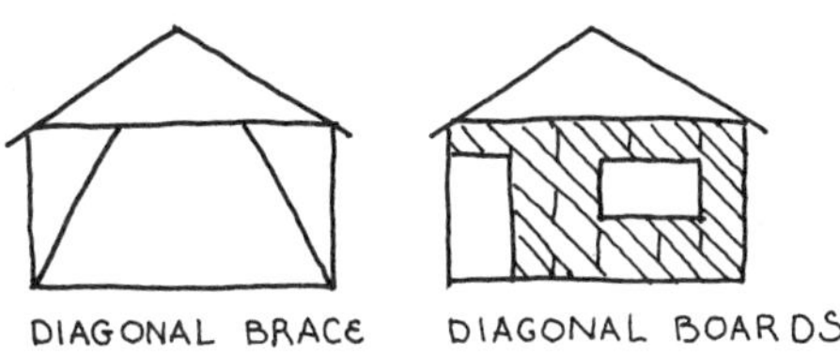

Figure 15-32. *Figure 15-33.*

In fact, even horizontal boards provide some bracing, because every three nails not in a row form a small triangle (figure 15-34). A house with horizontal boards will in effect be braced with hundreds of little triangles. The bigger these are, the more effective the bracing. That is why the most effective bracing is plywood or other materials that come in large sheets. They form hundreds of large triangles, which give the best possible bracing.

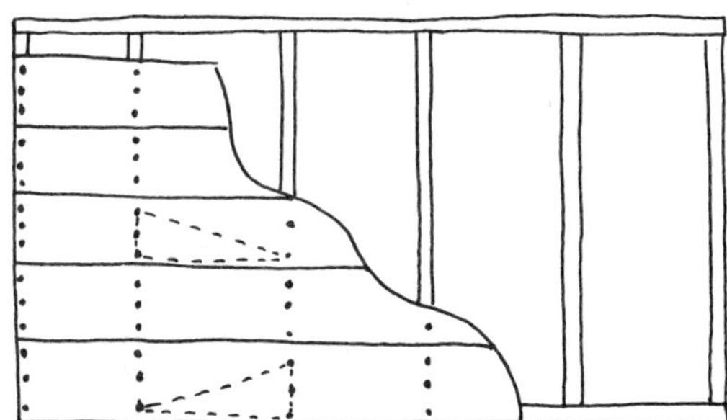

Figure 15-34.

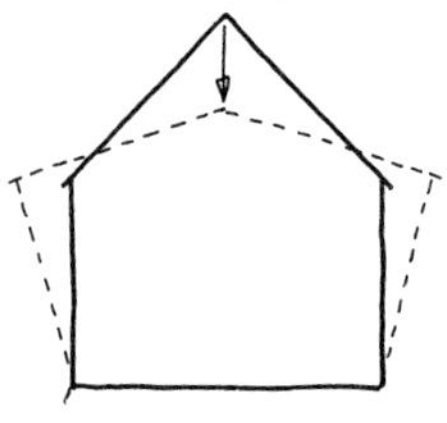

Figure 15-35.

You also need rigidity in the roof structure. An ordinary gable roof is shaped like a triangle with the bottom side missing (figure 15-35). If you put such a roof on top of the walls of a house, you will soon find (as many have found) that the roof tends to spread at the eaves as its weight pushes down. The roof will cave in, and the walls will be pushed out. Figure 15-36 shows three possible ways to solve this problem.

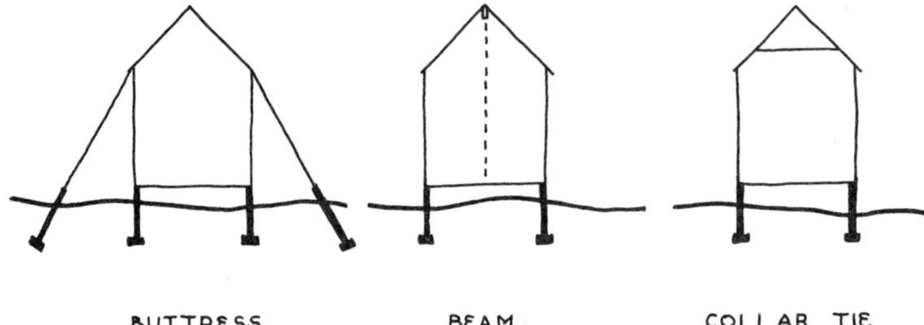

Figure 15-36.

The first solution is through some kind of exterior bracing or buttress. This is usually impractical, unless you are building a cathedral. The second solution is to support the ridge with a bearing wall or beam. This holds up the ridge of the roof and prevents it from moving downward. This is perfectly satisfactory, although it does require that the ridge beam (or wall) be properly supported. The most common solution is to install collar ties between the two rafters to hold them together. In many houses the attic floor joists will serve the same purpose. The roof becomes a stable structure exerting a downward force on the walls.

EVALUATING STRUCTURAL SOUNDNESS

Keeping these three principles in mind—strong pieces, strong joints, and rigidity—you can now evaluate your house design structurally. You can't make the precise calculations a structural engineer would make, but you can get an intuitive sense of whether the house is designed to stand up to the loads placed on it. Since most of the loads in a house occur in a downward direction, start at the top, following the load paths from the roof down to the earth. Every link in the structural chain must be strong. Every downward load must be transmitted all the way to the earth. As you follow the loads down, ask yourself: (1) whether the pieces can support the loads passing through them, (2) whether the joints connecting the pieces are strong enough, and (3) whether the structure can sustain the loads and still retain its shape.

For example, figure 15-37 shows a typical small two-room structure. To evaluate this design structurally, you would first figure out the live and dead loads on the roof. The dead load is the roof's weight. The live load includes snow and people occasionally walking about on the roof. These loads are supported by the roof sheathing (boards or plywood, probably), which in turn is supported by rafters. Your first task is to make sure the sheathing is thick enough so that it won't sag or break. How thick it needs to be depends on how wide the rafters are spaced. For example, on a roof ⅝-inch plywood will span 2 feet, but will sag over a 3-foot span.

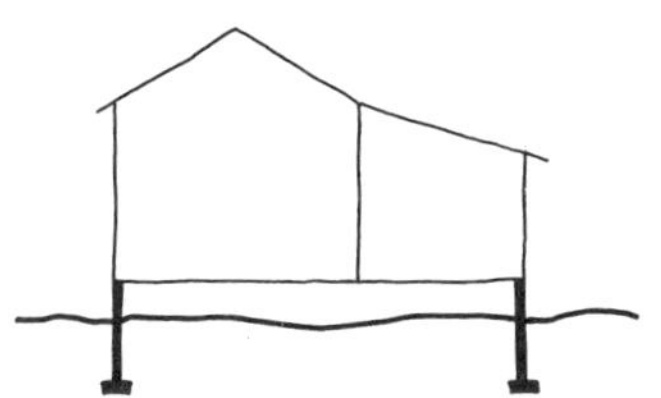

Figure 15-37. A very unstable structure.

Next, consider the rafters. They must be big enough to carry the dead loads of the roofing, the sheathing, and their own weight over the distance they span, plus the live loads of wind and snow. Chapter 16 shows how to calculate rafter sizes for specific situations. You also have to make sure your roof structure is rigid. The structure in figure 15-37 is not. The weight of the gable tends to spread the walls at the eaves. Using collar ties to fix this may solve the rigidity problem, but only if the collar ties are properly attached to the rafters. The outward spreading force of the rafters will put the collar ties in tension, so you must give special attention to fastening them. The collar ties have to be nailed, bolted, or nailed and gusseted to the rafters. The strong, stable structure created by the collar ties creates a clear vertical load path, with loads resting on the exterior wall to the left and the interior wall to the right.

Our next concern is the bearing surface where the rafters meet the walls. Usually a special notch called a bird's mouth is cut in each rafter to enable it to sit flat on the wall and transmit the load downward (figure 15-38). Often the rafters are positioned right above wall studs for a direct load transmission.

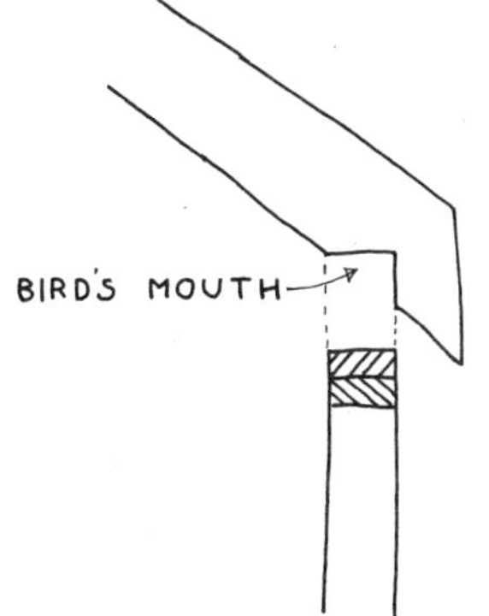

Figure 15-38.

Next, study the walls themselves. Walls are columns being compressed between the roof at the top and the floor at the bottom. Columns have

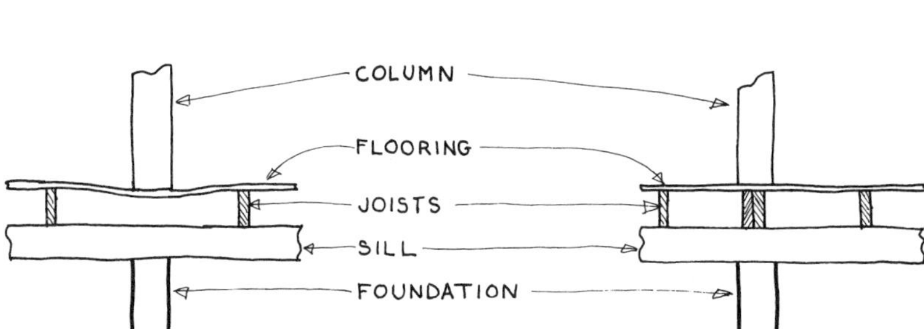

Figure 15-39.

tremendous strength in pure compression, so there should be no problem there. But whenever you have a column, you must make sure the structure it sits on is as strong as the column itself. The left-hand sketch in figure 15-39 shows a strong column over a strong foundation, with a structure between them that is not strong enough to transmit the load. In the right-hand sketch "blocking" has been added to provide a proper bearing structure.

Next, look at the right-hand side of the gable roof in figure 15-40A. This rests on a bearing wall, which in turn rests on the middle of the first-floor beam. In theory the beam could be large enough to carry the concentrated load, but such a beam would have to be absurdly huge. Any reasonably sized beam would sag as shown. The practical solution is a foundation post directly under the bearing wall (figure 15-40B).

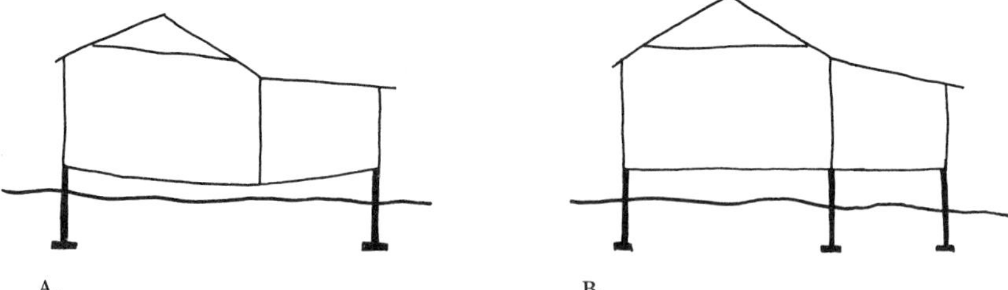

Figure 15-40.

At the base of the load path, make sure the bearing surface between the foundation and the earth is wide enough. On a rock base, a foundation column by itself would probably have enough bearing. But on a base of soil, the weight of the house might drive such a column right into the earth like a stake or cause uneven settling. A footing is needed to distribute the load. The softer the soil the bigger the footing has to be.

16

STRENGTH OF TIMBERS

THE SIMPLE MATH IN THIS CHAPTER SHOWS HOW TO DO load and timber sizing and span calculations. These calculations will probably cover ninety percent or more of the structural issues that arise in the case of a simple, small house.

However, there will still be times when some consultation with a professional builder, architect, or engineer makes sense. Long spans, large cathedral ceilings, or truss design are cases where a professional consultation is called for. Building codes also specify timber sizes for most applications, and they are in general a valuable source of construction information.

You can figure your timber sizes using span tables, like those in appendix B. But learning the math allows you to envision what forces are at work in any particular part of the house you are designing, and what your range of choices may be for managing them. If there is really no load above a window, you can leave out the header. If there is a beam supported by several columns, you can use the cantilever principle to span the column-to-column distances with smaller timbers, without sacrificing strength. With the math presented here you can imagine solutions to tricky problems, and do some mathematical comparisons to see which is the best way.

This saves lumber, money, and time. I'd estimate that on the average, I save about ten percent of my lumber costs using these ideas, compared to doing it the more obvious, standard way.

HOW MUCH WILL A BEAM SUPPORT?

You can get an idea of how the strength of beams varies by experimenting with a few small pieces of wood, say a 1 x 1 x 4-inch piece of oak, a 1 x 1 x 4-inch piece of pine, and a 1 x 2 x 4-inch piece of pine. Compare their strengths under different loads and configurations. Support them between two chairs. Orient them in different ways. Move the chairs closer together and farther apart (changing the span). Add or remove loads, watch how the boards react.

Species of wood vary greatly in strength. The oak 1 x 1 will be much stronger than the pine 1 x 1. The strength rating is called "fiber stress in bending," abbreviated F or Fb. Each species and grade of wood has an F number. Eastern spruce has an F of about 1,000, and southern pine an F of almost 2,000. That means that the southern pine is twice as strong as the spruce. Appendix A gives some of these values, and indicates where to obtain invaluable industry pamphlets giving values for most commercially available woods used in house framing.

Your comparisons will show that if the breadth (B) of a piece is doubled, keeping everything else constant, the strength also is doubled. Strength is proportional to breadth. A 2 x 4 laid flat will support twice what a 2 x 2 will. It's the same as having two pieces instead of one (figure 16-1).

If you double the size in the other dimension, in depth (d), the strength increases fourfold. Strength is proportional to the square of the depth. On edge, a 2 x 4 will support four times what a 2 x 2 will (figure 16-2).

This means a 2 x 4 is twice as strong on edge as lying flat (figure 16-3).

If you double the span (L), you halve the strength. Strength is inversely proportional to the span. A 2 x 4 that is 8 feet long supported at the ends can carry half the weight a 2 x 4 that is 4 feet long will carry.

Adding these facts together we get the formula used for computing the strength of a beam.

$$\frac{FBd^2}{9L} = \text{Maximum load in pounds}$$

where

F = the fiber stress (F) of the wood used
B = the breadth of the beam, in inches
d = the depth of the beam, in inches
L = the span, in feet

This form of the formula applies to conditions of uniform distribution of weight, or load. That is, we are assuming the load is spread approximately evenly over the length of the beam rather than concentrated in one place. The span will be the distance between supports. With sloping rafters, however, the span is measured horizontally and will be different than the rafter length. (See figure 16-4.)

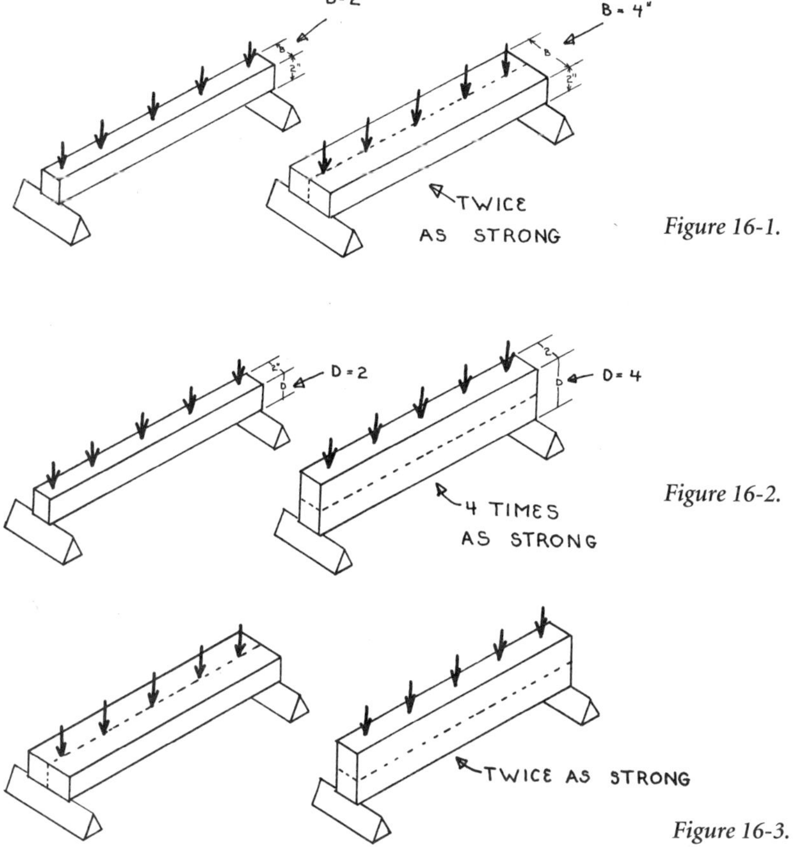

Figure 16-1.

Figure 16-2.

Figure 16-3.

If the load is concentrated in the middle of a beam, the formula is

$$\frac{FBd^2}{18L} = \text{Maximum load in pounds.}$$

For a cantilever with a uniformly distributed load, the formula is

$$\frac{FBd^2}{36L} = \text{Maximum load in pounds.}$$

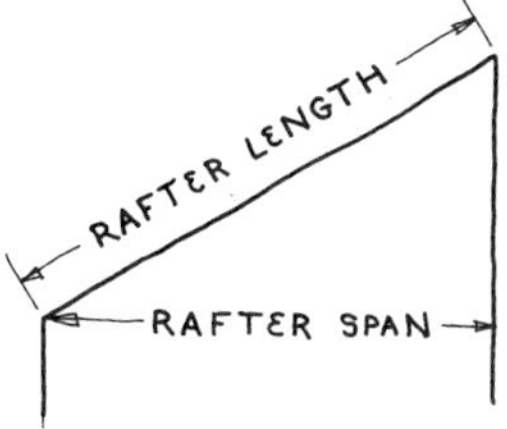

Figure 16-4.

For a cantilever with a load concentrated at the end, the formula is

$$\frac{FBd^2}{72L} = \text{Maximum load in pounds.}$$

I should mention here that cantilevers that have the maximum loads implied by these formulas will be springy. For this reason it is best to limit the length of cantilevers to at most two-thirds of the theoretical maximum extension. A review by a professional is advisable here. (Cantilevers are discussed more extensively in chapter 15).

Suppose you had a dressed native spruce 4 x 8, 12 feet long. How much weight would it hold loaded uniformly? From appendix A you would find the F value of 1,000. A dressed 4 x 8 is actually 3½ x 7½ inches. So B = 3½, and d = 7½, d^2 (rounded off) = 56, and L is 12.

$$\frac{\overset{(F)}{1{,}000} \times \overset{(B)}{3\tfrac{1}{2}} \times \overset{(d^2)}{56}}{\underset{(\text{constant})}{9} \times \underset{(L)}{12}} = 1{,}815 \text{ pounds}$$

This does not mean the beam would break if you put one ounce of weight on the timber more than the allowable load. A large safety factor is built into these formulas. Also, most beams are actually subjected to their loadings intermittently. The formulas are designed so that if you compute your timber sizes for the maximum loads they will be subjected to, they will be a good size for their routine and extra-large loadings.

ACTUAL LOADS AND LOAD ASSUMPTIONS

To use these formulas you must know the weight in pounds the timber will have to support. The timber will have to support itself plus part of the building structure. This is called the dead load on the beam. The beam will also be supporting furniture, people, fixtures, snow, and other contents of the building. These weights taken together are called the live load.

Dead Loads

You can find the dead load for a given beam by laboriously adding up the weight of the beam itself and the parts of the structure it will support. In figure 16-5, the frame weighs about two pounds per square foot. The plywood subfloor, the oak flooring, and the sheetrock ceiling also weigh about two pounds per square foot each. This is a total of eight pounds per square foot. A more precise calculation could be made using densities given in appendix F.

In fact, you do not need to figure dead loads piece by piece. Usually the total dead load of a floor, wall, or roof is assumed to be ten pounds per square foot in wood-frame construction. This assumption is conservative and avoids tedious calculations. If the structure contains very heavy materials, such as a slate roof or a concrete slab on a floor, make a detailed calculation using the information in appendix F.

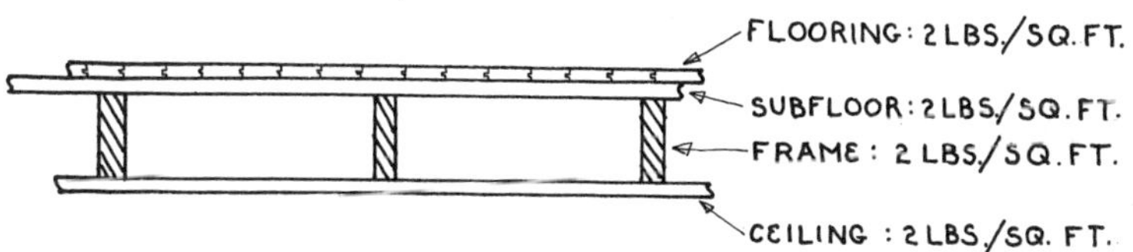

Figure 16-5. Dead load.

Live Loads

The snow, furniture, people, and other contents that constitute live load are continually changing, so it does not make sense to figure live loads piece by piece. Instead, calculations are based on assumed average loads. The first floor or living area live load can be assumed to be forty pounds per square foot. The second floor or study areas get lighter use and can be designed for a live load of thirty pounds per square foot.

Any roof should be designed to bear a minimum live load of twenty-five pounds per square foot, because even if a roof normally carries nothing, it should be strong enough to support workers during construction or repair. Twenty-five pounds per square foot will be a good standard in areas of little or no snow, or for roofs in snow climates that will shed snow. How well a roof sheds snow will depend on conditions of wind and weather and on how slippery the roofing surface is, but usually a roof with a rise of 6 inches per foot (a 6:12 roof) will shed deep snow.

Flatter roofs in snow areas can be designed by assuming accumulated snow to weigh about ten pounds per square foot of depth. In Vermont, where the maximum accumulation on a roof would be about 5 feet, a low-slope roof could be designed to carry a live load of forty or fifty pounds (see snow load map, appendix F).

Examples of Load Computations

Computing the load a beam will carry proceeds in several steps. First, figure out how many square feet the piece supports. Then figure what each square foot weighs, live and dead loads combined. To take a simple example, imagine a series of floor beams (joists) spaced 2 feet apart, each one 10 feet long and supported at each end (figure 16-6).

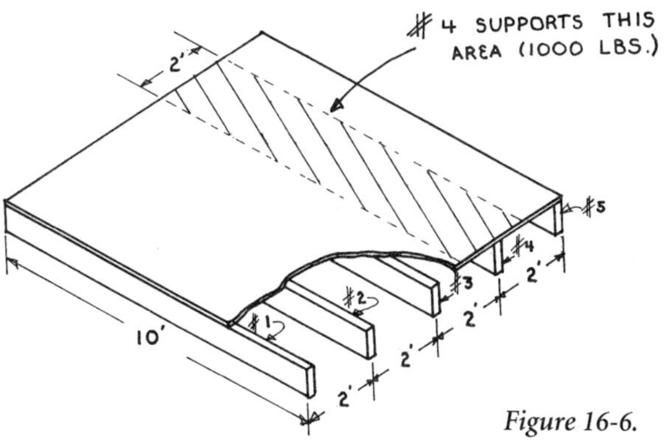

Figure 16-6.

Each one supports an area 10 feet long extending halfway to the next joist on the left and halfway to the next joist on the right. This is an area of 10 x 2 feet, or twenty square feet. If this were a first floor, the dead load would be ten pounds per square foot, the live load forty pounds per square foot, and the total load fifty pounds per square foot. Since each beam supports twenty square feet, the total load per beam is 1,000 pounds (twenty pounds per square foot x fifty pounds). Note, however, that the first and last joist, #1 and #5, support half the area, hence carry only half the load, or 500 pounds. The total load of the entire floor is 4,000 pounds.

Suppose this floor, in turn, was supported by two main beams, x and y, as in figure 16-7. These two beams span eight feet and share the weight of the ten-foot span equally. Each supports an area 5 feet wide and 8 feet long, or forty square feet. At fifty pounds per square foot, the total load is 2,000 pounds per beam.

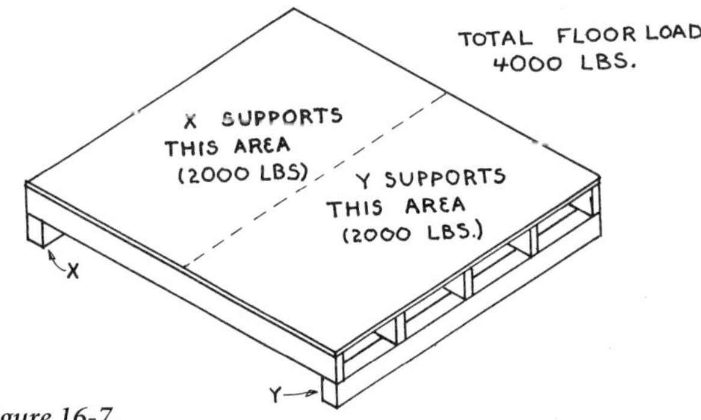

Figure 16-7.

Extra weights on the floor, such as partitions, could further complicate your calculations. Suppose there were a wall 8 feet high on the floor above beam y, as in figure 16-8 (wall a). This wall has an area of 8 x 8 feet, or sixty-four square feet. Since walls are assumed to weigh ten pounds per square foot, this one can be seen as weighing 640 pounds, 700 pounds rounded up. We add this 700 pounds to the floor load on beam y of 2,000 pounds to get a total load on beam y of 2,700 pounds.

Concentrated loads complicate things even more. In figure 16-9 we have another wall, wall b, over the middle joist. This wall exerts an extra uniformly distributed load on the joist directly below it of 10 x 8 feet x ten pounds, or 800 pounds total. But this wall also imposes a concentrated load at midspan on main timbers x and y.

Wall b is carried equally by x and y, so this concentrated load is one half of 800 pounds or 400 pounds on x and 400 pounds on y. For concentrated loads the formula will not be $FBd^2/9L$ but $FBd^2/18L$. In effect that means that a concentrated load counts double. We can convert our concentrated load of 400 to its equivalent as a distributed load, or 800 pounds on x and 800 pounds on y. This enables us to express the total loads on x and y in a single number.

Figure 16-8.

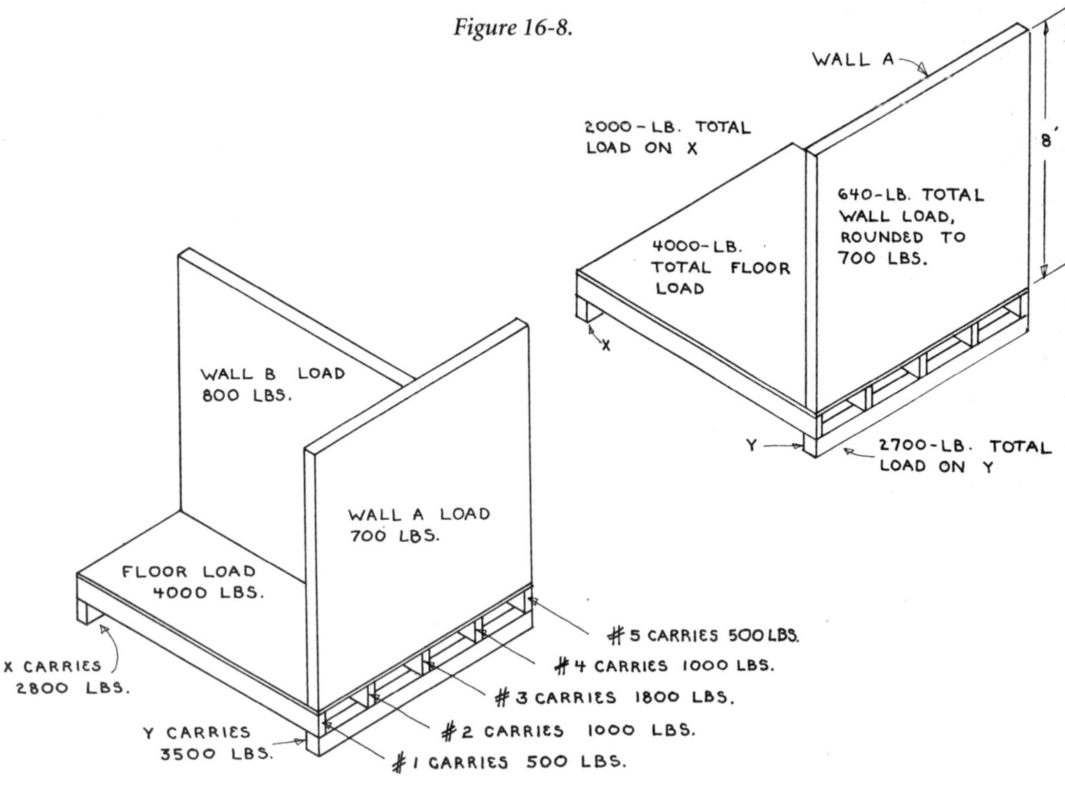

Figure 16-9.

To summarize, beam x carries:

2,000 pounds (half of the floor)
+ 800 pounds (half of wall b expressed as a uniform load) =
2,800 pounds total load.

Beam y carries:

2,000 pounds (half of the floor) + 700 pounds (all of wall a)
+ 700 pounds (half of wall B expressed as a uniform load) =
3,500 pounds total load.

EXAMPLES OF TIMBER SIZE CALCULATIONS

We can use the same imaginary structure to show how the formula is used to find the correct sizes of timbers under different conditions. Suppose we were using rough-cut spruce from New England, sawn to full dimensions so that a

2 x 6 would actually be 2 inches by 6 inches. Such spruce has an F of about 1,000. Let us compute what size to use for our 10-foot-long joists 2 feet apart. To make this calculation, guess what size timber would work and try it out in the formula. My guess is 2 x 8. The formula is $FBd^2/9L$. Here F is 1,000, B is 2, d^2 is 64, and L is 10.

$$\frac{\overset{(F)}{1{,}000} \times \overset{(B)}{2} \times \overset{(d^2)}{64}}{\underset{(L)}{9 \times 10}} = 1{,}422 \text{ pounds}$$

As shown in figure 16-9, joists #1 and #5 carry 500 pounds each, #2 and #4 carry 1,000 each, and #3 carries 1,800. Thus our 2 x 8 is more than strong enough for #1, #2, #4, and #5, but undersize for #3. In carpentry, a joist carrying a partition is reinforced simply by doubling up the joist, which will work here. So 2 x 8 would be an adequate joist size.

Since 2 x 8s actually give much more strength than needed, let's see if 2 x 6s would be sufficient.

$$\frac{\overset{(F)}{1{,}000} \times \overset{(B)}{2} \times \overset{(d^2)}{36}}{\underset{(L)}{9 \times 10}} = 800 \text{ pounds}$$

The 2 x 6s are sufficient for #1 and #5, but insufficient for #2 and #4 and also for #3 even if it is doubled up. The 2 x 6s are too small.

Of the beams that support the floor joists in our hypothetical structure, beam x, 8 feet long, carries 2,800 pounds. Again assuming wood with an F of 1,000, let's see if a 6 x 6 is strong enough.

$$\frac{\overset{(F)}{1{,}000} \times \overset{(B)}{6} \times \overset{(d^2)}{36}}{\underset{(L)}{9 \times 8}} = 3{,}000 \text{ pounds (more than enough)}$$

My guess for beam y is that it will be a 6 x 8. Since it is 4/3 as wide (B), it should hold 4/3 of the load, or 4,000 pounds. That is more than our design load of 3,500. It should be noted that the 6 x 8 will do more work placed with its narrow side down.

Other Factors that Influence Beam Calculations

There are several other factors that may influence your beam calculations. These include the depth of the beam, the duration of the load, and whether it shares the load with adjacent "repetitive" timbers, such as a group of floor joists. Lumber trade groups such as the Western Wood Products Association (WWPA) have researched how assigned F values can be adjusted to reflect varying load conditions, and developed excellent literature to show how these adjustments should be applied to graded lumber you use. Appendix A describes these factors further, and tells where the literature can be ordered. The WWPA's "Western Lumber Product Use Manual" is particularly valuable (see page 208 for ordering information.

Deflection

Until now we have been talking about calculations for strength. Sometimes you may also want to worry about stiffness or the wood's ability to resist deflection. Any timber, even one that has been properly sized, will deflect: It will both sag and spring when subjected to an impact. Usually this is nothing to worry about because a timber computed for strength will not noticeably deflect. A 12-foot floor beam or sill could sag ½ inch without being noticed. But in a few specific instances, you may want to make sure the amount of deflection is within acceptable limits. Consider these conditions: One, if there will be a real plaster ceiling the deflection should be limited to prevent the plaster from cracking over time. Two, it is a good idea to limit the deflection of floor joists 14 feet long or more, so the floor does not feel springy. Three, it is a good idea to limit the deflection of ceiling or roof timbers at or near eye level on spans over 12 feet. A timber way above your head could sag one inch without anyone being able to see it. But the same sag in a timber you could sight along, though harmless, might be irritating or look askew, particularly if you are fussy about such things (as I am). The timbers to worry about are beams or joists supporting low ceilings, or which can be viewed at eye level because of stairways.

In all of these examples a standard deflection limit used in design is 1/360 of the span. The deflection in a 16-foot beam, for example, should be limited to slightly more than ½ inch.

The stiffness of a particular kind of wood is indicated by its modulus of elasticity or E rating. The E of a timber varies with several factors. Some species are stiffer than others. Douglas fir is generally stiff, while cedar species are less stiff. A knot-free timber will be stiffer than a piece with large knots or knots along its edges. A dry timber will be stiffer than a green one. In general, stiffness is related to the density of the seasoned timber. Within a species, a heavy piece will be stiffer than a lighter one, and a heavy species will be stiffer overall than a light species.

Industry pamphlets you can order give E values for graded lumber (see appendix A). Values for ungraded species (the lumber from local sawmills) are also listed in appendix A. The ratings are actually averages. For stress-graded wood, the average covers a narrow range, because of the grading process. You can use the values given with confidence as long as there is nothing obviously wrong with the specific pieces of lumber you are using. With ungraded sawmill lumber, the average covers a much wider range, and you will have to temper your calculations by carefully looking over each piece of wood. For the long spans, choose the pieces with the highest (dry) weight and the fewest knots.

The formula for computing deflection is too complicated for nonengineers, such as myself, to use. You can use tables 4 and 5 in appendix B to find the minimum E values needed to limit deflection to 1/360 of the span under different circumstances.

The Western Wood Products Association makes a special little slide rule called a WWPA span computer that you can use to figure joist, rafter, and beam spans for both strength and stiffness. The span computer's only drawback is that it only makes calculations for planed timber sizes. It is available for $5.00 from the Western Wood Products Association, Yeon Building, 522 Southwest 5th Ave., Portland, Oregon 97204.

SHEAR

Figure 16-10 shows a beam failing as a result of shear forces. The timber is sheared off at its bearings, much as a paper punch makes a hole in a piece of paper. This kind of failure rarely occurs in small wooden buildings, because timbers computed for strength or deflection will be more than adequate against

Figure 16-10.
Shear failure.

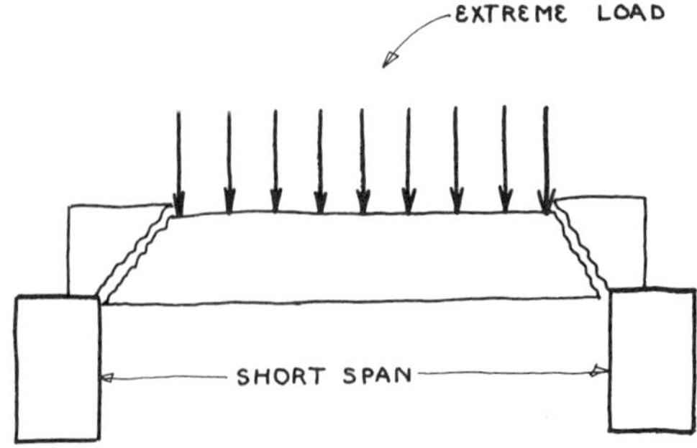

shear under most conditions. In general, you need not calculate beams for shear. But shear strength is important when the load on a timber is abnormally large over an abnormally short span. Under these conditions, a slender timber may be adequate according to calculations for strength or stiffness, but not stout enough to take the shearing forces at the joint. If you have such a situation, have your design checked by a structural engineer to make sure there is no problem.

COLUMNS

Wooden timbers used as columns are much stronger than they look. I don't think carpenters very often bother to compute posts in wooden houses. Often a big post will be used to make the various joints strong and easy to fit, even when a smaller column would theoretically hold the weight. A larger post also provides more bearing surface to the beams it supports. Often, also, a post will be large for the sake of appearances.

Appendix B, table 8, gives maximum loads for timber posts. I will briefly explain the formulas upon which these tables are based.

Columns can fail in two ways. Relatively slender columns will fail by bending sideways in the middle, like toothpicks you squeeze between your thumb and forefinger. Short stubby columns, however, will actually crush before they start to bend in the middle. Most columns in building are in the first category. The calculations for failure by bending in the middle are based on Euler's formula:

Maximum load in pounds = $.3 \times E \times \text{Area}\ (L/d)^2$

As figure 16-11 shows, area indicates the cross section of the post, in square inches. L is the height of the post, in inches; if a post is firmly braced in the middle, in both directions, L will be the longest unsupported height. The narrowest dimension of the column in inches is d. In the figure, the area is 12 square inches, L is 120 inches, and d is 3 inches. E refers to the modulus of elasticity of the wood.

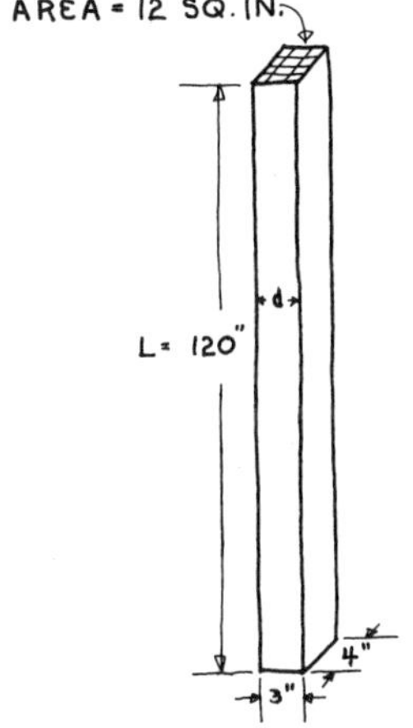

Figure 16-11.
Euler's formula.

A stiffer wood, or a larger cross-sectional area, will reduce the possibility of a column bending out in the middle. In the formula, the load is directly proportional to these two factors (E and Area). The other factor that affects the loading capacity is the slenderness ratio, L/d in the formula, which is really the relative skinniness of the column. If the height is greater in comparison with the least dimension, the post will be weaker.

Specifically, if you double the height of a column, it will hold only one fourth of the load.

That is why L/d is squared in the formula. This formula applies only to relatively slender columns that will fail first by bending out in the middle. If L/d for the column is less than about 20, you have a short column, and you should make sure it will not fail by crushing.

The compressive strength of a particular kind of wood (its resistance to crushing) is indicated by its "fiber stress in compression," or Fc. (This is a rating much like F in the beam formula.) Each wood has two Fc ratings, because wood can be crushed more easily when the force is applied perpendicular to the grain than it can when the force is applied parallel to the grain, as is the case in columns. The WWPA's "Western Lumber Product Use Manual: base values for dimension lumber" gives both values for graded lumber.

Fc ratings are in pounds per square inch. Thus, #1 hemfir has an Fc parallel to grain of 850 pounds per square inch. If you multiply the cross-sectional area in square inches of a short column by its Fc parallel to grain, you get the maximum capacity of the post in pounds. For example, a short hemfir post 3½ x 3½ inches has an area in cross section of 12.25 square inches. Fc parallel to grain is 850 pounds per square inch. The capacity is 850 x 12.25 or 10,412 pounds.

Here are two sample computations:

First, what will a #1 grade hemfir column 5½ x 3½ x 120 inches carry?

L/d is 120/3.5 or about 34. Since 34 is way over 20, the column will fail by bending out, so its capacity should be figured using Euler's formula. From the WWPA pamphlet we learn that E = 1,300,000.

$$\frac{.3 \times 1{,}300{,}000 \times 3.5 \times 3.5}{(120/3.5)^2} = 6{,}500 \text{ pounds}$$

Second, consider the same columns but 48 inches tall. L/d is 48/3.5 or about 13.7. Since this is substantially below 20, it is a short column and its capacity will be limited by the compressive strength of the wood itself, rather than by its tendency to bend out. Fc parallel to grain is 850 pounds per square inch. The area is 3.5 x 5.5 = 19.25.

850 lbs. × 19.25 sq. in. = 16,362 pounds capacity

If L/d is close to 20, say between 18 and 22, calculate the column capacity using both methods, and use the lower rating.

17

HEAT

HEATING IS THE MOST COMPLICATED PART OF A HOUSE design, and also the most fundamental and intuitive. There is a bewildering array of gear for heating your home, and many strongly held ideas about the best way. In the end, most people choose a heating system based on what appeals to their sensibility, and this is okay. There are many good ways to heat a house, particularly a well-built house.

Heat comes down to two questions: how to keep the heat loss low, and how best to provide the heat you do need.

KEEPING THE HEAT LOSS LOW

As you sort through heating options, it is well worth it to learn the basic heat loss calculation that is outlined in chapter 18. A heat loss calculation allows you to look at any version of your design, and figure the amount of heat you need for the heating season, the size of system you need, and what it will cost using different fuels. By mastering this bit of high-school algebra, you can figure out how various design choices will influence your heat loss, and which efforts are worth it to you. As you look at superinsulation, Russian stoves, or solar panels, you can compare design and installation costs to fuel savings, and make the decision that works for you.

The formulas quantify a few basic ideas. Build a tight house, because drafts of any kind can account for a huge proportion of your heat loss. Make the house compact, because heat loss is proportional to surface area. Use lots of insulation, because insulation value, or resistance to heat loss, is proportional to the thickness of insulating materials. Insulate everything. Put the big windows on the south, where their large heat loss will be offset by solar gain. Do everything you can do reduce the heat loss of the windows you do have, because windows lose much more heat than walls or roofs.

Figure 17-1. Finnish fireplace, with pizza oven. Baldwin house. Built by Doug Wood, Thetford, Vermont. Tiles by Ikuzi Teraki, Washington, Vermont.

Figure 17-2 summarizes the routine things you can do to minimize heat loss. These simple efforts require no special equipment or supplies. These measures include:

- 2 x 6 wall framing
- 12-inch-thick rafters
- Insulated foundation
- South exposure
- Modest size, simple shape to reduce area
- Small windows to the north
- Earth sheltering
- Wind protection
- Window overhang
- Dark basement floor
- No drafts at key points

By doing these things, you can build a house that will retain heat efficiently.

A 24 x 30-foot three-bedroom house as shown, even here in Vermont, would need only about 26,000 Btus per hour on the coldest day. That's about $420 worth of firewood, $475 worth of oil, or $800 worth of propane gas, at current prices, each year. Not that bad.

Yet some simple, inexpensive choices would make significant improvement in heating efficiency. For example, low E glass in the windows (most new windows have Low E glass) and more efficient heating units would bring the heating cost with wood down to about $250, $370 for oil, or $700 for gas.

It is possible to lower heat requirements even further. Figure 17-3 summarizes what some of these measures are. But at some point there is a diminishing return on your investment as you get a more and more efficient house. Learning the heat loss math allows you objectively to assess these choices.

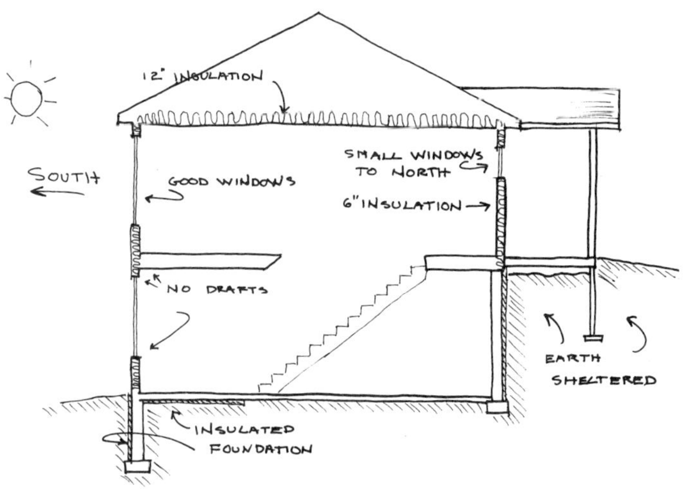

Figure 17-2. Basic measures for energy efficiency.

PROVIDING THE HEAT YOU NEED

Passive Solar

In a passive solar system, the house itself is designed to collect and retain solar energy, without any special equipment to move or store the heat. Although any well insulated house with south glass might be considered passive solar, a true passive system should have a method of storing the heat gained when the sun is out. The house in figure 18-2 has the rudiments of the simplest version of this—a masonry floor, in this case an insulated basement slab. Any dark, dense surface that receives direct sun can function this way. If this floor were painted a dark color, or tiled, or

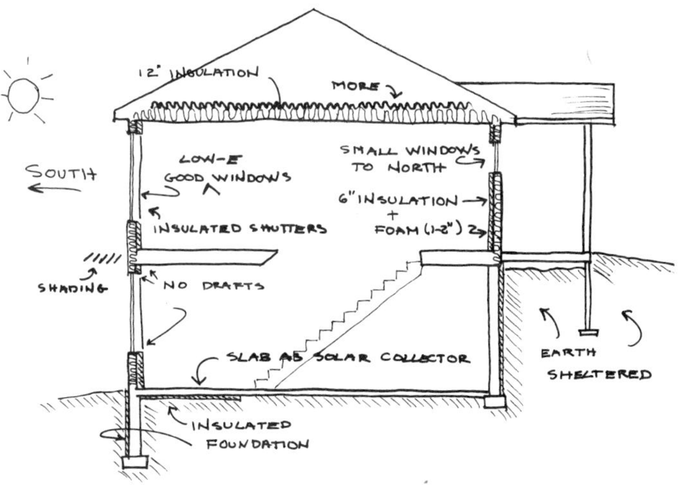

Figure 17-3. Additional measures to improve heat performance: more insulation in walls and roof; 2.5 air changes; air-to-air heat exchanger, and insulated shutters.

perhaps darkened with pigments mixed into the mortar and grout, it would make an excellent collector where daytime solar heat could be stored, and released back into the house when the sun is down.

Figure 17-4 shows other ways this has been accomplished.

Figure 17-4.
Storing solar heat.

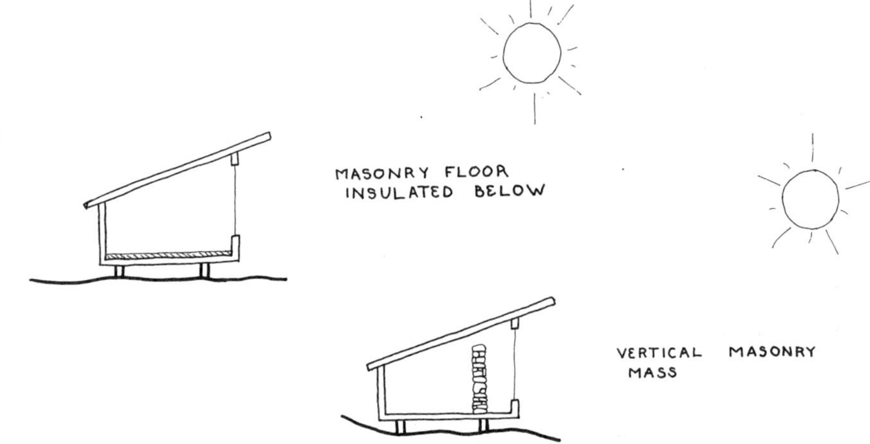

Active Solar

Figure 17-5 shows the basic idea of active solar house design. A collector—often mounted on the roof—has an absorber, a black surface. The heat collected by the collector is transferred to a fluid, typically water, and circulated through some massive storage medium, such as a tank or water or a rock bed. Another loop circulates this heat to the house when needed. The point is to capture the diffuse heat and concentrate it in some thermal mass for later use. Photovoltaic systems are also active systems. The solar energy is converted to electricity, and stored in batteries until needed. For an active system to work, you need enough sun, in comparison to your heating or electricity needs. In cold climates, solar panels are often used to heat water, but usually a heating system is needed for space heating. In milder climates, people can and do heat mostly with solar.

Space Heating Systems

In a space heating system, one or more small heaters heat the house. If more than one is used, the fuel, rather than the heat, is transported throughout the house. A small house may only need a single heater. A larger house may have two, three, or more. Usually a heater will not be needed in every room.

These systems are cheaper than central heat because they are simple. There is little or no ductwork or plumbing, perhaps only a gas line bringing gas to the heater, and some sort of vent.

Since the sources are widely separated, the heat is not that uniform; there will be hot and cold places. Only certain house layouts heat efficiently with space heat. A reasonably compact two-story house works well because the heat rises automatically to the second floor. A small cabin or house will heat fairly well because no part of it is far from the stove. But heat from a space heater will not travel far horizontally and will be blocked by closed doors. A large, horizontally spread out one-story house may need many heaters to maintain reasonably even temperatures. In such cases a central system may be more economical.

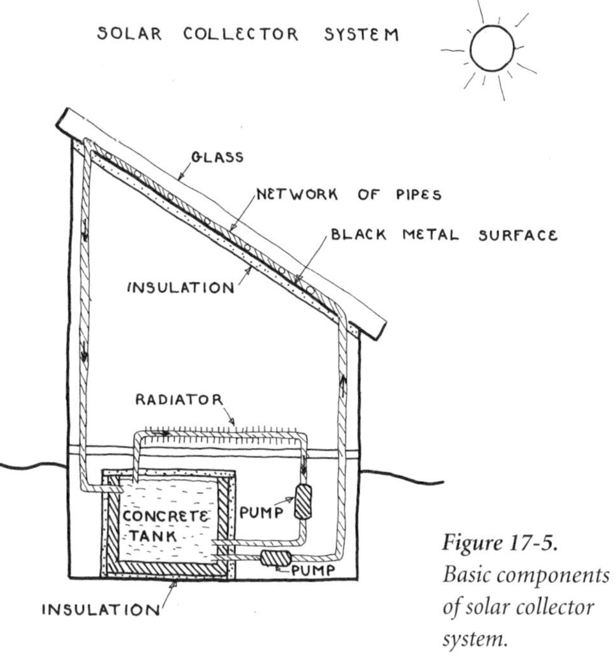

Figure 17-5. Basic components of solar collector system.

Wood stoves: Perhaps the most popular space heater is the wood stove, which comes in a great many types. We use a simple box stove. These are wonderful kitchen stoves that can heat and cook at once. Other stoves have glass in the doors and work like fireplaces. The newer stoves are much more efficient than older styles, and pollute less.

A popular system today is the Russian-style masonry stove (figure 17-1). A very hot, relatively brief fire is built one or twice a day, using small pieces of wood that burn quickly. A complex flue system works as a heat exchanger to transfer this heat to the stove's masonry mass, which releases the heat gently throughout the day.

Whereas a central heating system does its work quietly and automatically, a wood heat system requires more attention. The first thing you do each day is stoke the fire, and the last thing you do at night is prepare the fire. Over the course of the year, you handle the wood several times, and clean your chimney when necessary. You have to think about where or when to stack the wood, what sort of fire to build on a given day, and anticipate the weather.

All this activity can be seen as rituals that ground us. Or it can simply be viewed as chores to avoid.

A woodstove has an important effect on your design: The woodstove uses a lot of space. In addition to the stove and its chimney, there is usually a required "clearance to combustibles" of up to three feet around the stove where you can't put anything, although by adding heat shields this clearance can sometimes be reduced. When designing for wood heat, you have to draw your plan for a particular stove, because safety specifications vary.

Direct-vent gas space heaters: Direct-vent gas heaters are among the least expensive heating systems, and they work well. A 20,000 Btu unit might cost $300, and a 40,000 Btu unit—capable of heating a small house—might cost about $500, plus installation. The most efficient and quietest models cost almost double that, but they are still a cheap way to heat a house.

These units could be fired with natural gas, but in rural areas they use propane (bottled gas) stored in a tank outside. Direct venting means that instead of a chimney metal gas flue going up to and through the roof, the heating unit vents directly out through the wall of the house. The inner flue takes products of combustion out, and an outer pipe brings combustion air in. This is the air needed to feed the fire, which with other types of systems is normally drawn from inside the house.

Direct-vent units save the considerable cost of the chimney or insulated flue. Bringing the combustion air directly from outside is safer and more efficient, particularly in a tight modern house.

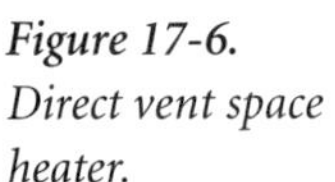

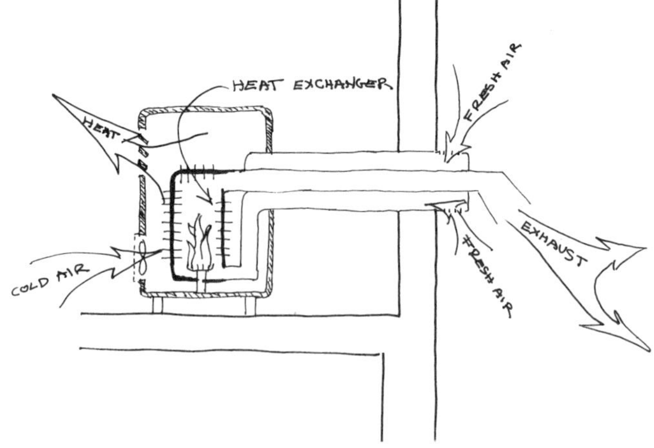

Figure 17-6. Direct vent space heater.

Kerosene direct-vent space heaters: The nicest type of space heaters are the new direct-vent kerosene heaters. The ones I have seen are imported from Japan. They are quiet, very efficient, and economical to operate because kerosene, which is a type of heating oil, is considerably cheaper per Btu output than bottled gas. Where gas units have to be located on an outside wall (and separated from opening windows that might allow the flue gases to re-enter the house), the kerosene heaters can be vented through relatively long tubes, which allow more flexibility in where the heaters can be located. They cost $1,000 to $1,500 depending on output.

Backup systems: My favorite space heating system combines wood stoves with one or two gas or kerosene space heaters. The wood stoves provide the bulk of the heat, but a gas heater or two keeps the house somewhat warm when no one is home, or when the woodstove dies down in the early morning. Without the backup, you are chained to the house: No winter trips to the islands.

Central Heating Systems

In a central heating system the heat is produced in one place and distributed throughout the house. In a "hydronic" or forced hot water system, water is heated in a boiler and carried by pipes to radiators or baseboards in each room. In a "forced air" system, air is heated by a furnace, and transported to grates in each room via ducts buried in the framing of the house.

A central heating system has two advantages. First, the heat can be uniform because it is so widely distributed. Second, the temperature is controlled automatically by thermostats. The main disadvantage is the higher installation cost. A hot water system for a small house will cost about $5,000 to $8,000 installed, and forced air perhaps $4,000 to $6,000.

Both hot air and hot water have their advocates. A hot air system is less expensive and can be combined easily with central air conditioning. A hot water system is quieter and less drafty, can easily be zoned, and lasts longer than hot air equipment. Air quality problems are sometimes associated with hot air systems, because the ducts create an inaccessible home for molds and dust, and provide a route to distribute them throughout the house.

Either type of system can be fired with wood, oil, gas, and even electricity. In most areas, electricity is the most expensive fuel by far. Gas is cleaner, and modern units can be direct-vented like gas space heaters, drawing their combustion air from outside. Oil is significantly cheaper to use. Though there are accessories called power vents that vent an oil furnace through the wall, they can be messy, and the combustion make-up air will usually come from the room. I would recommend a chimney rather than the power vent system until the gear improves.

Furnaces and boilers are much more compact than they were twenty years ago. For example, a typical boiler might be about the size of a small file cabinet, and a compact boiler as small as a back pack. It can hang on the wall in a closet. Air systems are more bulky because of the ductwork.

There are also multifuel units that can be operated with more than one fuel. Usually this is a wood furnace or boiler that automatically switches to gas or oil when the wood fire burns down. These units are expensive, and may not be as efficient as single-fuel heaters.

Distribution Systems for Central Heat

For forced air heating, the standard system for distributing the heat has been sheetmetal ductwork concealed in the house frame. Today ductwork is also made of insulated panels or tubes, which are quieter and conserve the heat better. One network of ducts carries the heat to the outside walls of the house, and another returns the cold air to the furnace to be reheated.

There are several popular distribution systems for hot water. The most familiar is baseboard radiation, which consists simply of copper pipe with hundreds of little fins attached. The fins dissipate the heat of the hot water flowing through the pipes. The fin tube is housed in a metal cover, the baseboard. The house can be subdivided into zones, each of which has its own loop of fin-tube piping, and its own circulating pump and thermostat. This is probably the most economical type of hydronic radiation.

The main disadvantage of the baseboard is that it eats up three or four inches of floor space at the perimeter of the room, usually along all or most outside walls. It makes it hard to place furniture up against the wall.

I prefer European style radiators. Unlike the old cast iron radiators, these are quite thin, and come in a great variety of shapes and colors. Also they can be located wherever they complement the design. They also come in special configurations, such as a bathroom unit with towel bars for warming towels, or a model to put by your front door, which has nice hooks to hang and warm your coat. These are expensive, unfortunately.

Another popular option today is to use the floor of the house, or of part of it, as the radiator, by embedding special coils of flexible heating tube in or under the floor. It's called "radiant floor heating." This works best with insulated concrete slabs or other masonry floors, but it is also done with wood floors that have been engineered to take the temperature changes. This is an old system, and many early versions of it failed with disastrous results, but new types of piping seem to be more reliable. This is also more expensive than baseboard, but some of the expense can be abated if you run the tubing yourself.

Complex Systems

So far I have surveyed the most common heating systems. Essentially, heating systems are just ways to transfer heat from one medium to another. There are infinite combinations. There are geothermal systems that extract heat from the earth. There are types of solar systems I didn't describe. Plus, there are many ways of combining systems. There are coal stoves, which have some advantages, pellet stoves, which use processed waste wood as fuel, and many other stove-type heaters.

HOW TO DECIDE ON A HEATING SYSTEM

Heat and fire carry a lot of meaning for people. How you heat is an expression of what is meaningful to you. I like a fire to sit around; I think I get to a little deeper level of relaxing and communicating if I can sit and look at the flames. I'm willing to do some work that might not be strictly necessary to get this feeling. I have friends with old kitchen stoves they wouldn't part with because the stoves are so evocative and cozy. The idea of a fire that cooks your food, warms the plates, and heats the house all at once is very appealing.

I have also had customers who have hired me to remove their fireplaces so that a central heating plant can keep them at the perfect temperature silently and automatically.

But choosing a heating system is not a completely arbitrary decision. First of all, look at cost. A system may have great appeal, but what does it cost? How much work will it take to build it initially and keep it running day by day? Is cutting firewood recreation to you, as well as work? Try to assign some quantities to both dollar and labor costs.

Choose a system that fits the features of your design. Space heaters function well in a smaller, compact house that isn't too spread out laterally. If you are heating with wood, you will need a source of wood, possibly some gear to split it if you can't buy it split. You will need a woodshed to store it in, and a convenient way to handle it when you get it, stack it, and move it from the shed to the stove or furnace. You'll need to plan a way to clean your chimney safely and easily. The wood heating will become a significant feature of your design.

A heated masonry floor fits some designs well. It works even better if it functions as a heat sink, warmed by lots of south glass.

Forced air works well if you have a convenient route from the central heater to the outside walls where the ducts can be located, and if the duct runs aren't too long.

Hydronic systems can generally work well anywhere, because of the variety of equipment available, but they are expensive.

In short, look for some synergy between the design of your house and the way you heat it.

18

HEAT LOSS CALCULATIONS & SUN PATHS

IN A SIMPLIFIED FORM, THIS CHAPTER SHOWS (1) HOW TO figure the hourly heat loss of your building design, which tells how much heating capacity you will need, and (2) how to predict approximately your annual fuel consumption and heating cost.

COMPUTING HEAT LOSS

All building materials have some insulative value, but some are much more effective than others. The effectiveness of a given material or combination of materials is indicated by its thermal resistance, or R rating. Here are a few examples:

ITEM	r(R/INCH)	R
Fiberglass insulation	3.3	
3-3½ inches		11.00
5-6 inches		19.00
Polyisocyanurate foam, foil faced		
1 inch		7.10
2 inches		14.20
Hardwood	0.91	
Softwood	1.25	
Drywall, ½-inch		0.45
Window, single glass		0.90
Insulating glass,		
3/16 inch of space		1.44

For most building materials, the R value is proportional to thickness. Thus, 2 inches of foam insulation has twice the R value of 1 inch. In the table (page 228) from which the examples above are taken, you will therefore see some materials rated for convenience in the form of R per inch of thickness. In this form, the value is called resistivity and is indicated by the lower case r. If you know that hardwood, for example, has an r of .91, you can easily multiply this by the thickness of hardwood you are using to find its R. Two inches of hardwood would be R 1.82, and ¾ inch of hardwood would be R .68.

There are a few cases where the R is not proportional to thickness. For example, ⅛-inch glass has an R value only slightly better than 1/16-inch glass, even though it is twice as thick. This is because window glass is a poor insulator that resists heat loss mostly because it seals off the window, creating an insulating air film on the interior surface. Its ability to seal off the window is not improved by making it thicker. Airspaces within a wall are another example of the R value not being directly proportional to thickness. A wide airspace will be somewhat more effective as an insulator than a narrow one, but not in direct proportion to the thickness.

To evaluate your design, first find the total R values for the types of floor, wall, roof, and window construction you are considering. Do this by adding up the R values of each element of the wall, roof, or floor. For example, a standard wall section might consist of ¾-inch softwood siding on the outside, a sheathing of ½-inch plywood, 6 inches of fiberglass between the studs, and ½ inch of drywall.

Siding	.79
Plywood	.62
Fiberglass	19.00
Sheetrock	.45
Total R	20.86

Typical Standards

At one time, buildings in cold climates were considered well insulated if the total R for the roof was 24 and 13 for the walls. These values were based on using 6 inches of fiberglass in the roof and floor and 3½ inches in the 2 x 4 walls.

I would say modern practice starts at 6 inches in the walls and 9 or 10 inches in the ceiling, or about R 21 and R 30 or 35, respectively. If the design allows it, it makes sense to increase the roof insulation to 12 inches, or about R 40.

In superinsulated buildings, still higher levels can be attained—perhaps R 30 in the walls and R 50 in the roof—by adding yet more insulation, often in the form of rigid foams.

U values

Total R values tell you whether individual parts of your design are well insulated. To get a picture of the total heat loss of a design, the R values can be converted into another unit, U, which is the coefficient of thermal transmission. R tells you how well something insulates. U is the opposite side of the coin and tells you how much heat gets through, which in turn tells you how much fuel you will need to replace the heat lost. Mathematically, U is the reciprocal of R:

$$1/R = U$$
$$1/U = R$$

U is expressed in a very specific unit: Btu/hour per square foot/degree difference.

Btu stands for British thermal unit, the unit of heat used in this kind of engineering. Square foot means square foot of building surface. Degree differences refers to the difference in temperature inside and out, in degrees Fahrenheit. If a certain wall has an R value of 20, it will have a U value of 1/20 or .05. This means that .05 Btu of heat will pass through each square foot of the wall every hour, for each degree of temperature difference.

If it is 20 degrees F outside and 70 degrees F inside, that is a temperature difference of 50 degrees F. Thus, our U value of .05 times 50 degrees F temperature difference means that 2.5 Btus will pass through each square foot of the wall. If the whole wall were 10 feet square (100 square feet) the heat loss through it would be 100 x 2.5, or 250 Btu/hour. To maintain the inside temperature of 70 degrees F, you would need to replace these 250 Btu/hour with your heating system output, the sunlight, and other marginal sources of heat such as cooking and body heat.

In the case of basement walls, the degree difference will be less, because it is based not on outdoor temperature but on the temperature of the earth. Below the frost line, earth temperature will be near 50 to 55 degrees F. Above the frost line in cold climates, it could be at or below freezing temperature. Estimate an average earth temperature to find degree difference for basement walls.

HOW TO FIND THE HEAT LOSS FOR A SPECIFIC DESIGN

First, find the area in square feet for each part or section of the exterior surface of the building.

SECTION	AREA (IN SQ. FEET)
Floor	800
Roof	1,000
Walls, excluding windows and doors	1,200
Windows	300
Doors	50

Next, find the total R value for each section. Then take 1/R to get the U value for that section. Make a chart as follows:

Heat loss per hour

SECTION	AREA (IN SQ. FEET)	R	U	U × AREA DEGREE DIFF.		U × AREA × DEGREE DIFF.
Floor	800	17	0.059	47.06	50.0	2,352.94
Roof	1,000	22	0.045	45.45	50.0	2,272.73
Walls	1,200	18	0.055	66.67	50.0	3,333.33
Windows	300	1.44	0.694	208.33	50.0	10,416.67
Doors	50	1.75	0.571	28.57	50.0	1,428.57
				Total Btu/hr.		19,804.24

INFILTRATION

Number of air changes		Volume in cubic feet			
	×		× .018 × Degree Difference = Infiltration		
1	×	8,000 cubic feet	× .018 ×	50 degrees F	= 7,200 Btu/hour
				Total Heat Load	27,004.24

Now, multiply the U value of each section by the area of the section. Then multiply the resulting figure by the degree difference. The degree difference chosen will be the desired difference between the inside and outside temperatures on the coldest day. Choose a figure that represents the maximum performance you will expect from your heating system. The final product is the Btu/hour needed to replace the heat lost through that section of the building exterior, assuming the specified temperature difference.

Finally, total the Btu figures for each section to find the conductive heat loss for the entire building. In the example, the total is 19,804 Btu/hour.

Infiltration

This total conductive heat loss does not take into account the heat loss due to infiltration: leaks, drafts, and open doors and windows. The infiltration can amount to twenty-five percent or more of the heat loss.

That figure is based on the fact that it takes .018 Btus to raise the temperature of one cubic foot of air one degree F. If you can approximate the cubic feet of air that will infiltrate each hour, you can calculate the added heat needed.

Modern “blower door tests” can be used to measure the infiltration for an existing house.

The air in many older homes is entirely replaced one to one and a half times per hour. The house I live in probably has around one air change per hour. It's insulated and has plywood on the walls, but there are probably voids in the insulation, and many of the windows are old and a bit drafty. Not long ago, this kind of house would have been considered quite tight and efficient.

By using modern windows, carefully weatherstripped, and by taking pains to install a vapor barrier, it is not difficult to get down to .4 or .5 air changes per hour. This is the level I shoot for in my designs.

It is possible to get even lower by using even better windows, providing airlocks (double doors) at entries, and other measures.

To illustrate the calculation, suppose the example house had a volume of 8,000 cubic feet, with approximately one air change per hour. Each of these cubic feet of air will require .018 Btu, per degree of temperature difference.

If the degree difference was 50 degrees F, the heat loss per hour due to infiltration would be 7,200 Btus/hour. This calculation is shown above in the chart "Heat Loss per Hour."

Add the calculated loss from infiltration amount to your previous total. In the example, at one air change, we'll be adding 7,200 Btus/hour to get to about 27,000 Btu/hour.

You now have the total heat loss per hour for a given design under the coldest conditions. This tells you what the maximum output of your total heating system should be. Most heating systems are rated in Btu output per hour, so you can find the right-size unit.

This Btu total we come up with is the output needed for your total heating system. If you are using a gas or oil system in combination with a wood stove or fireplace, the central heat by itself will not need an output equal to the maximum Btu/hour needed. Furnaces and boilers work most efficiently operating somewhere near their maximum output, and if you pick a furnace to match the most extreme weather, it may not have to work to that output under normal conditions. It is better to assume that under extreme cold the wood stove or fireplace will be supplying some fraction of the required heat, and with that in mind to choose a smaller furnace. The more you emphasize wood heat, the more you can lower the capacity of your boiler or space heater. However, it is a good idea to have a capacity big enough to keep the house reasonably warm (perhaps 40 to 50 degrees F) when you are away and the wood fires are out, at least in those parts of the house where pipes may freeze.

All the calculations assume your house is designed to take maximum advantage of its heating system. If you have very high ceilings and have provided no way to prevent the warmer air from collecting at the roof peak, you will need much more heat than the calculated amount of fuel. With a wood heat system, you might need extra fuel to heat remote spaces far from the stove, unless these rooms can be used for activities that can be done comfortably at a lower temperature.

YEARLY HEATING COSTS

The next step is to extend the calculations to find out what your approximate yearly heat needs will be. You do this by comparing the building heat loss with local weather data to find the Btus needed for the heating season. Then you can find the cost with different fuels.

You need to know the number of "degree days" in the area where you live. Degree days is a measure of the average difference between the outside temperature and 65 degrees F on a given day. For example, if on January 27 the average temperature is 35 degrees F, that is thirty degree days for that day, because the difference between 65 and 35 degrees F is 30 degrees F. Weather bureaus keep records of the average temperatures from year to year and use this information to figure an average number of degree days for the entire heating season.

Figure 18-1 shows the average number of degree days in different regions. The more degree days there are, the more severe the weather.

Here is the procedure for finding your annual fuel consumption:

1. Take the total heat loss figure and redivide it by the degree difference you originally used to figure it. This takes us back one step, to convert your heat loss total into a form you can use for this computation. In the above example, we divide 27,000 Btus/hour by 50 degrees F to get 540, because the figure 27,000 assumed a degree difference of 50 degrees F. This number (540) represents the total Btu loss per hour, per degree of temperature difference.
2. Then multiply this figure by 24 (representing hours). In the example, 539 × 24 = 12,960. This is the Btu loss per day, per degree of temperature difference.
3. If you multiplied this second result times the degree days for any given day, you would get the total heat loss in Btus for that day. For a thirty degree-day day, our heat loss would be 12,960 Btus × 30 = 388,800 Btus. Similarly, if you multiply the figure times the total degree days in your region you get the Btus needed for the entire heating season. For example, if the figures we have been using above were for a house in southern New Hampshire, we would multiply 12,960 Btu × 8,000 degree days to get the 103,680,000 Btus needed for the whole heating season.

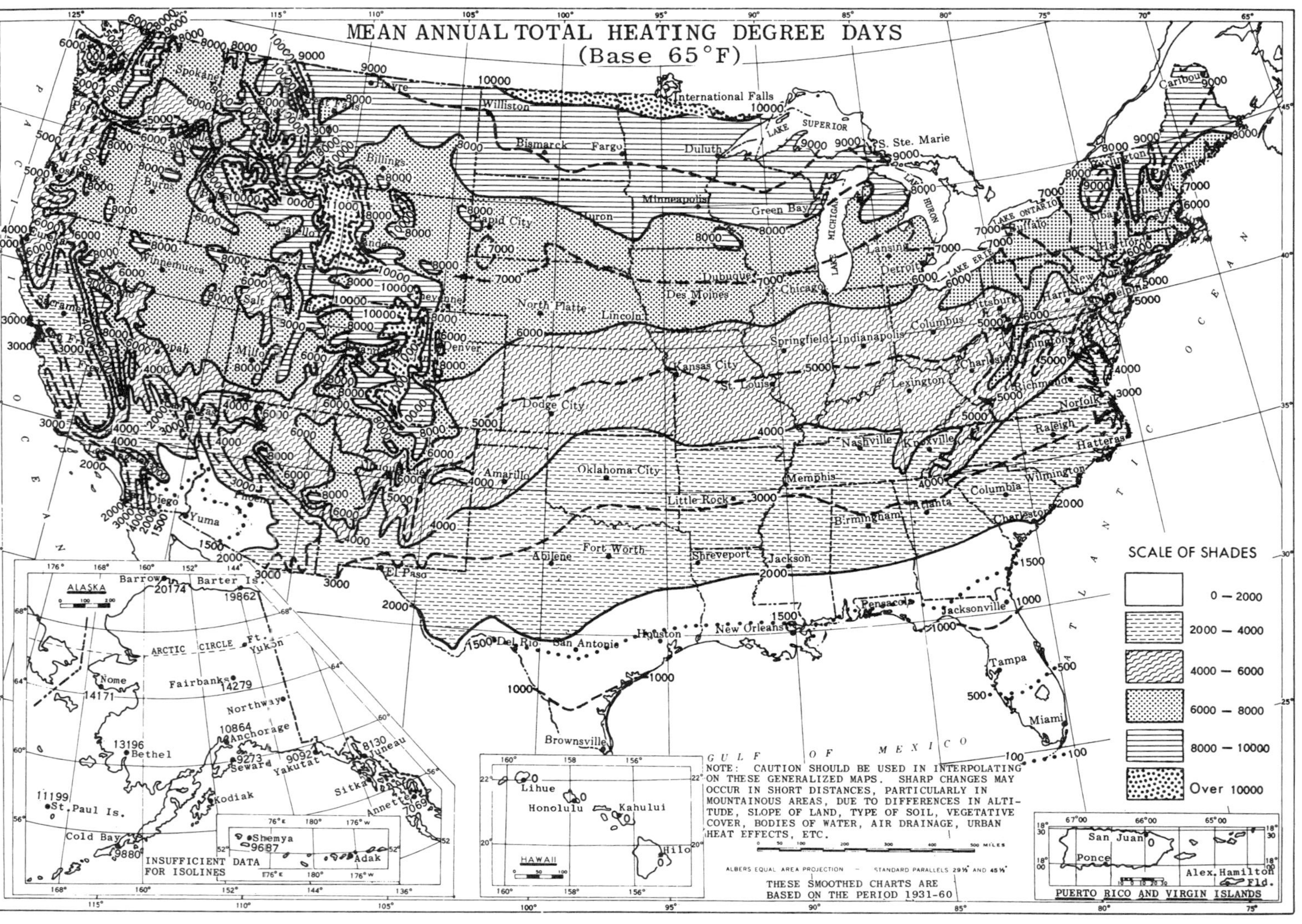

Figure 18-1. Degree day map.

4. Before figuring the approximate fuel costs, subtract from your Btu total the free solar heat you get through the south windows of the house. In most parts of the continental United States, a south window will contribute about 100,000 Btus per heating season, per square foot of unshaded glass area. If your climate has clear, hot weather, the figure might be more like 120,000, and in more overcast climates, such as Portland, Oregon, around 80,000 Btus per square foot of unshaded south window.

When you figure your south glass area, subtract for any shading due to trees, other buildings, hills, curtains, overhang, or shades. This rule of thumb assumes that your house is designed with the sun in mind. If the heat from south windows really only gets to one or two rooms, or cannot circulate to the colder parts of the house, the usable heat gain will be proportionally less. In the example I have been using, if the south glass area were about 150 square feet, solar heat gain through south windows would contribute about 15,000,000 Btus per heating season:

$$150 \times 100{,}000 = 15{,}000{,}000 \text{ Btus/year}$$

This represents about fifteen percent of the total heat needed. This leaves 89,000,000 Btus to be supplied by the heating system. The percentage contribution from the sun may be more or less depending upon your design.

Figuring Heating Costs

5. To find the cost of the remaining Btus needed for any fuel, compare the Btu output of the fuel with its cost in your locale. Make sure the price figures you obtain are expressed in the same unit as are the outputs for the fuel. With all fuels except electricity, a lot of the heat goes up the chimney, so the outputs for each fuel must be adjusted for the degree of efficiency of the burner used. The table below gives the figures for the most common fuels.

   To find the cost of a particular fuel for the heating system, divide the adjusted Btu output per unit of the fuel into the total Btus needed (Step 4 above). Then multiply the result times the price per unit in your area. In our example, 89,000,000 Btus are needed per heating season. Elm wood puts out about 24,000,000 Btu per cord (a cord is a pile 4 x 4 x 8 feet), but since as much as half of this heat goes up the chimney, the usable product is about 12,000,000 Btu. The annual need divided by the output per cord is

$$\frac{89{,}000{,}000 \text{ Btu}}{12{,}000{,}000 \text{ output per cord}} = 7.4 \text{ cords needed per year}$$

At $100 a cord, the annual heating bill would be $740.

R values for common building materials

MATERIAL	r(R/1 INCH)	R
Fiberglass	3.3	
3½ inches		11.00
6 inches		19.00
Foams		
Polyisocyanurate/foil	7.1	
Extruded Polystyrene	5-5.4	
White "Beadboard"	4.2	
Plywood		
½ inches		0.62
¾ inches		0.93
Drywall, ½ inches		0.45
Siding		
wood shingles		0.87
bevel (clapboards)		0.79
1-inch boards		1.00
Wood		
softwood	1.25	
hardwood	0.91	
Roofing		
Asphalt shingles		0.44
Wood shingles		0.94
Roll roofing		0.15
Floorings		
Carpet and fibrous pad		2.00
Carpet and rubber pad		1.23
Vinyl tile, etc.		0.05
Masonry materials		
Concrete Block, 4 inches		0.71
Concrete Block, 8 inches		1.11
Face brick		0.44
Windows (varies with unit)		
Single glass		0.90
Insulating Glass		1.4-2.10
Low E argon-filled		2.80
Triple glass, Low E		4-6.00

Sources of data: I have compiled these R values from various sources, including industry literature. R values differ from source to source. Usually these discrepancies are small. Standard sources for this kind of data include the *ASHRAE Handbook of Fundamentals* (New York: American Society of Heating, Refrigerating, and Air conditioning Engineers, Inc., 1977), and *Architectural Graphic Standards*, 9th edition, by Charles Ramsey (New York: Wiley, 1994).

Efficiency of fuels

FUEL	BTU PER UNIT ADJUSTED FOR BURNER EFFICIENCY	ASSUMED EFFICIENCY OF BURNER	BTU/ UNIT
Oil	140,000/gal.	75–85%	105,000/gal.
Natural gas	1,000/cu.ft.	80–90%	800/cu.ft.
Manufactured gas	525/cu.ft.	80–90%	420/cu.ft.
LP gas	92,000/gal. or 21,700/lb.	80%	73,600/gal. or 17,360/lb.
Electricity	3,412/kwh	100%	3,412/kwh
Hickory, oak, sugar maple	27–30 million/cord	50–80%	13.5–15 million/cord
Birch, ash, elm, red maple, paper birch, cherry, Douglas fir	21–27 million/cord	50–80%	10.5–13.5 million/cord
Eastern white pine	15.8 million/cord	50–80%	8 million/cord

R values for air spaces

The insulating value of an air space depends on its thickness, the reflectivity of the surfaces creating the air space, and on temperature conditions. The indicated R values reflect average conditions and are accurate plus or minus about twenty percent. For more precise figures see *ASHRAE Handbook of Fundamentals,* from which this table was computed by the author.

	Foil Both Sides of Air Space		Foil One Side of Air Space		No Reflective Surface	
	¾" AIR SPACE	4" AIR SPACE	¾" AIR SPACE	4" AIR SPACE	¾" AIR SPACE	4" AIR SPACE
Flat roofs	1.8	2.2	1.6	1.9	.9	1.0
Floors	3.4	8.1	2.6	4.6	1.2	1.4
45° Sloping roofs	2.2	2.5	1.9	2.1	1.0	1.0
Walls	3.0	2.8	2.4	2.3	1.1	1.1

SUN PATHS

The location of the earth in relation to the sun, which varies with the season, creates a shifting relationship between a speciWc building site and the "path" of the sun across the sky. Calculating these sun paths for your site will allow you to consider how heat from the sun will impact your house as "solar gain," which is usable heat for space- and water-heating. The quantity of available sunlight also profoundly aVects the viability of using photovoltaics at your site.

In summer, the sun rises in the northeast, follows a high arc across the southern sky, and sets in the northwest. In winter, it rises in the southeast, follows a low arc across the southern sky, and sets in the southwest. The exact pattern depends on the latitude. In northerly areas, the arc will be lower in the sky. For example, in Canada (latitude 52 degrees north) the sun never gets above 62 degrees off the horizon, while in the Caribbean (latitude 24 degrees north) it reaches almost 90 degrees.

Figure 18-3. Where the sun rises and sets.

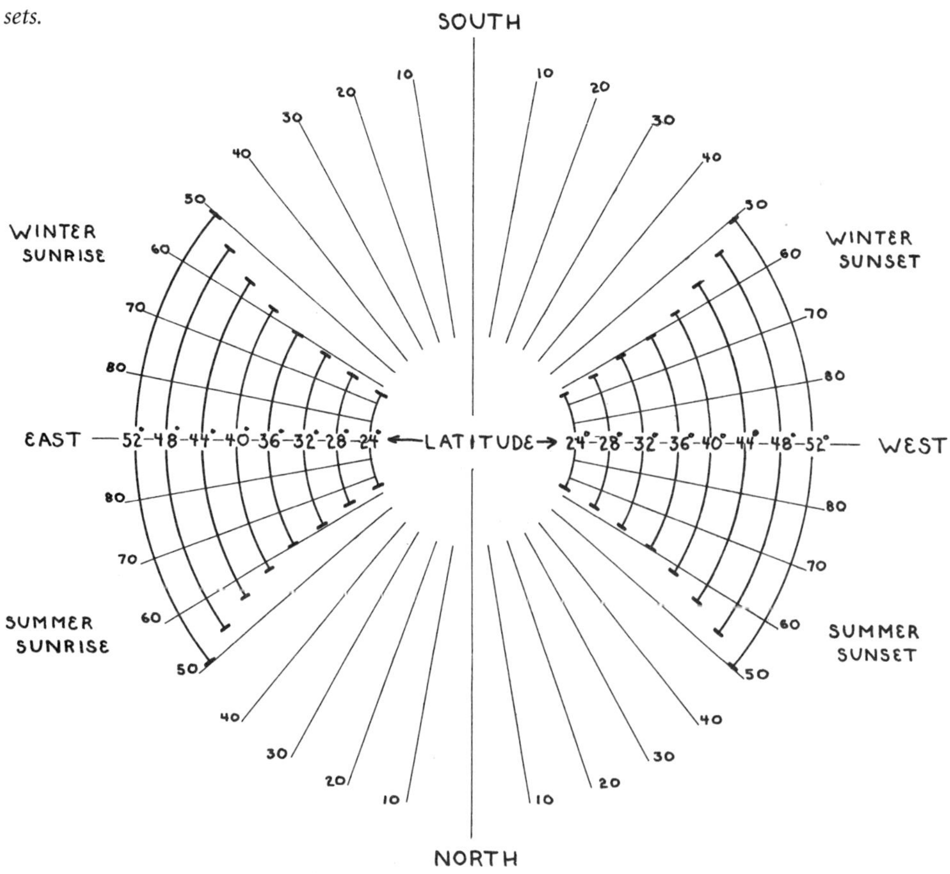

Similarly, in northerly areas, the sunrise will move farther into the north in summer and farther into the south in winter than in more southerly areas.

In any northern latitude, December 21 is the shortest day, and the sunrise and sunset will be in their most southerly locations for the year. The altitude of the sun above the southern horizon at noon will be the lowest for the year. On June 21, that altitude will be at its highest point and sunrise and sunset will be at their most northerly locations.

Halfway in between these dates (September 21 and March 21) are the only two days when the sun actually rises in the east and sets in the west. These days are the fall and spring equinoxes.

Figure 18-3 shows how the sunrise and sunset vary with the seasons at each latitude. Figure 18-4 shows how the sun's altitude varies. In both cases, the diagrams show the range of variation and how the sun's location varies gradually between the extremes.

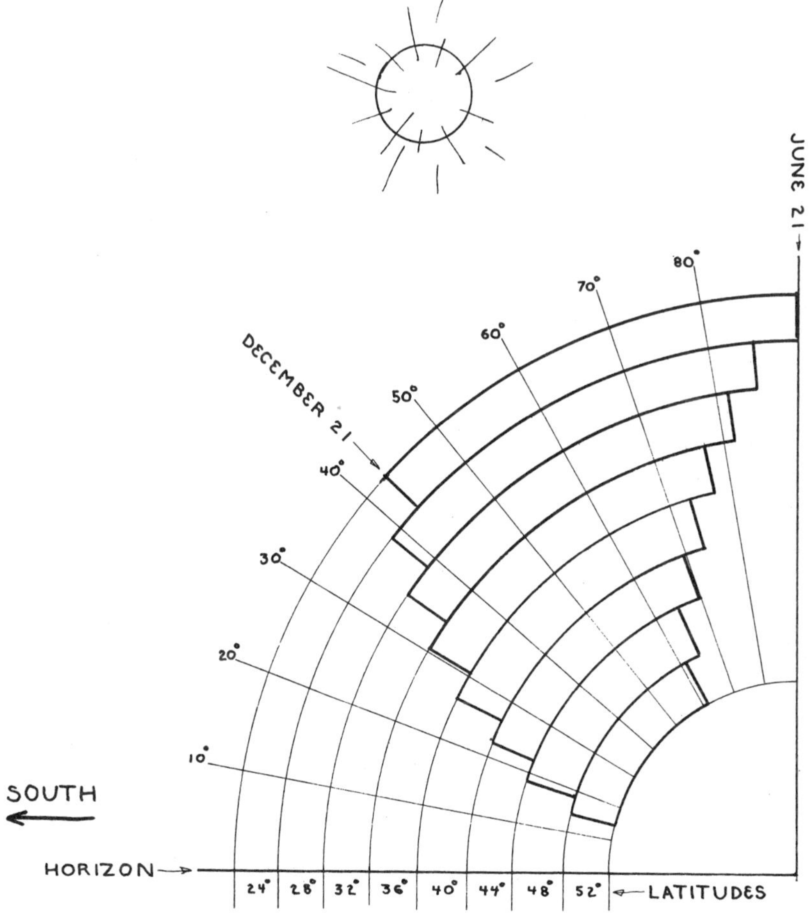

Figure 18-4. *Altitude of sun above horizon.*

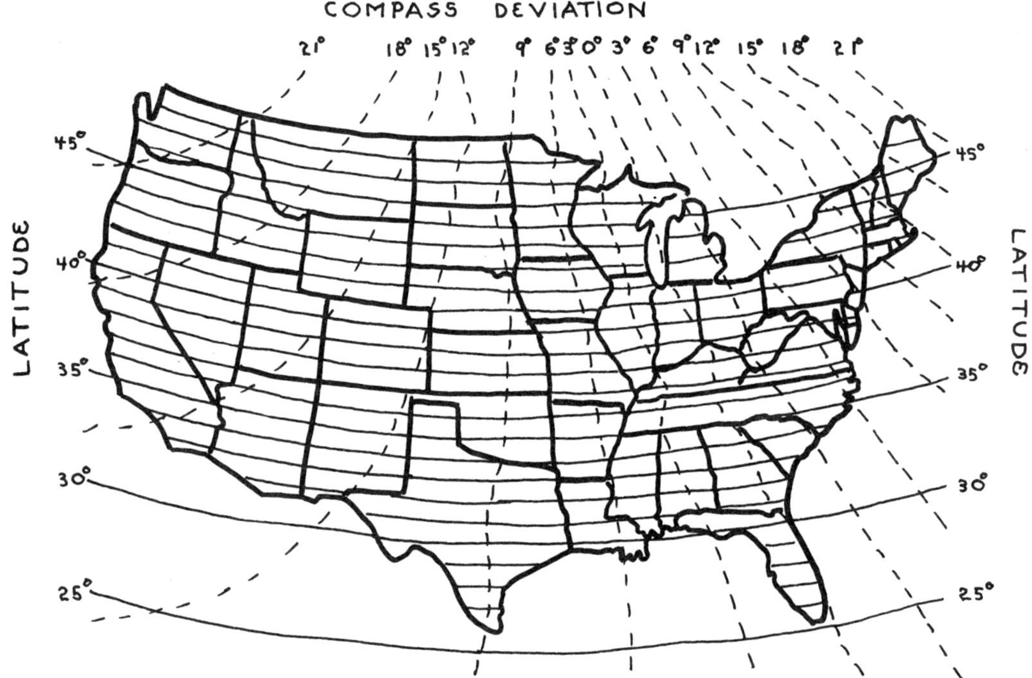

Figure 18-5. Isogonic chart of the United States.

Use figure 18-5 to find the latitude and the relationship of magnetic north to true north where you live. Magnetic north matches true north along a line that is south of the North Pole and runs through Indiana, across Appalachia, and follows the east coast of Florida. This is marked "0" on the map. East of this line, true north will be east of magnetic north by the number of degrees shown. West of the line, true north will be west by the amount shown.

Find the band on figures 18-3 and 18-4 for your latitude. If you live in central Wisconsin, latitude 44 degrees north, you are on the 3 degree line west of "O," so true north will be 3 degrees west of compass north. The sun on December 21 will rise at a point 56 degrees east of south and set a point 56 degrees west of south. On June 21, the sun will rise at a spot 56 degrees east of north, and set at 56 degrees west of north.

From figure 18-4 you can see that on December 21 the sun at midday will only rise about 23 degrees above the horizon. On June 21 the sun will reach its maximum height of about 70 degrees.

19

SUMMER COOLING & VENTILATION

In northern climates you can keep your house cool by opening one or two windows, but in most of the United States you will need to do more.

SITE CONDITIONS AND ORIENTATION

Since most houses face the sun, you will need shade trees. Shade trees are broad-leaved trees—hardwoods—that shed their leaves in winter when you need maximum sun.

Summer breezes are equally important. Often the prevailing wind direction varies seasonally. By consulting neighbors, and by making your own observations, you can find the wind patterns for your site. Build your house where hills, plants, and outbuildings will obstruct the winter but not the summer wind. Of course you may have to remove or add plants and trees to do this. You might want to live in your house for a season before removing too many large, irreplaceable trees.

The temperature of the wind is affected by what it passes over as it approaches the house. A street or driveway acts like a solar heater; it warms the adjacent air. A pond or vegetation will cool the air and therefore help cool your house.

The insulation that keeps the winter heat in keeps the summer heat out. Effective insulation, especially in the roof, is as important for summer cooling as for winter heating.

South Window Area

You need enough south windows to collect the sun's heat, but too much south-facing glass will make the inside intolerable in summer, and increase your winter heat loss too. Use common sense to determine the right amount of glass. Observe other houses in the area with similar orientation.

Ventilation

Leaks admit cold air. This is called infiltration. The leaks you do want in summer are called ventilation. Ventilation can consist of open doors and windows, but you can also add supplementary screened vents, particularly if much of your glazing is fixed. Whatever type of opening you use, the principles are the same (figure 20-1). The air should enter on the windward side where possible and exit on the opposite leeward side for cross-ventilation. The wind is the fan driving the air through. If the inlet is low and the outlet high, the house acts like a chimney. A natural draft is set up, so air will move even without the wind. Finally, the air flow will be more effective if the outlet size is larger than the inlet size. Chapter 28 discusses the construction of windows and vents.

Figure 19-1. Natural ventilation.

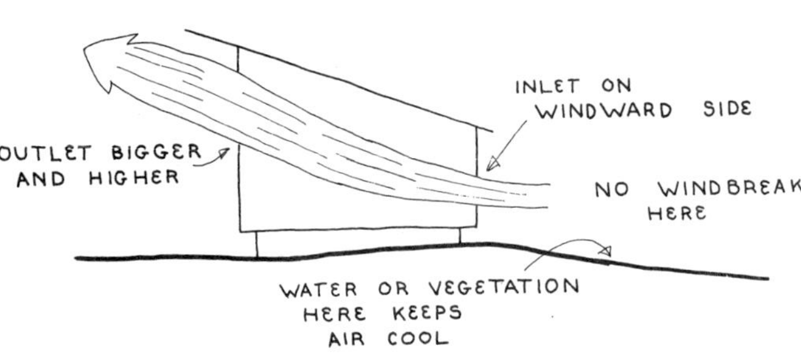

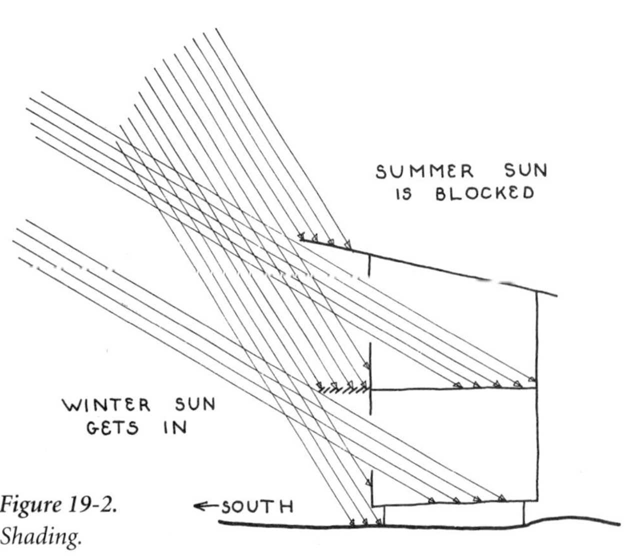

Figure 19-2. Shading.

Architectural Shading

In some climates a strategically located shade tree will be all the shading you need. In hotter climates you may need to build the shade into your house. Figure 19-2 shows shading provided by extending the overhang of the roof on the south side. The sun is higher in summer than in winter. The goal is to find an amount of projection that lets in the winter midday sun but blocks out the high midsummer sun. You can determine the amount of overhang you want by studying the sun angles in your latitude (see chapter 18). An overhang of this type could project as much as four feet.

When a roof does not provide convenient shading you can devise a projecting sunshade like the one shown in figure 19-2. If such a shade is spaced or louvered, the summer breeze will be more free to flow. Make the outer portion adjustable so you can vary the overhang with conditions.

MECHANICAL VENTILATION AND AIR QUALITY

The subject of mechanical ventilation involves several issues. It is important to manage moisture in the house, and avoid moisture damage to the house. It is important to provide fresh air. And it is important to have air that is as free as possible of toxic components.

I don't think there is agreement among builders about the best way to provide clean, fresh air for a house. There are new challenges because houses are so much tighter than they used to be. There are also new kinds of heating equipment that influence air quality, and air handling systems to control air flow. Some experts insist that costly ventilation systems are essential. So far, I'm not convinced. In this discussion, I will start with the simplest issues, and work toward the more complicated ones.

The Bathroom

A bathroom can generate huge amounts of moisture, particularly if a family takes a lot of showers. A few long showers every day can lead to mildew, staining, and worse problems in the bathroom or elsewhere in the house. Codes specify that bathrooms either have fans or opening windows, but often people don't bother to open the window. A bathroom exhaust fan is usually a good idea, unless people will open the window consistently.

The Kitchen

Many kitchen designers insist that a hood or other kitchen vent is essential. I always found this a perplexing position, since I had never lived in a house with one, and rarely experience the need for one. How could it be essential? However, after years of building kitchens, both with and without fans, I have come to believe that there are situations where fans make good sense.

Kitchen fans help control smells, grease, moisture from cooking, and possibly carbon monoxide emissions from the operation of gas stoves. Perhaps if you cook fish a lot, the first reason makes sense. Today most people use less grease in their cooking, but I have always recommended that people put in some sort of hood if they fry a lot. A hood also makes sense, perhaps, if there is often a pot of spaghetti, a pot of stock, or similar dishes on the burner generating lots of steam. My general conclusion is that a kitchen vent, located above the stove, makes sense if what you cook tends to throw off grease or excess moisture, or simply if you cook a great deal. But if you don't perceive that you need a fan, you may really not need one, in spite of what your kitchen designer says. There is more discussion of this in the kitchen and bathroom chapter (chapter 10).

The carbon monoxide issue is a little more complicated. Any gas flame creates carbon monoxide. In quantity, it's toxic. Normally, gas stoves don't have integral vents, and aren't required to have hoods. I think this is because most houses, until recently, had enough fresh air coming in that the stove emissions didn't present a problem. But in very tight houses, stove emissions can be problematic. If so, the simplest solution is a hood. One possibility for determining if you have a problem is to purchase a carbon monoxide detector.

The Laundry

Any dryer puts out both heat and moisture as it dries your clothes. I would be happy to have the heat in winter. Sometimes, also in winter, the moisture would help humidify my dry, woodheated house. But I don't want the heat in summer. When there is a lot of laundry use, or in the summer, the drier moisture contributes to excessive house moisture.

A dryer also puts out a lot of lint, which in my experience is impossible to completely filter out. A gas dryer, which is much cheaper to operate than an electric dryer, also throws off carbon monoxide from incomplete combustion.

On balance, it makes sense to vent your dryer to the outside.

Make-Up Air and Backdrafts

When in use, a dryer, kitchen fan, or bath fan sucks warm air out of your house. Most traditional wood stoves, space heaters, and furnaces draw their "combustion air" from the house. This air has to come from somewhere.

In older conventional building, this wasn't a problem. The house was so drafty that all the replacement air easily came in as infiltration, from ordinary drafts around doors and windows, and sometimes, right through the walls. The result was wholesome, but cold, air.

Of course, in a tight, energy-efficient, house, this doesn't work, particularly if the house has air locks at the doorways or other features that frustrate or defeat normal air changes.

The worst situation in tight houses is when make-up air is taken from one of the few openings you aren't able to plug: your heating system chimney. It is possible for the make-up air demand in the house to pull toxic combustion air back down your flues. This is called backdraft. With a clean-burning gas heater, you might be completely unaware of this happening. Another possible outcome would be that your fans and heaters, starving for air, couldn't efficiently do their job of forcing out toxic exhaust.

Direct-vent heaters: One of the most important recent developments in heating technologies is direct venting of heating appliances. Figure 17-6 shows the concept. First, whether it's a water heater, a boiler, or a free-standing space heater, the heater is made very efficient by recycling flue gas Btus. So flue gasses in these units aren't as hot as with less efficient units. Therefore, perhaps with the aid of a small fan, they can be expelled directly through the sidewall of the house. This approach is much cheaper than building a chimney out through the roof. Just as important, the combustion air is brought directly in through the same hole in the wall. An inner pipe is the vent for the exhaust, and an outer pipe provides fresh air directly to the flame of the heater. As result, the backdraft hazard is eliminated, and the total amount of suction inside the house is reduced.

This technology is essential in modern building. It is now standard in gas heaters of all kinds and in the new kerosene space heaters. Oil central heating plants often provide sidewall venting (referred to as power-vent systems), and some now provide a direct air inlet for combustion air. Some woodstoves have intake vents to the outside, and some homebuilders run a small air duct from outside to the front of their wood stoves, though I doubt these ducts really supply much of the air the woodstove uses. I can't think how the wood stove would know to take that air, when there is so much more air available from the room.

Other mitigating factors: Though direct venting is a major piece of the puzzle, there are many smaller measures that can reduce the suction effect, and save energy at the same time. Well-insulated houses use less heat, requiring less combustion air. More efficient heaters and appliances use less fuel and therefore less air. Low-flow shower heads use less water and may make less steam. Some cooking practices cut down on stove-top emissions. Using lids on pots or pressure cookers are examples. Drying clothes on a line or rack saves the fuel and eliminates one source of air outflow.

Supplying fresh air on purpose: Even if your heater were direct vent and your stove electric, and all your household practices were ideal, you wouldn't want to eliminate air changes. Our bodies are heating appliances, cranking out body heat Co_2 as we breathe. And our houses are home for airborne dusts, microorganisms, and other substances. If the house is sealed up, fresh air has to be provided.

The chances are that adequate replacement air comes from the suction created by kitchen or bath fans. The replacement air could also come by way of cracking a window in the bedrooms. It is also possible to add small wall vents that admit just the right amount of air. These are often located in bedrooms.

This air flow dynamic depends on the fans being turned on frequently. A bath fan wired with bathroom lights might be sufficient to guarantee this. Another approach is to hook the fan to a timer that runs the fan a preset amount based on calculating the amount of ventilation needed given the house size and the number of occupants (see page 239).

Another approach to assuring fresh air is to use a separate whole-house ventilator. These are systems that use a central fan to draw air through ducts from each major space. The problem with this is that exhaust fans take heat along with stale air. Another approach is the air-to-air heat exchanger. These devices basically use one fan to take warm stale air out, and another to bring fresh air into the house. Fins of some sort transfer some of the heat from the exhaust air to the fresh air, so as to reduce the net Btu loss.

What We Do: Half-Way Measures

If you took all these lines of discussion to a logical conclusion, you might, after worrying about it for a while, go to great lengths to achieve perfect solutions. You might use special sealants and gaskets and housewraps to build a very tight house, buy the most expensive and efficient equipment, and install a whole-house heat recovery ventilation system, with air quality monitoring of some sort. It might make sense for some people to do this.

We don't do that. It seems odd that anyone would go to extremes to build a superinsulated, tightly sealed house, then spend money to force air into it.

We build tight houses, but not supertight. Using ordinary plywood, no house-wrap except at plywood joints, and careful work, we can achieve around .4 or .5 air changes per hour. This is tight enough for good heating, but loose enough to admit some air naturally. Ventilation is also provided by opening windows, running bath fans, and sometimes installing quiet, exterior mounted kitchen fans or range hoods. We use direct-vent heaters for reasons of both efficiency and safety. We rely on the intelligence of the people who live in a house to be attentive about the air they breathe and about the products they bring into the house.

TOXIC INDOOR AIR

Houses, and the air inside, are of course made of chemicals, some organic, some inorganic, some non-toxic, some toxic, some that are toxic only in large quantities, and some toxic to some people more than others. Some substances become more toxic or irritating with increased exposure. I used to put up fiberglass insulation with very little itching; now I start scratching as soon as I place

Sizing Fans

VENTILATION FANS are sized to provide specific air flow rates in cubic feet per minute (cfm). The size of the fan needed for ventilation depends on the size of the house and the number of occupants. The ASHRAE Standard recommends a minimum of 0.35 air changes per hour (ACH), but not less than 15 cfm per occupant during the time the house is occupied. You should calculate minimum whole-house fan size based on both house size and number of occupants, then use the higher number.

Here's how to calculate 0.35 air changes per hour:

1. Multiply the exterior square footage of the house or apartment by the average ceiling height to calculate total volume. Then multiply by 0.85 to account for the wall and partition thicknesses.
2. Multiply the volume by 0.35.
3. Divide by 60 (minutes per hour) to get the required cfm.

So, for example, an 1,800-square-foot, three-bedroom house with 8-foot ceilings would need a ventilation rate of 71 cfm to ensure 0.35 ACH.

Now compare this with the other ASHRAE recommendation of 15 cfm per occupant. Since the number of occupants in a house or apartment changes over time, the assumption is made that there are two occupants in the master bedroom and one occupant in each of the other bedrooms. So a three-bedroom house requires 60 cfm of ventilation by this measure. Because the ACH measure—71 cfm—is higher, that is the one to use. In this case, one Panasonic FV-08 will meet the required ventilation needs.

Most fans' airflow ratings are usually listed at 0.1 and 0.25 inches of water. When choosing a fan, use the CFM rating at 0.25 inches of water, which approximates the static pressure of a typical duct installation.

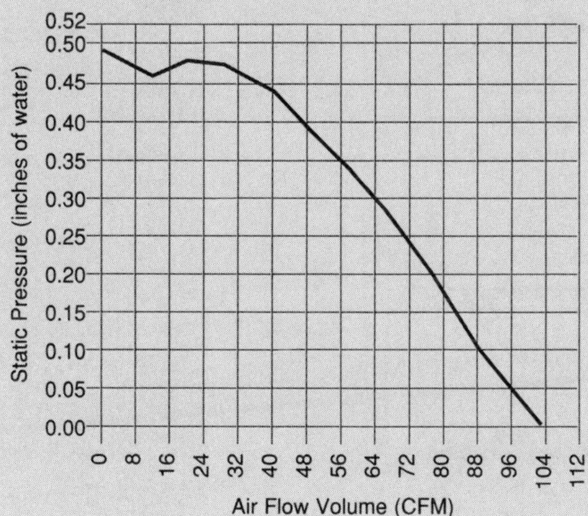

From the Journal of Light Construction, *(August, 1995), by Andy Shapiro. Reprinted courtesy of* JLC *and Andy Shapiro.*

the order for it at the lumber yard. Some materials are toxic for a while, but eventually "outgas" their toxic elements. Some of the toxic or irritating materials, like fiberglass, are buried in the walls.

There are entire books for people who are particularly sensitive, or think they may be, to common toxins. Careful research of materials' toxic properties may become a big part of building a house. But there are also a range of products that have received attention recently. Adhesives and binders, particularly urea-formaldehyde adhesives, which are in building products such as plywood, particle board, and carpets, are the most well known. Finishes have received a lot of attention. Varnishes and paints are examples of products that give off the most volatile organic compounds (VOCs) during application and disposal, but may also have some toxicity in use. There are now companies that market non-toxic finishes, some of which are similar to natural finishes used 100 years ago.

Figure 19-3. Radon removal system: Find out from local officials your radon hazard level. Consider including radon venting in your foundation plan.

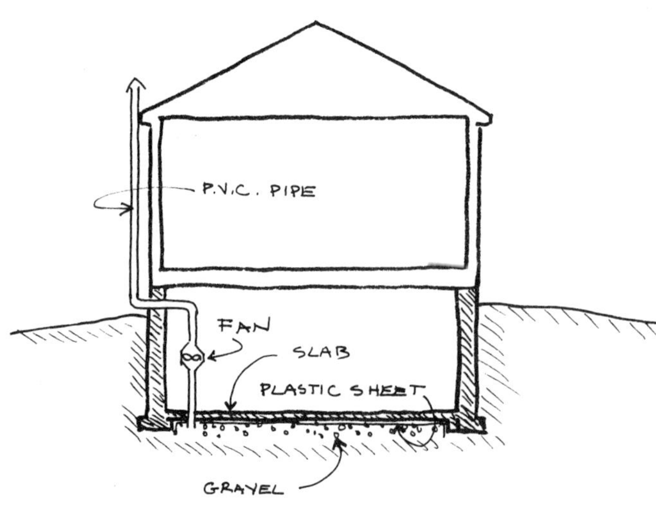

Though your own needs could require special measures, and more is learned about air quality each year, there are several basic steps that can be taken to reduce toxics.

Use fewer materials, particularly new materials, that consist of fibers glued together, unless you know what the glue is. I don't use nearly as much particle board or new carpeting as I once did. If shopping for carpet, I would be willing to pay a little more for a type that seemed safer. If I were thinking about kitchen cabinets made with particle board, I would make sure it was sealed up well, and I would ask what was in it. Better yet, I would use natural wood cabinets.

Seal the walls carefully to keep what's in the wall inside, particularly insulation particles.

Think about the finishes you use. We are always trying out new, more natural, less toxic finishing products. We use a lot of water-based polyurethanes, though they don't function as well as traditional solvent based-types. We use the new non-toxic finishes at times, and sometimes we use no finish at all.

Old things may be less toxic. An old, recycled piece of plywood will have long ago given off its fumes. An old stove will be made of simpler materials, perhaps, than a new one.

Be careful of all the little things you bring into the house. You could have built your house of totally benign, natural materials, but if you fill it with all manner of stuff made of materials you don't know the name of, you might be adding in all the toxic risks you so carefully excluded. As I look around my little office, with its copier, computer, blueprint machine, myriad objects made of myriad plastics, and thousands of sheets of dozens of types of paper, I realize I haven't the slightest idea what this stuff is made of. This could be a death trap!

20

PLUMBING & WATER SYSTEMS

Water supplies and the laws that govern them vary greatly from place to place. When you construct your water system, follow local practice, but don't merely investigate those solutions popular with contractors. Check out how farmers get their water. Farmers of necessity are experts at cheap but reliable water systems. If you don't find a certain type of system in local use, the chances are it has been tried and found impractical.

A water system has four parts: the source, the pipe that goes from the source to the house, the plumbing in the house, and the sewer system.

SOURCES OF WATER

When there is no town water, most people have a well drilled by a huge truck-mounted drill that looks something like a crane. This rig cannot maneuver on soft ground or in confined areas such as dense woods, so make sure the machine can get close enough to your house before planning on a well. The machine drills a 6-inch hole and keeps drilling until it hits an adequate supply of water. Two hundred feet is standard in New England. Drilling costs around $8 per foot, plus another $8 for a casing, which is a steel pipe used in those segments of the well that penetrate sand or soil rather than rock. The pump is usually a submersible pump hung from the water pipe. The pump and other gear might cost another $1,000. The depth at which water is found is unpredictable, although you can get some idea by finding out how deep other wells in the neighborhood are.

Less expensive water sources could be investigated before deciding on a drilled well. Dug wells are still used in some areas. This is the old-fashioned hole in the ground lined with stone, wood, concrete, or steel. Usually such a well will be at least 3 feet in diameter. This kind of well works when the water-bearing soil or rock is within about thirty feet of the surface. Another inexpensive type of well is the driven well, which consists of a perforated pipe with a special tip called a "well point" that you drive into the ground. In the right soils, this type of well can be fifty or sixty feet deep.

A source at or near the surface is usually more economical than any type of well. The simplest of all is an unpolluted stream, river, or lake. If you have such a source nearby, have the water tested. Usually some branch of the state government will test the water for a small charge.

Springs are a more common source. A spring is a place where water surfaces from underground. Sometimes its location will be obvious because it creates a swamp or stream. Other times the ground will just seem damp. Often ferns, cedar trees, or other wet-ground vegetation will signal its location.

When you are looking at a spring, bring along someone familiar with local water systems to help you evaluate what you find. Begin by digging out the spring with a shovel to see how much water there is, unless there is obviously enough. Have the water tested. Find out if it runs year-round.

Springs can be developed as a water system in a variety of ways. A temporary method is to dig out a hole and put in a garbage can full of holes as a liner. Surround the can with rocks and cover the can with its regular lid, making sure the top of the can is several inches above ground level. A more permanent arrangement is to dig a large hole at the spring, 4 to 10 feet deep, and install spring tiles, round sections of concrete pipe 2 feet long and typically 4 feet in diameter. The top tile sticks a foot or so above the ground and is covered with a concrete lid.

A surprisingly effective secondary source in some places is a rain barrel that catches the water from the roof gutter. This can amount to a lot of water as an emergency or supplementary supply, or a main source for cabin or outbuilding. With a smooth roof surface such as sheetmetal, the water is surprisingly clean and often fit for washing or for animals, though not always for drinking.

MOVING THE WATER TO THE HOUSE

Whatever the source, the water will usually be brought to the house in flexible black plastic water pipe. For a summer house, the pipes can simply run across the surface of the ground or be buried in a shallow trench. In a freezing climate the pipes must be buried in a trench below the frost line so they won't freeze up. The frost line is the maximum depth to which the ground freezes. Usually there will be a local rule of thumb that you should follow (also see appendix D). The trench will be dug by a backhoe, which is a tractor with a digging shovel on the back.

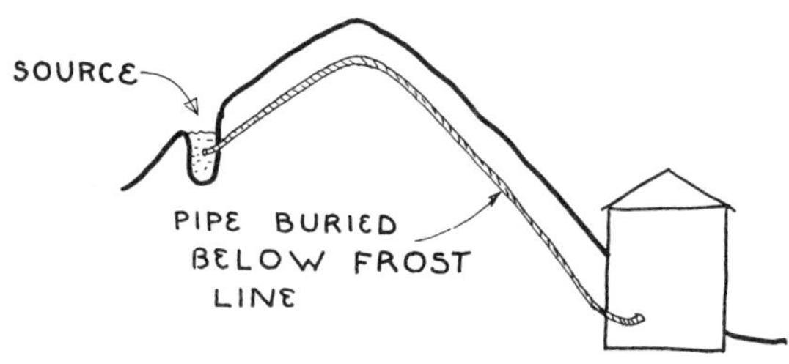

Figure 20-1.
Gravity flow with siphon.

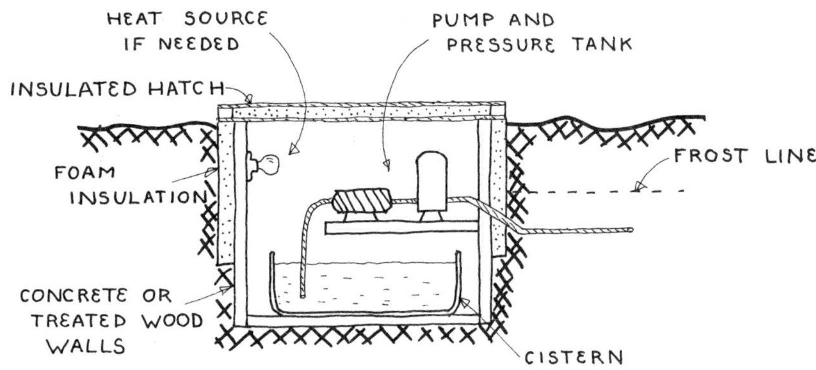

Figure 20-2.
Types of pump houses.

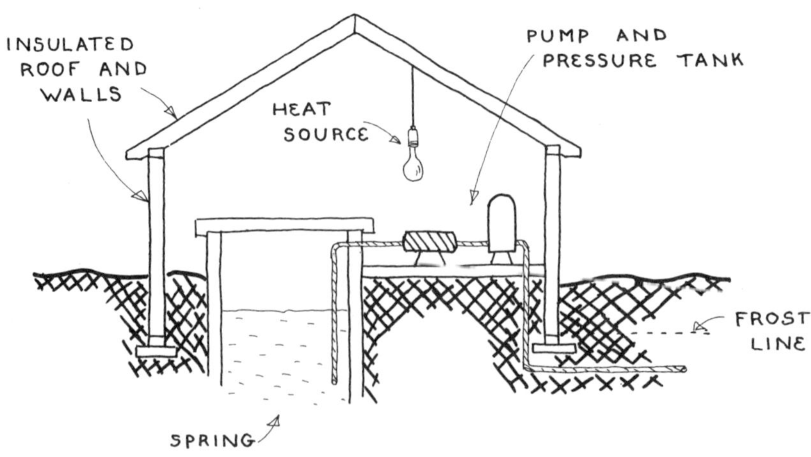

A backhoe operator can look at your site and estimate the cost of a trench. A trench across open fields may not be too expensive. In the woods you will have to clear a road wide enough for the machine to work in. If a trench is too long or too hard to dig, the spring may not be worth using.

Water will flow by gravity as long as the source is above the house. If segments of the pipeline are above the house, the water will still flow by gravity; this is a siphon.

If the source is below the house, it must be pumped.

Pump Location and Protection

Pump location will often determine the type of pump used. A pump can be submerged in the water at the source or placed above, but within a few feet of the source. It can also be located in or near the house, as long as the house is not too far above the source. Here we come to the characteristic problem with pumps: Though they can push water hundreds of feet uphill, they can only suck water from about twenty feet below themselves. That means the pump can only be about twenty feet in elevation above the water source, though the horizontal distance can be much greater.

In cold climates the system must be protected from freezing. If the water freezes, your water supply is of course plugged up, possibly for the season. If the pump freezes the water inside it will expand as it crystalizes, and crack the pump.

There are four ways to protect a pump. First, you can locate it deep underground, submerged in the water source. Second, you can put it in the house, in a heated basement or utility room. Third, you can build a pump house. This can be an insulated box on top of the ground built much like a wooden house, or an underground box with an insulated lid. The underground type will tend to be warmer, since the earth is warm. An above-ground pump house will be more accessible, which may be important if the pump is a type that doesn't operate automatically. In either case, you will probably need some sort of heat source hooked up to a thermostat to keep the pump house warm in extreme weather. This heat source may be an electric heat tape, an electric bulb, or a small electric heater, depending on the severity of the weather. With a buried pump house covered with snow, a light bulb will provide plenty of heat even in a very cold climate.

The fourth method of protecting the pump is to drain it of all water after each use. This may be workable if you only use the pump every few days to fill a large tank in the house, but I think most people will find the process too aggravating to put up with indefinitely.

Types of Pumps

Pumps can be operated by hand, by wind, by the hydraulic power of a stream, by a gasoline engine, or by electricity. The type you choose will depend on the power sources you have, the type of water source, and the lay of the land.

The old-fashioned hand pump is probably the simplest and can be quite practical if the water source is fairly near the house. One common type has the pump mechanism located right at the handle. Below this is a pipe leading to the source. Since a pump can only pull water by suction from about twenty feet below itself, this type of hand pump must be no more than twenty feet above the water. At the handle there will either be the familiar spout, or a second pipeline leading upward from the pump to a cistern (holding tank) from which it can flow into the plumbing (figure 20-3).

A type of hand pump called a deep-well pump can be used to retrieve water from a source more than twenty feet down, as long as the source is directly below the handle. The handle at the top is connected by a long shaft within a pipe to the pumping valve so that the valve itself is within twenty feet of the water. Because of the shaft, this type of pump is only for use in wells (figure 20-4).

Figure 20-3.

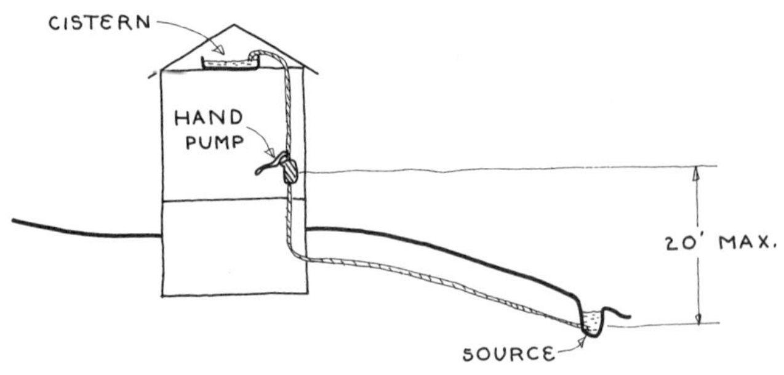

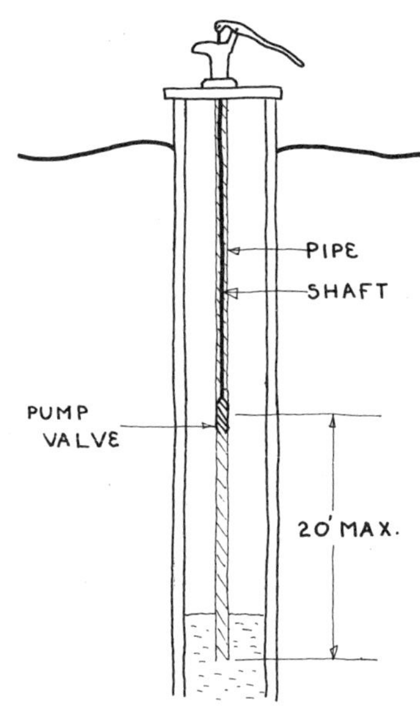

Figure 20-4. Deep well pump.

If your source is a flowing stream or river, water can be pumped by the stream's own power. You do this with a special pump called a hydraulic ram. (For information about one such product, write the Rife Hydraulic Engine Manufacturing Company, P.O. Box 367, Millburn, N.J. 07041.) A ram uses the flow of the stream or river to pump a small fraction of the water up to the house.

At one time, wind used to be a pump power source. A windmill seems like a simple, practical way to pump water, and before electricity came along windmill pumps were common, particularly in the Midwest, where abandoned windmills can still be seen. Because of their cost, they won't be a practical choice in many situations.

Pumps can also be powered with gasoline motors. If you had a gas-powered pump, periodically you would start up the pump and fill a large 200- to 500-gallon tank in the attic. You might also fill an even larger water tower with thousands of gallons. In either case, the water would flow from the tank to your plumbing by gravity. You will have no freezing problem if the pump is in the house. Outside, you would have to protect the pump as described earlier.

The most convenient and perhaps most economical solution is usually an electric pump. Deep wells normally necessitate a submersible pump located right down at the bottom of the well (figure 20-6).

For other water sources, the jet pump is the popular type.

Usually an electric pump is combined with a small tank called a pressure tank, which stores a small supply of water under pressure so the pump doesn't have to go on every time someone gets a glass of water.

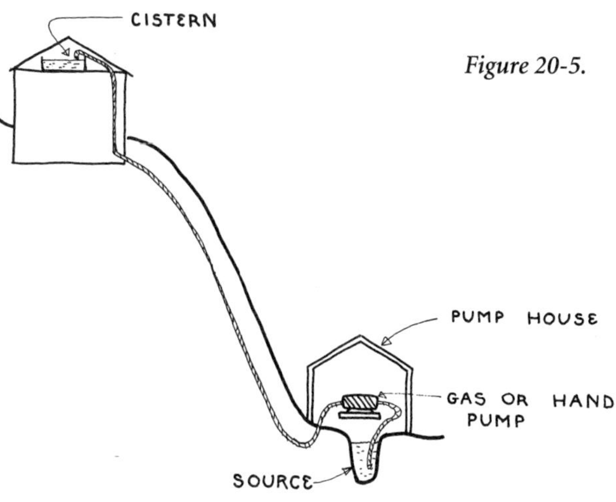

Figure 20-5.

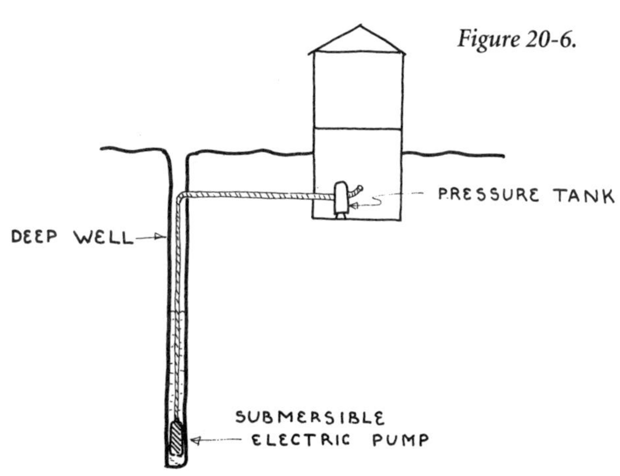

Figure 20-6.

If the water source is nearby, the pump and pressure tank can be right in the house (figure 20-7). Otherwise you may need a pump house. This is usually located right at the source (figure 20-8), but it occasionally happens that you can't economically get an electric line to the source. If so, one solution is to have the water flow

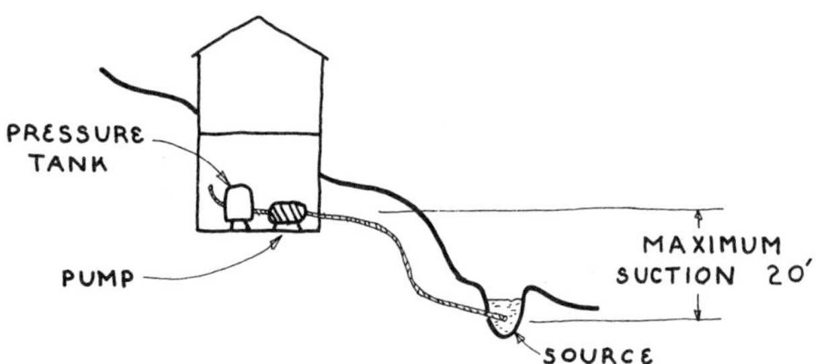

Figure 20-7.

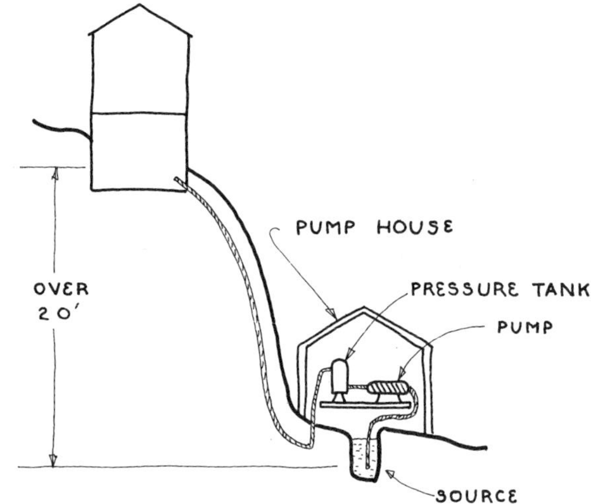

Figure 20-8.

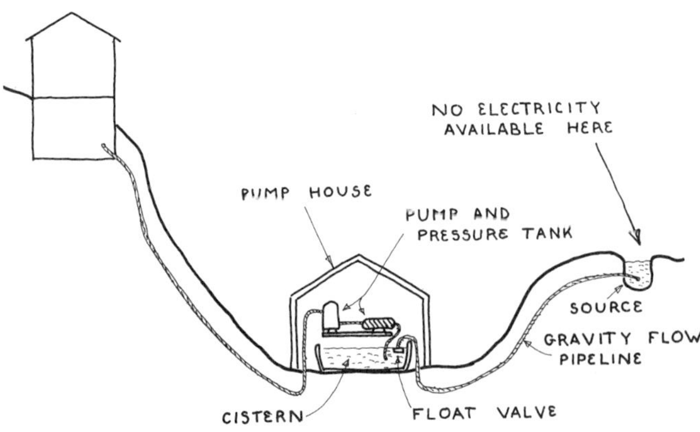

Figure 20-9.

from the source by gravity to a lower elevation to which you can more easily run an electric line. The water collects in a cistern consisting of a galvanized metal stock-feeding tank or a more permanent concrete tank. To prevent overflowing, the cistern has a float valve which, like the valve in a toilet tank, cuts off the water flow when the tank is full. The cistern is located in a pump house with an electric pump and pressure tank. These send the water on to the house when needed. If your house is above the water source, but the cellar is slightly below it, all this hardware can be in the basement (figure 20-9).

Constant Flow and Runback Systems

Sometimes none of the arrangements described above will work because distance, expense, or bedrock make it impossible to dig a trench deep enough to avoid freezing. There are two ways to solve this problem, both based on the principle that water moving continuously in a pipe will not freeze. When water flows to the house by gravity, you can use a constant flow system, as shown in figure 21-10. The water flows to the cellar, and spurts constantly into a cistern through a small hole, perhaps as big as a pencil lead, in the end of the pipe. Though the flow is small, it is enough to prevent freezing, so the pipeline can be buried in a shallow trench. Even a small flow will amount to

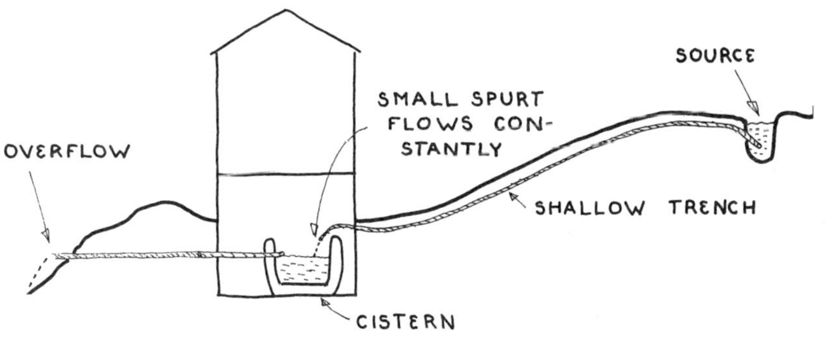

Figure 20-10. *Constant flow system.*

thousands of gallons over time if it never stops, so you will need an overflow pipe to carry the excess water over the edge of a nearby hillside.

When the water is being pumped uphill in a pipe, you can use a "runback" system, as shown in figure 20-11. Here there is a hole in the bottom end of the pipeline, either in the pump house or (with a deep well) down in the well just above the pump. Each time the pump stops running, the water in the pipeline flows back into the well or cistern, so there is nothing left to freeze. There must be no low points in the pipeline, because water would be left behind there to freeze. The runback system is commonly used with wells where bedrock is near the surface. The problem with both of these methods is that if the flow stops for even one minute in winter, the line will freeze and you will have no water until the ground thaws.

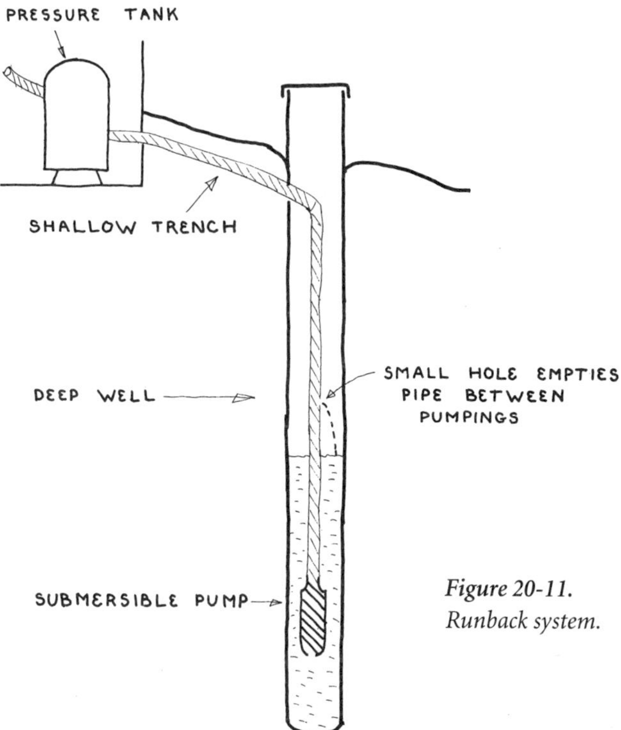

Figure 20-11. *Runback system.*

HOUSE PLUMBING

Supply Pipes

Plumbing in the house includes the supply system, which consists of the hot and cold water lines, and the waste system (technically called the drain-waste-vent or DWV system), which usually leads to the town sewer or a septic tank. The supply begins with the pipe that enters the house from the water source. You must get this pipe into the house without exposing it to freezing temperatures. If you live in a warm climate or have a furnace in the basement, this is no problem. If you live in a moderately cold climate and do not have a heated basement, it may be sufficient to wrap the supply pipe in thick insulation and firmly attach an electric heat tape to the vulnerable portion of pipe. A heat tape is a heating element with a built-in thermostat.

Sooner or later this system will fail in very cold climates and you will find yourself with no water and perhaps a burst pipe. In such a climate the supply pipe should emerge from the earth and enter the heated part of the house through a fully insulated and heated chamber called a "hotbox" (figure 20-12).

For a house on a post foundation, this may mean building a small, insulated basement of treated wood or concrete block just for the pipes' entrance. The hotbox goes below the frost line. It should be protected on the outside with rigid foam insulation and on the inside with rigid foam or fiberglass. The heat source can be one or two light bulbs, located at or near the bottom of the hotbox, or a

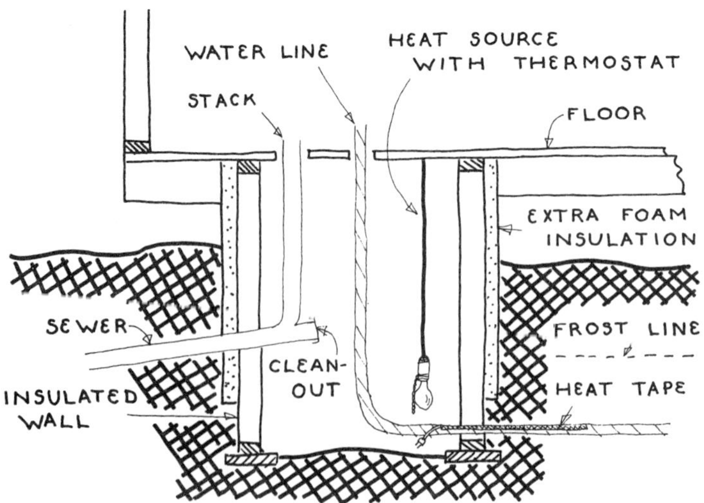

Figure 20-12. *A hotbox protects water lines where they enter the house.*

short section of electric baseboard. Leave a peek-hole into the house so you can make sure the bulb has not burned out. I like to install an indoor/outdoor thermometer, with the sensor in the hotbox, and the readout upstairs where I can check the temperature below without opening the hatch.

Another system is to use an appliance such as a furnace, boiler, or hot water heater to heat the hotbox. In this case, you have a mini-basement.

Even with these precautions, water pipes may freeze, as they might during a power failure. Therefore, the box should be at least big enough to permit you to work inside it, perhaps thirty inches square, even bigger. A box this big also allows the sewer line to exit through the hotbox, protecting the sewer from the cold and giving you access to the sewer cleanout. It is a good idea to install a section of heat tape on the first few feet of pipe that enters the earth, as shown in figure 20-12. If a freeze-up happens, the inaccessible section is likely to freeze along with the section in the hotbox itself, and the tape gives you an emergency method of thawing it. (Normally, this heat tape would be left unplugged.)

Once the water is in the house and past the pump, it will divide into two lines, one heading directly to the cold water taps, the other leading through the hot water heater to the hot water taps. These supply lines will usually be copper, which is good but expensive, or PVC plastic, which is cheaper and easier to install. Plastic is not legal everywhere, so consult the local plumbing code. Copper is more durable, and easier to fix if the pipes burst.

Supply lines are usually run within the walls and under floors for cosmetic reasons. They are installed after the house is framed, but before surfaces are covered with finish materials. The actual tap fixtures are installed after the finished wall and the floor surfaces go in. In cold climates, it is critical to locate pipes where they won't freeze. Usually that means avoiding concealing any pipes in exterior walls. If they must run within an outside-facing wall, they should be exposed to interior heat, or enclosed in a pipe box or cabinet that can be opened in case of emergency.

Drain, Waste, and Vent Lines (DWV)

Though plastic is not always legal for supply lines, it is standard for waste lines. The waste system of a house is centered around one or more stacks. A stack is a 3½- or 4-inch pipe running from the sewer underground up through the framing in the walls and on up through the roof. Where it protrudes through the roof, it is the vent that lets out sewer gas.

Each fixture (sink, toilet, etc.) has a drainpipe that pitches downhill all the way to a stack so that waste can flow away by gravity. A toilet drain will be 3½- or 4-inch pipe, and a sink drain will be a 1¼- to 2-inch pipe. An S- or P-shaped trap should be positioned between the fixture and the drain. It will stay full of water all the time. (In a toilet, the trap is built into the toilet itself.) Traps provide a seal to prevent sewer gas from entering the room through the fixture.

Every fixture needs a vent, which is a pipe that starts from the drain below the trap and goes upward and out the roof. Vents allow sewer gas to exit into the outside air instead of into the house, and they prevent the traps from being emptied by suction. Fluids flowing down the drains act like a pump to create suction in the system. In a vented system, the suction pulls outside air in through the vents instead of through the traps at each fixture. If a fixture is very close to the main stack, the stack itself will be the vent. But any fixture even a few feet away from the stack should have its own vent pipe.

Figure 20-13. *Drain–Waste–Vent or DWV system.*

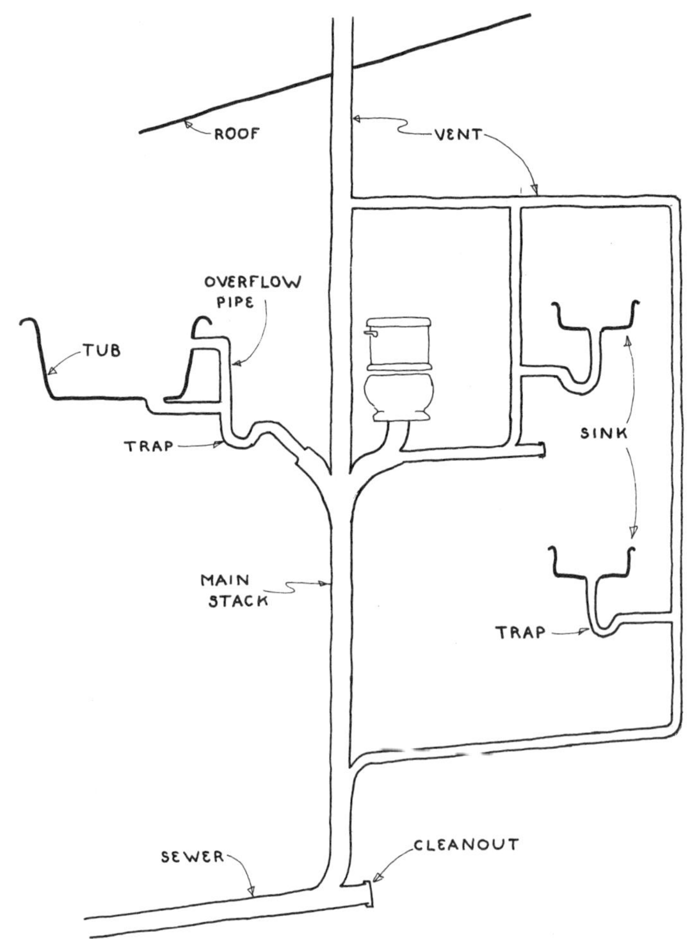

The maximum distance between a fixture trap and the nearest vent depends on the type of fixture, the drainpipe diameter, and the floor the fixture is on. Local plumbing codes will give standards to follow. Usually a sink, tub, and perhaps other fixtures will have a common vent pipe, as in figure 21-13, which will reenter the main stack above the top-most drain or go through the roof separately. In an economical house, the toilet will be located adjacent to the main stack, because large-size pipes are expensive and you want to avoid having to provide a special vent.

Lay out plumbing location to minimize materials and plumbing labor. Ideally this means locating all the pipes in one wall with the kitchen and bath back to back or one on top of the other. Then all fixtures can feed into the same stack. Another good arrangement is to have a centrally located room containing most of the house machinery such as pipes, electrical panels, furnace, hot water heater, and laundry equipment. The bathroom and kitchen would be adjacent to the machine room. This machine room has no interior wall covering, so the plumbing pipes will always be accessible for repair or remodeling.

Waste Pipes

If there is no town sewer, the waste system will consist of a pipe leading from your stack to the septic tank nearby, the tank itself, the pipe leading to the drainage field, and the drainage field. The pipes are all pitched to flow by gravity, and they do not need to be below the frost line.

The sewer pipe will be a 4-inch pipe with sealed joints, usually 4-inch plastic designed for this purpose. The septic tank is a large sealed tank with an inlet near the top for the sewer line, an outlet on the opposite end leading to the drainage field, and a removable hatch for inspection and pumping out. The most common and practical septic tank is concrete, which will be delivered and placed in its hole by a special truck with a crane.

The tank should be at or below the same level as the house, since the sewage normally will flow by gravity. It can be within twenty or thirty feet of the house to economize on the cost of the sewer line, but if possible put it where heavy vehicles will not drive over it. The concrete tank can take the weight, but you want to avoid disturbing the alignment of the pipes.

The sewage flows into the tank, where, if the system is working right, bacteria will attack the sewage and transform it into a less noxious liquid, which flows out of the tank through another pipe into the drainage field. This drainage field is a network of nearly level pipes with rows of holes in the sides to let the treated sewage percolate back into the soil. The drainage pipes rest in a seepage bed of coarse gravel covered with filter fabric. Topsoil is placed over the pipes. The filter

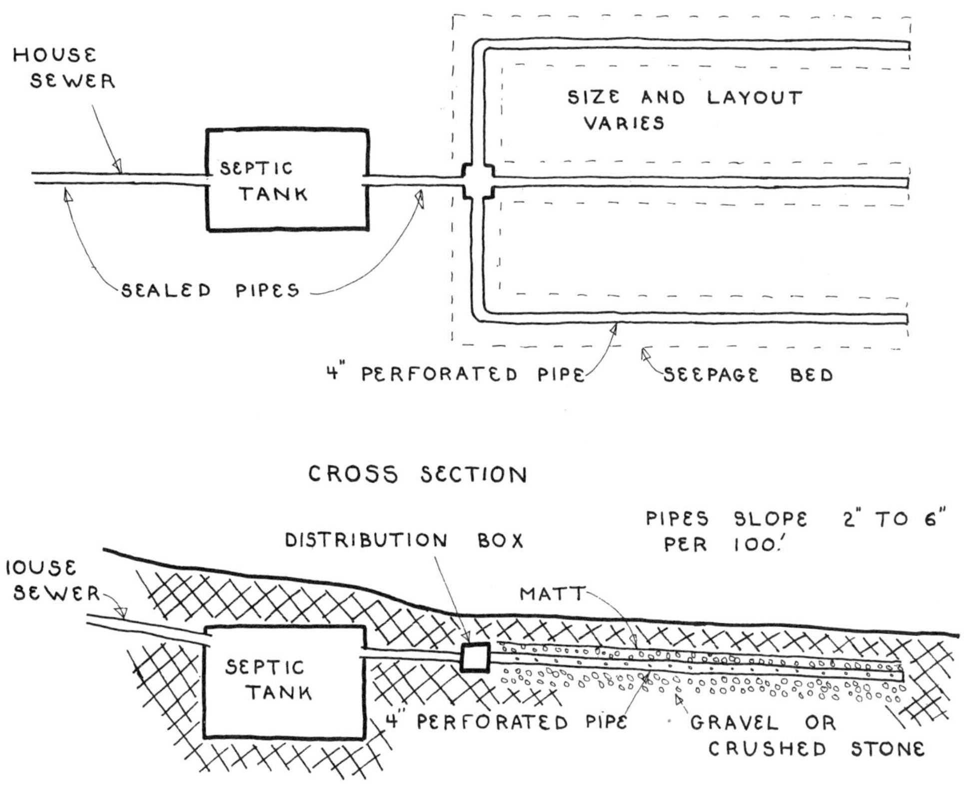

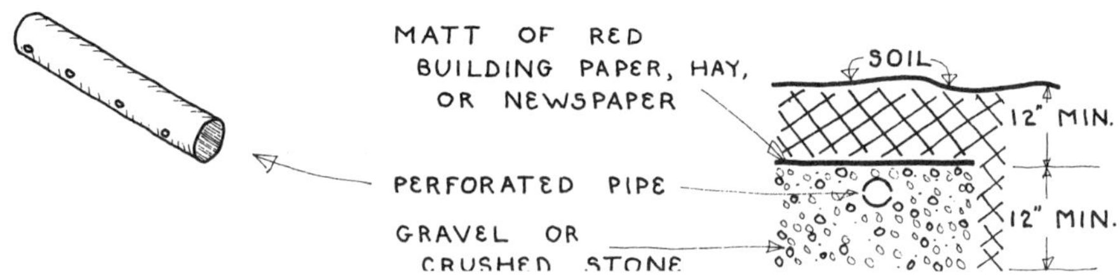

Figure 20-14.
Septic systems.

fabric keeps the topsoil from clogging up the system, without preventing ground water from sinking into the earth.

Some systems have two sewers coming from two stacks. The toilets feed one pipe, which goes to the septic tank. Everything else—known as graywater—feeds into the other sewer line, which bypasses the septic tank and goes directly to the drainage field. This prevents nondegradable detergents from ruining the bacterial action in the septic tank.

The drainage field should be level with or downhill from the house and septic tank, because everything flows by gravity. Like the septic tank, it should be located where it won't be driven over, if possible. The field should be below the water source and at least 100 feet away from it. Equally important, the soil must be porous enough so that the wastes can percolate back into it.

You have to make sure the soil is suitable for a septic system when you plan your house site. At one time, anyone could and did build septic systems, but today they are engineered, based on an actual survey of soil conditions called a percolation test or perc test. The engineer will usually also designate an area for a replacement field that can be developed in the future when and if the first system fails.

A simple system like the one I've described will cost $3,000 to $5,000. But often wetlands, bedrock, or clayey soils make it difficult for the engineer to locate the system. If the system ends up at a higher elevation than the house, a pumping station will be needed to pump sewage up to the system. If there is no suitable soil anywhere nearby, a mound system may be needed. A mound system really means building up an area of land with decent drainage. These options will add thousands to the cost of the system.

In general, the suitability of a piece of land for a septic system will influence its market value. That is why it often makes sense to have the perc tests done before agreeing to buy any piece of land.

Alternative Waste Disposal Systems

There is a lot of interest today in alternative systems for waste disposal that pollute less, use less water, and are otherwise considered ecologically sound. One such system is the composting toilet, such as the Clivus Multrum, which was developed in Sweden. This is a tank that fits under the house. Excrement and garbage go directly into the tank, but waste water from sinks, tubs, and washing machines does not. The Clivus is designed so that the excrement and garbage are turned into a rich humus over a period of months through the action of various microorganisms. Water use is cut in half, since there is no toilet flushing and the septic tank is eliminated.

The graywater from sink and shower would go to a conventional septic system or a simple drywell. There are also systems that use this nutrient-rich water to irrigate specially designed planting beds in a greenhouse.

The appeal of these systems is that they recycle human waste for human use, where septic system cycles waste back to the earth. But they are usually much more expensive than a septic system. It may be that an alternative system is worth considering if the soil conditions are such that a septic system would be very costly.

If you have a small budget and want to avoid a septic system, the best solution is the old-fashioned outhouse, which is still legal in some rural areas. A properly ventilated outhouse is hygienic, uses no water, does not pollute the soil, and can be built with scrap materials and a minimum of labor. You can use a covered potty chair or chamber pot, emptied daily, for nighttime or cold weather use.

PART 5

CONSTRUCTION METHODS

21

BUILDING IN THE RIGHT ORDER

PART 5 DESCRIBES EACH STEP IN THE CONSTRUCTION OF a house, usually with several variations. I do not try to give every possible method but just a variety of straightforward and economical choices. In each chapter in this section, I look at a part of a house in three ways. First I discuss relevant design considerations. Second, I describe and compare a variety of construction methods. Third, I give examples of the kind of working drawings you may need, and tips on ordering and selecting materials.

Figure 21-2 is a floor framing plan, which shows how a floor frame is arranged. Figure 21-3 is a detail drawing, a cross section that clarifies how some elements of a house fit together. These working drawings help you figure out how the house will be constructed. They are also the easiest tool for making the materials lists complete and accurate. You can quickly count the number of joists of each size needed to frame the floor shown, or other elements you might forget.

Also, the drawings help you keep track of your design while construction is underway. When things get really busy on a building site, and different people are doing several different tasks, it's easy to get confused. The working drawings keep things clear on site, and provide a way to communicate with your crew.

You may want to consider building systems that aren't explained here. If so, don't simply get the latest book on the subject and start building. Talk to people in your area who have used these systems themselves. Find out what the costs were, how much time was spent, how the system worked, and where supplies came from. As you do this research, you'll be gathering the information to move ahead if you decide a particular system is really for you.

This book in general focuses on "light timber framing," or "stick-building," the most common wood framing system. In a stick-built house, the load-carrying frame consists of small, closely-spaced framing members concealed inside floors, walls, and roof. Though it is often criticized, it is an economical, quick, strong, and flexible system that works well for owner-builders. Most of Part 5 describes this system.

But at times it makes sense to borrow from timber framing methods to solve particular design problems, or to give a house some of the comforting, hefty feel of an exposed timber structure. For example, we often build a second story floor in heavy timbers; it looks great, and saves money. For this reason, the first chapter of Part 5 begins with a survey of basic timber framing approaches.

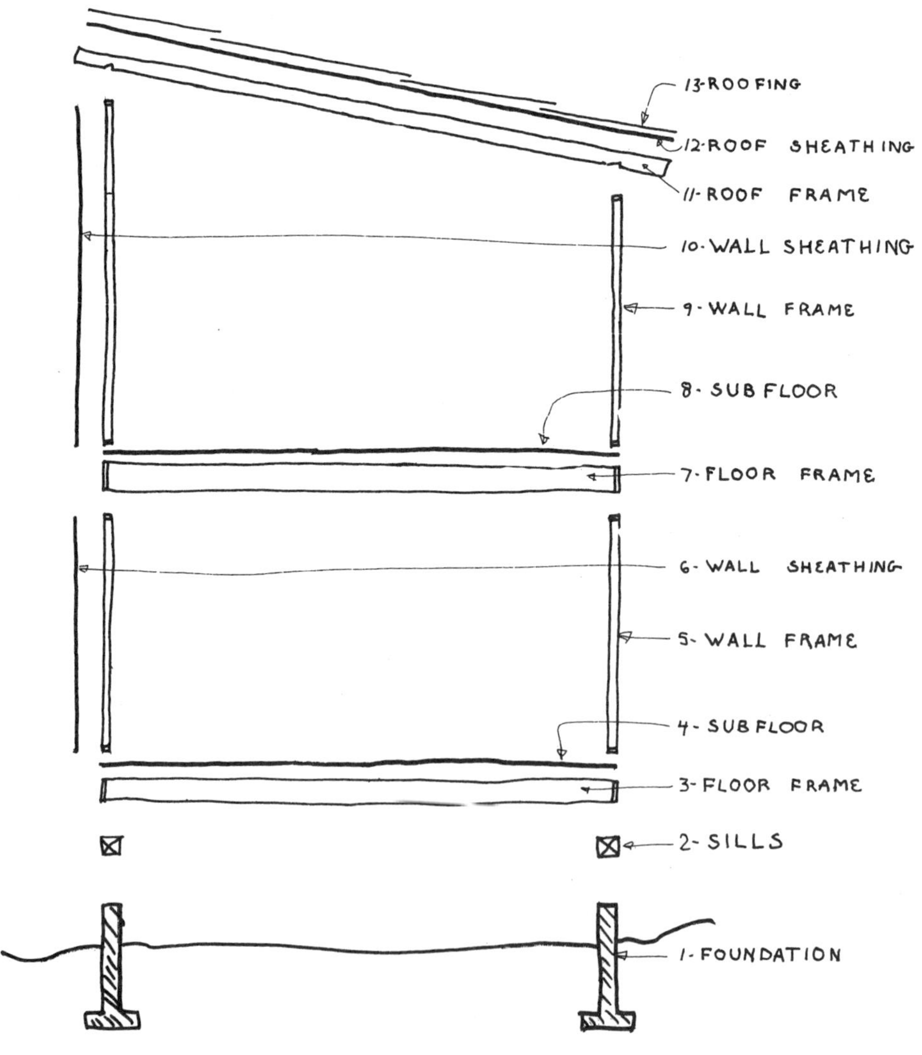

Figure 21-1. Typical steps in building.

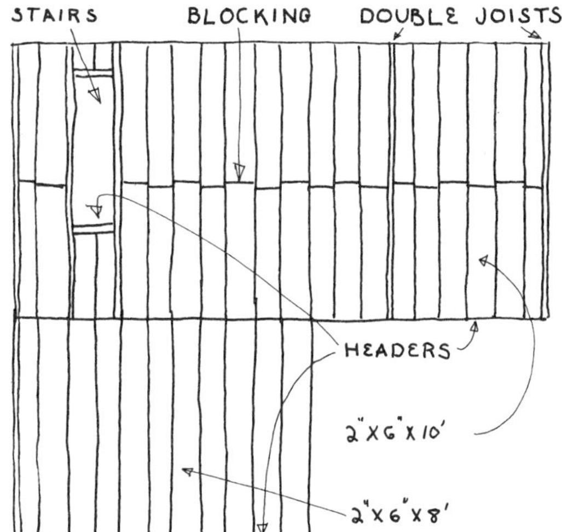

Figure 21-2. Floor framing plan.

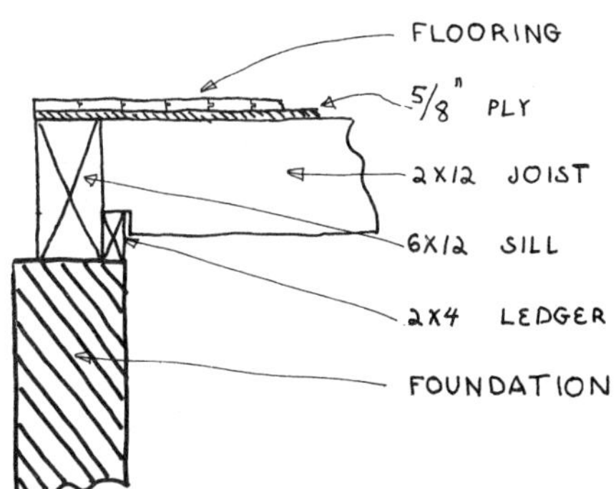

Figure 21-3. Detail.

INSPECTIONS

Your building sequence will probably be interrupted at several points by inspections. Communities vary greatly in how closely they supervise your building.

Normally the septic system must be inspected and approved before it is covered, so that the inspector can see the array of pipes. Typically there will be two inspections of the house itself, the rough and the finish inspections. The rough inspection occurs when the frame is up and covered on the outside, and before insulation goes in. The plumbing will be "roughed in" but the fixtures will not be installed. The electrical wiring will be completed except for the actual outlets, switches, and light fixtures. At this stage, the inspectors can see that everything is up to code. If you have hired a plumber or electrician, they will prepare the plumbing and electrical permits, and arrange the rough inspections; the general contractor will pull the overall building permit and arrange the structural inspection. Owner-builders are responsible for scheduling these inspections. Give inspectors a couple of days notice when scheduling inspections. These inspections may be cursory or exhaustive, depending on local law and on the personalities of the inspectors. If the site looks orderly, complete, and under control, inspectors are less likely to look for something to criticize, which you must alter before continuing. Many building inspectors are flexible concerning carpentry practice, but strict when it comes to fire safety, egress, or safety in general.

The finish inspection can be scheduled when the building is substantially done, and plumbing and electrical fixtures are installed. This inspection may be more cursory. Then the inspectors will sign off on the job, and the building inspector will issue a "certificate of occupancy."

22

POST-AND-BEAM CONSTRUCTION

At one time, most wood frame houses were built using the post-and-beam system, in which the loads are carried by a few large, widely spaced, and usually exposed heavy timbers such as 8 x 8s. Colonial houses are probably the most-well known example of post-and-beam construction. Today many owner-builders are turning to modern versions. Figure 22-1 shows one example.

When I started building I believed, as many new builders do, that post-and-beam construction was the key to economical building. I liked the feeling of a big visible frame, and I believed that with intelligent design a post-and-beam system could save enormous amounts of money and time. After building several post-and-beam houses, observing many more, and discussing the subject with many other builders, I have come to the reluctant conclusion that, though post-and-beam methods are economical in particular cases, and specific post-and-beam techniques are widely usable, a post-and-beam structure will almost always take more work and cost extra money. As most builders have known for years, I discovered that nothing goes up faster than studwalls and plywood; nothing is stronger either, despite the heft of those big timbers.

So this book does not contain an exhaustive manual on post-and-beam construction. Instead, you will find instructions on conventional framing methods and on particular post-and-beam techniques that can easily be combined with conventional wall framing. Since post-and-beam framing is very practical for certain specific designs, and since some people will simply prefer to build post-and-beam, I provide here a general introduction to the subject and some suggestions about how to go about designing a post-and-beam-style house.

Post-and-beam framing has characteristics that can be advantageous with good planning. The frame, having few parts, goes up fast. You can put the roof on early, to provide a covered place to work. The walls that fill in between the

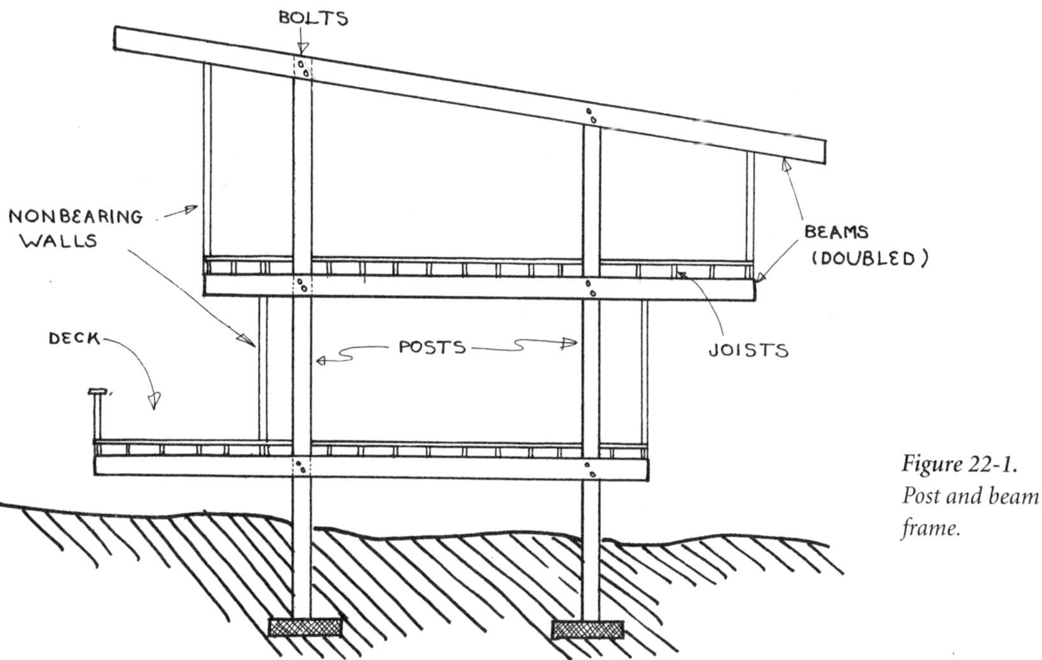

Figure 22-1. *Post and beam frame.*

heavy timbers are not load-bearing. Window openings can be put anywhere and can be any size, and the walls can be easily altered or removed when your needs change, as long as the major posts are undisturbed. With an exposed frame you can eliminate a lot of time-consuming finish work with clever detailing.

However, I think the major appeal of post-and-beam is aesthetic. The frame is big, obvious, and exposed. This exposed frame looks good and gives a house a reassuring feeling of strength and permanence. The simple frame is easier for many people to understand and visualize. You can easily imagine extending the posts upward or the beams outward to create interesting and useful nooks, alcoves, lofts, cupolas, or overhangs that might be harder to conceive of with regular framing.

Post-and-beam construction has disadvantages too. Those big timbers are heavy. You often need a real crew to move them around. Framing details can be tricky. Members are cut down on the floor and have to fit right when they are put up. Joints may be more complex, with various notches often replacing simple butt-nailing. Framing errors are costlier. If you cut a 2 x 8 too short you can go to the pile and get another, and use the short one somewhere else. If you cut a 14-foot 6 x 8 too short you may be out of luck.

Figure 22-2. *Post-and-beam looks neat inside.*

The main design objection to post-and-beam is that the simplicity and economy are often canceled out by structural duplication. The walls you build in between the framework, though they carry no important loads, must be designed for insulation. If you are using fiberglass, you will end up building a conventional 2 x 6 studwall just to create bays to hold the insulation. But this wall could carry the weight of the house by itself, so the heavy posts become superfluous. You have built a strong structure to hold up the house, and then an equally strong structure just to hold insulation.

Post-and-beam construction demands more thorough planning. Studwall construction is additive; you build one part at a time and what you do in one place doesn't necessarily affect what you do ten feet away. But post-and-beam framing is systematic, everything is interrelated. You must detail everything in advance. You have to plan where your wiring and plumbing will go, because they can't necessarily run through the walls just anywhere. You can't cut joists as freely to accommodate staircases. In general, your layout and structure must be unified and orderly if post-and-beam methods are to be a help rather than a hindrance. There is room for improvisation, but it must be done within a system.

Post-and-beam will also demand more from you as a designer. There are as many ways to build post-and-beam as there are ways to attach two pieces of wood together (figure 22-4). You have to find your own system, and your own

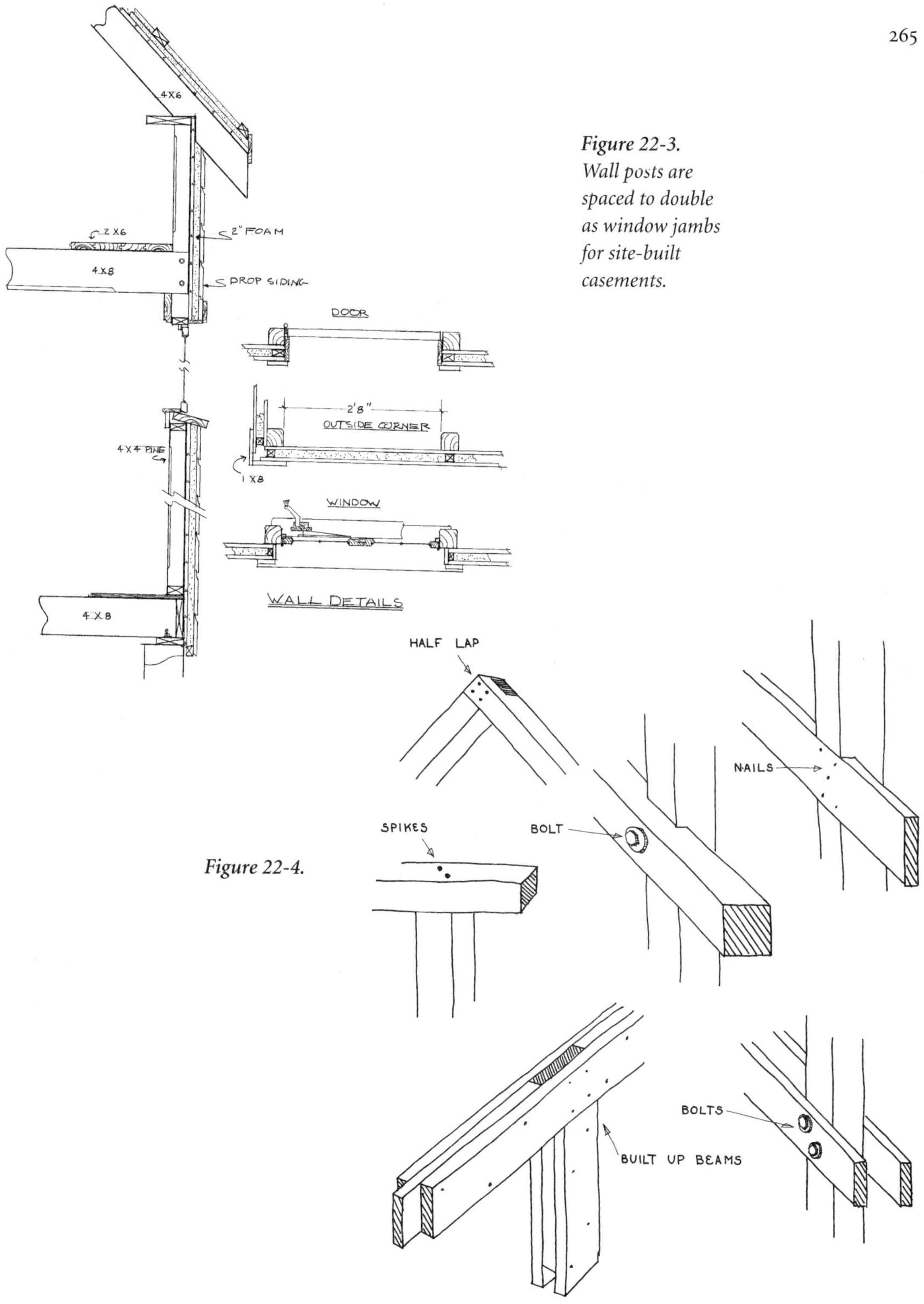

Figure 22-3. Wall posts are spaced to double as window jambs for site-built casements.

Figure 22-4.

solutions to problems. For some people, the inventiveness and resourcefulness that post-and-beam calls for is part of its fascination.

Some people settle on post-and-beam construction before they even start designing. If you find post-and-beam systems easier to understand, or if, like me, you are simply a post-and-beam fan, this is all right. But from the standpoint of economy and efficiency, do not automatically assume that post-and-beam is the right approach. Some house designs lend themselves to post-and-beam; others don't. A house on a hilly site with lots of large windows and no basement, and which you plan to add to, will be a good candidate for a pole house, one type of post-and-beam. A traditional mortised timber frame goes well with a traditional gable roofed building form. A one-story house with an irregular floor plan might well be easier to build conventionally. I have often found that a combination of post-and-beam and conventional methods works well. Sometimes just a few heavy posts or beams will solve special problems, such as an extra-long span. Familiarize yourself with post-and-beam methods so that they become part of your repertoire. Then use them to the extent they suit your layout, site, preferences, and the local building code.

POST-AND-BEAM FRAMING METHODS

As I have explained, post-and-beam framing must be systematic and unified, and everyone has to work out his or her own system. I have chosen to summarize briefly a few of the most common and practical post-and-beam approaches. Elsewhere in the book you will find most of the basic principles and technical information you need to work out your own system.

Frame Built One Story at a Time

Figure 22-5 shows a heavy timber frame built in stages. The first level consists of concrete foundation posts with sills joining them on top. Over the sills goes a floor frame as in a conventional house. Then a heavy timber wall frame, one story high, is built on the floor. The second floor rests on this frame and then a second wall frame is built on that. The roof can be framed in any fashion. The walls can be filled in with horizontal or vertical studs, whichever are more convenient.

This is a modified post-and-beam frame. It combines the overall strategy of post and beam with some features of studwall construction. The joints are simple. The timbers are smaller and easier to handle than ground-to-roof posts would be. Each floor provides a platform for doing the next story.

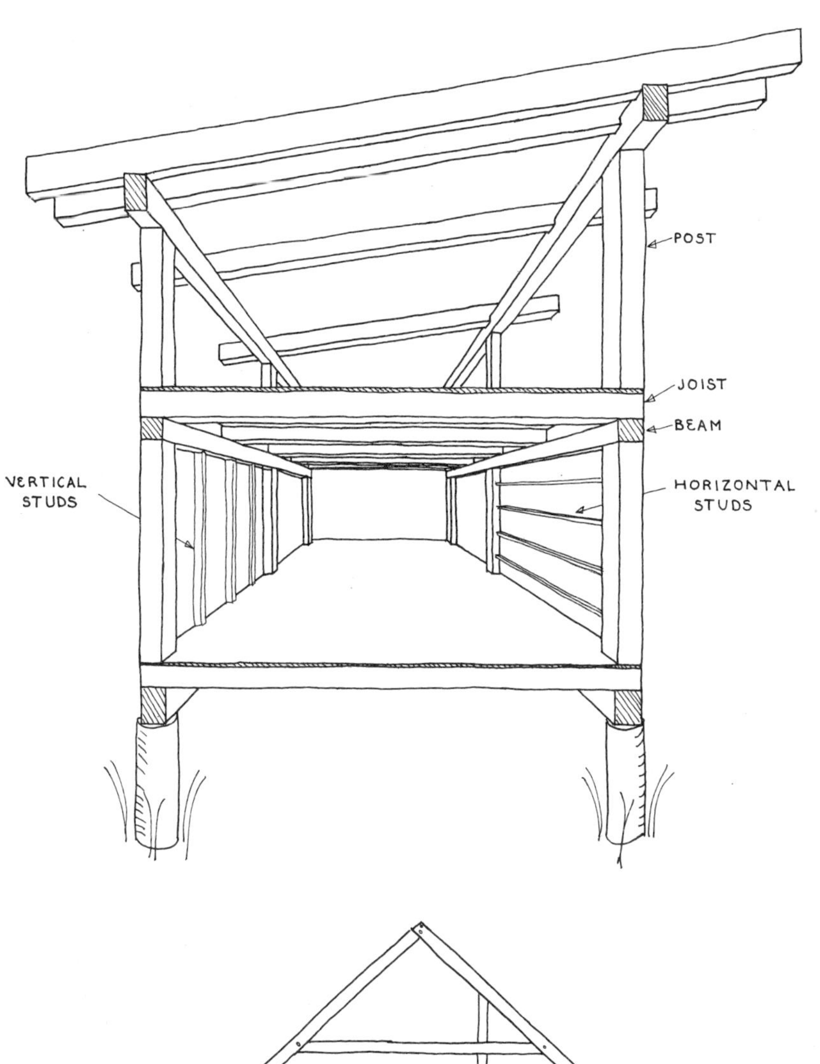

Figure 12-5. *Post and beam frame built one story at a time.*

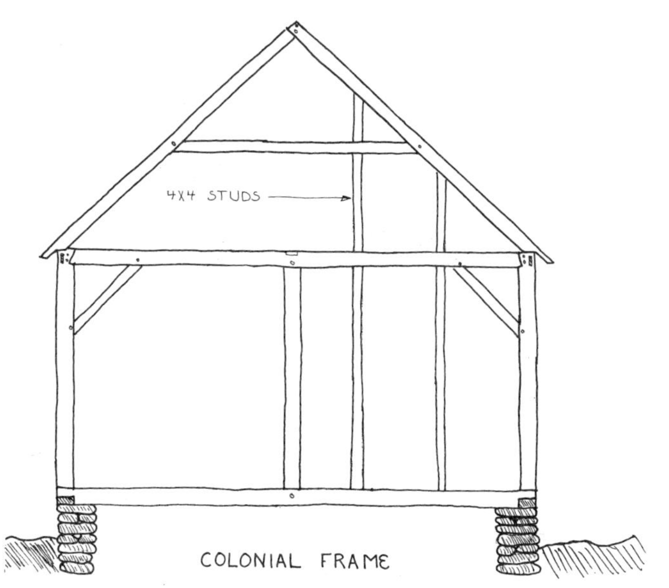

Figure 22-6. *Traditional heavy timber house frame.*

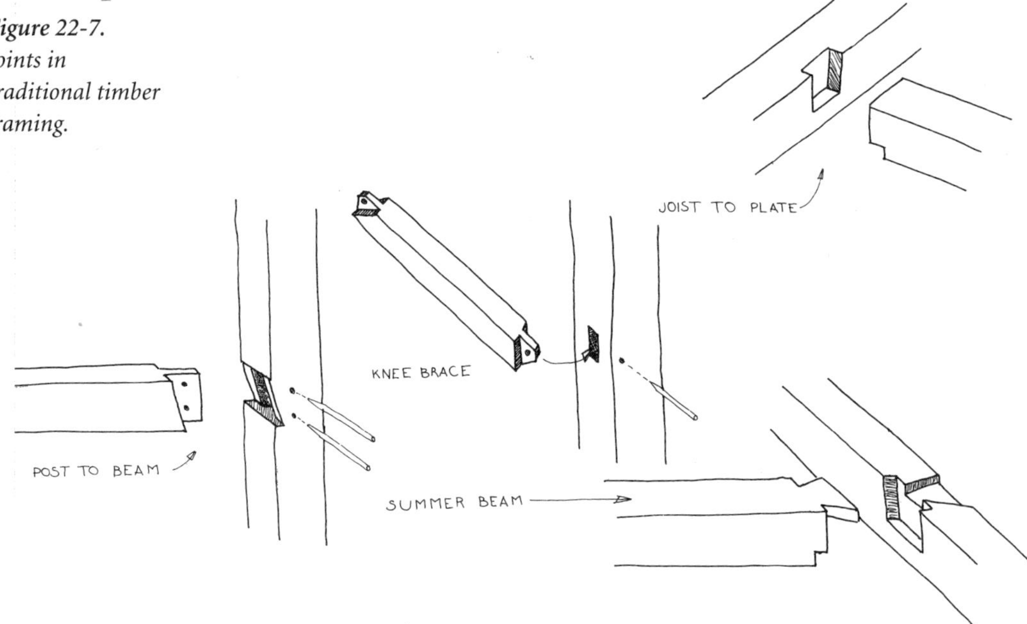

Figure 22-7. *Joints in traditional timber framing.*

Colonial Frame

Figure 22-6 shows a traditional house frame, framed as most wooden houses were before the mid-ninteenth century. There are no nails, only mortises and tenons, dovetails, notches, and dowels. Some of these joints are detailed in figure 22-7. This is the most ambitious kind of post-and-beam frame and assembling it takes great skill. It is one of the most satisfying ways to build a house, but I would not suggest it to anyone who is in a hurry.

Pole House

Another popular method of post-and-beam construction is the pole house (figure 22-8). In pole house construction, pressure-treated posts extend from the roof all the way down into the ground, where they serve as the foundation. If the holes are deep enough, the posts will be held so rigidly by the earth that they will brace the building. The economy of this system lies in the fact that a few heavy posts do the jobs normally done by many different elements: foundation posts, anchor bolts, wall studs, and diagonal bracing.

Depending on the size of the beams running across them, the poles can be spaced up to sixteen feet apart. Eight feet on center is a convenient spacing for any modular material such as plywood. The beams are often 2 x 12s notched into the posts. A horizontally applied member on top of the posts, called a stiffener, keeps the wall from flexing outward between posts. The stiffener's size depends on the span between posts and the roof span.

Figure 22-1 shows another type of pole house. Here, instead of a 2 x 12 beam notched into each post, you have pairs of beams with a post sandwiched between. The two beams are bolted to the post with no notches. The joint is detailed in figure 22-4, lower right. This is probably the simplest and fastest kind of post-and-beam construction.

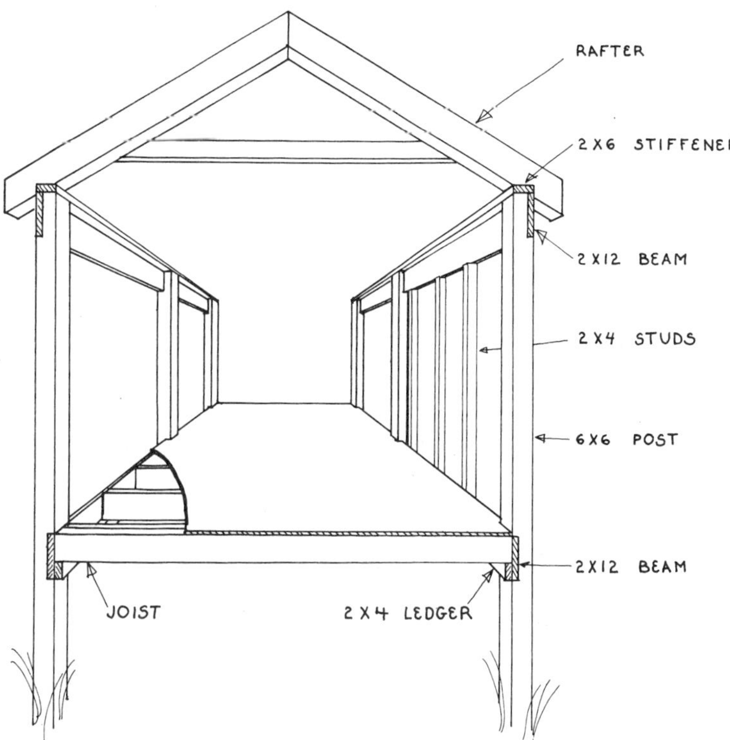

Figure 22-8. Pole house.

FILLING IN THE WALLS

The walls that enclose a post-and-beam frame have to be well insulated, support siding and interior finish, and be able to carry wires. They also have to conveniently hold doors and windows.

Since a conventional wall frame within a post-and-beam frame is structurally redundant, and can at least partially cover up the beautiful frame you've just built, the idea of a light, non-structural panel has a lot of appeal. Many companies now make wall panels specifically for timber frames. They come in various thicknesses, but a panel with about 4½ inches of expanded polystyrene foam insulation is common. The panels give an R value of about 26. The foam is laminated between a sheet of ½-inch oriented-strand board (OSB), which will become the sheathing of

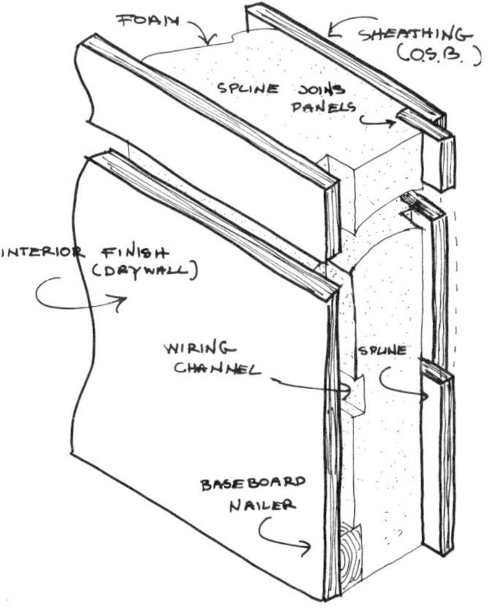

Figure 22-9: Foam insulated wall panels.

the house, and ½-inch drywall, which will become the interior finish wall. The panels are simply nailed to the timber frame. They are partially self-trimming because the timbers cover many joints that would otherwise have to molded, trimmed, or taped. The panels provide an unbroken blanket of insulation in

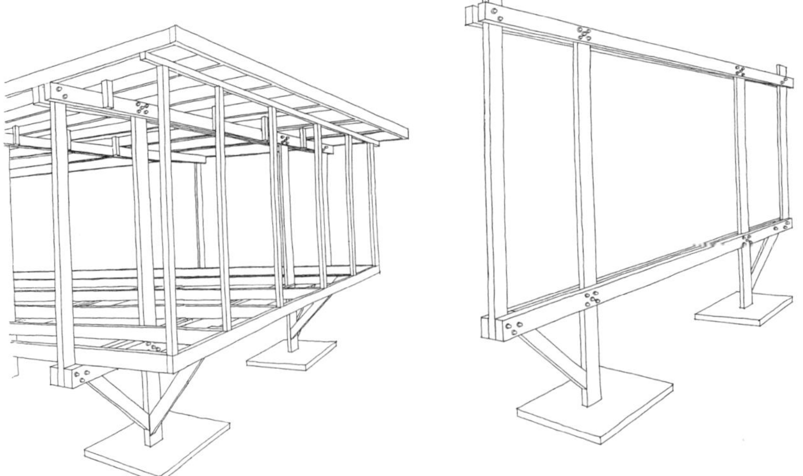

Figure 22-10. *Segal Method house, a post-and-beam system popular with owner-builders in England and Europe and described in* The Self-Build Book *(see book list below). (Drawings and photo from* The Self-Build Book. *Photo by the authors, reprinted courtesy of Green Books.)*

comparison to studwalls, in which the insulation blanket is interrupted frequently by the 2 x 6s.

Wiring can be surface mounted, but panel manufacturers now provide built-in chases to accommodate wires. The panels cost about $2.50 per square foot.

Panel makers and timber framers argue that the timber and panel systems are cost effective, if you factor in the energy savings from the high R value of the walls. As I said earlier, our experience is that conventional building is less expensive. If you are drawn to this approach, it would make sense to cost out your design both ways.

Though the panel industry is growing up around the timber frame industry, I think the greatest potential lies in the development of structural foam panels. These are engineered panels with plywood or oriented strand board on both the inside and outside. The panels are capable of supporting wall, roof, or floor loads by themselves. This type of panel could make the studwall and timber frame unnecessary. A house could be built with a very minimal frame, which would actually only serve to stabilize and contain the panels. This type of design can be truly economical compared to conventional building.

Resources on Timber Framing

— *Building the Timber Frame House,* Tedd Benson. Simon and Schuster, New York, N.Y., 1980.

— *Timber Frame Houses,* Fine Homebuilding Great Houses, Taunton Press, Newtown, Connecticut, 1996.

— *The Timber Framing Book,* Stewart Elliot and Eugene Wallas. Housesmith Press, Kittery, Maine.

— *Practical Pole Building Construction,* Leigh Seddon. Williamson Publishing Co., Charlotte, Vermont, 1985.

— *Early Domestic Architecture of Connecticut,* J. Frederick Kelly. Dover Publications, New York, N.Y., 1952.

— *The Self-Build Book,* Jon Broome and Brian Richardson. Green Earth Books, distributed by Chelsea Green Publishing Co., White River Junction, Vermont, 1995.

23

FOUNDATIONS

There are many types of foundations, but they have critical design features in common. At the top, an anchor bolt or some form of strong fastening ties the foundation to the building. At the bottom, there is a footing that distributes the weight on the foundation to prevent it from settling. The size of the footing depends on how much weight it holds, and upon the hardness or softness of the the material under it. The footing should be well above the water table (don't build in a swamp, or a seasonal swamp).

In any foundation, good drainage is critical. Water damages the foundation itself, and if it gets into a basement or crawlspace, it can cause severe moisture problems. Some soils, particularly sandy or gravelly soils, drain naturally. More dense soils, particularly clay soils, retain water; mechanical drainage or soil modification may be needed to carry water away. This could be simply a trench filled with crushed stone that leads to "daylight," but the standard method is to imbed a perforated pipe in a gravel bed, and cover it with a filter fabric that keeps the soil from silting up the stone bed.

In a cold climate, there must be some strategy to limit damage from "frost action." Wet soils that freeze can grab and lift columns or even heavy foundation walls. These "frost heaves" result from the way water expands with enormous force when it crystalizes. The most common preventive strategy is to have the footings be deeper than the "frost line," which is the depth to which the ground freezes in the locale, and to minimize the amount of water near the foundation with good drainage.

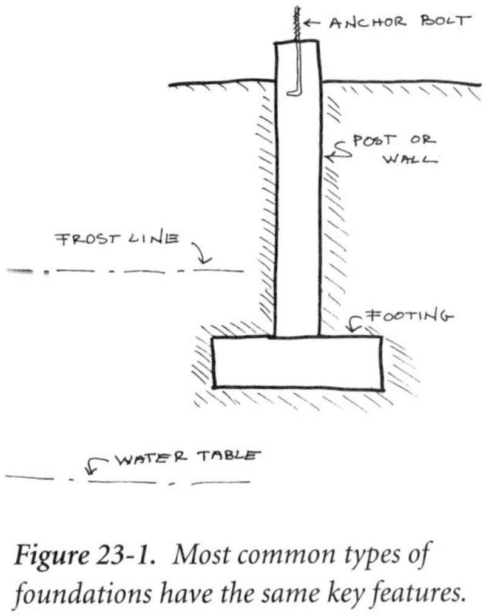

Figure 23-1. Most common types of foundations have the same key features.

Common Types of Foundations

There are a few common foundation types. A column foundation consists of rows of columns sunk into the ground, resting on footings, and connected at their tops with wooden sills. The columns can be of either pressure-treated wood or concrete.

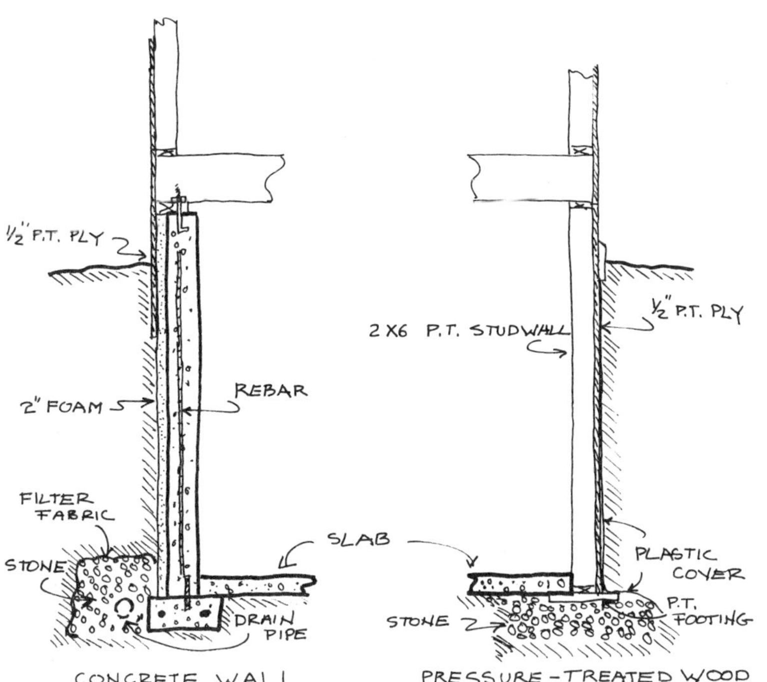

Figure 23-2. Common types of foundation.

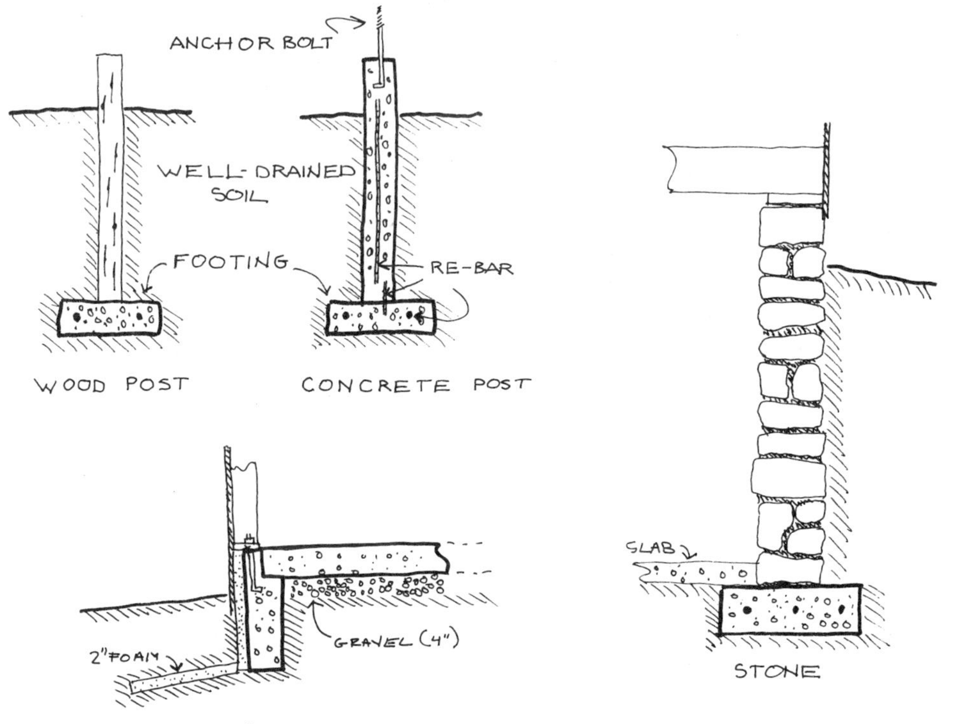

A basement foundation consists of a continuous wall of block, concrete, stone, or treated wood resting on footings set below the frost line. A variation on this is the crawlspace foundation, which has the same walls but no basement, just a crawlspace about 3 to 5 feet high. In moderately cold climates, the basement walls can be insulated.

A "slab on grade" is a concrete pad on "prepared ground" which serves as foundation and floor in one. Slabs are often built on conventional "frost walls," no different than the walls that would have been built for a basement. But even in cold climates slabs are built with shallow footings at their perimeters. Frost heaves are minimized by eliminating the water in the ground with good drainage (perimeter drains and gravel), and by using foam insulation to keep the cold from getting under the foundation.

See appendices C and D for specific information on bearing capacity of soils and frost lines. For more information on shallow foundations, send for the booklet "Frost Protected Shallow Foundation Development Program, Phase II," from the NAHB Research Center, 400 Prince George's Blvd., Upper Marlboro, Maryland 20772-8731.

A foundation can combine elements of more than one type. For example, there can be a small basement for utilities, with most of the house resting on posts. The garage can be on a slab. A house need not rest on a single foundation as long as each part of the house is properly built.

Choosing a Type of Foundation

Column foundation vs. basement walls: A column foundation is quicker and more economical to build than a basement foundation. The materials for a column foundation for a small house can cost as little as $400, compared to many times that for a concrete cellar. A column foundation can be put in by two or three people in a few days. No bulldozer will tear up the landscape, because the digging can usually be done quickly by hand.

But the column foundation gets more complicated when you factor in insulation and plumbing. Figure 23-3 shows one common approach. Here, the floor gets heavily insulated, and the insulation is covered underneath to keep out "critters." Then a mini-foundation or hot-box is needed to bring plumbing into the house. This is a workable plan, although animals will still try to make a home in the floor, and may succeed.

In a cold climate, the wind can make the floor cold. It helps to add a pressure-treated plywood skirting to keep out the wind, but this skirting has to be vented to prevent moisture buildup which would eventually rot the floor. The earth floor gets covered with plastic to keep moisture down as much as possible.

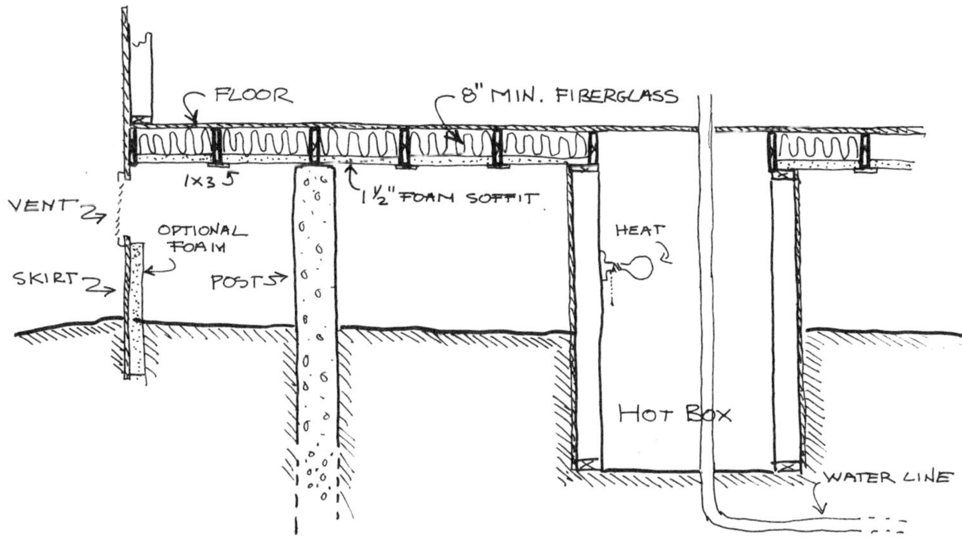

Figure 23-3. Column foundation with insulated floor.

Another approach is to insulate the skirting rather than, or in addition to, the floor. We use this approach where it is impossible to get high levels or insulation in the floor, or where wind or the height of the skirting increases exposure to the cold.

Though a post foundation is simple and cheap, it is not as great a bargain as it appears. It may not make sense in wet soils, particularly wet soils in cold climates. Unprotected by positive drainage, it is subject to frost damage. But if you go to the trouble to do the excavation for drainage, and building an insulated skirt, you might have been better off building a conventional frost-wall of concrete or perhaps a continuous pressure-treated wood foundation. You would have lost most of the savings and advantages of building on posts.

Full foundation: If a full foundation is an expensive way to hold up a house, it is an inexpensive way to provide space. If a basement can satisfy some of your space requirements, you should definitely consider building one. Anyone planning central heat, a laundry, a shop, and/or a darkroom, for example, should find a basement economical.

On sloping sites, part of the basement can be at grade. The south side can be living space, with windows and an exit door, while the back half houses utilities and storage.

In cold climates, a full basement or deep crawlspace solves many of the problems of building on posts. It keep out the wind, water, and critters, and provides an insulated space for heating equipment and other utilities.

If you live in a cold climate, plan on central heat, and a basement fits in with your overall design, then you should probably build at least a partial basement if you can afford it.

Basement walls cost roughly $35 to $45 per running foot of wall, including the form work, plus the cost of the excavation, drainage pipe, stone, filter fabric, and foundation coating. These items might add up to $1,500 to $3,000, depending on the size and characteristics of the site.

Slab foundations: In a slab foundation the slab doubles as the floor. Heating pipes and other utilities can be cast right into the slab, making the entire floor a radiator. Then the slab can be covered with carpet, tile, or similar flooring, either installed directly on the concrete, or over a subfloor built on 2-inch spacers called sleepers.

In warm climates, slabs are commonly used as an economical floor foundation combination. In cold climates they are not usually used under living areas because the floor tends to be cold near the perimeter. This need not be a problem if the slab is totally isolated from the cold with rigid foam insulation. In fact, a well-insulated floor slab can become a storage bank for solar heat.

Concrete can be made into a finish floor with a simple paint job, but I have seen beautiful concrete floors that have had pigment mixed into the concrete. I have seen others that also had grooves struck into the concrete to make it like a tile floor. Of course, the floor can be also be tiled.

This is a tricky design area, so if you are considering a slab foundation in a cold climate I suggest you get professional advice.

Wood vs. masonry: Modern pressure-treated wood products make it possible to build a wooden foundation that is virtually permanent. Today even full basements are being built of wood.

Wood foundations have several advantages. They cost less than masonry foundations. They're faster and easier to install, especially for a novice with no experience in masonry work. The components are light, which means less back-breaking labor for you and less weight on the ground. The diagram in figure 23-2 shows the basic elements of the scheme.

Unlike reinforced concrete foundations, the components of a wood foundation are changeable; you can remove, repair, or alter one without much trouble, whereas altering a reinforced concrete foundation is almost impossible. In that sense, wood foundations sit lightly on the land.

Many people don't think treated-wood foundations will last. They may not be accepted by every building code. Our experience so far has been that treated-wood foundations stand up well in well-drained soils. I do not think a wood foundation should be used in soil that tends to stay wet much of the year, because the wood may break down sooner. If I were doing the work myself, I'd use the wood foundation system. But if I were having a foundation built, I would have it done in concrete the conventional way.

HOW TO LAY OUT A FOUNDATION

Whether you use columns, a basement, or a slab, the foundation will have to be laid out precisely. If a contractor is building your basement foundation, he or she will lay it out using a transit (figure 23-4), which is a tool for establishing consistent elevations.

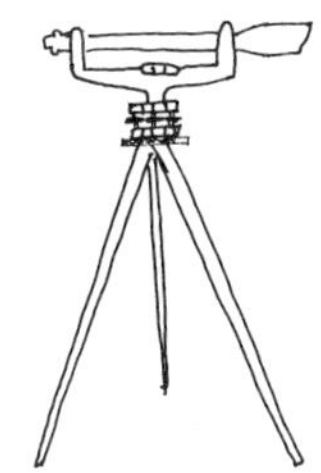

Figure 23-4. Builder's transit.

Otherwise you will do your own layout using batter boards and strings. Usually this is the first thing you do at the site. However, if a backhoe or bulldozer will be used for excavation, you do the accurate layout after the digging, since the heavy machinery would knock down the stakes and strings. Excavation can be done with an approximate layout made on the ground, typically with an outline made with lime striping or spray paint.

Figure 23-5 shows batter boards and strings. The string locations are marked on the batter boards with nails so that the strings can be removed when they are in the way and put back in exactly the same positions.

Here are the steps for laying a foundation:

Step 1

Rough layout: Using a compass and a fifty-foot tape, lay out roughly where the house will be. Mark the corners with rocks or stakes.

Step 2

Put up stakes: About three to four feet outside of these stakes, plant the tall stakes that will support the batter boards. These should be 2 x 3s, 2 x 4s, or similarly strong poles from the woods. They must be as high or higher than the foundation, because the batter boards are set at the level of the foundation top. The foundation height will often be about 8 inches above the ground at the highest corner. You can point the stakes with a hatchet and drive them into the ground with a sledgehammer. If you are building a column foundation with a row of columns down the middle of the house, provide stakes to support batter boards to mark this row of columns.

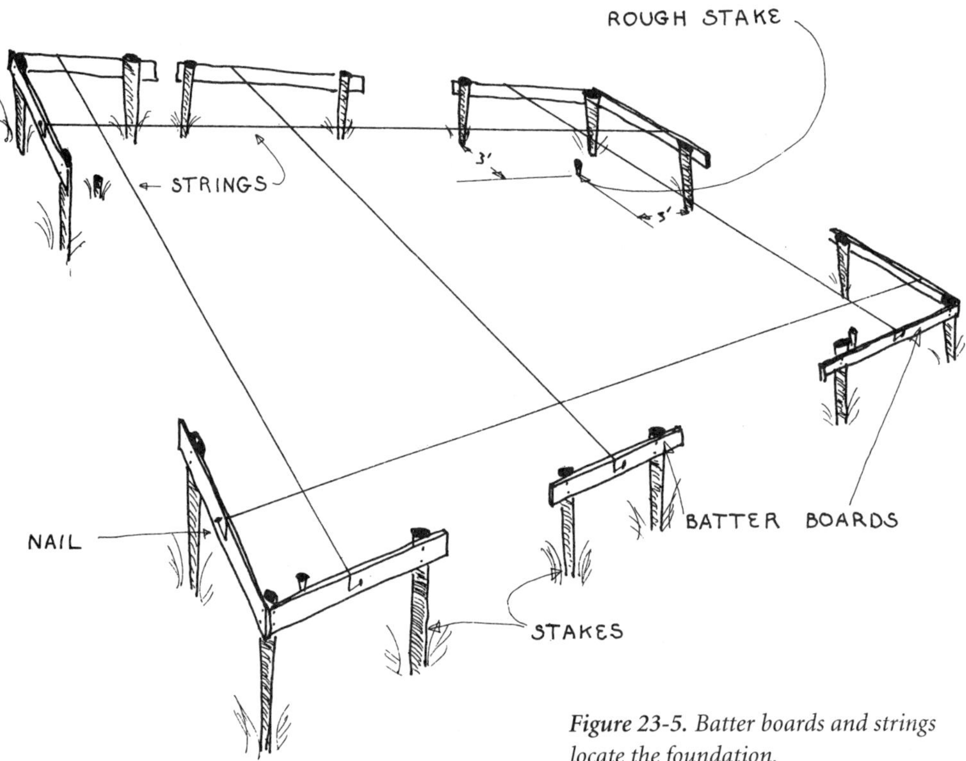

Figure 23-5. Batter boards and strings locate the foundation.

Step 3

Attach batter boards: When the stakes are in, nail 1 x 6 rough boards onto them as shown in figure 23-5. The top edge of these boards should be at the level of the top of the foundation. Level the batter boards with a water level, which you make yourself with a gallon jug and clear plastic tubing from the hardware store (figure 23-6). Chapter 35 shows exactly how to use a water level.

When the boards are nailed on at the right height, sight across them to make sure they are all in a single plane.

Step 4

Attach strings (figure 23-7): If your house is 16 x 24 feet, the strings you attach to the batter boards will form a rectangle 16 x 24 feet. Your foundation will be located exactly within this rectangle.

Using eyeball or compass, set up the first string, the one representing the first side of the rectangle. Pull it tight, and tie it off to nails in the top edge of the batter boards.

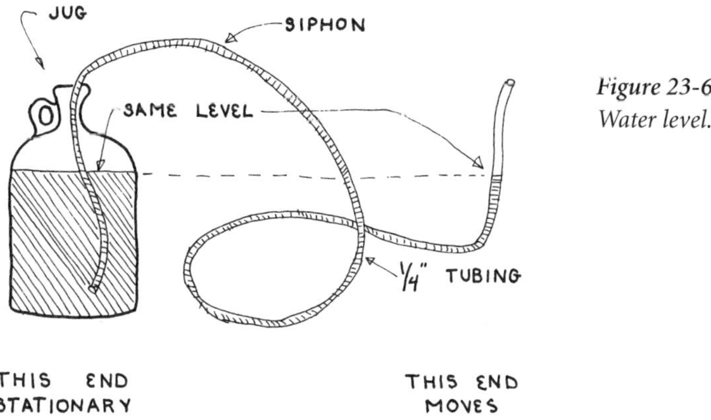

Figure 23-6. *Water level.*

Next, attach the opposite, parallel string in the approximately correct spot. You will need a fifty-foot tape and someone to help you. Make the second string equidistant from the first at both ends. The string should be drawn tight over the board and tied to a nail on the bottom outside edge of the board. This allows you to adjust the string back and forth until it is exactly parallel to the first string and exactly the right distance away. When the second string is just right, mark the batter board by putting a nail into the top edge where the string comes across.

Put up the third string approximately square (90 degrees) to the first two. Use a framing square to make a right angle. Do not set the final location of this string yet. Figure 23-9 shows the relationship of all four strings.

The fourth string will be parallel to and opposite the third, just as the second was parallel to and opposite the first. Measure and mark where the fourth string should go by tying marker strings onto the first and second strings (figure 23-8).

Figure 23-7.

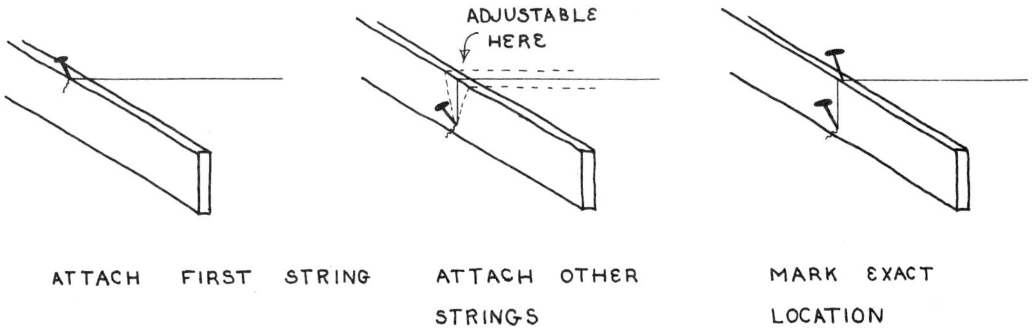

ADJUSTABLE HERE

ATTACH FIRST STRING

ATTACH OTHER STRINGS

MARK EXACT LOCATION

You can slide these little markers back and forth until they are exactly the right distance from the third string.

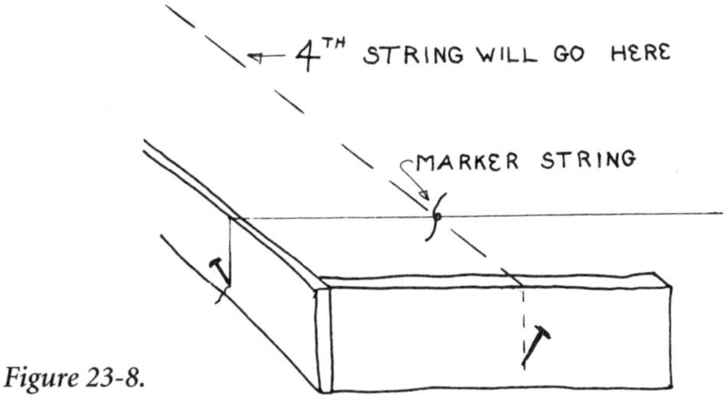

Figure 23-8.

STEP 5

Squaring the strings: The dimensions should now be accurate, but the squareness of the shape may be off. Squaring is done by measuring diagonals. In geometry, when the diagonals of a parallelogram are equal you have a rectangle. Equalize the diagonals by moving strings three and four back and forth together from the same ends until the diagonal measurements are within ⅛ inch of one another. When you get it right, fix the locations for the third and fourth strings with nails at the batter boards.

You can now lay out strings for interior rows of columns by stretching other strings parallel to the outside ones. If your layout is L-shaped, U-shaped, or some other odd shape, divide the shape into rectangles and lay out one rectangle at a time. When all the strings are right, mark the positions permanently with nails at the top of the batter boards .

Figure 23-9.

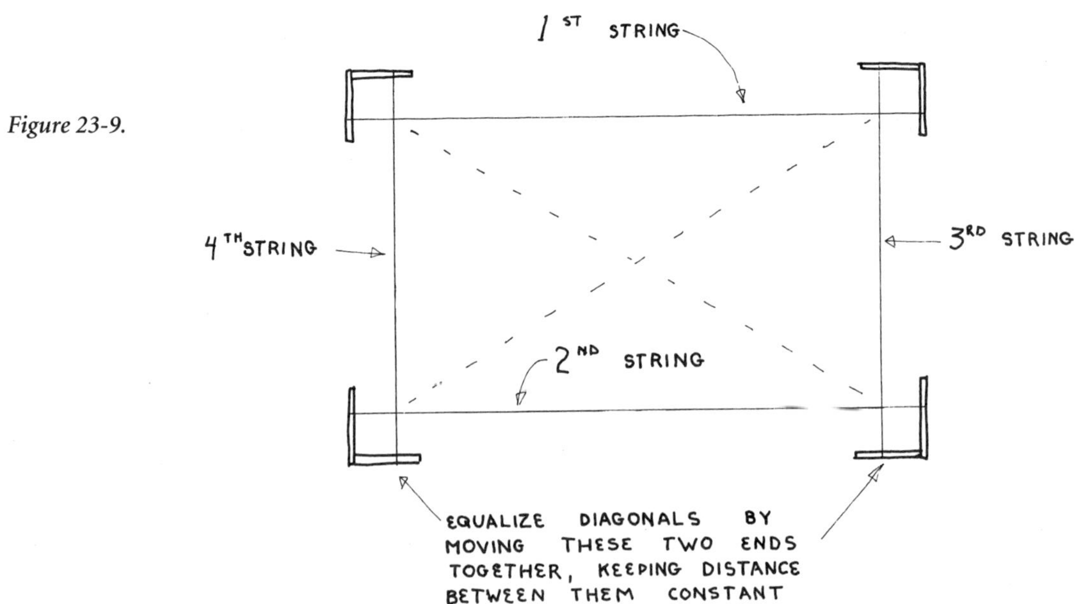

BUILDING FOUNDATIONS

Column Foundation

Wooden columns will consist of 6-inch or bigger pressure-treated posts. You could use a round column, but a square section such as a 6 x 6 is easier to work with.

Concrete columns are poured in place in special cylindrical cardboard forms, often known by the trade name Sonotubes. One-story houses, with each column supporting an area of 150 square feet or less, can be built on 8-inch columns. Use 10-inch columns if the columns are less frequent, if the house has two stories, or if the columns stick more than 3 feet above the ground.

Layout and design: Foundation posts are always arranged in rows. On top of each row is a sill, a heavy timber such as a 6 x 8 that bridges the posts and is a main carrying timber of the house. The floor beams or joists rest on the sills and span between them (figure 23-10). For purposes of illustration the column height is exaggerated in the drawing. Usually the columns will be as low as possible. It is also possible to hang the joists between the sills.

Figure 23-10.

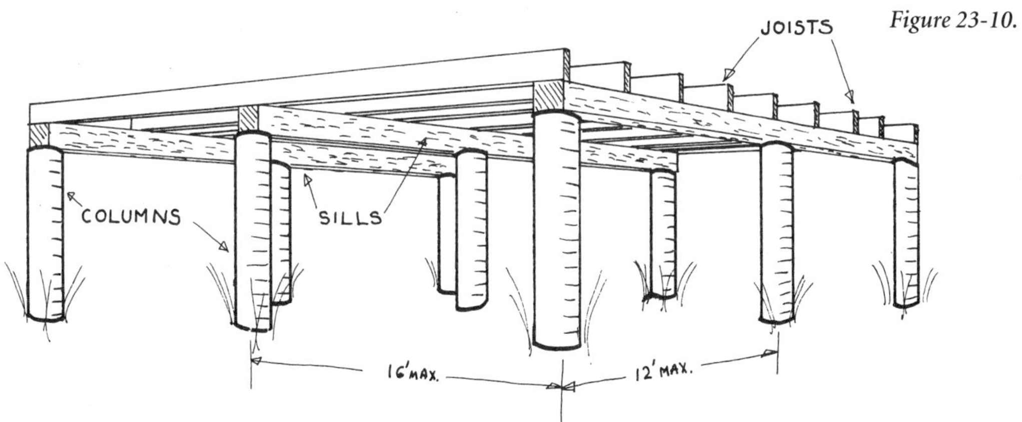

Figure out how many rows of posts you will need and how many columns should be in each row. If there are too many columns, the work and cost of putting them in is too high. If there are too few and the distance between them too large, the timbers that connect the posts will have to be impractically large.

For most designs, column rows should fall about every 12 to 14 feet, which is a reasonable length for floor joists to span. Sixteen feet should be the maximum distance between column rows. Within a given row, columns should be spaced perhaps 10 to 12 feet apart.

A foundation plan must accommodate any special loads that come down inside the building. If part of the roof or second-story floor is resting on walls or posts inside the house, these walls or posts should be directly supported by strategically located sills or posts (figure 23-11).

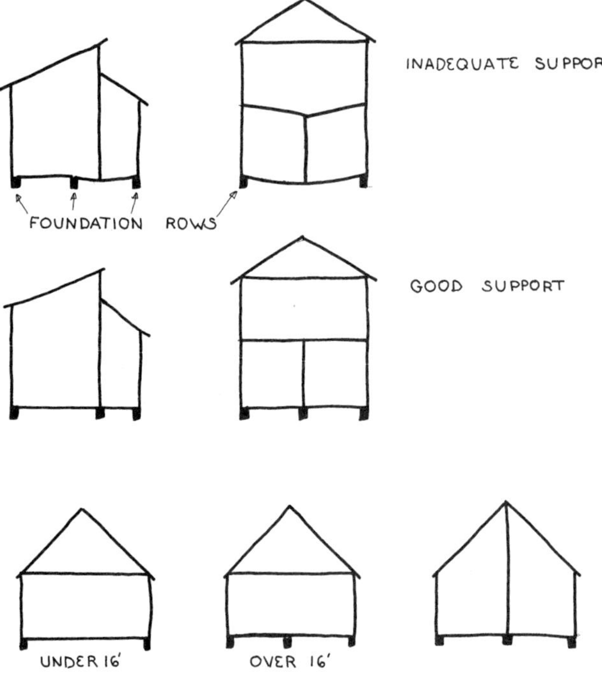

Figure 23-11.

Finding the sill size: Once the foundation is laid out, you can figure out what size sill will be necessary. The longer a sill is and the more weight it supports, the larger the timber has to be. A two story house will need bigger sills than a one-story house. Chapter 16 explains how to compute what size timber is needed for any situation. I recommend that you spend a few hours mastering that material, because it will give you a better understanding of how a house structure works and enable you to make your own design more efficient. The sill size table in appendix B also gives sill sizes for different cases.

Figuring out how the sill will sit on the column: The sill should be centered as much as possible on the column without sticking out where the rain can collect and rot it. With a narrow sill and a wide column, centering the sill on the column will sometimes cause a water trap. If so, move the column over until the siding just covers it (figure 23-12).

Figure 23-13 shows how to calculate where the center of the column should be with a 4-inch sill on a 10-inch column. The siding and sheathing add up to 2 inches. The strings mark the outer edge of the sill. If the column is positioned with its outside edge flush with the siding, the center of the column will be three inches inside the string. Use a plumb bob (figure 23-14) to locate the center of the column on the ground, and mark the spot with a small stake.

Digging: You can dig the holes yourself, or hire a backhoe at $50 or more per hour to do the digging. I suggest you dig a test hole. If it takes more than an hour or two by hand, the backhoe will save a lot of time and aggravation. If a backhoe is to do the

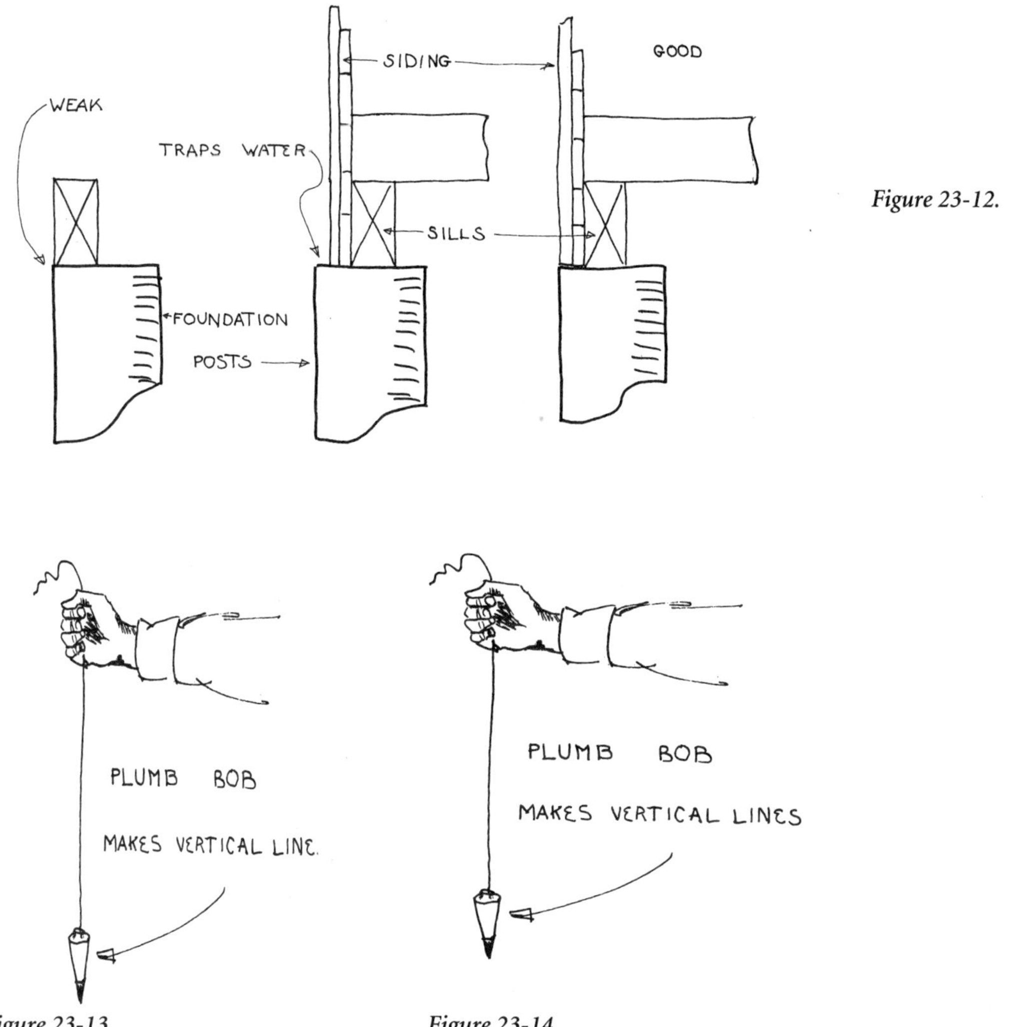

Figure 23-12.

Figure 23-13.

Figure 23-14.

digging, put up the batter boards as described earlier after the backhoe finishes. The backhoe digs a big hole, so it can work from an approximate, quick layout done with stakes and a fifty-foot tape. The accurate layout can be done later.

If you dig yourself, you will need:

- Shovel
- Posthole digger (rentable)
- Mattock for roots
- Pick for rocks
- Five-foot crowbar for prying out boulders

Make your holes 1½ to 2 feet in diameter at the top; the sides should go straight down or, if possible, splay out at the bottom. Dig the holes below the frost line. You can find out where this is by asking local builders or by referring to appendix D. If you hit bedrock, you can pour the footing on that. The diameter of the hole at the bottom is determined by the size of the footing.

Footings: The footings consist of reinforced concrete puddles poured in the bottom of the holes. Their diameter and depth depend on the amount of weight each column must hold and the bearing capacity of the soil. In general, a one-story house can have footings 20 inches or 24 inches in diameter, and a two-story house can have footings 30 or 36 inches in diameter. The footing thickness depends on the diameter, as follows:

DIAMETER (IN INCHES)	THICKNESS
20	6
24	8
30	10
36	12

The steel reinforcing can be four ½-inch reinforcing bars placed in a tic-tac-toe shape about 2 inches up from the bottom of the footing. For a 36-inch footing, use six bars.

You can have ready-mixed concrete delivered by huge cement mixing trucks. Use it if possible to save a lot of backbreaking labor, and because the economy made by mixing your own is not great. You may need to mix your own concrete if footing volume is smaller than the minimum you can get delivered, or if you are building in an inaccessible place.

Mixing concrete: Prepared concrete mix comes in bags, but it is much cheaper to make your own. Here is a good recipe.

One part Portland air-entraining cement
Two parts fine aggregate (clean sand)
Three parts coarse aggregate (gravel)

A lumberyard can sell you the cement in ninety-four-pound bags, which is one cubic foot of cement. The aggregates can be bought from a sand and gravel company. The fine aggregate must be clean sand. The coarse aggregate ideally should be a mixture of particles ¼ inch to 1 inch or 1½ inches in diameter. But often peastone is used. This is uniformly about ¼ of an inch or ⅜ of an inch in diameter. What you want is a mixture that contains an even gradation of particle sizes. Tell the person at the gravel pit what you are doing, so he or she can give you the right materials. Sometimes concrete is made with a single, ungraded bank-run gravel dug right out of the hillside and used as is. If you use bank-run gravel, make sure it contains an even gradation of particle sizes and not mostly small or large pieces. Eliminate pieces over about 2 inches in diameter, because they will make the mix harder to work with.

You can mix concrete in a rented electric mixer or make your own mixing pan out of a 4 x 8 sheet of plywood and some 1 x 4 or 1 x 6 boards, as shown in figure 23-15.

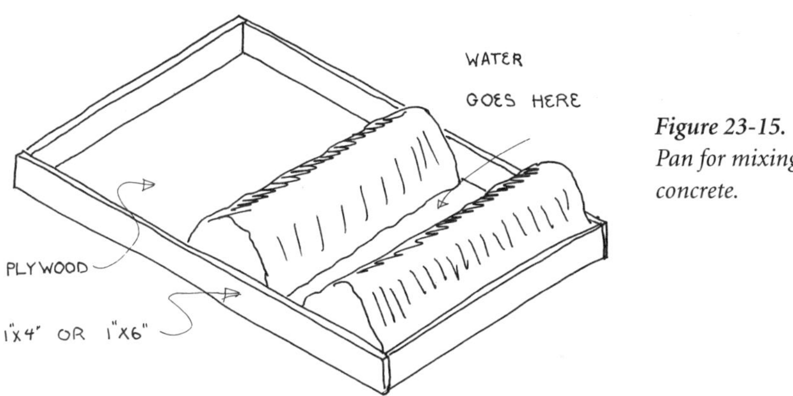

Figure 23-15. *Pan for mixing concrete.*

If you use an electric mixer, first mix the ingredients dry, then add water until the mix reaches the right consistency. It's hard to describe what the right amount of water is, but the mix should be stiff without being dry. Every particle should be wet, and the mix should get shiny when you whack it with the back of a shovel. It should not be runny because a wet mix makes weak concrete.

If you mix in a wooden pan, the trick is to mix the ingredients as you fill the pan. Sprinkle a shovelful of cement lightly around the pan. Then sprinkle two shovelfuls of sand and then three shovelfuls of coarse aggregate. Do this ten times. After a few passes with a shovel or hoe, the dry ingredients will be mixed. Then use a hoe or shovel to form two mountain ranges with a deep valley between them. Pour some water in the valley. Take the hoe and gently pull part of the mountainside into the water, spreading it out with a pulling stroke. Keep this up until your mountains have been eaten away and the water is used up. Add water if needed. Then keep mixing until the batch is well mixed. Keep track of the amount of water you use to reach the right consistency. On the next batch, put in all the water at once, which will make the work easier.

Pouring concrete footings: Concrete trucks run on a tight schedule and have to dump their loads quickly. If you are using ready-mixed concrete, you will have to be completely prepared with a crew of three or four people all set to work. Line the bottom of each footing hole with old newspapers, felt, paper, or plastic to prevent the water in the mix from leaching into the soil and weakening the concrete. If the chute on the truck won't reach a hole, put the concrete into a wheelbarrow or buckets and transport it to the hole. As you pour the footing, try to keep dirt from the sides of the hole from falling into the concrete. You can stretch the concrete by throwing in rocks you have dug up, but rinse them first. When you have put about 2 inches of concrete into the hole, put the reinforcing bars, which you have cut in advance with a hacksaw, in the tic-tac-toe shape.

Finish pouring the concrete to the right depth. Use a 2 x 4 to pack it tight so there are no air pockets. If the columns will be concrete, reinforcing bars should be put in to tie the column to the footing. On a relatively flat site, a single ½-inch rod 12 to 18 inches long will do. It can be put in after the concrete has set enough for it to stand up straight. On a sloping site, use three L-shaped rods about 2 feet long, as shown in figure 23-16.

These must be placed immediately after you finish pouring the concrete. Wire them together or use sticks to hold them up while the concrete sets. Whether you use one bar or three, set up the layout strings to make sure they emerge from the footing within the circle where the column will be.

Figure 23-16.

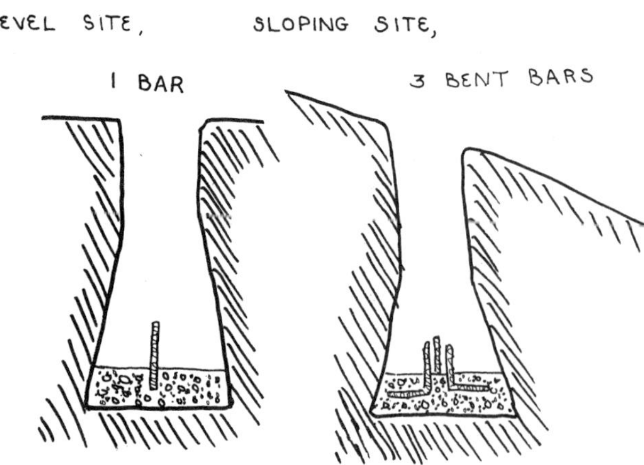

Cutting the columns: You must cut your wooden columns (or cardboard forms if your columns are concrete) so that each of them will come to exactly the same elevation. Since the bottoms of the holes will not be at the same elevation, each column will have a different length. One method is to set up your strings, then measure from the strings down to the footing.

You can also use the water level. (Consult chapter 35 to see how the water level is used.) Set the reservoir of the level at the column height, which is the top of the batter boards minus ½ inch. Get a stick, such as a 1 x 2, at least as tall as the tallest column, to use as a ruler. (You will not need a tape measure at all.) Take the stick to the first hole, set its bottom on the footing, and hold it as plumb as you can by eye. Put the free end of the level's tube next to the stick and make a mark opposite the water level in the tube. Write "1" next to the mark. Take the stick over to the second hole, put it on the footing, straighten it, mark it in the same way, and put a "2" next to the mark. Continue the procedure until your stick has a mark for each column.

You are now ready to mark the columns or cardboard forms. The stick is the ruler. Make sure the columns or forms are cut square. If you've used cardboard tubes for concrete, make several measurements around the perimeter and connect them freehand; then cut along that line with a hand saw or saber saw (figure 23-17).

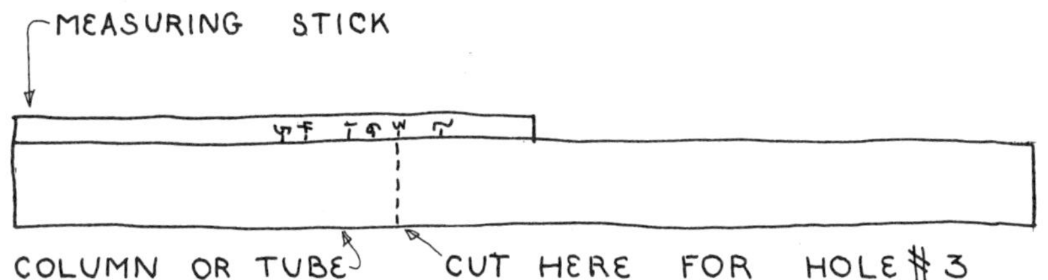

Figure 23-17.

Sonotubes come in 10-foot pieces. You can avoid waste by taping the scraps together with duct tape. Put the taped joint within 1 foot of the top or underground so that the weak place at the joint won't cause trouble.

Setting the columns: Cardboard forms or wooden columns are set in the ground similarly. One person holds the column or cardboard form at the right place relative to the layout strings at the top. Another person uses a 2-foot, or longer, level, held plumb, to adjust the bottom of the column or form until it is plumb (exactly vertical). Take readings on two different sides of the column.

When the column or form is vertical and lines up with the strings, place a few inches of dirt in the hole, and pack it down carefully with a 2 x 4 without moving the bottom of the column in the process. When the bottom is positioned, one person continues to hold the top in place while the other fills in more dirt.

About every 6 inches, pack the dirt down with a 2 x 4. Rocks or other rubble may be added to fill the hole since the dirt will run short.

Pouring concrete columns: If your foundation is wooden, it is now done. If you are using concrete columns, these can now be poured the same way that the footings were. Any tubes that stick above ground more than 30 inches will have a tendency to wave back and forth when they are full of heavy concrete, so the tall ones should be braced to the ground or to the batter boards to keep their tops positioned.

Any column that sticks more than 3 feet above the ground should be reinforced with three pieces of ¼-inch #2 reinforcing rod. The rods can be cut to fall just short of the top of the tube, so they won't be in your way later. Put them in after you have poured the concrete about 2 feet deep.

Agitate the concrete as it is going in to prevent air pockets that will weaken the column. This may be done several ways. A 2 x 4 can be used to pack down the concrete as it goes in, shovelful by shovelful. You can drum on the sides of the tube with your fists, which will free a lot of air. If the tube is reinforced, you can wiggle the reinforcing rod vigorously to settle the concrete.

When a tube is full, install a special L-shaped bolt called an anchor bolt. This will later hold the sill firmly to the column. The trick is to hold the anchor bolt in the right position long enough for the concrete to set (figure 23-18). Put the L-shaped end down. The bend in the bolt prevents it from being ripped out of the hardened concrete by a violent event. The anchor bolt should sink at least five inches into the concrete and should be centered.

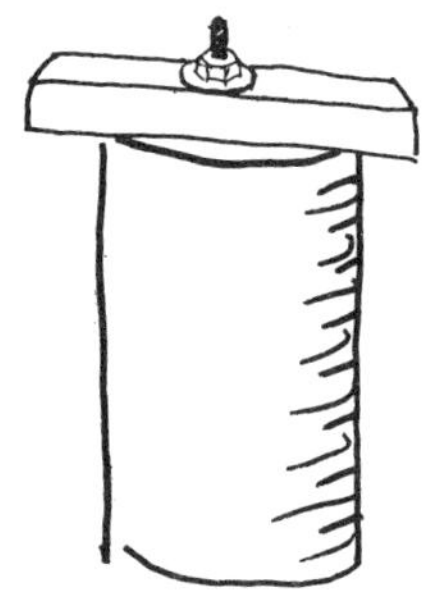

Figure 23-18. *Use a scrap 4 x 4 to position the anchor bolts while concrete sets.*

Setting the sills on the column: Sills which are within 8 inches of the ground should probably be pressure-treated lumber.

You may be using large timbers for your sills, but they are often constructed by nailing together several 2 x 8s, 2 x 10s, or 2 x 12s. If a sill will span over several posts, span it across at least two posts if possible, or stagger the joints as shown in figure 23-19.

Nail up these compound sills on sawhorses. The ideal sill has a slight crown upward. It will settle to a flat floor when loads are applied. If the lumber you have for your sill is quite warped, you can position the warps to cancel each other out as you nail them together. Begin nailing at one end, then use a large bar clamp or sliding clamps to align the pieces as you work your way down. As the sills get assembled, use a rafter square to make sure the face of the sill stays square with the side.

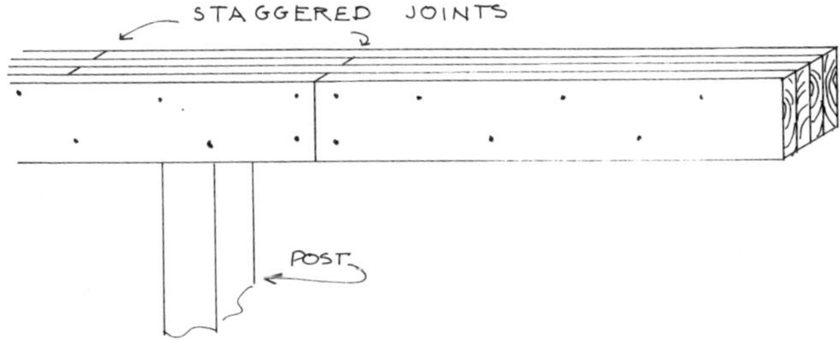

Figure 23-19. Building up a sill from 2-inch lumber.

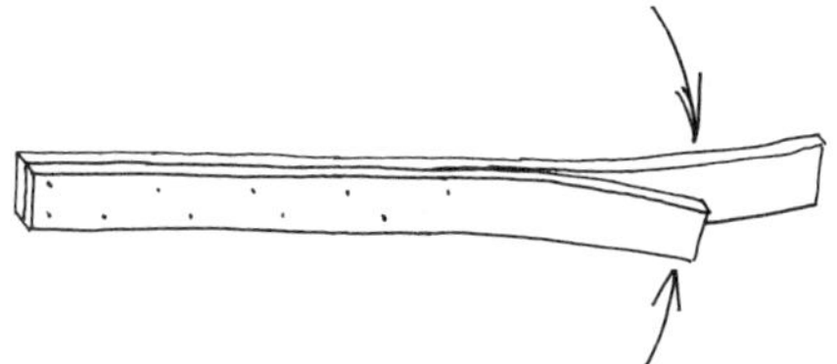

Figure 23-20. When building up sills from crooked stock, oppose the warps, then force them into alignment to create a straight sill.

Locate the resting place of the sills. Even if there is some inaccuracy in the location of the columns, set the sills exactly. Often the outside edges of the sills should fall right on the layout strings. Set up the strings, and mark on the column exactly where the sills should go. If the columns are wood, the sills can be placed on the marks and toe-nailed in with 20d nails. Also tie the sill to the column with steel strapping or plates (figure 23-21).

If your columns are concrete, wait until the next day to place the sills, since the concrete must set. Even then be gentle with the concrete, which only gradually cures to its full strength over the course of twenty-eight days. Locate the sills by marking the columns exactly where the strings pass over them. Then drill holes in the sills for the anchor bolts. To locate the holes you will have to measure from the end of the row to the center of the bolt and from the layout string to the bolt, since the anchor bolt may not be perfectly centered. Since it is hard to measure precisely where these holes should be and the anchor bolts may not be perfectly straight, drill oversize holes to allow a little play. Use a 1-inch hole with a ½-inch anchor bolt, and then put a large-size washer under the nut.

Since the spade bits are only 5 or 6 inches long, they will not drill all the

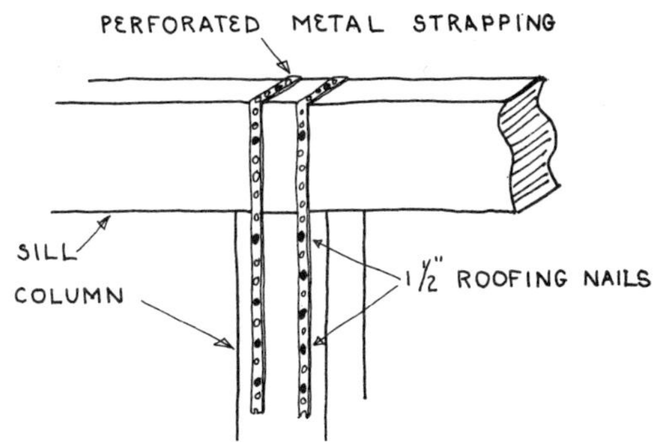

Figure 23-21.

way through the sill. You can drill from both top and bottom of the sill and hope the holes will meet in the middle, or get a drill bit extension. In either case, have a friend hold a square on the sill next to the drill to make sure that you are drilling straight, or duct tape a torpedo level to the top of the drill.

The sill should not be installed in direct contact with a concrete column, since masonry attracts dampness. Before you put the sills on, cut a piece of sheet aluminum bigger in radius than the column, make a hole in it for the anchor bolt, and place it on top of the column as a shield against moisture. Put the sills on loosely and check them for level. Check too that they are exactly the right distance apart and that the diagonals are equal within ¼ inch if possible. Make adjustments, and bolt the sill in.

What to do if the foundation is inaccurate: Your foundation can be out of whack in two ways. First, the column heights can be a little off. If a column is too high, notch the sill above it by the same amount. If the column is too low, add to the column with wood scraps or plastic shims of the right thickness. Separate the wood from the masonry with a layer of aluminum flashing.

Second, the columns can be out of line. If the discrepancy is an inch or two, line the sills up where they should theoretically go and let them sit a little incorrectly over the columns. If a column is more than 2 inches out, you need advice from an experienced person.

Building a Basement Foundation

A basement foundation is a more ambitious undertaking than a column foundation. I describe four types of basement foundations: poured concrete, block, stone, and wood. Readers planning to build any of them should work with an experienced person, and consult the references in the bibliography. Many of the steps are the same for all types of foundations.

Layout: It is usually good practice to insulate the outside of the foundation walls with 2-inch foam board, sometimes known as "blueboard," which is made for this purpose. The outside dimensions of the building will then be to the outside of the foam, so the foundation wall will be inset 2 inches. If the basement is finished living space, an alternative would be to do the insulating on the inside using regular 2 x 4 or 2 x 6 wall framing.

Excavation: A cellar hole is usually dug by a backhoe, at about $50 per hour. At such rates your cellar hole will cost a few hundred dollars. There should be a corridor 2 or 3 feet wide all around the foundation, to make room for building the foundation walls. A 16 x 20-foot house needs a hole about 22 x 26 feet.

Put up batter boards and strings after the hole is dug. Because the hole is oversized, these must be set further back than they would be for a column foundation. Otherwise, the layout strings are set up exactly as for column foundations (pages 281–82). If you're having a contractor do the foundation, he or she will do the layout and no batter boards will be needed.

Building footings: Any masonry foundation wall will need the type of footing shown in figure 23-22. The footing should be as deep as the wall is wide and its width twice that. Here is a step-by-step approach.

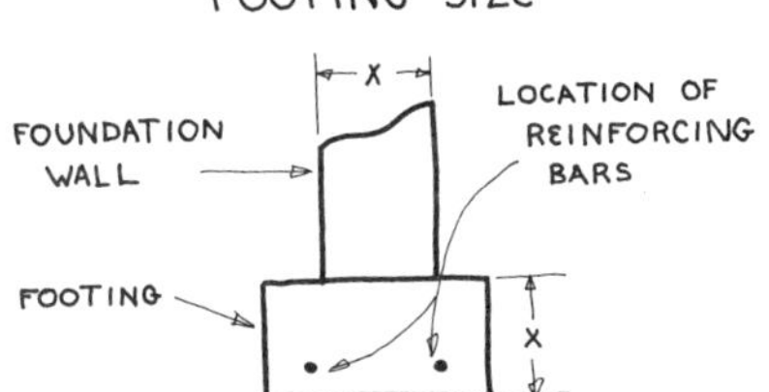

Figure 23-22.

Step 1

The simple forms for a footing are shown in figure 23-23. An easy way to locate them is to place an extra string on the batter boards right above where the outside edge of the footing will be. Measure down with a plumb bob to locate the outside of the footing. The first plank of the form will go there. Measure down from the string to find the correct height of the footings. Since it is difficult to get the planks perfectly level at the top, the planks are set a little high. Then a chalk line on the inside of the planks establishes the height the concrete should be poured to.

The planks are held in place with stakes and stabilized at the top with spreaders as shown in figure 23-23.

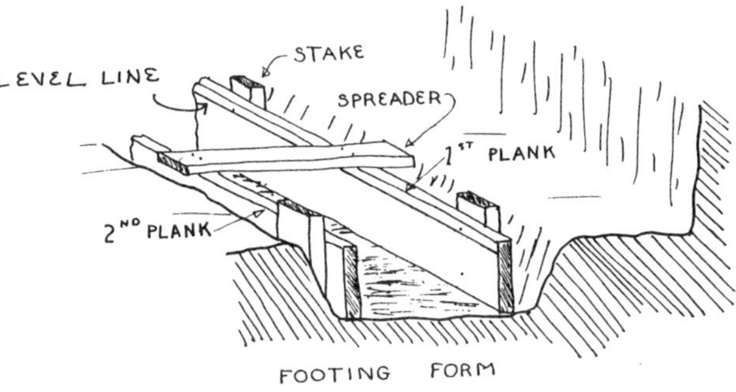

Figure 23-23.

Step 2

Before pouring the footing, pack down the loose earth inside the footing form with a tamper (figure 23-24).

Step 3

If the soil is dry, it will absorb water from the concrete too fast and weaken the mix. Sprinkle water in the bottom of the footing to dampen the soil.

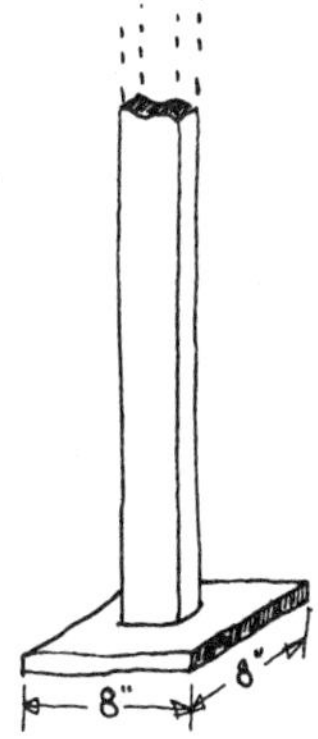

Figure 23-24. Tamper.

Step 4

Next, mix and pour the concrete. Pages 284–86 explain how to mix your own concrete. Ready-mixed concrete, delivered to you all ready to go, costs about $70 per cubic yard. The cost for the footings of a small house might be around $300. This expense is small compared to the labor of mixing it by hand.

When a concrete truck comes, be ready. These trucks are on schedule and charge extra for any time they spend idle. Have a crew of at least four people prepared with shovels and a good strong wheelbarrow. The truck will need a good, fairly flat road right up to the hole. Plan exactly how the mix will be brought to the forms from the truck. Fill the part of the form farthest from the truck first, and work back toward the truck.

Step 5

Reinforcing: Though many footings are put in unreinforced, ½-inch reinforcing bars laid in the concrete will strengthen the footing against shifting and hold it together if it ever cracks. When the concrete is 2 inches deep, lay in two bars parallel with the footing (figure 23-22) and then continue pouring.

Step 6

As the concrete approaches the chalkline on the inside of the form, use a shovel first and then a short board (called a "screed") to level the concrete. The concrete should be level, but it need not be smooth. In fact, if you pour concrete walls above, the roughness will help the wall adhere.

Step 7

If you pour concrete walls on your footings, use short pieces of reinforcing bar to tie the two parts together. Cut pieces of ½-inch reinforcing bar about 16 inches long with a hacksaw. Place the bars sticking up out of the center of the footing every 2 feet. They can be stuck into the concrete as soon as it has stiffened enough to hold them up.

Step 8

In hot weather, the sun and heat can cause the concrete to dry too fast, which weakens it. Keep the concrete damp until it has fully set at least twenty-four hours. Cover it with plastic between soakings.

Poured Concrete Foundation Walls

Concrete full-wall foundations are cast in one piece in special forms. Since there are no joints in the forms to break or leak, they are both stronger and tighter than any other type of foundation. Reinforced with steel, they are nearly indestructible.

If you have chosen a poured concrete foundation, I recommend having all of the work done by a foundation contractor. If you do it yourself, you can save half of the cost, assuming you can reuse your form material, but I don't think it is an efficient use of your time. The plywood forms you build must be super-strong, and you must build them piece by piece, a job you have never done before and probably will never do again. The plywood you make the forms from will be coated with a thin film of concrete and will dull any tool you use to cut it. The contractor, who has done it all before, has special forms that can be put up in a morning and reused again and again. A labor that will take you weeks will take the contractor two days. Concrete foundation work is one place where a contractor's services are worth their premium cost. If you want to build your own foundation, I suggest you build a block or treated-wood foundation.

Even if you hire your foundation work out, there are several important tasks that will keep you busy. First is to make sure at each step that the foundation is going in as you have planned, particularly that the height is as designed. Figure out where sewer pipes, floor drains, or other pipes or wires will penetrate the foundation. Holes for a sewer line are often made by casting a paint can into the wall; it's the right size.

Block Foundation

A block foundation costs about sixty percent as much as a poured concrete foundation you build yourself and about twenty-five percent as much as a concrete foundation a contractor builds for you. Block is definitely the most inexpensive way to build a masonry foundation in a reasonably short time. These foundations are not as strong or watertight as reinforced concrete because they are not monolithic. They are more likely to develop leaks and are more susceptible to damage caused by settling or frost. But they work well in a well-drained soil.

If you wish to build a block foundation, get someone experienced with block construction to help you design your foundation and to show you how to lay block. The Audel's guide on the subject has an excellent, detailed description of block construction.

Block construction is modular. Every dimension in every direction is, or should be, a multiple of four inches. The standard block, called a stretcher, is 7⅝ x 7⅝ x 15⅝ inches. With a standard ⅜-inch mortar joint, each block occupies a space 8 x 8 x 16 inches (figure 23-25).

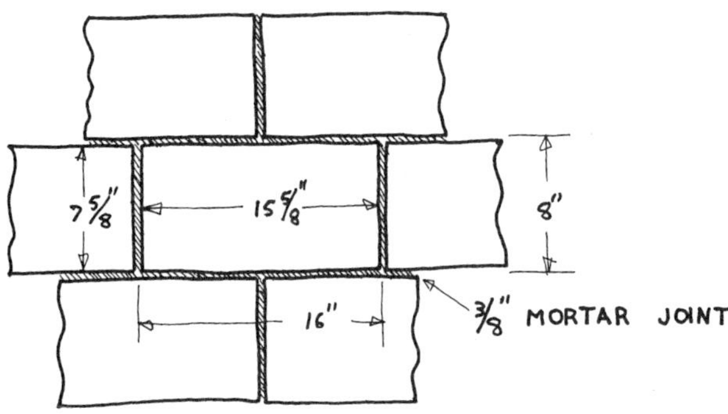

Figure 23-25. Standard block dimensions.

Eight inches is a standard thickness of a block wall. Other common thicknesses are 4 inches (used for nonbearing partitions), 10 inches, and 12 inches. These extra-thick walls are used where extra strength is needed or desired. For each thickness of wall, a variety of blocks, including one-half and three-quarter length blocks, are available for special situations, such as corners and window locations (figure 23-26).

A block foundation should be designed to avoid cutting any blocks. A wall 16 feet long will be 12 blocks long and easy to build. A wall 16 feet, 2 inches long will require a lot of block cutting and make the work longer. Usually the corners are laid first in a bed of mortar on the footing and are staggered back as shown in figure 23-27.

Strings are then stretched from one end of the wall to the other to line up the block courses (rows) in between. All the blocks are carefully plumbed and leveled to keep the wall straight and accurate. Anchor bolts are embedded in the top course to hold down a 2-inch-thick sill.

A modern variation on block construction that is becoming popular is surface bonding. In a surface-bonded wall, the first course of blocks is set in mortar. After that the courses are stacked with no mortar at all. Then both sides of the wall are plastered with a special cement such as "Block-bond," which is reinforced with glass fiber. Though this construction might not seem strong, tests have been conducted that show that a surface-bonded wall is stronger than a conventional block wall. The mortar joints in a regular wall do not actually hold the block together very strongly. But the glass fibers in the surface bonding make the plastered surface into a continuous film with considerable tensile strength. The surface is also more waterproof.

A surface-bonded wall is faster to put up than a conventional block wall, particularly for the amateur, and is probably one of the best methods for building your own basement.

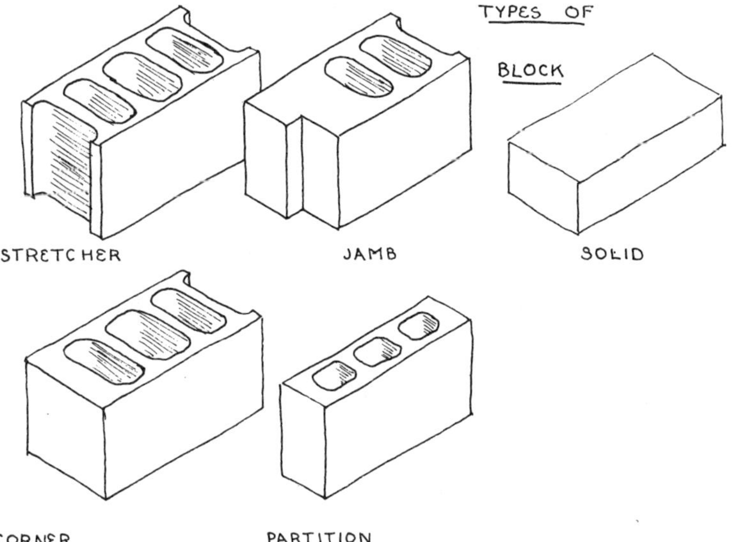

Figure 23-26. *Standard block shapes.*

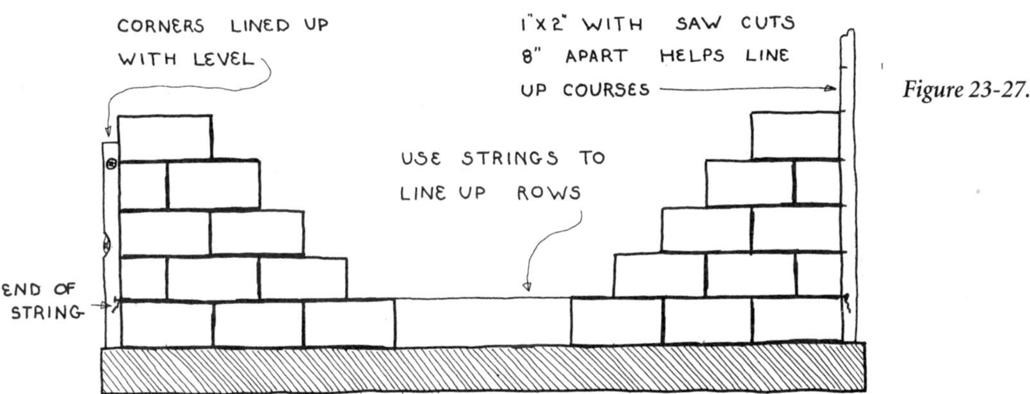

Figure 23-27.

Stone Foundations

A stone foundation, like a block foundation, is susceptible to frost damage and uneven settling. On a good footing and in well-drained soil, this will not be a serious drawback. The major disadvantage of stone construction is that it is the slowest possible way to build. I do not want to discourage anyone from learning stonework. To me it is one of the most satisfying kinds of building, perhaps the most satisfying. And it is a skill you can learn without being a professional. But

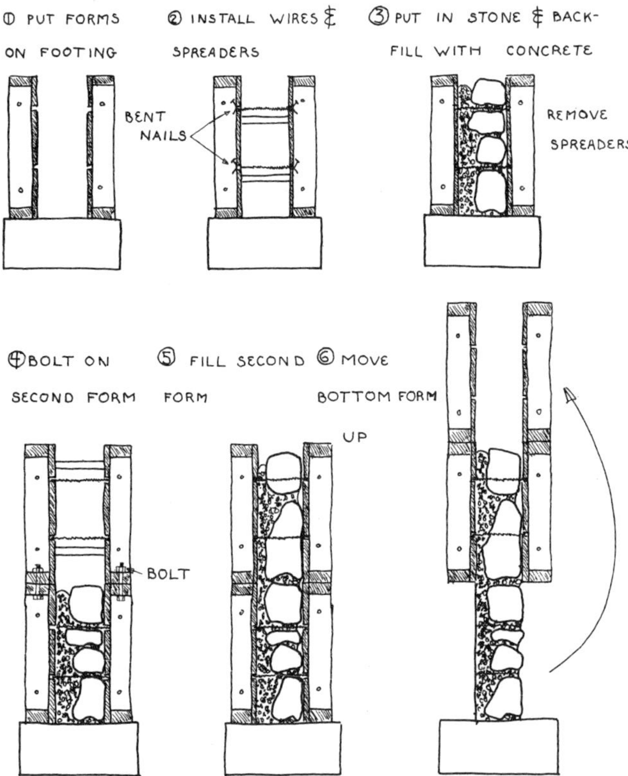
NEARING METHOD
① PUT FORMS ON FOOTING
② INSTALL WIRES & SPREADERS
③ PUT IN STONE & BACK-FILL WITH CONCRETE
BENT NAILS
REMOVE SPREADERS
④ BOLT ON SECOND FORM
⑤ FILL SECOND FORM
⑥ MOVE BOTTOM FORM UP
BOLT

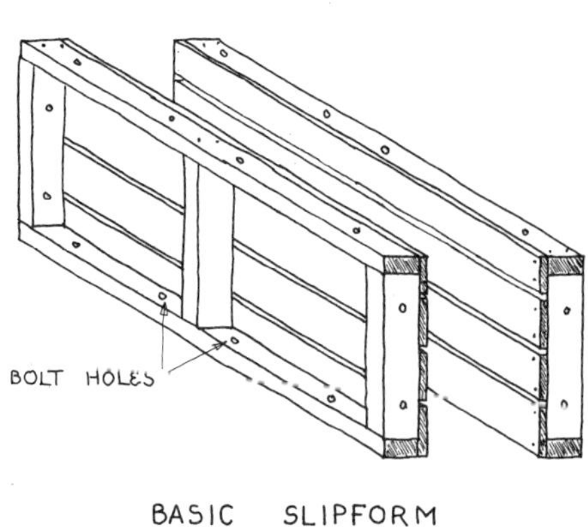
BOLT HOLES
BASIC SLIPFORM

Figure 23-28. The Nearing method.

do it for the pleasure in the work and in the result, not because you think it will save you money. Should you take up stonework, hire a local mason to teach you the art.

One type of stonework suitable for an amateur is the slipform or the Nearing method, pictured in figure 23-28. Stones are laid against the wall of a movable form (figure 23-29) on the side of the wall that will be exposed. Concrete is poured behind the stone. When the concrete sets, the forms are removed and repositioned farther up the wall. The bibliography gives references that describe the Nearing method in detail.

Sills for Masonry Walls

Figure 23-29 shows the standard pressure-treated sill that goes on top of any type of masonry wall. This sill ties the frame to the foundation and separates the concrete from untreated wood. If the outside of the foundation is being insulated, as is often the case, the sill will sit flush with the outside of the foam.

Before setting the sills, measure the dimensions of the foundation (including exterior foam) to see how accurate it is. Also measure the diagonals of the the major rectangles of the foundation to see how square they are. You are in great shape if things are within ¼ inch, and okay shape within say ¾ inch. While you could make some adjustments here to correct for any error, it is probably best to simply bolt the sills so they line up with the outside of the foundation (including the foam), and make further adjustments when you build the floor.

The anchor bolts are usually ⅜ inch diameter. Lay the pressure-treated 2 x 6 sill stock on the top of the foundation to mark where along the length of the 2 x 6 the holes will be drilled for the anchor bolts. Measure in from the outside of the foundation or foam to find out where the hole goes. Drill a ¾ inch hole to give a little room for error. Then apply the sill seal, which comes in rolls, and bolt down the 2 x 6s tentatively.

Next, check each foundation wall for alignment. Start by eyballing each line from the end. Some people can eyeball such a line quite precisely. If it looks dead straight, it probably is.

Most prudent carpenters, though, will also check with a string. The standard procedure is to apply a 2 x 4 (which is 1½ inches thick) spacer block at each end of the sill, as shown in figure 23-30, and stretch the line tight around that. Then you can measure between the string and the sill anywhere along the line, and the distance should be 1½ inches. I actually prefer to tie the string to a nail at each end, and bend the nail til it is right above the sill line. The main point is to make sure nothing is nudging the string; it has to run free. Adjust each sill until it lines up. Then tighten the bolts down hard.

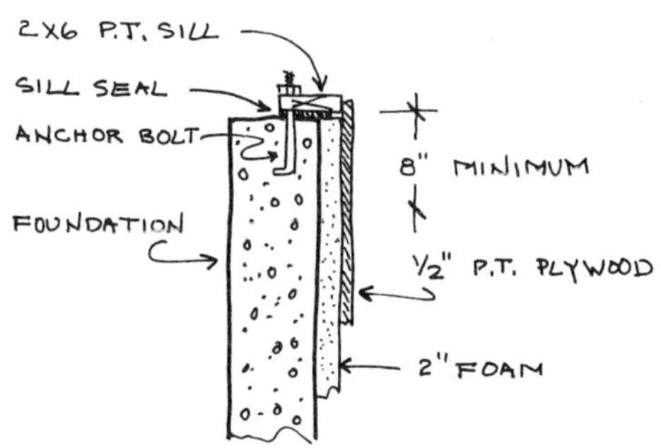

Figure 23-29. Sills on masonry walls.

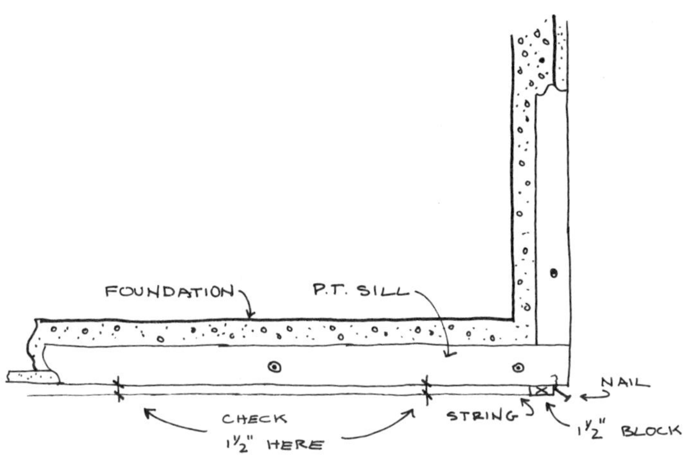

Figure 23-30. Checking with a string.

Treated Wood Foundations

The American Plywood Association has developed the design for a full basement foundation built of pressuretreated wood. The basement consists of an ordinary studwall built with treated studs, covered with treated plywood on the outside, and resting on a treated 2 x 8 used as a footing. The outside of the wall is

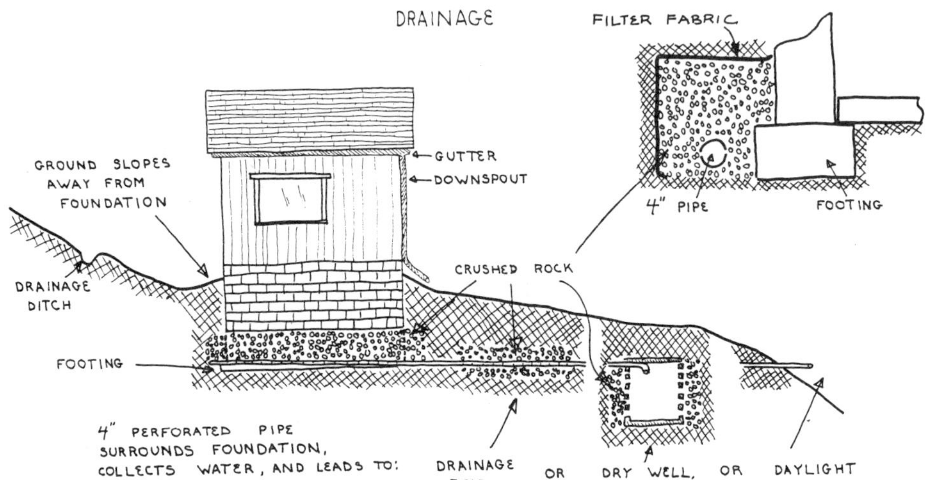

Figure 23-31. Strategies for drainage.

sealed with plastic film. A bed of crushed rock, covered with plastic film, fills the cellar hole. On that rests a concrete slab floor, or possibly a pressure-treated wood floor, which holds the foundation in position at the bottom. A sump or drain within the slab area provides drainage for the cellar itself and for the perimeter of the foundation walls (figure 23-32).

The wood foundation costs almost exactly the same amount for materials as a block foundation, and I suspect it is as durable. It's probably the fastest full foundation to build, particularly for people wood to whom wood is a more familiar material than stone.

The address for the technical manual is listed below.

FOUNDATION DRAINAGE

When you build any basement, you must keep the water out of it. Figure 23-31 shows common ways to do this. The first objective is to keep as much water as possible away from the foundation in the first place. If there is a hill behind the house, you may need a drainage ditch of some kind to divert water coming down the hill toward the house. The earth near the house should slope away from the foundation. Finally, gutters and downspouts can be installed. Your house may need all or none of these provisions, depending on the contour of the land and the porosity of the soil.

The second objective is to remove water that does find its way to the cellar walls. Special 4-inch perforated pipe is placed in the bottom of a trench just beside the footing. This pipe surrounds the house and is carefully laid at a pitch of about 1 inch in 12 feet. The pipe is covered with crushed rock to gather the water. The crushed rock is covered with a filter fabric, which lets water through but keeps soil from clogging the pipe. The pipe continues at a downward pitch, till it emerges from the ground to daylight or, if that is impossible, flows into a dry well or a drainage field similar to the leach field of a septic tank. Exactly how you provide drainage will depend on the conditions at the site. Seek reliable advice. Often a bulldozer or backhoe operator will know a lot about drainage systems.

To keep the remaining water out of the basement, coat the outside of the foundation with a special tarlike foundation coating.

If water still finds its way into the cellar, you can provide a sump, which is a concrete tank below the basement floor, where the water can collect. From the sump, water can flow to a dry well, drainage field, or daylight, or if absolutely

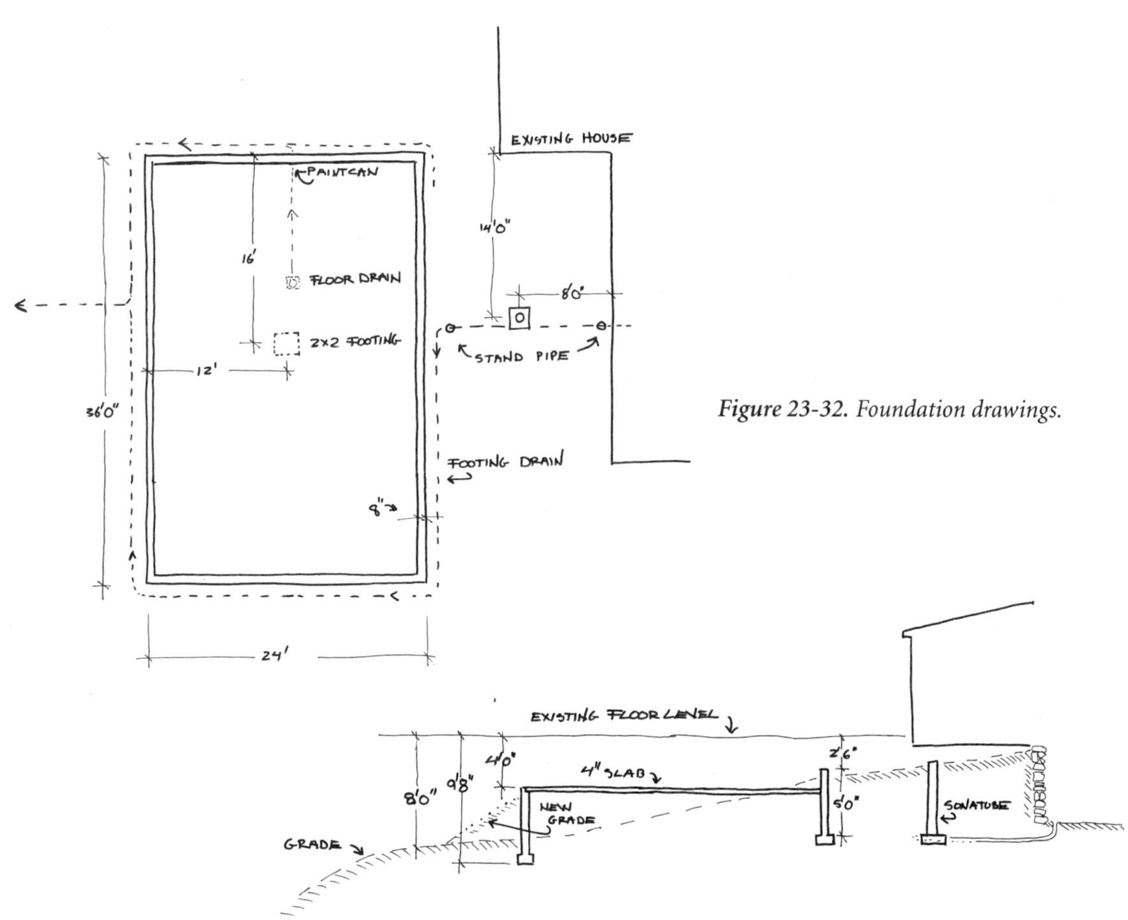

Figure 23-32. Foundation drawings.

necessary, be removed with a sump pump. However, if the drainage in your site is so bad as to require a sump pump you should consider a different site, or perhaps a column-type foundation.

WORKING DRAWINGS

For both design and estimating purposes you will need a plan and a cross section of your foundation. Show the dimensions of the foundation, plus any special details such as anchor bolts and reinforcement. For a post foundation, list the estimated heights of the various columns. Figure 23-32 shows an example of foundation drawings. The paint can refers to the practice of sliding a paint can, which is the same height as the wall is thick, into the wall where a hole is required so drain pipes can exit the foundation.

FIGURING CONCRETE MATERIALS

Step 1

Figure concrete volumes in cubic feet: To figure out the volume of foundation walls, multiply the area times the thickness in feet. An 8-inch concrete wall 10 feet wide and 8 feet high will have a volume of 10 x 8 x ⅔ cubic feet because 8 inches is two thirds of a foot.

Volume of round column footings:

DIAMETER (IN INCHES)	DEPTH	VOLUME IN CUBIC FEET
20	6	1.1
24	8	2.1
30	10	4.1
36	12	7.1

Volume of round columns:

DIAMETER (IN INCHES)	VOLUME IN CUBIC FEET PER FOOT OF HEIGHT
8	.35
10	.55

Step 2

Volumes for ready mix: Divide cubic feet by twenty-seven to convert to cubic yards. Use the cubic yard total for ordering ready-mix concrete.

Volumes for mixing your own concrete: Computing the volume of materials is complicated by the fact that one cubic foot of cement, plus two cubic feet of sand, plus three cubic feet of peastone will not add up to six cubic feet of concrete. To some extent, the smaller particles will occupy the spaces between the larger particles.

For estimating purposes you can roughly figure that the sand in your concrete formula will disappear into these spaces. One cubic foot of cement, two cubic feet of sand, and three cubic feet of stone, for example, will make approximately four cubic feet of concrete. Cement is sold by the cubic-foot bag, which weighs 94 pounds.

Stone and sand are sold by the cubic yard.

For a mix of one part cement, two parts sand, and three parts peastone (1:2:3), order:

Cement: .25 bag/cubic foot of concrete
Sand: .5 cubic feet sand/cubic foot of concrete
Peastone: .75 cubic feet peastone/cubic foot of concrete

For a mix of one part cement, two parts sand, and four parts peastone (1:2:4), order:

Cement: .2 bags/cubic foot concrete
Sand: .4 cubic feet/cubic foot concrete
Peastone: .8 cubic feet/cubic foot concrete

Convert sand and peastone figures to cubic yards by dividing by twenty-seven, and add fifteen percent for waste.

Books on Foundations

—*Mason's and Builders' Library: Concrete, Block, Tile, Terazzo* (Audel's guides). Louis M. Dezettel. Howard W. Sams and Co., Indianapolis, Indiana, 1972.
—*Concrete and Masonry.* T. W. Love. Craftsman Book Company of America. (The best general masonry book.)
—*Foundations for Farm Buildings* (Farmer's Bulletin #1869). Northeastern Region Agricultural Research Service, U.S. Dept. of Agriculture, Washington, D.C., 1970. (An excellent pamphlet.)
—*The Forgotten Art of Building a Good Fireplace.* Vrest Orton. Yankee, Inc., Dublin, N.H., 1969.

—*Building Your Own Stone House Using the Easy Slipform Method.* Karl and Sue Schewenke. Garden Way Publishing Company, Charlotte, Vermont, 1975. (Another good book on the slipform method.)

—*How to Plan and Build Fireplaces.* Sunset Books, Lane Books, Menlo Park, California, 1975.

—*Building Stone Walls.* John Vivian. Garden Way Publishing Company, Charlotte, Vermont, 1976.

—*"Permanent Wood Foundations: Design, Fabrication, and Installation Manual,"* American Forest and Paper Association, available from AWC Publications, P.O. Box 5364, Madison, Wisconsin 53705-5364 ($6.00 plus $5 shipping and handling). Phone: 800/890-7732.

24

FLOORS

PARTS OF A FLOOR

Figurc 24-1 shows the most common type of floor framing, used with both column and basement foundations. Figure 24-2 shows floor details that go together to complete the floor, including insulation if the floor is over an unheated space.

Joists: Floor joists are the main beams that support the flooring. They span between the sills either by resting on top of them, or by being hung between them on a ledger strip or joist hangers. Joists are usually spaced 16 inches or 24 inches center to center (o.c.) to match insulation, plywood, and other materials. Joists are doubled where they

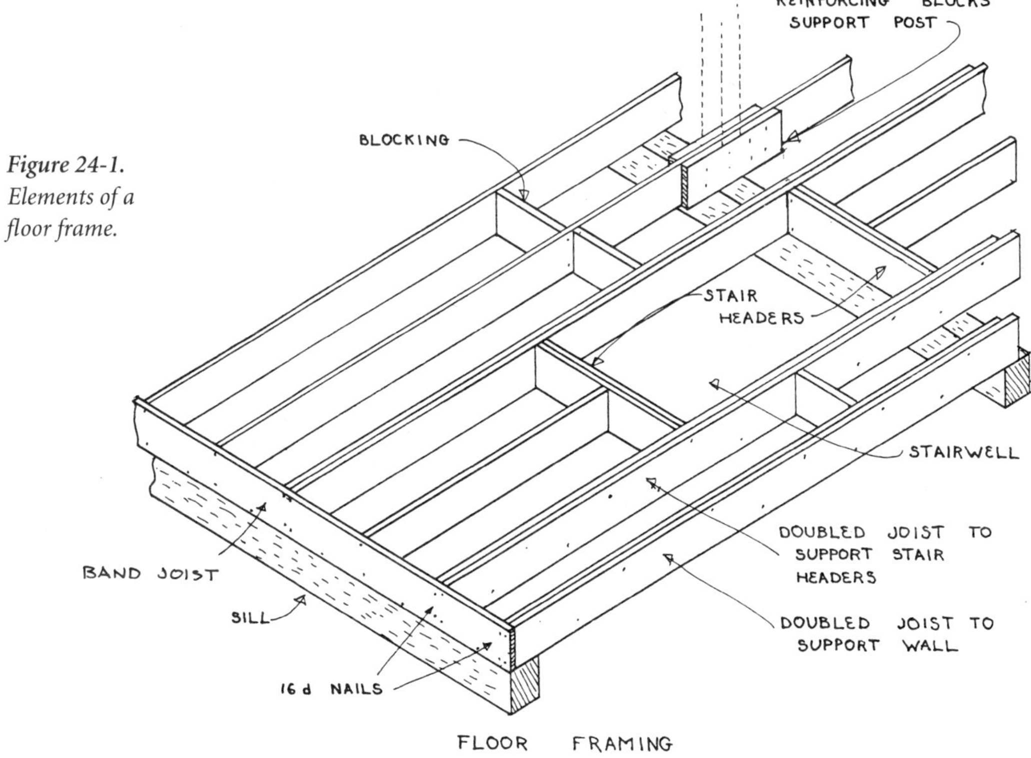

Figure 24-1. Elements of a floor frame.

carry extra weight around stair openings, or under walls, partitions, and bathtubs.

Band joist: At the outside walls, the joists usually butt into a piece of the same joist material known as the band joist. The joist layout is marked out on these band joists.

Headers: Headers also run perpendicular to the joists. They support the joists where they are cut for stairwells, chimneys, or other openings.

Blocking: These are short pieces of joist material that run between joists. They are used with long joists over 12 feet to stiffen the floor in midspan and space the joists, or to provide a nailing surface for subflooring. I usually omit blocking unless the joist span is being stretched to the limit.

Reinforcing blocks: Reinforcing blocks are scraps of the joist material nailed to a joist under the area in which a major load-bearing column will be located. These transmit the column load firmly down to the foundation or sill.

Insulation: A floor that is exposed below or is over an uninsulated crawl space should be insulated. In the drawing, 8 inches of insulation plus 1½ inches of foam gives an R value of about 30; that's a good amount in a cold climate. If the joists allow room for more insulation, add more. The soffit serves to keep the insulation in place and keep out animals, who often enjoy tearing down or burrowing in fiberglass.

The floor is sometimes insulated even when there is a full basement. If so, the soffit isn't necessary. The insulation can be supported periodically with stiff wires called "insulation supports." However, our practice in this situation is usually to insulate the basement walls instead.

Subfloor: The first layer to go down on the joists, the subfloor, is made of ¾-inch tongue and groove ("T&G") plywood, ⅝-inch plywood, or cheap 1-inch boards. Subflooring stiffens the joists and provides a surface to work on while building the rest of the house. The ¾-inch T&G underlayment is probably the first choice. It is very strong, and the tongue and groove support the joints not located over joists. It is smooth enough to serve as a finish floor with a nice coat of paint or varnish.

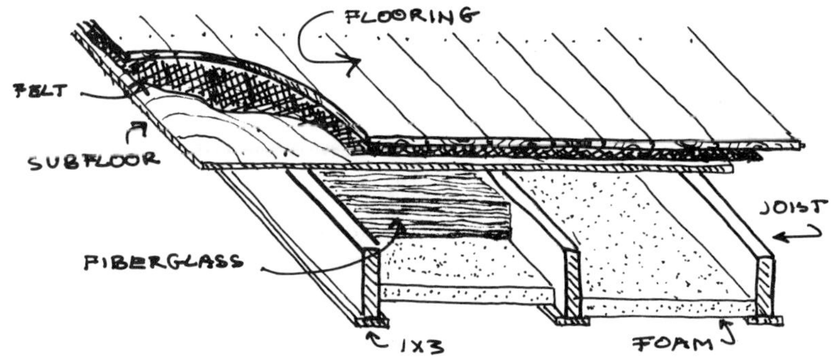

Figure 24-2. *Floor over unheated space.*

Vapor barrier: The plywood itself is a decent vapor barrier, but often a layer of .004-inch plastic film will be used in addition, either under the subfloor or between it and the finish floor.

Finish flooring: This is the final flooring layer. Installing the finish floor is usually delayed as long as possible to minimize the amount it gets scarred up by the building process. The finish floor can be good-quality dry softwood or hardwood boards laid over a layer of 15-pound felt (tarpaper) to prevent squeaking and reduce drafts. If linoleum, vinyl tile, or ceramic tile is used, there is usually an additional layer of ¼-inch or ⅜-inch plywood underlayment added to the subfloor to make a stiffer and smoother base. Sometimes the thickness of the additional layer will be selected to allow the tile or linoleum to come flush with an adjacent wood finish floor.

HOW TO BUILD A FLOOR

Designing the Frame

Your first job is to determine the joist size. The common sizes are 2 x 6, 2 x 8, 2 x 10, and 2 x 12. The size will be influenced by the joist span and spacing. A wider spacing may require a deeper joist. If you are using a subfloor plus wooden flooring, 24-inch spacing will be most economical. A 16-inch spacing should be used when there is no finish flooring layer, or when the finish flooring is tile, carpet, or another nonstructural material. The tables in appendix B show what size joists to use for different situations, and chapter 16 shows how to make these computations for yourself. When you are planning the floor, remember all the extra joists, headers, and blocking shown in figure 24-1.

Usually a single joist will not span the house: Sets of joists will butt together over a beam in the middle of the house, or perhaps over a load-bearing partition. When such a beam is located in a cellar, it is called a girder. Joists meeting this way should be tied together in some way. A plywood subfloor will do this effectively, but with a subfloor of boards, tie the pairs together as shown in figure 24-3.

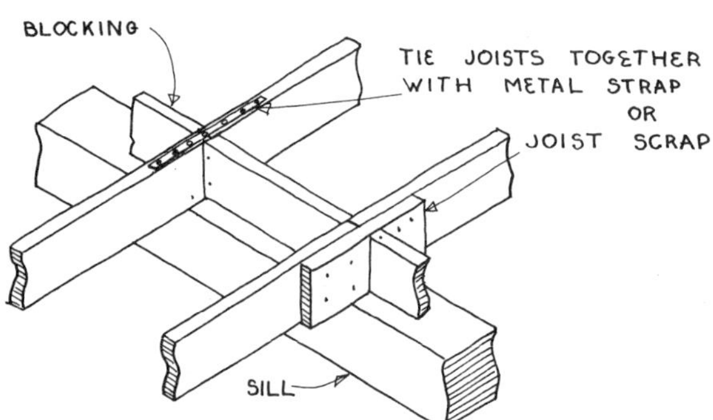

Figure 24-3.
Joining joists over a sill.

When a house is built on posts and large timber sills, the floor may be framed much as it would be on a masonry foundation, with a band joist as shown in figure 24-1. But an alternative is to hang the floor joists between the sills. This makes sense when you are trying to keep the floor frame as low to the ground as possible. Joists made of planed lumber can be hung on joist hangers.

Rough-cut joists can be supported by a 2 x 4 ledger strip nailed firmly to the sill. Often the latter method will require notching the joists (figure 24-4).

Framing the Floor

Checking the sills and foundation: Ideally, every stage of your house framing will be exactly the same size as your drawings indicate, and each part of the house will be level and square. That way, when you get to the roof, the rafters will be all the same length, and will install easily.

But each phase of construction can introduce some inaccuracy, particularly the foundation work. There is a good chance your foundation is a bit out of square and a bit out of level. Floor construction provides an opportunity to make adjustments.

Figure 24-4.
Nailing joists.

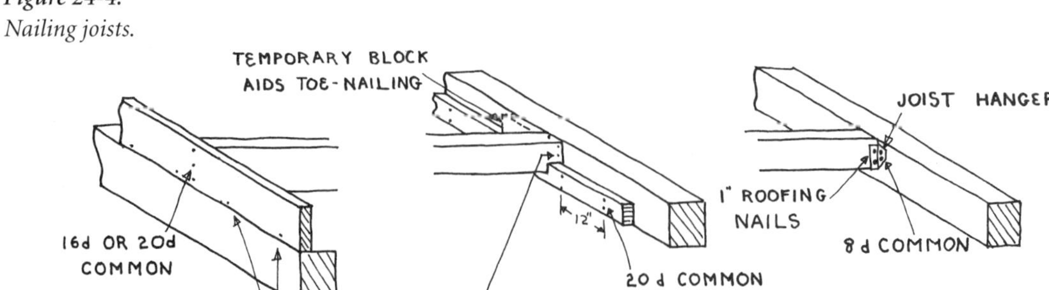

If you are building on posts, and the joists are hung from the sills, the strategy will be to make corrections now by shimming or repositioning the sills. If the floor frame sits on top of the sills, the strategy is to build your floor frame to the exact size it should be, with only minimal fastening to the structure below. Then you square it, position it, and level it with shims so that it is nearly perfect.

Before beginning to frame, do a careful survey to see where your foundation may be askew. First check the outside dimensions to see if the basic rectangles of the house are the size they should be. It is probably better if the foundation is a little small than if it is a little big, but either are manageable. Second, check diagonals. If the diagonals of the major rectangles are equal, the floor is square. I like to get within ⅛ inch, though ¼ inch or more is not bad at all. An inch off square is worth doing something about. Third, check how level the top of the sills are using a water level. In the case of a basement, the easy way to do this is to make a series of marks on each wall at the same height (any height will do). Connect these marks with a chalk line. Then you can measure up from this line to the top of the sills to gauge how consistent the level is, and if things are off, where the high point is.

For a house on posts, with the framing between the sills, make any adjustments you can to get the floor correct. It is worth it to recut things, re-drill bolt holes, or do some shimming to get it right. It will save time later.

For the type of floor frame that sits on the sills, the corrections get made once the frame is assembled.

Figuring cutting lengths: As you figure out the lengths to cut the floor framing members, you have to allow for "build-up," as shown in figure 24-4. Framing members are bound to be a little warped or cupped. As they are nailed together, the frame may end up bigger than the measured lengths of the pieces added up. To compensate, cut the joists a little short. Often the joists are cut ⅛ to ¼ inch short to get the proper total. If you have a built-up beam such as shown in figure 24-5, you can measure it after assembly to see how much it has built up.

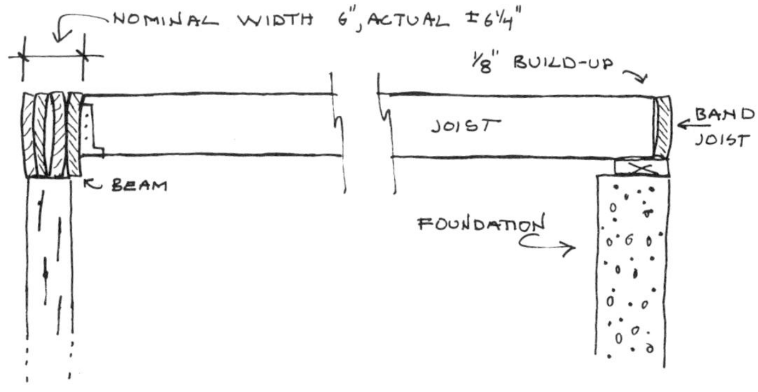

Figure 24-5. Build-up: you have to cut things a little short to get the floor frame to end up the right size.

If the frame is at all complex, it makes sense to make a floor framing plan, which gives the adjusted sizes of all the pieces. Figure 24-6 is an example.

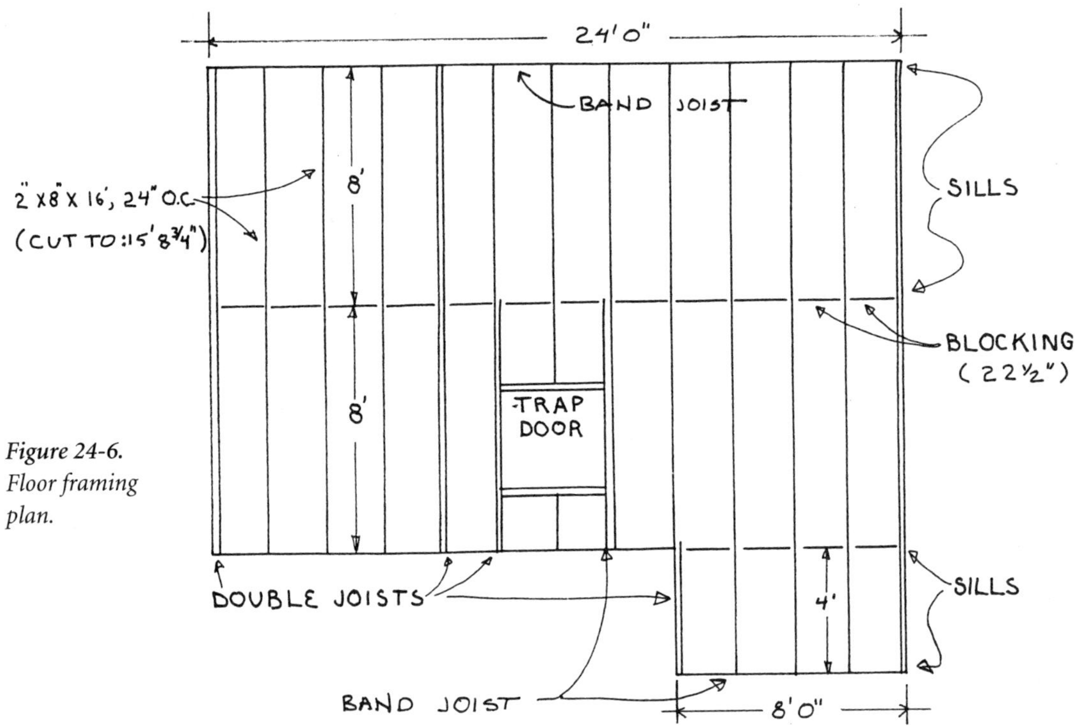

Figure 24-6. Floor framing plan.

Cutting the joists: You can cut all the joists to the same length all at one time. Chapter 35 describes cutting methods.

As you handle each joist, sight down it to check for straightness. If there is a slight crown (convex bow), plan to place that up, as in figure 24-7. A joist that bows as much as the one shown should be discarded or cut up for blocking. The bigger or more numerous knots should be on the top (compression) side if possible.

If you are using rough lumber and your joists sit on top of the sills, you may have to trim some joists and shim up others to make the floor come out flat. To do this, first find the average or typical size. Variations of ⅛ of an inch or less either way from the average width can be ignored. You can trim oversize joists on the bottom where they will rest on the sills by using a hatchet or saw. Narrow joists can be shimmed up with a shingle or scrap of the right thickness. The headers have to be flush too, so make them from pieces of average width. The headers should also be pieces of uniform thickness, so that the joist lengths can be constant.

If the joists run between the sills, your problem is to make them come out flush with the top surface of the sill. This is easy if you are using joist hangers. But

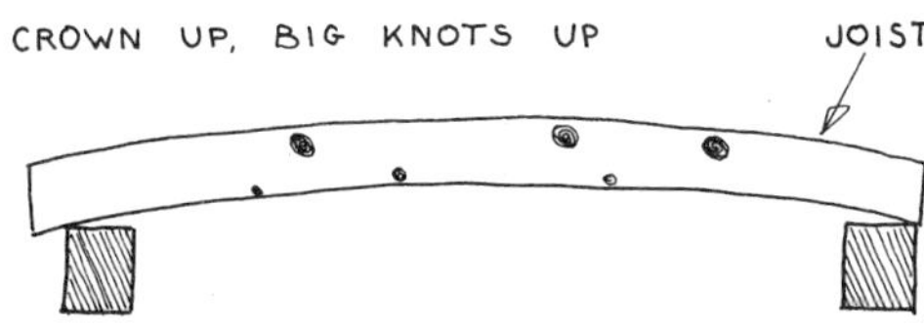

Figure 24-7.

with ledger strips, the trick is to match the notch location with the ledger strip location accurately. Figure out the proper notch size by making a scale cross section through the sill. As figure 24-8 shows, the notch should remove no more than one-third of the joist depth or the joist will be weakened.

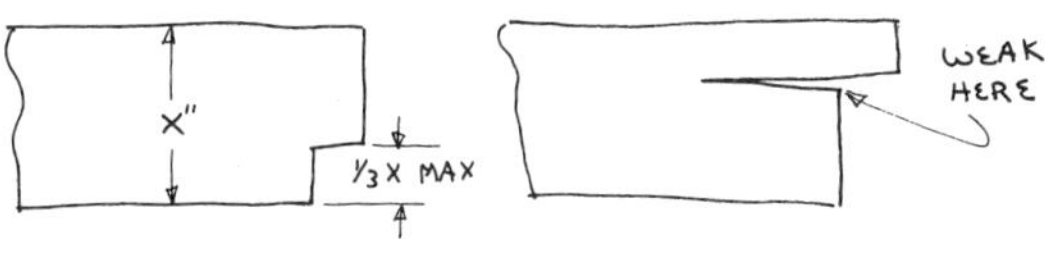

Figure 24-8.

If your joists are 2 x 8s, for example, you might have a 2-inch notch with 6 inches of wood remaining. When you mark the joists for notching, do not measure 2 inches up from the bottom. Measure down 6 inches from the top. This will automatically compensate for any variation in the size of the joists. I find it easiest to cut a marking template. In this case it would be a 6-inch-long 1 x 1 or 1 x 2. The template makes measuring mistakes almost impossible. Use the same template to position the 2 x 4 ledger strip on the sill. That will guarantee that the joists come out flush with the sill.

Installing band joists: If you have band joists, select straight pieces from your pile to cut them from, and cut them to length exactly. Tack them in place with a few toenails, leaving the heads up a little. I often find that it is easiest to toenail something first with the thinner 8d nails to establish the right position. Then add strength with 16d nails.

Installing joists and headers: Your first step is to lay out the joist locations (figure 24-9). If the joists are resting on top of the sills, mark the band joists. If the joists are hung between sills, the marks will be on the sills themselves. For 24-inch centers, hook the end of a long tape measure on one end of the header (or sill) and make a mark every 2 feet. Do not measure 2 feet, make a mark, measure 2 feet more from that mark, and so on. You will get a cumulative error. Hook the tape on one end and make a mark at 2 feet, at 4 feet, 6 feet and so forth.

If your interval is 16 inches instead of 24 inches, you will probably find that the 16-inch intervals on the tape are numbered in red or boxed for convenience. Your marks indicate the centers of the joists, except for the first and last in the row. For these two, the edge of the joist will be flush with the end of the header or sill, as in figure 24-9.

Once you mark the centers, measure to either side of the marks to indicate the edges of each joist. If you're using standard 1½-inch-thick joist, you'll measure ¾ inch either way from your center mark. Then use your big rafter square to carry those lines square across the band joist. The joist will go between these lines.

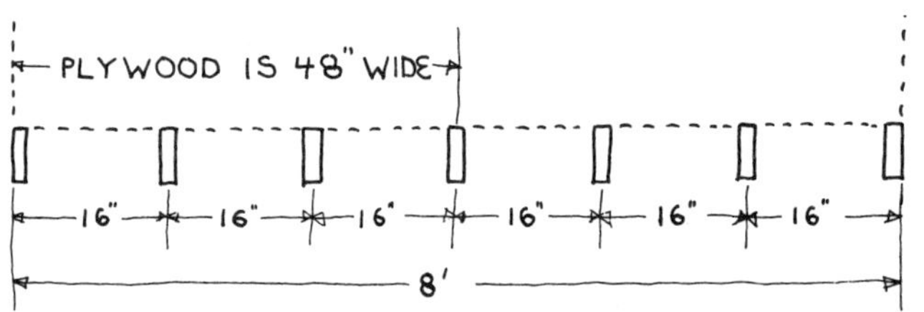

Figure 24-9. Joist Layout.

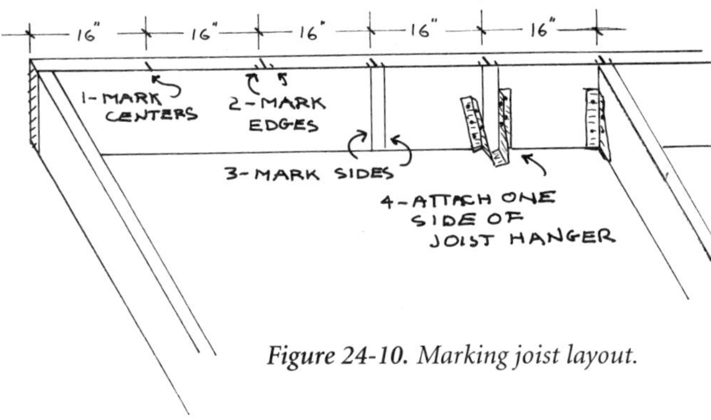

Figure 24-10. Marking joist layout.

If you are using a 2 x 4 ledger strip you can now attach it to the sills. Use a ruler or template piece to position the ledger strip just the right distance down from the top of the sill. Nail the strip in with two 20d common nails, one right over the other, every 10 or 12 inches.

Joist hangers: Use a scrap of the joist material to mark where the bottom of the hanger goes. Nail one side of the joist hanger home on the line with three 8d common nails. Position the other side, and put in one nail halfway.

If the floor will be insulated, now is the time to nail 1 x 3 strapping to the bottom edge of the joist to form the ledge which will support the foam soffit. These strips can be attached with almost any 5d, 6d, or 8d nails (figure 24-2). Leave the strapping about 3 inches short at the ends if you are using joist hangers.

The joists are now ready to go in. At band joists, nail through the band joist into the end of the joist with four 16d common nails, making sure the joist is square and flush at the top. Where joists rest on top of timber sills or girders, toenail first with one 8d nail, then two 16d nails. If your joists rest on 2 x 4 ledgers, toenail everything together in the same way. At joist hangers, slide a joist in, make adjustments to get the top of the joist flush, then nail the other side of the hanger to the sill with 8d common nails. Then tack the hanger to the joist with joist hanger nails or 1-inch roofing nails.

As figure 24-1 shows, stairwells and other large openings in the flooring will require doubling the adjacent joists (called "trimmers") and putting in doubled headers to support the cut joists. If one joist will be cut, and the header will therefore be 4 feet long or less, install the header before you double the joists. End-nail the headers in position through the joists with numerous 16d or 20d common nails, and then add the second joist. If the header is over 4 feet long, the joist should be supplemented with a ledger strip or joist hangers. Large joist hangers that are made to fit a 4-inch beam are simplest to use.

Adjusting the frame: If your joists are hung between the sills, you've already made adjustments to get everything square, straight, and level.

But if the floor sits *on* the sills (as in figure 24-3), make the adjustments next. It is easiest if you do this in the correct order. First, with a helper, check the diagonals of your new floor to see if it is square. You can adjust it for square with the aid of a sledgehammer, but it works better sometimes to "walk" it into place with a pry bar, slid between sill and joists. Once square, see how the floor aligns with the sills. You may be able to get a better match by repositioning the floor on the sills, and re-squaring it. Once it is both square and as well-aligned as possible with the sills, nail down the high corner (which you have already identified) to the sills, toenailing with 16d nails. You have a grade line on the foundation below. Measure from this line to the top of the floor frame. Shim the other corners of the frame up to this height (if needed), and nail them home.

Next eyeball the frame, sighting down each side, and across the entire floor. You can also stretch strings (see figure 23-30). Probably you will have to use your pry bar again to force the centers of each side into perfect line, and also do some shimming with shingles to get to the perfect height above your grade line. When you have made these adjustments, firmly nail the whole floor frame to the sills below.

If it takes all day, get this floor as square and level as possible. It will make life easier later. On the other hand, if foundation problems or mistakes make this impossible, there is sure to be some sort of way to fix things. This would be the point where a professional carpenter, who has had many years of experience correcting mistakes, can help you develop a fix for the situation.

Putting in the soffit: If the floor is to be insulated with its underside exposed below to the weather, now is the time to put in the soffit. You can use 1-inch, 1½-inch, or 2-inch foam board, or any moisture-resistant cheap material that comes in sheets.

Cut the foam to size and just drop it in place on the ledgers you already put on.

Blocking: Add blocking on long spans, or where joists meet over a girder or sill. End-nail the blocks in place where possible, and toenail otherwise. You can precut the blocks to the standard size—often 14½ inches or 22½ inches, but make sure as you put them in that small cumulative errors or "build-up" aren't distorting the layout.

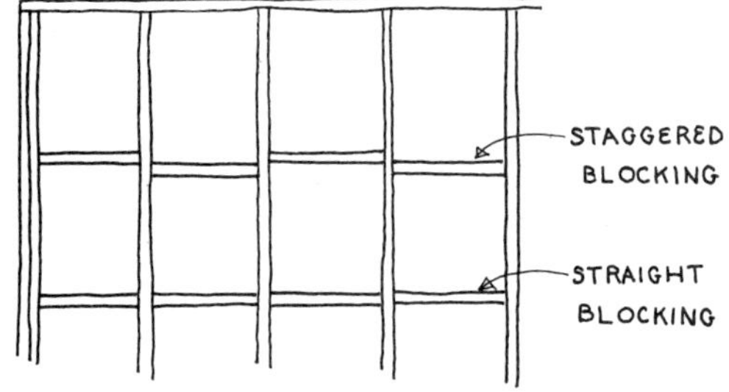

Figure 24-11. Blocking.

Reinforcing blocks: Add reinforcing blocks wherever load-carrying posts will be located in order to transmit the column load to the sills and foundation. This reinforcing should be at least as wide as the base of the post. The blocking for column support can consist of blocking between joists, or of pieces of scrap wood nailed flat to the joist, as shown in figure 24-1.

Wiring: Any wiring that will be run in the floor should be put in before the floor is insulated and before the subfloor is put down.

Insulation: Fiberglass insulation comes in widths for 16-inch and 24-inch spacing, and in a variety of thicknesses. It also comes with various backing materials. The unfaced kind is much quicker to install.

Cut the insulation to length with a utility knife. It helps to use something like a 1 x 6 scrap as a straight-edge to minimize contact with the material. Compress the fiberglass as you cut it.

Fiberglass is itchy to work with. Wear gloves, and cover your whole body with old clothes. Wear a respirator designed for use with such materials; it is worth the investment.

Make sure the insulation fits in neatly. Fill any voids or pockets.

Subfloor: Any subfloor will go down quickly, but a plywood subfloor will go down particularly fast because the pieces are big. The trick is to get the first few sheets down flush and square with the frame. Nail them tentatively with two or three nails until you are sure they are well located. Then use 7d plywood nails or 8d galvanized box nails every 8 inches or so.

If the first row starts with a full sheet, start the second row with a half-sheet to stagger the end joints.

It is common practice to glue plywood sheathing down with construction adhesive, which minimizes floor squeaks. If you want to do this, put down the adhesive with a caulking gun one sheet at a time so the glue doesn't get all over.

If you are using T&G plywood for the subfloor, there are a couple of tricks that will help. Align the first row so that the tongue of the second row tucks into the groove of the first. When you start the second row, it may take some force to seat the tongue in the groove. Place the sheet as close as possible, then place a 5-foot-long 2 x 4 or 2 x 6 on the joists, against the groove of the second sheet; this piece protects the groove as you can tap the sheet into place with a sledgehammer.

If your T&G plywood will be your finish floor, even temporarily, you could forgo the adhesive, and put a plastic vapor barrier under the plywood. Alternatively, seal the cracks with a resilient caulk. If you are varnishing the floor, use silicon caulk applied after the varnish.

Otherwise, put the vapor barrier in later between subfloor and finish floor.

A subfloor of 1-inch boards is cheaper, but slower to install. Inexpensive green boards can be used for a subfloor, as long as they are uniform in thickness (plus or minus 1/16 of an inch). The really thin or thick boards in your pile should be used somewhere else. If available, buy rough boarding that has been planed on one side to a uniform thickness. Sometimes this is sold with a shiplap on the edge, which will result in a better quality job.

Nail the boards perpendicular to the joists. Only a fair fit is required, since they will be covered up later. Usually the ends of sawmill boards must be squared. Often subflooring boards are allowed to hang out over the edge of the floor frame as much as is necessary and trimmed off all at once later with a skillsaw. Use two or three 8d nails at each joint (figure 24-12).

Some of the boards will be warped and will have to be forced into position. Nail down one end, as in figure 24-13, then force the warped end over. Sometimes if you flip a board

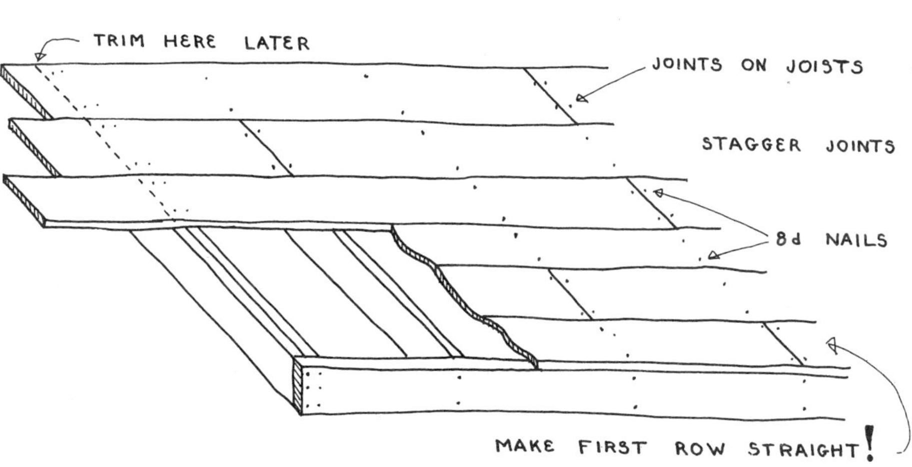

Figure 24-12. Nailing boards to joists.

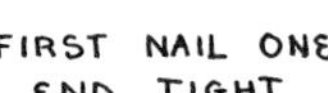
FIRST NAIL ONE
END TIGHT

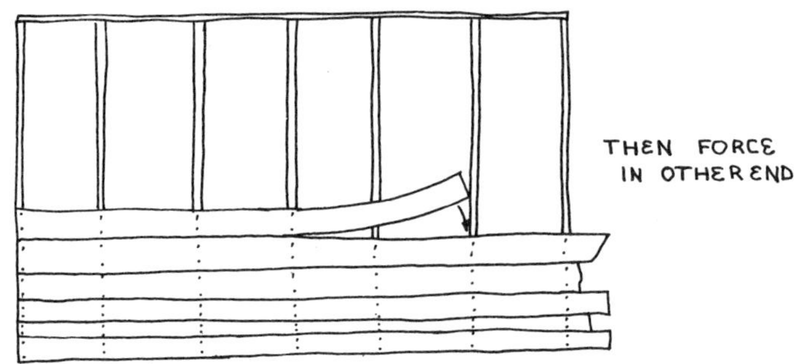
THEN FORCE
IN OTHER END

Figure 24-13.
Nailing warped boards.

over, or reverse it end-for-end, it will fit much better.

Figure 24-14 shows two other handy ways to draw boards tight. In the first method, you place a chisel next to the board at the joist with the bevel of the chisel touching the board. Tilt the chisel to one side and pound one point of it into the joist about ¼ of an inch. Then lever the chisel over to force in the board.

If the board is only ⅛ or 3/16 inch out of line, you can usually bring it over just by toenailing the board to the joist as shown in the picture.

A plastic vapor barrier can be installed later between the subfloor and finish flooring.

Rain cover: From this point on you have to worry about the rain soaking your floor (and everything else), particularly if there is insulation in the floor.

Cut a piece of plastic larger than the floor to use as a rain cover when needed. Hold it down with a few 2 x 4s. If one piece of plastic will not cover, slide a 2 x 4 under the lap joint to prevent water running in.

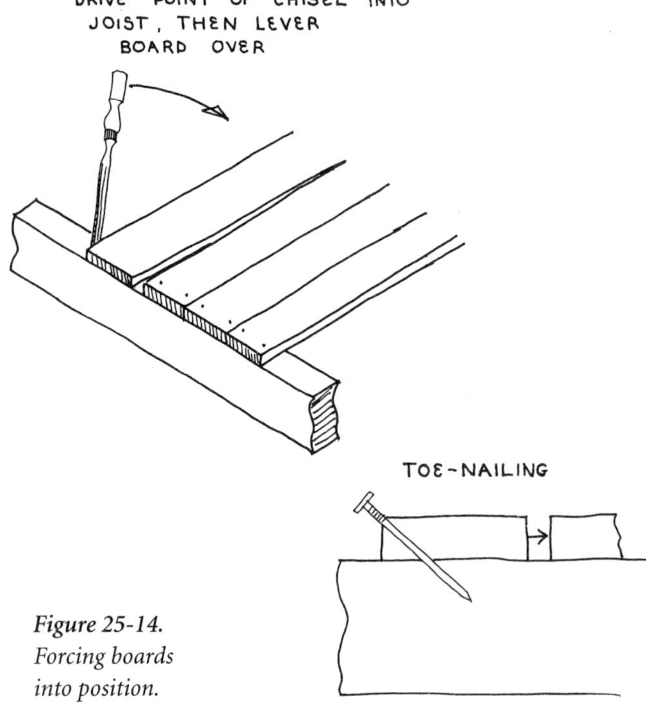
DRIVE POINT OF CHISEL INTO
JOIST, THEN LEVER
BOARD OVER
TOE-NAILING

Figure 25-14.
Forcing boards into position.

Building A Second Floor

The second floor is usually framed just like the first floor. Joists span from one wall to the other and rest on top of the walls. A floor is nailed down on the joists from above: first a subfloor, then a finish floor. A ceiling (often wallboard) is installed from below. There is no insulation within these joists, since both spaces are indoors and heated. If you plan to have this kind of closed-in ceiling, just build your second floor exactly as you did the first floor. If your floor joists are on 24-inch centers, you may want to add "strapping" later. Strapping is cheap 1 x 3, applied perpendicular to the joists at 16 inch centers. Strapping provides better support for the sheetrock, and is easy to shim if the joists are wavy.

Long spans: If the house is over 16 feet (or at most 20 feet) wide, the joists will have to be supported in the middle by a partition or beam. If the house is 24 feet wide, the space might be divided by a partition into an 8-foot room and a 16-foot room. You would then have a 16-foot joist meeting an 8-foot joist over the partition. These would be joined just as first floor joists would be joined over a sill, using one of the methods shown in figure 24-3. Since such a partition would be a load-bearing wall, it must be located over a sill. If the second-floor joists meet over a beam, the posts that support the beam should be located over a foundation wall or column, and the space under the post between the subfloor and sill should be filled with reinforcing blocks.

Another alternative for longer spans is to use engineered timbers of various kinds. These materials can be used for quite long, unsupported spans, but they are expensive. It is also possible to push your spans by using higher grades of lumber, or stronger species. Yellow pine or Douglas fir will span longer distances than spruce or other common framing lumber. Check your span tables.

Exposed frame ceilings: Figure 24-15 shows exposed frame ceilings. This is one place where timber framing is both beautiful and cost effective.

The joists consist of rough sawmill timbers 3 or 4 inches thick and 3 or 4 feet on center. No ceiling is nailed to the underside of these joists. The ceiling is the bottom side of whatever material you put over the beams. This could be a layer of rough boards. It could be a layer of sheetrock, good side down, which is later screwed to the layers above.

The traditional detail is simply two layers of boards, both perpendicular to the joists. Even spanning 4 feet, they will be stiff enough to walk on. Sometimes I've put a layer of sound board or other soft material (or at least building paper)

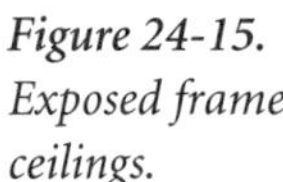

Figure 24-15. Exposed frame ceilings.

between the two layers of boards to suppress sound to some extent (figure 24-16). A modern approach is to use 2 x 6 T&G decking. This creates the ceiling below and the floor above all in one step.

These systems are inexpensive because no extra finish ceiling needs to be built. Everything is nailed down from the top, which is faster than nailing up from the bottom, and all the joints are hidden above the joists.

To build this ceiling/floor, first find the correct joist size by using the formula in chapter 16. Or use the same depth of joist you would use in a conventional floor and increase the width as you increase the spacing.

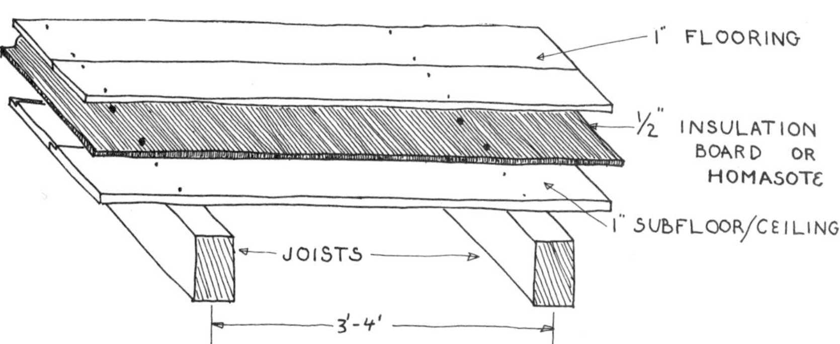

Figure 24-16. Traditional 2- or 3-layer floor.

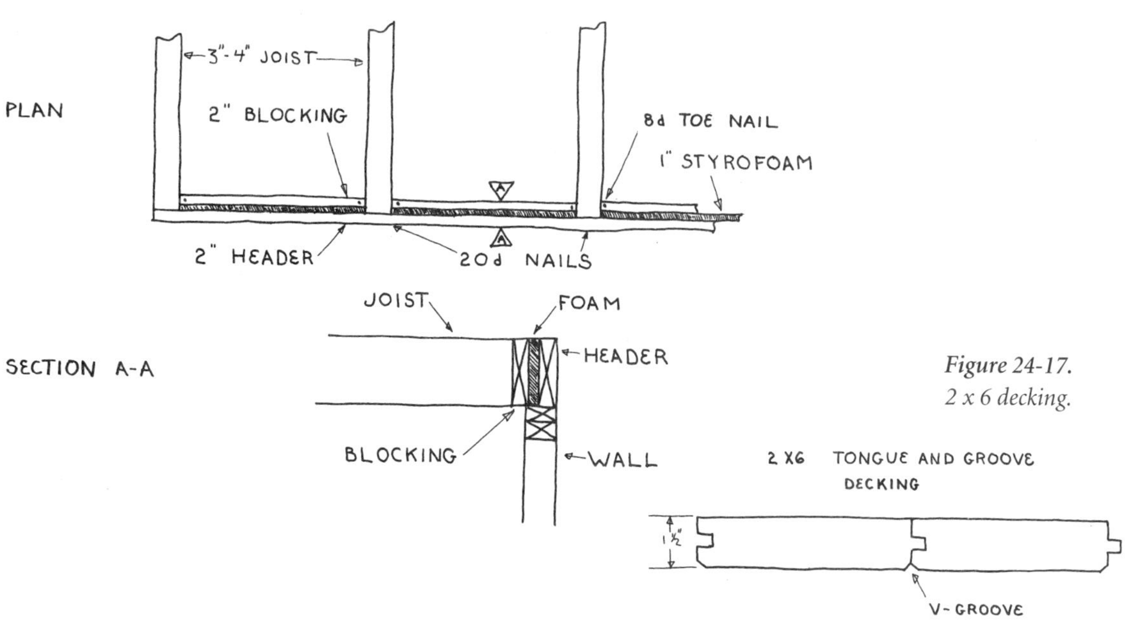

Figure 24-17. 2 x 6 decking.

The floor is put together much like a conventional floor. The joists will butt into a 2-inch thick header flush to the outside of the wall framing. Nail the header to the joists with 20d nails, and toenail both joists and header to the top of the wall. Put 2-inch rigid foam insulation board against the inside of the header. Cover this with an extra row of blocking (figure 24-16), which can be nailed through the foam into the header and carefully toenailed to the joists. Do this nailing neatly, since these nails will show. The blocking conceals the foam, stiffens the joists, and trims the top of the inside wall surface.

If you are using nominal 1-inch boards, choose good looking boards, keep them clean, and get them as dry as possible.

Make sure the first row is straight by aligning it with a chalk line or by sighting down the row before it is finally nailed. Use two 8d common nails into each joist, or three if the board is over 6 inches wide. Fit each board as tightly as possible to the previous row.

Boards vary somewhat in width from one end to the other, and often the flooring will be out of parallel with the house frame by the time you get halfway across. Every few rows, measure the remaining distance across at both ends to see if you're still parallel. If the rows are not parallel, make corrections by putting the wider ends of the next few rows at the end that needs to be expanded.

A layer of black insulation board or Homasote could go on top of the subfloor. Black or gray Homasote will reduce the visibility of any shrinkage cracks that appear from below. Wherever you can, nail the Homasote through the subfloor to the joists with 8d nails, and elsewhere into just the subfloor boards with 1-inch roofing nails, which have large heads. The finish floor will go on later, parallel to the subfloor.

Two-by-six decking from the lumberyard is as fast or faster to install than the floor system just described. Decking is flat on top and has a V-groove pattern on the bottom.

If you buy decking, look over the stock carefully, because quality varies. Decking must be dry to work well. Make sure it has not been sitting around for several months outdoors in a lumberyard picking up moisture.

Notes on Ordering Materials

I always order one or two extra joists, so that a really crooked or split one can be culled. Also, if one is miscut, it is cheaper to have bought an extra ahead of time than to run to the lumberyard while everyone stands around.

When ordering boards for a subfloor, calculate the amount needed using the actual width of the boards, not the nominal width, and add ten percent for waste.

25

WALLS

THERE ARE MANY WAYS TO BUILD A WALL. IN FACT, MOST new building systems that get introduced or revived are really wall-building systems. The roofs and other parts get built conventionally. This chapter outlines the simplest and most common type of wall building, which is 2 x 6 studwall construction. You build a wall frame on the floor, as in figure 25-1, then tilt it into position in one piece. This is called platform framing.

Done thoughtfully and with care, as detailed below, this produces a strong, energy-efficient wall very quickly. It gives an R value of about 21. There are a variety of super-insulated wall designs that achieve higher R values. Most involve adding layers of foam insulation either inside or outside of a basic fiberglass insulated wall. The sidebar on pages 322 and 323 shows one wall system that achieves an R value of about 28 in a simple and elegant way.

PARTS OF A STUDWALL

Figure 25-1 shows the parts of a typical studwall.

Studs: The studs are the vertical 2 x 6s that support the house and provide nailing surfaces for the exterior and interior wall coverings. In the old 2 x 4 walls, studs were 16 inches on center, but with 2 x 6s, 24 inches on center is plenty strong, and with wider stud bays more of the wall surface is insulated.

Plates: The plates are the horizontal 2 x 6s to which the studs are nailed. The top one is doubled for extra strength and to make it possible to interlock the corners of adjacent walls (figure 25-2).

Rough openings: The openings left for windows and doors are called rough openings. They are usually about 1 inch larger than the window or door frames that will be used.

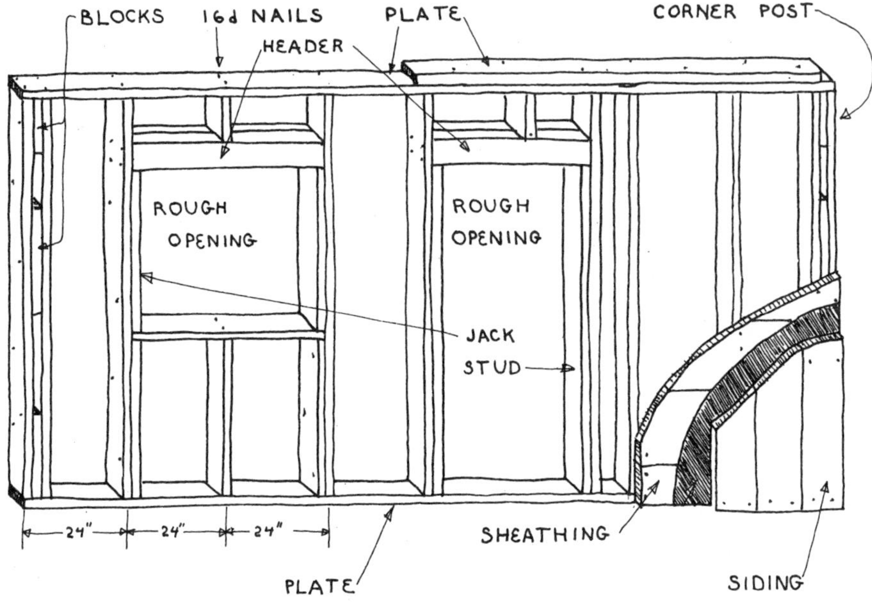

Figure 25-1. Parts of a studwall.

This allows the window or door to be adjusted exactly plumb and level when it is installed.

Headers: The headers support the top plate above wall sections where studs have been removed to create windows and door openings. Headers usually consist of a pair of 2 x 6s or 2 x 8s on edge, one flush with the outside wall and one flush with the inside wall. The gap between them gets stuffed with insulation. Though many builders put headers above every window or door, they are only needed in bearing walls, which support a roof load or heavy floor load.

Sill: A piece of lumber at the bottom of window rough openings is called the sill. It is the same length as the headers.

Jack studs: The jack studs support the headers over doors and windows.

Sheathing: Sheathing is the outside covering of the house under the siding. It strengthens and braces the house and makes it tight. Traditionally it consisted of boards, preferably T&G boards or shiplap boards. But ½ inch CDX (a cheap grade of exterior plywood) plywood goes up faster, and is stronger.

Air barrier: It is critical to keep the wind from blowing through the wall from the outside. The traditional detail was to cover the sheathing with either felt paper or red building paper, which come in 3-foot wide rolls. A few years ago house wraps such as Tyvek and Typar were introduced. They come in wide rolls capable of covering a whole floor of a house in one piece. These materials are designed to keep the wind out while allowing moisture to escape. House wrap is a great idea over a sheathing of boards. It is less obvious to me why it's necessary over carefully applied plywood, which is itself a great air barrier. Our practice is to put strips of house wrap over plywood joints and corners, or simply caulk the joints.

Siding: It is possible to let the plywood be the outside finish of the house, if the joints are properly sealed or flashed. Special plywoods such as T-111 are designed to be used as a sheathing and siding all in one. But most people add a layer of siding material to the house. Shiplap boards, board and batten, clapboards, and shingles are common choices.

Corner post: Where two walls meet at an outside corner, an extra stud is often added to one of the walls. It provides nailing on the inside for wallboard or other interior finish materials, and combined with the other two, makes a strong corner post.

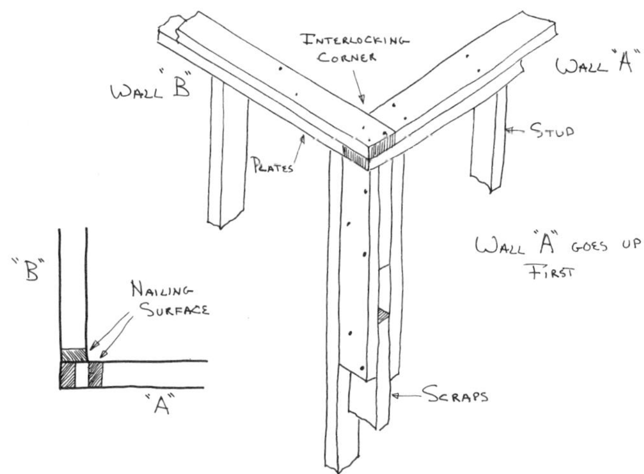

Figure 25-2. *An extra stud in the A wall makes the corner stronger, and provides nailing for interior finish.*

DESIGNING WALLS

When you design a wall you must (1) find the header sizes, (2) figure out what combination of siding and sheathing you will use, (3) plan how the wall will be braced, (4) figure out the rough opening sizes, and (5) make a drawing.

Finding Header Sizes

Headers above windows and doors in bearing walls must be designed to hold the loads they will carry, which may include a second story or roof. Consult the chart in appendix B, or make the computation yourself using the method in chapter 16.

Choosing Sheathing and Siding

Though there are many possible variations, two types of sheathing make sense to me. The standard is to use ½-inch CDX plywood, which seals the house, provides diagonal bracing, and provides nailing for sid-

Best Energy-Efficient Wall Design

Clayton DeKorne

Thomas Brown, an architect from Stevens Point, Wisconsin, won the best wall design at the 1991 Quality Building Conference held in Springfield, Mass., and sponsored by the New England Sustainable Energy Association. Brown's entry for an energy-efficient strapped wall was judged by an architect, an engineer, and a builder for "energy efficiency, buildability, simplicity, and innovation."

Brown's design combines conventional framing methods with several energy-conserving construction practices (see illustration). Brown claims his design "is very forgiving . . . and allows flexibility in the completion of the project." Most of the energy-conserving measures have been added to a standard 2 x 6 wall, he explains, and don't require a whole new method. "In fact," he adds, "it is possible to revert to more standard practices if budget or other constraints prevail."

The basic wall consists of an insulated 2 x 6 studwall covered on the inside with ½-inch foil-faced rigid insulation and strapped with 2 x 2s before the drywall is secured. With batt insulation, Brown claims the R-value of the wall materials adds up to about R 28, including an air space next to the foil-faced insulation. But Brown is quick to point out that high R-value is not the key to the wall's good performance, but rather tight construction.

A tight envelope depends on following through with several sequences. Most important is careful attention to the air/vapor barrier. Here Brown urges a few departures from conventional practice:

First, the vapor barrier must form a continuous envelope around the entire house. To achieve this, Brown calls for (1) running the sub-slab vapor barrier up the foundation wall before pouring the slab; (2) draping a wide strip of 6-mil poly over the top of the sill plate before the joists are laid down; and (3) running strips of poly around the rim joists and over the top plates on the upper floor. These steps leave tabs that can then be taped to the vapor barrier on the walls and ceiling.

Second, Brown calls for overhanging the 2 x 6 wall plates 2 inches so the rim joist can be inset to accept rigid foam insulation on the exterior.

Third, rough openings for windows are oversized by 3 inches in width and height, and the window is installed on added 2 x 4 nailers, creating a step for recessed wood trim

Brown has designed about fifteen houses with some variation on this wall, and in many cases, he says, the builder liked it enough to adopt it on other projects. Brown attributes the success of the design to its versatility. The strapping, he says, is a high performance feature that is not absolutely necessary, but which adds an extra level of efficiency. Without the air space and the electrical raceway, the wall can still perform better than average if care is taken around electrical boxes and other penetrations. And even if the interior foam is eliminated, the continuous vapor barrier and rim joist detail make a difference.

Brown has worked up a full set of construction notes and drawings of his wall. For a free copy, send a self-addressed stamped envelope to Wall Detail, *JLC*, RR#2, Box 146, Richmond, VT 05477.

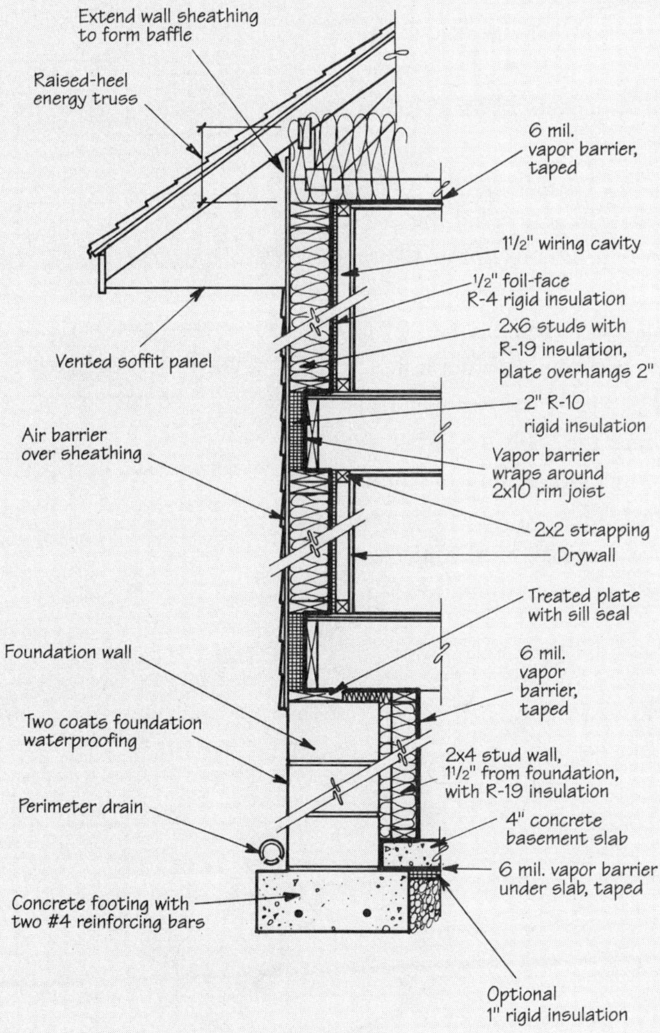

The basic wall consists of an insulated 2 x 6 stud wall covered with ½-inch foil-faced foam insulation and 2 x 2-inch strapping on the inside. Note the indented rim joist to accept 2-inch rigid foam on the outside, and the continuous vapor barrier. (Article and illustration first appeared in Journal of Light Construction, *August, 1991. Reprinted with permission.)*

ing in one quick operation. A somewhat cheaper version of this choice is to us a ½-inch manufactured oriented-strand board, commonly called OSB. These sheathings provide adequate nailing for shingles or clapboards, and thicker horizontal boards can be nailed directly to the wall studs. Vertical boards require more nailing, which can be provided by adding horizontal blocking in the walls every four feet. As I discussed above, the joints should be sealed in some way.

Depending on your local lumber sources, it may be somewhat cheaper, though more time consuming, to use 1-inch boards for sheathing, preferably shiplap or T&G boards. This provides better nailing, but a housewrap will be necessary. You will also have to brace the walls by installing let-in bracing in the wall frame itself. This bracing is provided by notching diagonal 1 x 4 or 2 x 4 pieces into the wall frames, as shown in figure 25-3.

There should be two braces in each wall. If possible, braces should run going from the bottom to the top of the wall. Create bracing triangles as large as the window and door locations will permit.

Figure 25-4 shows some possible arrangements.

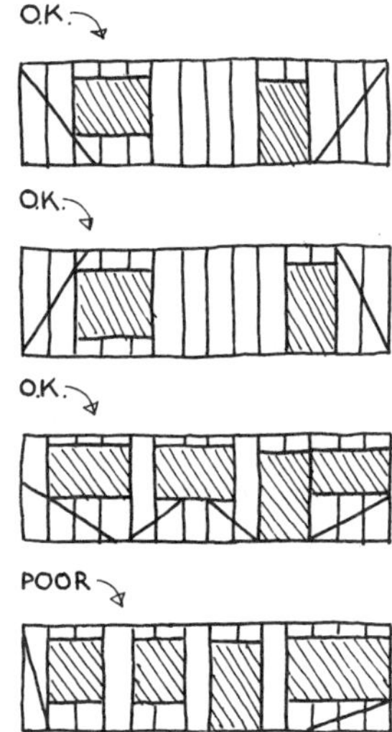

Figure 25-4. *Let-in bracing locations.*

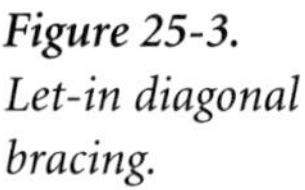

Figure 25-3. *Let-in diagonal bracing.*

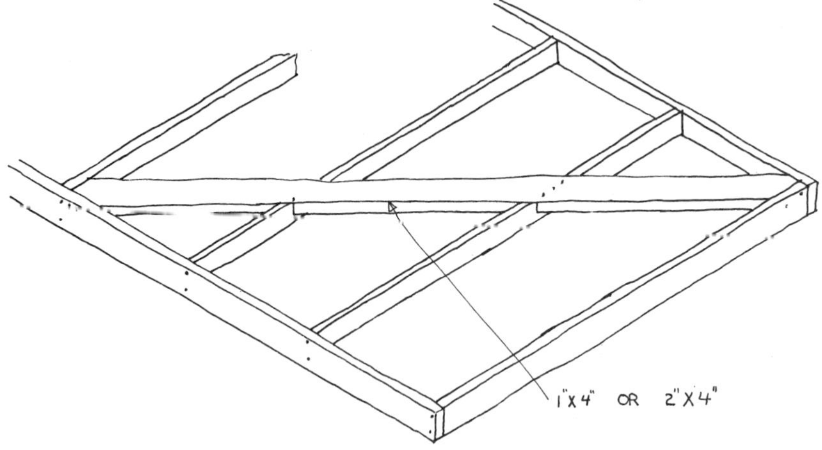

Determine Rough Opening Sizes

If you are buying new windows and doors in their frames, each unit will have a specified rough opening (R.O.) dimension that you can get from the supplier. Usually rough openings can be about 1 inch bigger than the window or door frame. If you are building your own windows or doors, see chapter 17.

Making a Layout Drawing

To clarify your wall design and make ordering materials easier, make a scale layout drawing that shows the parts of the wall frame. First determine for yourself how long each wall frame will be.

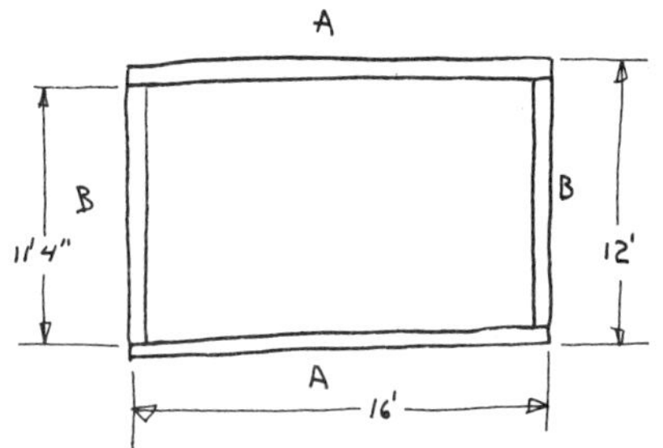

Figure 25-5. Wall A goes up first.

Figure 25-5 shows the wall lengths of a 12 x 16-foot house. The two long walls, marked A, will be built exactly 16 feet long. The short walls, marked B, will be less than 12 feet because they fit between the long walls. Standard 2 x 6s are 1½ x 5½ inches, so the B wall will be 11 inches shorter than the total wall length. You could subtract another ⅛ inch or ¼ inch for build-up.

Get the basic outline of each of your wall framing plans by making a tracing from your exterior elevation drawings. Then draw the 24-inch intervals where studs will be centered. The intervals should always start at the corner of the building for the convenience of installing plywood sheets. On A walls, this will also be the end of the wall frame, but for B walls, which are erected after A walls, the intervals should be offset by the stud width, as shown in figure 25-6. Finally, mark all the doubled studs, diagonal braces, jack studs, and headers on your drawing (figure 25-7).

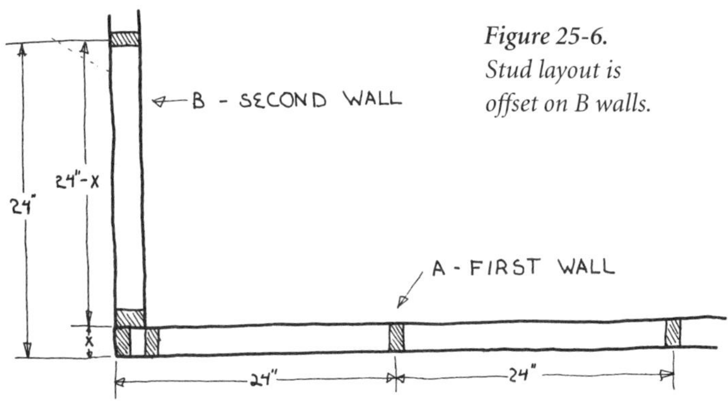

Figure 25-6. Stud layout is offset on B walls.

Figure 25-7. Studwall plan.

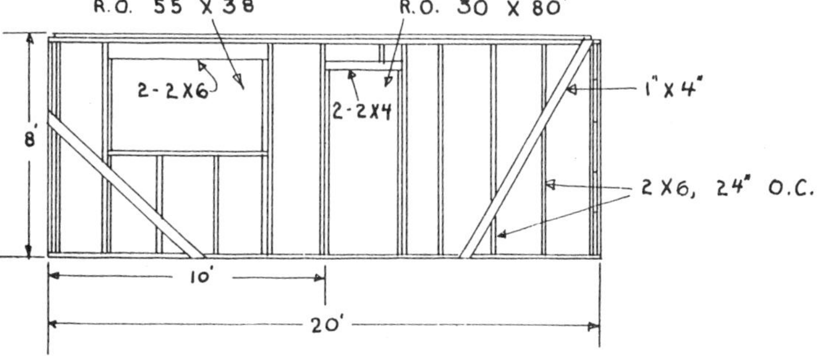

BUILDING WALLS

STEP 1

Pre-cut studs. Figure out how many vertical full-length studs there are, and cut them all at once with a power or hand saw. Usually they will be 4½ inches less than the total wall height. You can also pre-cut many of the parts that go around windows and doors. The horizontal member (the headers and the sill stud below windows) will be the width of the rough opening, plus 3 inches. Window jack studs will be the same height as the window R.O. Door jacks will be the height of the door R.O., less 1½ inches. It helps to give each set of these special studs a code letter that should also appear on your floor plans. For example, all windows with a 3 foot, 2 inch-by-5-foot R.O. would be marked A, the next size B, and so on. Always give the horizontal dimension first when writing down a R.O.

STEP 2

Strategize. Before you build the walls, strategize about which walls will be easier to put up first, and exactly how you will maneuver each wall into place. Usually there is a logical order for building the walls. Walls are heavy if they are sheathed with plywood. It takes several people to put them up. It might make sense to put long walls up in parts. For example, I would do a 24-foot wall in two parts.

STEP 3

Plates. Cut the top and bottom plates to length from the straightest 2 x 6s you have.

Lay the two plates next to each other on the sawhorses with the ends flush. Mark the upper one "top" and the lower one "bottom." Lay out the 24-inch intervals on the plates, remembering to set the zero hook of the tape at a point representing the corner of the building, which may or may not be the end of the plate itself.

If you're laying out a B wall, and the wall sits 5½ inches in from the corner of the building, have someone hold the tape 5½ inches out from the end of the plate as you do the layout (figure 25-8).

As you work down the plate, make two additional marks at each 2-foot interval, one ¾ inch on each side of your center mark. These marks represent the two edges of the stud.

Take a large framing square, and place it on the plates with the 1½-inch-wide blade of the square crossing the plates. At each stud location, put the square between the outer marks, then mark along both sides of the square. These

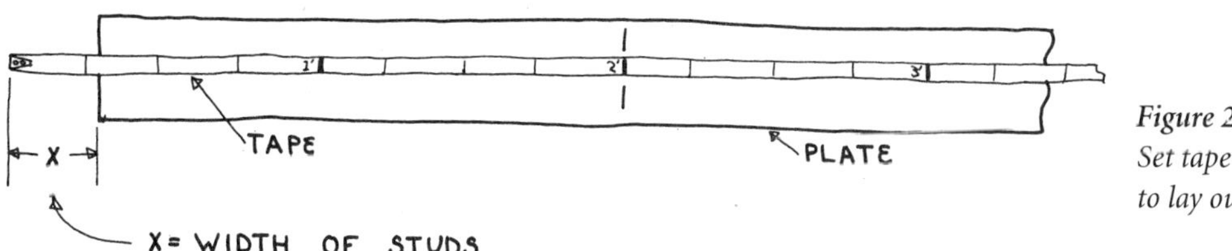

Figure 25-8.
Set tape like this to lay out B walls.

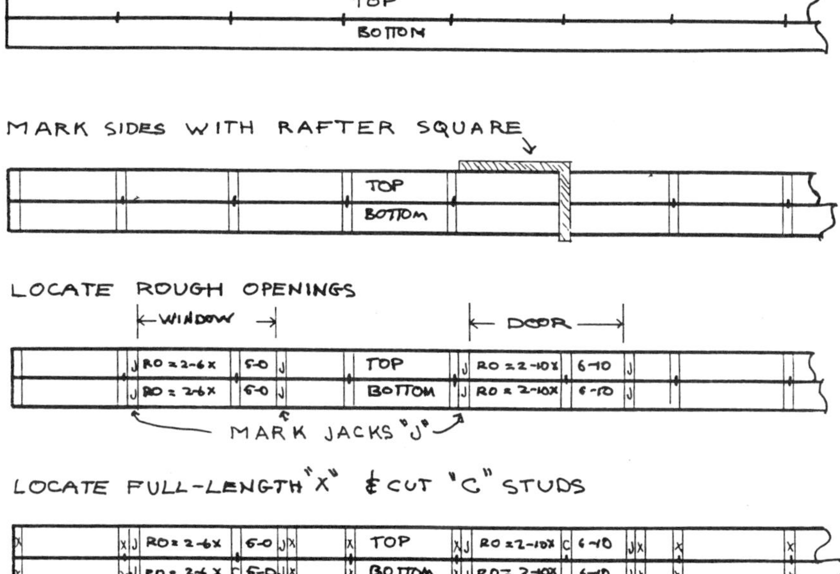

Figure 25-9.
Laying out wall plates.

double lines make it easy to see exactly where the stud goes when you nail the frame together.

Next locate the rough openings for doors or windows. If it is convenient, use the stud locations you have just defined at one side of the rough opening. To avoid confusion, I label R.O.s with a marking crayon, giving the letter designation and the R.O. At each side of the R.O. will be a jack stud, marked J. Outside of each jack is a full-length stud. Mark these "X."

The studs within the boundaries of the R.O. are not full length. They are cut studs, marked "C" on the plates. At window R.O.s, there will be "C" studs below the window, and there may be "C" studs above the window if the header doesn't completely fill the space between the top of the R.O. and the top plate. On doors, there will be cut studs at the top only.

Finally, mark Xs for all the remaining studs where there aren't R.O.s to indicate full length studs.

Step 4

Frame the walls. To assemble the walls, locate the plates on the floor where you can conveniently tilt the assembled wall frame right up into position. Arrange the studs and headers in place. As you position the full length X studs, sight down them, and put the bows up.

When the studs are in position, nail through the plates into the studs with three 16d nails at each joint. Make the joints square and flush by eye. The trick to nailing studwalls together is to prevent the studs from jumping around every time you hit the nail. The simplest system is to hold the stud down with one foot while you nail through the plate. If the stud and plate are not flush on top, shim

Figure 25-10.

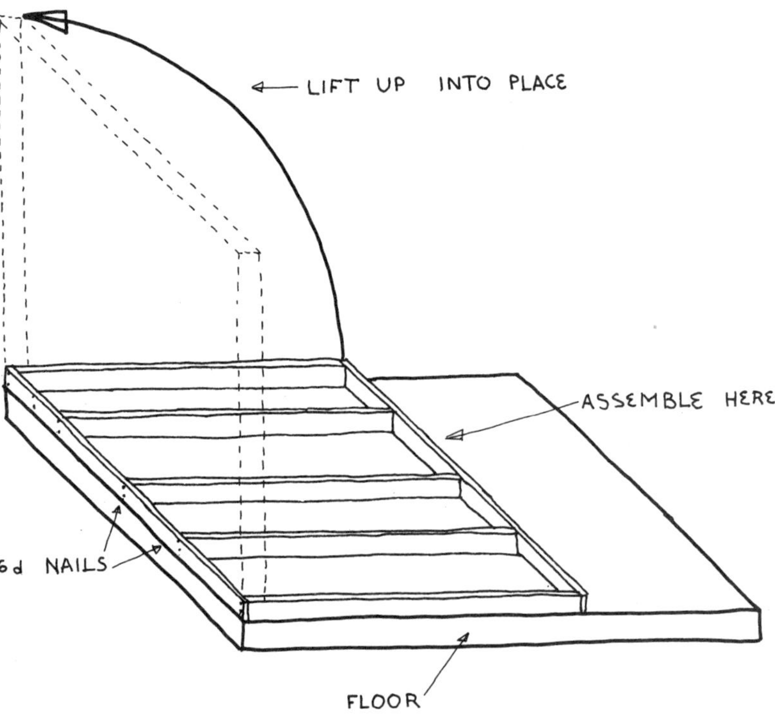

the lowest one up on a shingle while you nail. You can also nail cleats to the floor around the frame to keep it from sliding.

Working from your drawings and the marks on your plates, nail in headers and other R.O. parts. With doors, it's easiest to start by nailing on the jacks, then the headers, then cut short "C" studs to fill in above the headers. With windows, it may be easier to work from the top down, particularly if the headers fit right up against the top plate. I don't bother to make a full length J stud to go under the sill piece. A scrap block provides plenty of strength.

Step 5

Diagonal bracing or plywood sheathing. If you are sheathing with plywood, it goes on now. Board sheathing, which is heavier, should probably go on after the wall is up. In that case, the necessary let-in braces are installed at this point.

In either case, make sure the wall is sitting straight and square on the floor. Sight along the plates to see if the perimeter pieces are straight, and then adjust the wall square by making the diagonals equal. If necessary, lightly tack the frame to the subfloor to hold it straight.

It pays to spend a few minutes thinking about how the sheets of plywood can best be positioned. There are several considerations. Structurally, the plywood serves to tie the various floor and wall frames together, so it should lap over joints where floor and wall frames join. This makes the house strong, and excludes drafts. You also want to use full sheets to minimize cutting and speed the work. Carpenters often let the plywood overlap the wall frame at the bottom by 3 to 6 inches to establish this lap. Then later, another band of plywood is added to finish covering the floor frame. This band in turn will lap over the foundation one or two inches.

However, this overhanging plywood makes it more awkward to tilt the wall up, and carpenters have a variety of tricks to make this tilting-up sequence safe and convenient.

My suggestion for first-time builders would be to align the edge of the plywood parallel with the bottom plate, and about six inches up from the bottom plate. Later you can add more plywood at the bottom to cover the floor frame and lap onto the foundation by about one to two inches.

Also figure out where the plywood should go to at the top. If you have raised the bottom sheet by 6 inches, uncut sheets will often stick up around 6 inches at the top. If there will be a floor above, install the sheets leaving this projecting flap of plywood at the top. When you build the floor above, this flap will tie the lower walls to the second floor structure, and exclude drafts where they meet. When there is no floor above, cut the plywood to project 1 inch at the top. This 1-inch flap will tie the wall frame to the top "doubling" plate which you will be adding later.

If you are framing a B wall, the plywood will stick out 5½ inches on one side. This flap will tie the B wall to the A wall it butts into, making a strong corner.

Overall, the goal is to cut as few sheets as possible, while using the plywood to tie all the structural elements together.

In earthquake zones, coastal areas, or other areas with risk of high winds, there will probably be building codes that specify how this must be done.

Position the first sheet of plywood with the proper overlaps, as accurately as you can, holding it with four or five tentative nails. Tack down the second sheet the same way. Make adjustments if needed.

Nail the plywood every 8 inches with 8d galvanized box nails, or special 7d threaded plywood nails. Cut out the window openings before erecting the wall, because it make the wall unit lighter, and provides more hand-holds.

If you are going with board sheathing instead of plywood, install let-in bracing now. Take two 2 x 4s or 1 x 4s and lay them on the studwall in the best location you can find (figure 25-11). Using the brace piece as your ruler, mark where the studs must be notched out to accept the brace. Without moving the brace piece, kneel down, reach under with your pencil, and mark the brace piece where its ends meet the plate or corner stud. A 2 x 4 brace can butt into the plates or corner post, as shown in figure 25-3, but a 1 x 4 brace should be notched in everywhere, as shown in figure 25-11.

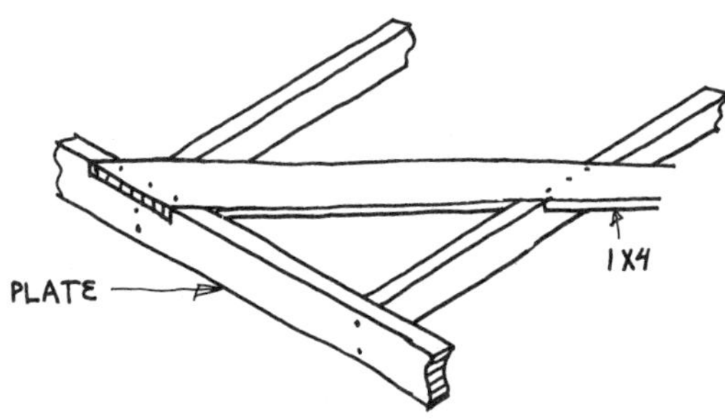

Figure 25-11.

Cut the necessary notches, using the method given in chapter 35. Then tap the braces into place. Two-by-four braces can be nailed in with 16d nails, and 1 x 4 braces can be nailed with 8d nails or 10d box nails.

Step 6

Putting up the wall. The wall is now ready to tilt into place. Strike a chalkline on the floor, 5½ inches in from the outside of the floor frame. This will help you line the wall up.

I also like to clamp or nail blocks to the outside of the wall, as shown in figure 25-12, to keep the wall from sliding off as you tilt it up. Make sure

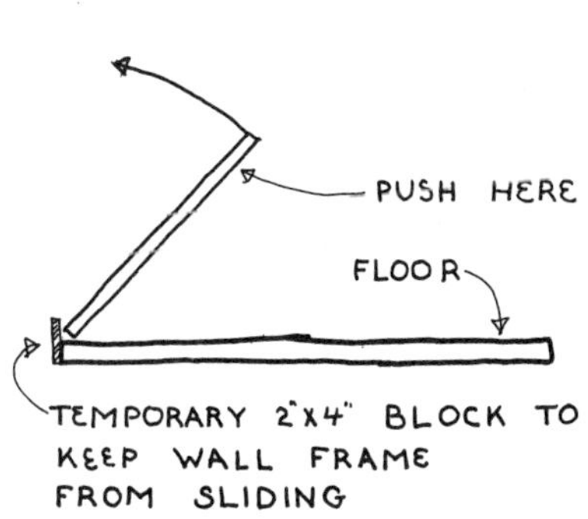

Figure 25-12. Lifting the wall.

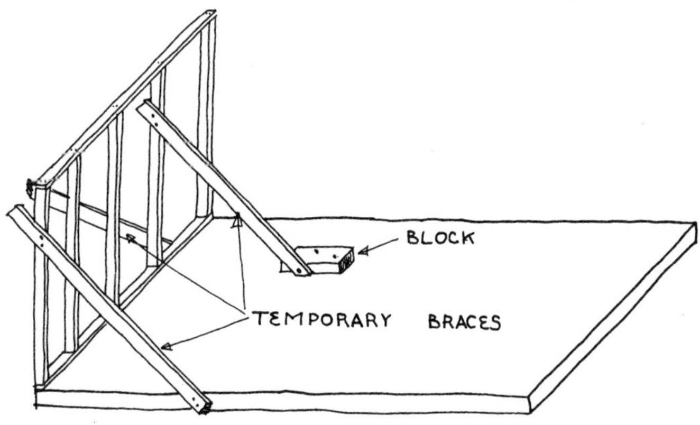

Figure 25-13.

they don't interfere with the plywood.

Get several people to help with the lifting, particularly for a long wall. Have some 2 x 4s, nails, and a 4-foot lever handy, and a sledgehammer. I would advise having an experienced carpenter with you the first time you do this. He or she can help you find the safest and most efficient procedures for your situation.

Slide the wall to its location, then tilt it up.

When the wall is vertical, use the sledgehammer to tap it into the exact location, on the chalk line and flush at the ends. Nail the bottom plate down using 16d nails, leaving the nail heads up a little so that minor adjustments can be made later if necessary. Take the longest level you have and plumb the wall in and out. When it is just right, use the 2 x 4s to temporarily brace the wall in place, as shown in figure 25-13.

Step 7

Doubling plates. By now, all the walls are up. Next add the doubling plates, nailing them down with 16d nails. Pick the straightest, longest pieces you can find.

Step 8

Benchmarking. This is a good point to make sure everything is as it should be. In theory, because you squared your walls so carefully, the rectangle defined by the tops of the walls will be square and level.

It is good to check everything for plumb, level, and dimensional accuracy. Check the top diagonals with the fifty-foot tape, and if you have any doubts, set up the water level as before to make sure all walls are the same level at the top. You might be able to force the walls into better alignment with diagonal braces, as in figure 25-13. If they aren't level within

⅛ or ¼ inch, you could lift the low spots by shimming between the top plate and the doubling plate.

Step 9

Complete sheathing. Usually at this point, the plywood sheathing will be on, but it may be missing in some spots. You could complete this now, or come back to it after the roof is on. Make sure the plywood is nailed particularly well at stress-points such as the horizontal joints between floors.

If you are using boards for sheathing, you must choose whether to put them up now or later. Sometimes, particularly in rainy weather, it seems better to roof the house, and then come back to the sheathing. If you do this the walls should be braced with strong temporary braces so they do not get out of plumb.

Sheathing boards should be cut so that every joint comes at a stud, and then nailed at each stud with 8d common or 8d galvanized box nails. Start from the bottom and work up. When you get to a height that you cannot reach from sawhorses, build a scaffold to work from (see chapter 32).

VARIATIONS ON WALL FRAMING FOR SPECIAL SITUATIONS

The other walls usually go up one at a time in the same way. There are also some special situations that are handled in slightly different ways.

Studwalls Constructed in Place

Sometimes it is inconvenient to build a studwall on the ground and erect it in one piece. This is particularly true of irregular walls, such as those that meet a sloping rafter at the top (figure 25-14).

They can be built after the rafter is up. First put the bottom and top plates in place with 16d nails. Then lay out the stud locations on the bottom plate. Place a stud a little longer than necessary on a layout mark and plumb it with a level. The stud should be offset as shown in figure 25-15.

With the stud plumb, mark it for cutting at the correct angle, and at the same time mark the top plate to show where the stud will fit when cut. Cut the stud to size, and toenail it in (figure 25-16).

Studwalls for Houses with Kneewall Design

Some houses have a gable roof sitting on a short 3- or 4-foot-high second-story wall called a kneewall. This kneewall cannot be built as a separate short studwall. Part of its job is to help hold in the roof, which has a tendency to spread. To do

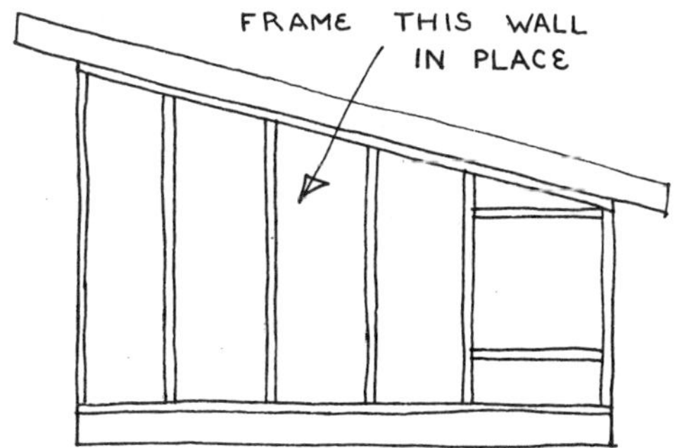

Figure 25-14. Odd walls are often framed in place.

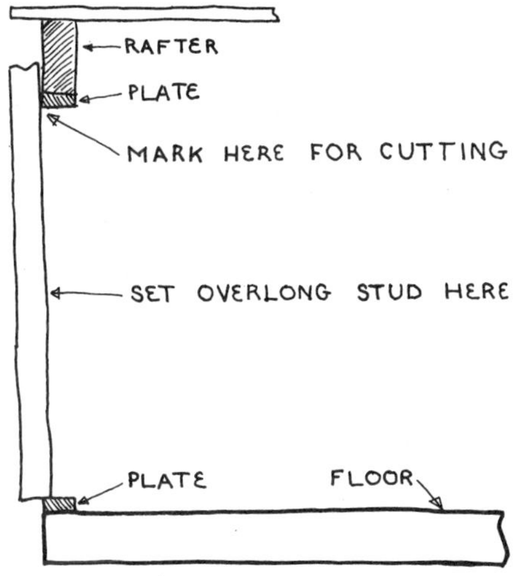

Figure 25-15. Marking studs in place.

Figure 25-16. Toenailing.

this the kneewall and the wall below it are built in one piece; the studs go from the first floor right up to the roof (figure 25-17).

These continuous side walls are held together by the second floor joists, which can be built using one of two methods, as shown in figure 25-18. In either method, the strength of the structure depends on a strong joint between the wall stud and the floor joist.

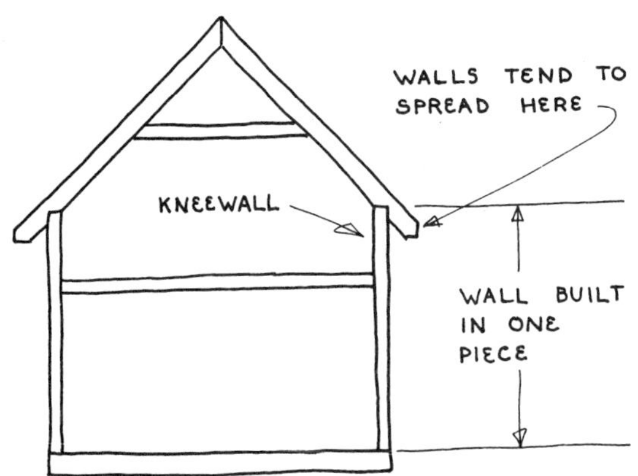

Figure 25-17. *Kneewall design.*

Interior Partitions

An interior partition is sometimes merely a room divider. It bears no loads. Such a wall is not heavy and can go anywhere. If it runs parallel to the joists underneath it, double the joist right under it. If it is perpendicular to the joists and is near the middle of their span, consider its weight when you compute the size of the joists. A partition that carries roof, rafters, or second-floor joists (load-bearing) should be located over a sill. If such a

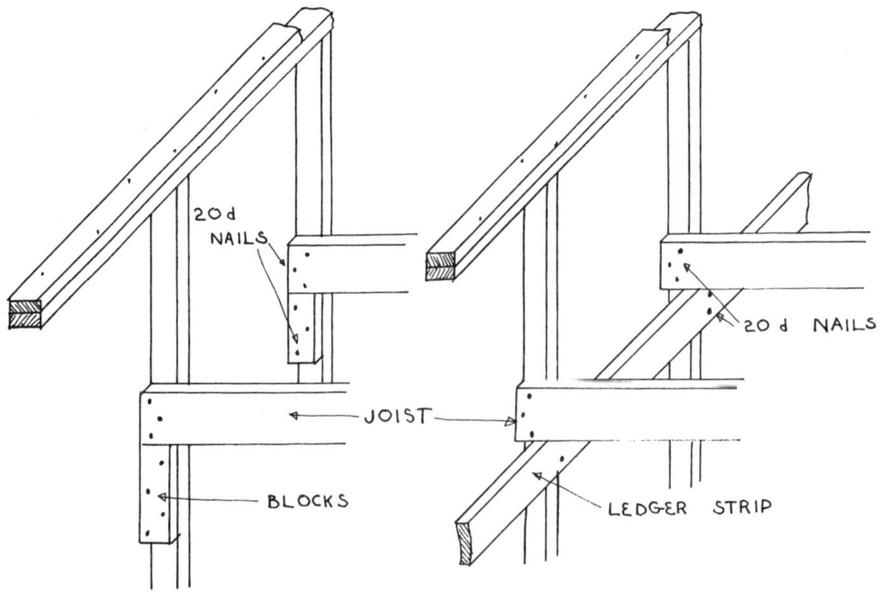

Figure 25-18. *Two ways of attaching joists in kneewall design.*

partition is perpendicular to the sills, the joist beneath it can be doubled or tripled to create in effect an extra sill. These interior partitions are built like exterior walls, but out of 2 x 4s instead of 2 x 6s.

Supporting Inside Loads on Posts and Beams

Sometimes the second floor or roof needs support inside the house, but a partition would be in the way. Such a load can be carried on a horizontal beam, with the ends carried on posts, just like a sill or girder (figure 24-15).

Compute the size of such beams using chapter 16. The posts can be exposed 4 x 4s, 4 x 6s, double 2 x 4s concealed in a wall, or for a very large load, triple 2 x 4s concealed in an existing studwall. Locate these posts right over the foundation. There should be solid wood at least six inches wide between the post and the foundation.

NOTES ON ORDERING WALL MATERIALS

Your framing plans are a good guide for ordering wall framing material. Try to pick lengths that will minimize waste. Usually, however, there are so many different sizes of wall pieces that few short pieces over one foot long will go to waste. There is a rule of thumb that says to order one stud for each foot of wall length; the extras go for plates, headers, etc. That is a reasonable guide.

But 2 x 6s are also invaluable for staging, bracing, and many other purposes. It always pays to order extras. Also order fifty-pound boxes of 8d and 16d nails.

26

ROOFS

TYPES OF ROOF STRUCTURES

The roof is the most complicated and tricky part of a house to frame. Yours should be as simple as possible.

Structurally, most roofs are either shed roofs or gable roofs. Figure 26-1 shows the basic structure of each. A shed roof is a sloping roof in a single plane. A gable roof is the common A-shaped roof, which can be built in almost any size and pitch. More complex looking roofs are usually variations or combinations of these basic types (figure 26-2).

A shed roof consists of rafters laid across from one wall to another. It's simpler and faster to build than a gable roof. It has fewer pieces and they are easy to figure out, cut, and fit in place. Often a shed roof is pretty flat and therefore safe and easy to work on.

But a shed roof has drawbacks. It provides no attic space. It does not look good on every house. And it does not span distances much greater than 16 feet without intermediate support from a wall or beams inside the house.

A gable roof can span longer distances. It can enclose additional living or storage space, and its traditional shape suits some houses better than a shed design.

Where a shed rafter simply spans across a house as floor joists do, a gable roof must be triangulated to be stable. Without a crosspiece to hold the bottom ends of the opposing rafters together, they will spread apart from the weight of the roof and the loads it carries (figure 26-3). These crosspieces are called collar ties.

There are several main types of gable roof, each of which solves the problem of triangulation in a slightly different way. These are shown in figure 26-4. In the attic design, the collar ties are located where the rafters meet the walls. Collar ties double as attic floor joists. Often the rafters are put up individually against a ridge board at the top. The collar ties/attic floor joists are put down first, and then a

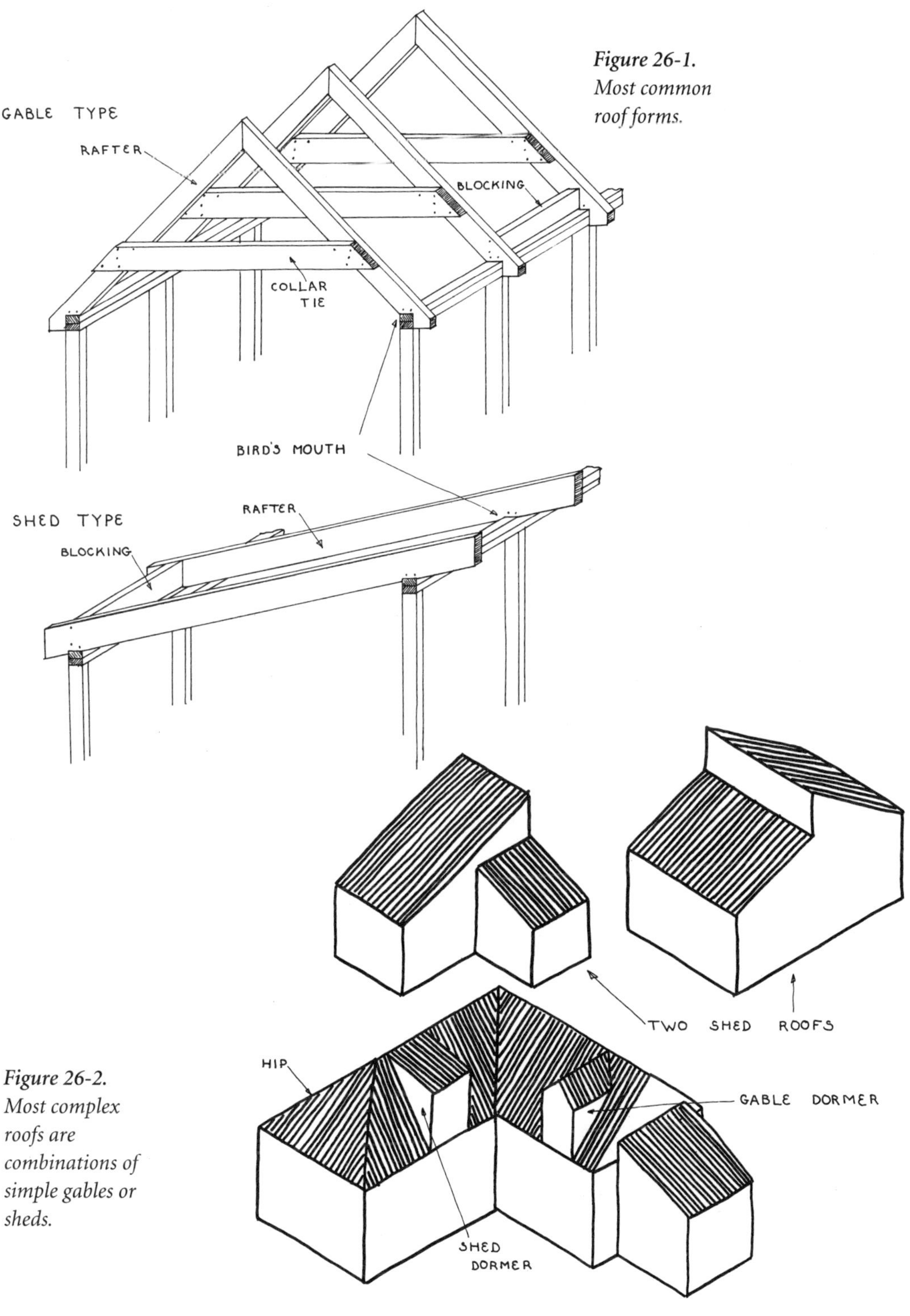

Figure 26-1. Most common roof forms.

Figure 26-2. Most complex roofs are combinations of simple gables or sheds.

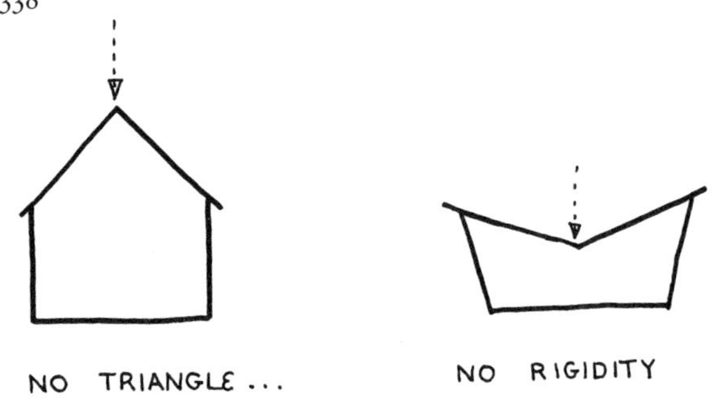

Figure 26-3.

subfloor is laid on them as a work surface for putting up the rafters. In a small house the attic may be useful only for storage because the headroom is poor. But in larger houses the attic can be living space if the roof is steep enough.

In a small house the area under the roof can be made usable by using the story-and-a-half or kneewall design, so the collar ties are located farther up the rafter to create headroom. They cannot be above the rafters' midpoint, however, or they will be ineffective in defeating spread load. If you are designing a roof of this type, get some help with the engineering.

The little walls extending up past the second floor are the kneewalls. The added height they yield makes the space usable as a second floor. In this type of design, the rafters and collar ties are often assembled into units called rafter units, which are hefted into place in one piece. The second floor provides a convenient work platform.

Sometimes an attic-type design is assembled in rafter units, rather than one rafter at a time. This works particularly well for smaller roofs.

A third type of gable-roof frame is the truss. A truss roof consists of a series of triangles assembled into a frame, generally bought ready-made from lumberyards. You lose the storage space, but trusses can span long distances without interior support, and can be installed quickly. Most companies that market trusses can deliver them on lift trucks that allow you to unload the trusses at roof level.

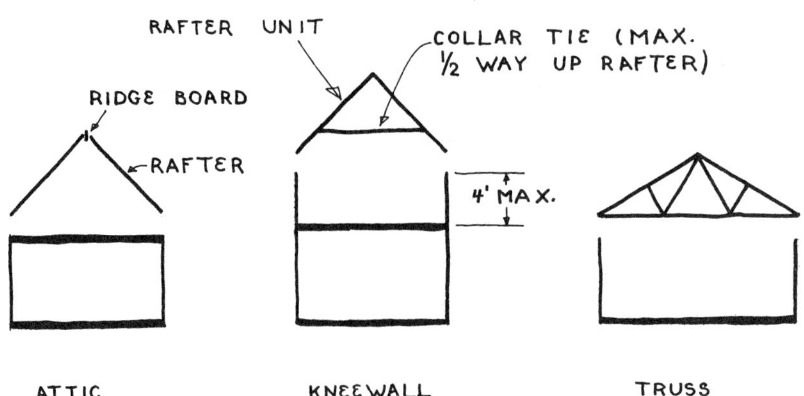

Figure 26-4. Gable roofs.

DESIGN DECISIONS

Each of these basic roof types can be built in many sizes and at many angles or pitches. How do you decide which to make for your house? The best method, I think, is to let the roof type be determined by your space requirements, the general character of the site, the sun orientation of the house, and similar nonstructural considerations. Find the roof shape that fits your overall plan. Chapter 5 describes how to choose a roof shape this way.

You must also decide how the roof will be covered, how to vent it, and how overhang details will be handled. Chapter 9 discusses some basic considerations. Chapter 27 describes advantages of different roofing materials, and how they influence the roof structure.

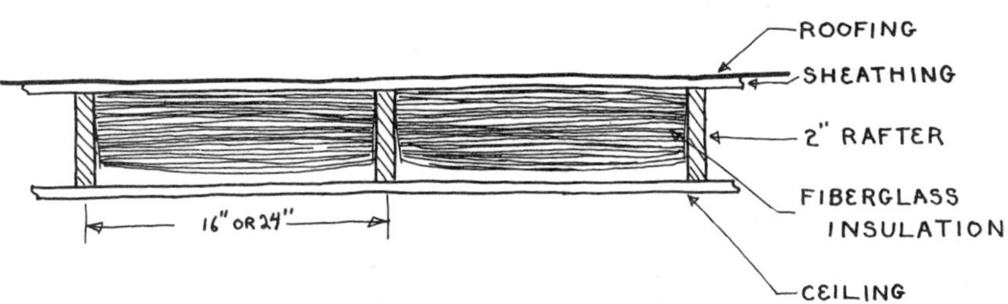

Figure 26-5.

Figure 26-5 shows the most conventional way roofs get covered, with plywood on the outside, sheetrock or some other ceiling material on the inside, and fiberglass insulation between the rafters. This works well with any kind of shingles, roll roofing, or metal roofings.

Galvanized metal roofing is probably the quickest type of roofing to put up, and it is very durable. It now comes in nice colors. Although it can be put down over plywood, it is often installed over spaced boards, which saves money, and provides an integral ladder for moving about the roof during construction.

Chapter 27 discusses roofing options and methods.

Roof Overhang

Overhangs protect the walls and windows, and provide necessary shading from the sun on the south side. The amount of overhang and how it is detailed also have a large effect on how your house looks. As you work through these decisions, it is very useful to make detailed elevations or sections, or even mock up different options. An inch difference in the fascia board, or a few inches difference in an overhang, can change the look of a house.

Figure 26-6. Sheets of galvanized roofing being installed over spaced boards.

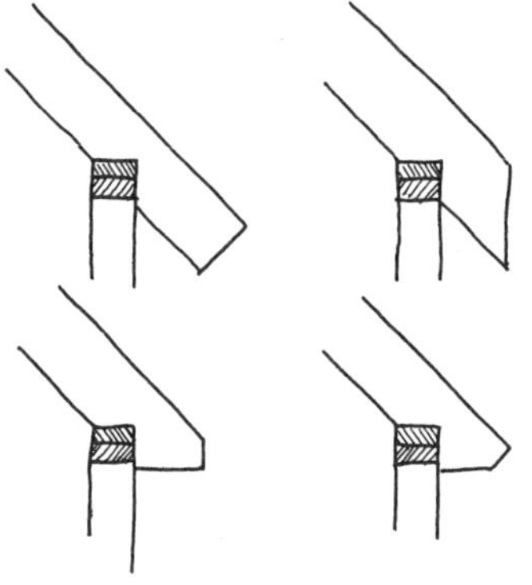

Figure 26-7. Roof-overhang shapes.

A starting point for your decision might be how the rafter overhang is shaped, and this affects how the rafters are cut. Figure 26-7 shows some options. The next question might be whether to box in the rafters or leave them exposed. Exposed rafters can give a cottage look, or a modern, timbered style. Leaving rafters exposed probably saves some time, by eliminating most of the exterior roof trim.

Boxing in the rafters gives a different look, and provides better protection to the rafter ends.

The details of the overhang come next, starting with the eaves. With an open rafter, you need to specify the amount of overhang, and the shape of the rafters at the ends. It makes sense to install a small trim strip, as shown in figure 26-9. This keeps some of the water away, and covers the edge of the roof sheathing.

With a boxed rafter, the goal is a detail that looks good, is proportional to your house design, and uses standard lumber sizes for easy installation.

Similarly, think about the rake, which is the overhang and trim at the gable end that is parallel with the rafters. With exposed rafters, you can just add an "outboard" rafter on top of the sheathing. You might snip off

Figure 26-8. Commerford Maracek house. The roofing is "Onduline," an asphalt roofing that comes in sheets.

the overhang on the end rafter for looks; it might look funny having two exposed rafters only an inch apart (see figure 26-11).

In the case of a boxed rafter, the outboard rafter might be a narrower size than other rafters. This creates a little slot that the siding can be tucked up into. With either type of rafter, a little more overhang can be created by extending the outboard rafter out with a 2 x 4 or 2 x 6 spacer.

A larger overhang can be framed as in figure 26-12.

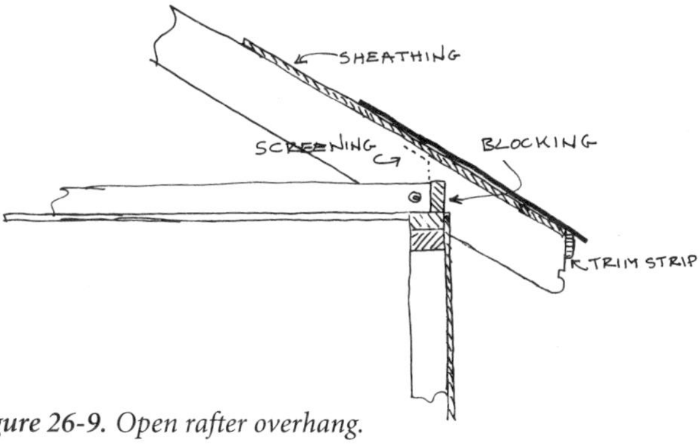

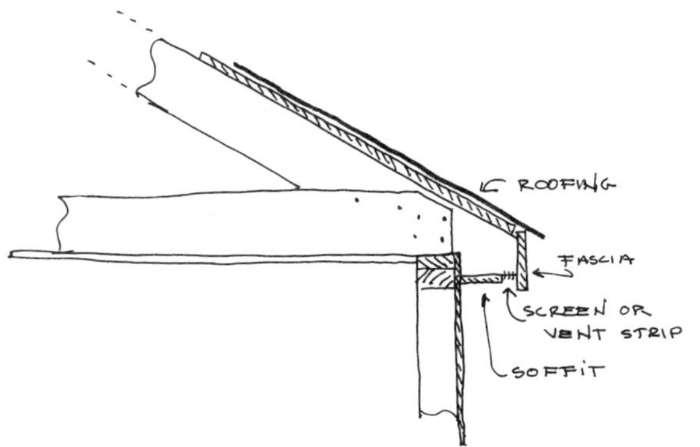

Figure 26-9. Open rafter overhang.

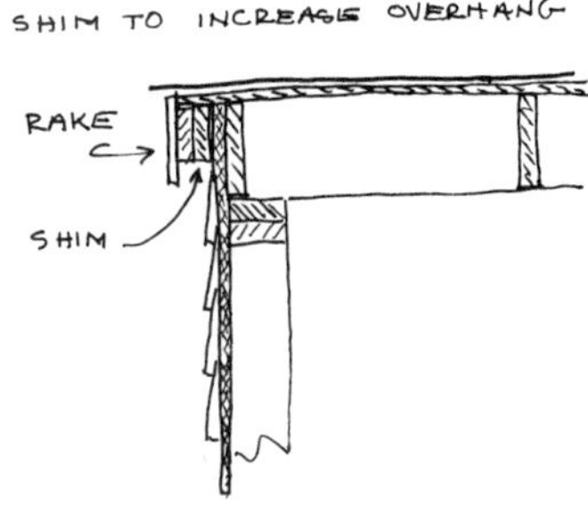

Figure 26-11. Rafter details at rake.

Figure 26-10. Boxed rafter details.

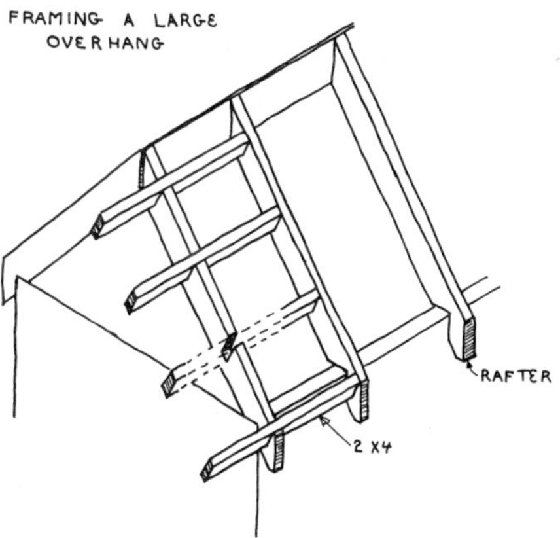

Figure 26-12. Framing a large overhang.

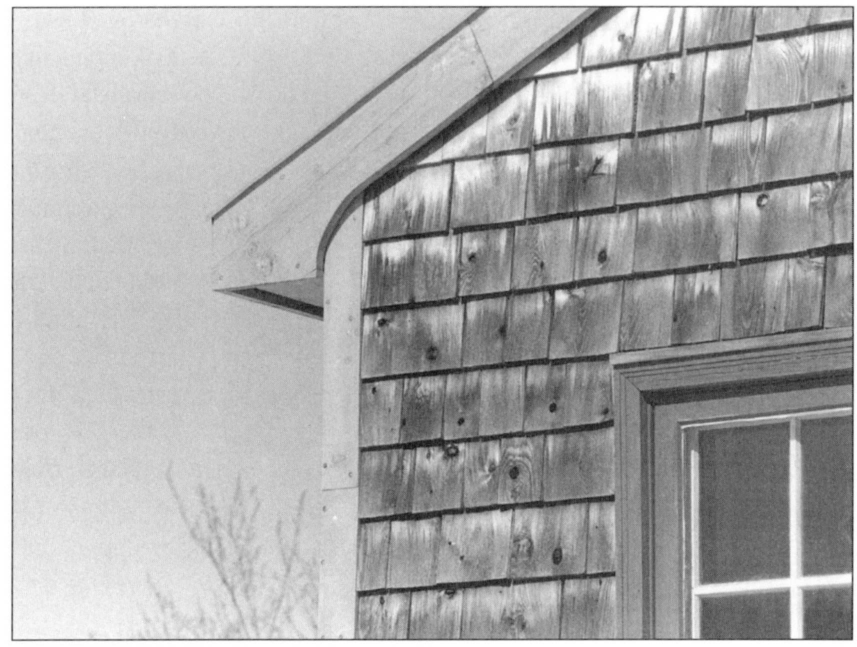

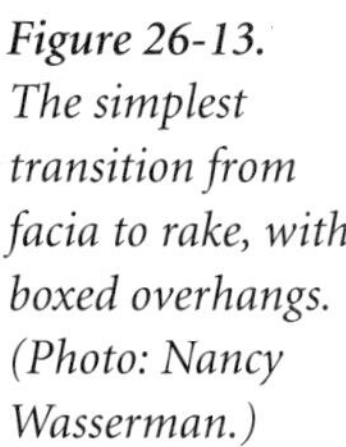

Figure 26-13. The simplest transition from facia to rake, with boxed overhangs. (Photo: Nancy Wasserman.)

With boxed eaves, the eaves' trim boards will usually continue around the corner and up the rake. Figure 26-13 shows one standard way this is done.

It's worth it to spend some time over this detailing. You want something weatherproof, in proportion, and easy to build. Draw in your trim detail on your elevations and section drawings. It may be helpful to make a large-scale detail of these things, or even mock up alternatives. The simplest way to get this right is to look carefully at the buildings you see, and notice how the overhangs that look right were constructed.

Rafter Sizing

Finally, find the rafter size using chapter 16 or appendix 1. We often use a 2 x 10 or bigger rafter to make room for thick insulation, even though a 2 x 8 or 2 x 6 would handle the load. The collar ties can usually be 2 x 6s in terms of the function of keeping the rafters from spreading.

Next, you also need to plan how the roof will be vented, and how excess moisture inside the roof can escape. You need to create airflow above the insulation. In a shed roof, there are often vents either in the soffit, or in the blocking between exposed rafters. This approach works at the bottom of a gable roof, but the upper vent will usually be in the end wall (if there is an attic or attic-like space), or by using a continuous ridge vent.

Figure 26-14 shows standard approaches to creating this needed airflow. For a soffit vent, you can use standard aluminum soffit vent sold by lumberyards. Gable-end vents with screening are sold in various sizes, in both wood and aluminum. Ridge vents are made in several materials. But the goal with each type is to permit airflow along the ridge while keeping rain or snow from being blown in through the gap. They are designed to be at least partially covered by asphalt shingles.

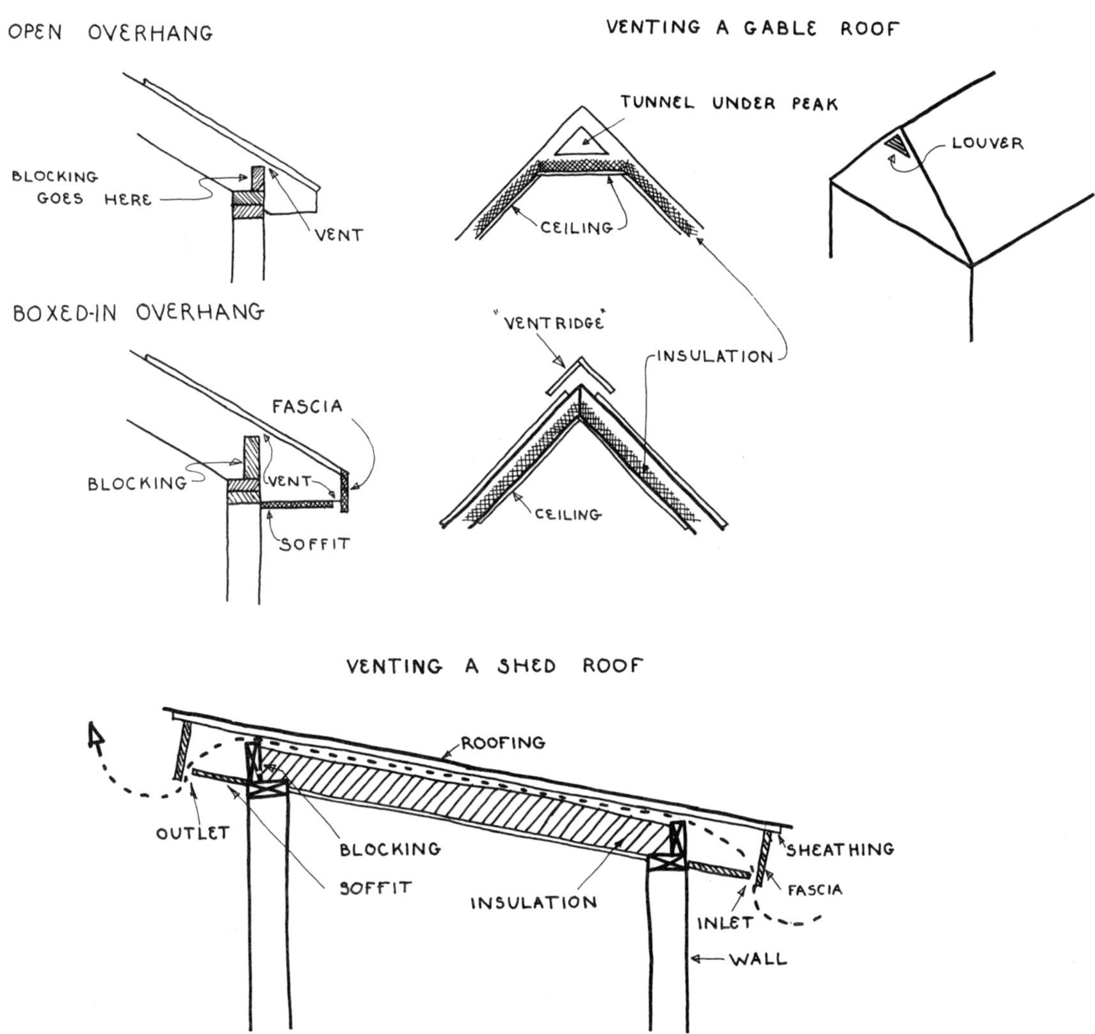

Figure 26-14. Venting options.

The stock vents may not look quite right for every design. For example, a ridge vent can look quite out of place on a very traditional colonial-style building. We often make our own soffit vents using ordinary household insect screen, neatly stapled to the rafter ends before installing the soffit trim (with a 1-inch gap for airflow). We make gable-end vents ourselves when we want to match a particular roof pitch. We have also fabricated ridge vents using rot-resistant wood and metal flashing.

The total vent area should at least equal 1/300th of the area of the roof in plan view. A greater vent area will help with summer cooling. However, both insect screening and louvers impede airflow by about half. Therefore, with screening or louvers, double the vent area to 1/150th of the roof area. With screening and louvers combined, triple the vent area to 1/100th of the roof area. Ridge vent products will come with specifications indicating the amount needed.

It is important that insulation in the roof doesn't impede the airflow. If the rafters are insulated, a gap of 1 inch or preferably 2 inches should be left above the insulation. If the ceiling joists or collar ties are insulated, make sure the airflow is not obstructed at the eaves. Foam baffles, stapled to the bottom of the roof sheathing between the rafters, are sometimes used to maintain the integrity of the airflow.

The most important point, however, is to have a well-sealed interior vapor barrier in the ceiling, to minimize the flow of moisture upwards into the roof insulation in the first place.

LAYING OUT AND CUTTING RAFTERS

The notch in a rafter where it meets the top of the house wall is called a bird's mouth. Any cut that will be vertical when the rafter is in place is called a plumb cut. Any cut that will be level when the rafter is in place is called a level cut (figure 26-15).

In carpentry, roof angles are measured not in degrees, but in "rise and run." In figure 26-16, for example, the roof pitch is a rise of 8 feet in a run of 12 feet, or 8:12. You rarely need to know the roof pitch in degrees.

There are many ways to lay out rafters for cutting. Two are given here. Pick the one that will be simplest for you, or adapt one to suit your own requirements. If your roof is too complex for these methods, learn to use a rafter square, the traditional carpenter's computer for roof framing and other complex framing problems. Stanley rafter squares and Swanson Speed Squares both come with excellent introductory booklets on this subject.

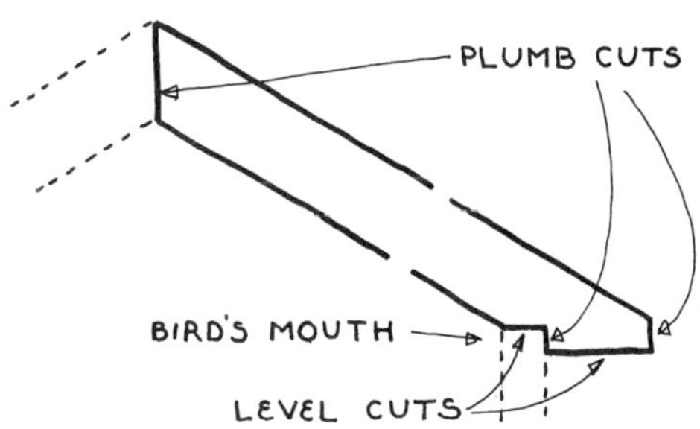

Figure 26-15. Rafter cuts.

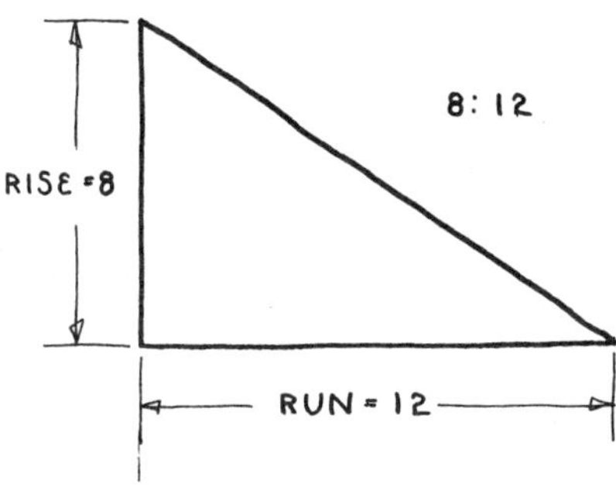

Figure 26-16.

How to Lay Out Moderate Pitch Shed Rafters

When a shed roof has a pitch of about 4:12 or less, the rafters can be marked for cutting with no measuring at all (figure 26-17).

Step 1
Lay out the rafter positions on the top of the wall plates. They will normally be 16-inch or 24-inch centers for a roof with an inside ceiling and 3-foot to 4-foot centers for an exposed frame. Then take an uncut rafter and hoist it up onto the two wall plates. Put it on two corresponding marks where it will actually sit.

Step 2
Take the level to the lower end and mark a plumb (vertical) line on the rafter even with the outside of the wall. Make another vertical line at the top even with the inside of the wall.

Step 3
Rest a piece of 1-inch board on the wall plate, right against the rafter. Use it to mark a level or horizontal line on the rafter 1 inch above the plate. Do this at the lower end, as well. These lines show where to cut the bird's mouth. While the rafter is there you can decide how far out at the top and bottom you want to let it overhang. Mark the rafter at those spots. If you want that overhang cut plumb, mark a vertical line with your level while the rafter is up in position.

Step 4
Take the rafter down and cut along the lines you have made. The bird's mouth cuts can be started with a circular saw but must be finished with a hand saw.

Step 5
If your wall framing is uniform, this rafter (once you are sure it fits) can be used as a template to mark all the others. Simply lay it on the next rafter you must cut and trace. Use the origi-

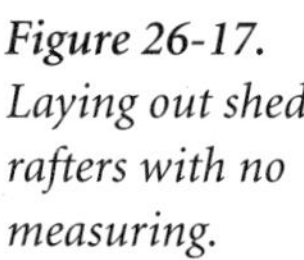

Figure 26-17. Laying out shed rafters with no measuring.

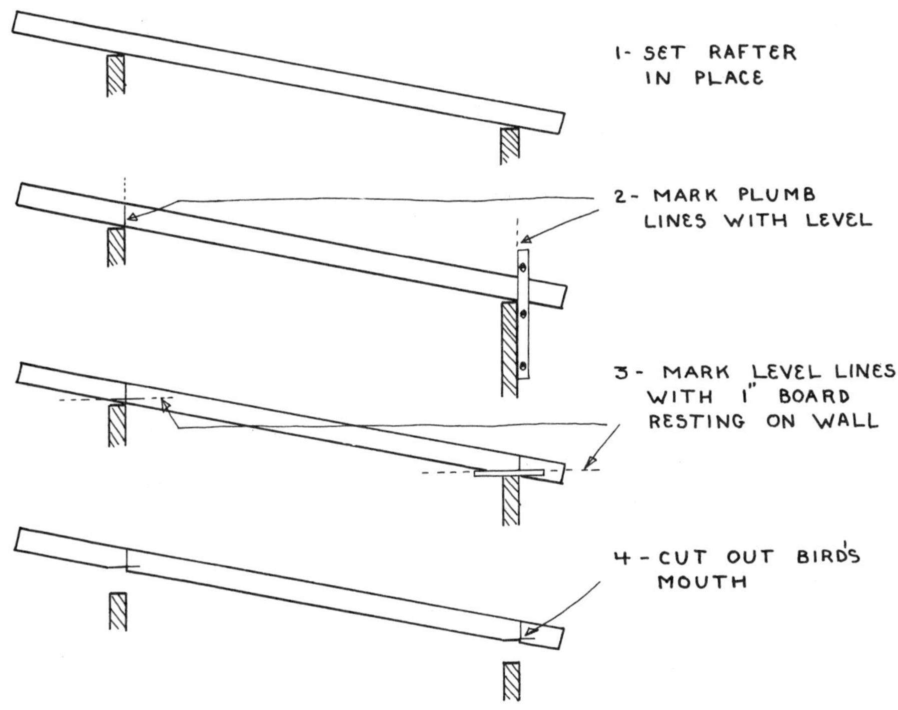
1- SET RAFTER IN PLACE
2- MARK PLUMB LINES WITH LEVEL
3- MARK LEVEL LINES WITH 1" BOARD RESTING ON WALL
4- CUT OUT BIRD'S MOUTH

nal rafter as the template each time, to avoid a cumulative error. If rough, non-uniform lumber is used, line up the template piece with the rafter flush on the top side. This way variations in the width of rafters won't affect where their top edges sit, so the roof will be flat.

Sometimes your template rafter will not fit everywhere, because the wall framing is not uniform. If this is the case, first try to make the framing uniform so that all the rafters can be cut identically. Sometimes a slight warp in a wall can be forced back into line as you install the rafters. Sometimes a wall will be out of plumb, and you can readjust temporary braces to cure the problem. Try to find a way to straighten out the walls and make them parallel on top.

If a correction can't be made, measure each rafter as you did the first by hefting it up, marking it, and bringing it down again to cut it. Or, to avoid this extra hefting, measure each rafter length with a tape measure. Measure from point A to point B as shown in figure 26-18, which represents the distance between the plumb cuts of the bird's mouths. Similarly, measure between the corresponding points of your first rafter, which has already been cut. Then use the first rafter as a marking template, but after marking the top bird's mouth, slide the template up or down as needed to adjust the length from point A to point B.

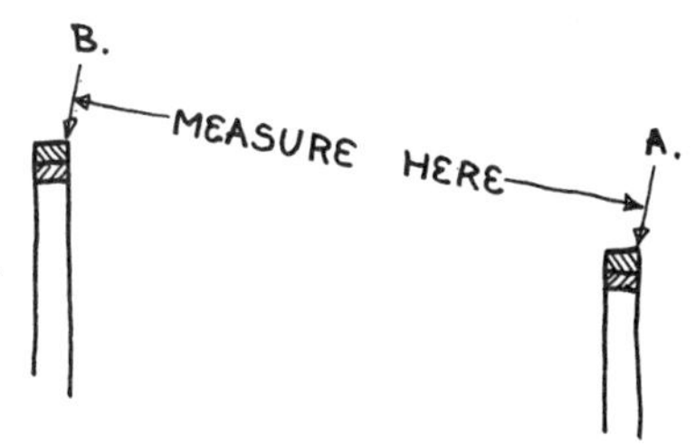

Figure 26-18.

How To Lay Out Steep Shed Rafters and Gable Rafters

For most other roofs, the simplest way to mark rafters is by making a full-scale drawing of the roof frame right on the new floor (figure 26-19). A gable roof built with rafter units is used in the example, but the identical method can be used with almost any roof.

Step 1

Using a chalk line, make a base line representing the level of the top of the walls. On this line, lay out the width of the house, from the outside of the framing on one side to the outside of the framing on the other side. Below the line, at the appropriate place, draw the studwalls themselves. If the sheathing covers the framing at the top, include that, too.

Step 2

Find the midpoint of the base line you have made, and draw a line perpendicular to your first one. From this midpoint to the outside of either studwall represents the run of the roof. The vertical line will represent the nominal rise of the roof. Measure up the rise line an amount equal to the rise you want and make a mark. The distance from this mark to the left end of the base line (the first line your drew) should equal the distance from the mark to the right end of the base line. If not, the rise line is not square to the base line and should be adjusted.

Step 3

Make a chalk line going from the left-hand end of the base line up to the mark at the roof peak. This chalk line should begin at the outer top corner of the left-hand studwall, since that is where the run is measured from. In figure 26-19, this line is marked "1st" line. Make another similar line on the right-hand side.

Step 4

Inside each of the two lines you just made, make a second and parallel line intersecting the inner top edge of the wall. The bottom edge of the first rafter you mark will rest on these lines. This line is marked "2nd" line in figure 26-19.

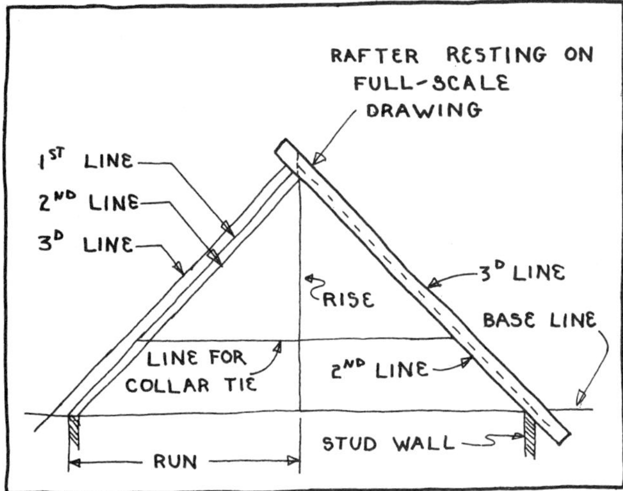

Figure 26-19. Full-scale drawing of rafter unit, drawn on subfloor.

Step 5
Make a horizontal line to represent the bottom edge of your collar tie. In some designs the collar-tie line will be the base line itself; in others it will be another line higher up.

Step 6
Place an uncut rafter with its bottom edge on the line marked 2nd line in figure 26-19. You will mark this rafter for a plumb cut at the top, a bird's mouth, and the overhang shape by transferring lines up from the chalk lines on the floor. The line for the plumb cut at the top will be the rise line on the floor. Place a straightedge on the rafter above the rise line. You can use a rafter square on the floor to make sure the straightedge is directly above the chalk line. When the straightedge is aligned, draw the plumb cut on the rafter. Take your straightedge and use the same method to mark the bird's mouth and the overhang shape.

Step 7
Cut the rafter on the lines you have made. Flip it over, and place it on the corresponding position on the other side of the roof. It should fit perfectly. The whole procedure may seem a little confusing to read about, but when you have a full-scale drawing made and the wood and tools in hand, it will be fairly easy. If you are building the type of roof frame that has a ridge board, use the same method, but draw the ridge board on your full-scale drawing and adjust the rafter length accordingly.

Step 8
To avoid a cumulative error use the first rafter as a template to mark all the rafters. With rough lumber, line up the top edge of the template with the top edge of the piece being marked to keep the roof flat.

Step 9
When you have marked and cut two rafters, place them on the drawing. When they look just right, and fit right at the bird's mouth, make a third parallel line on the floor along the outer edge of each rafter. This line is labeled "3d" line in figure 26-19. It represents the outside edge of the roof frame. Later, when you assemble the

rafter units, they will be assembled in this position on these two lines to guarantee that all units are identical.

Step 10

While the first two rafters are in position on the floor, take the piece of stock you are using for a collar tie, usually a 2 x 6, and place it on top of the rafters, right above the line on the floor that represents the collar tie. Mark where to cut the collar tie to length, and mark on the rafters where the tie will go. This piece can be cut, tested, and then used as a template for marking the other collar ties.

When you make all the rafter units identical, you presuppose that the wall framing is accurate. It sometimes happens that by the time you get up to the roof, the house will vary an inch or two in width. The goal then (without spending too much extra time) is to frame the roof in a way that disguises these inaccuracies. If the variation is under 2 inches, the rafters can be made to fit the widest dimension. This will mean that although the walls vary, the roof is uniform and straight. Since each bird's mouth will sit slightly differently on the wall, you may find that the interior and exterior trim near the eaves will be tricky to fit, but the framing will proceed quickly. If the discrepancy is much over 1 inch, I suggest you get a skilled carpenter to help you find the simplest solution.

Improvising a Roof Framing Method

Sometimes neither of the systems mentioned here will be convenient. The roof may be too steep to heft the rafters into place, or there may not be enough room on the floor for a full-scale drawing. In that case you can improvise some way of finding the angles of the cuts and the distances between the cuts. The essential trick is to find or make a line along which to measure and from which to read the angles. This might mean stretching a string from one point to the other so that you can find the angles of the cuts with your angle bevel. Alternatively, you can use a lightweight piece of lumber such as a 1 x 6 in place, mark it as you would a real rafter, and then use it as a template. The advantage of the latter approach is that if you make an error you have not ruined valuable rafter stock.

Sometimes the trick is to find the right place to measure from so that you can measure along a line parallel to the rafter. In figure 26-20 you cannot measure from the top wall to the bottom because the line would not be parallel with the rafter. However, you can find the angle of the top plumb cut by making a scale drawing, or simply by reading the angle from an existing scale drawing with your angle bevel. Take a scrap of the rafter material, or any scrap of the same width, and cut it to that angle. Place it in the position shown, and mark the wall at the bottom of the scrap. This point establishes the top end of a line that will be parallel to the rafter and that you can measure along.

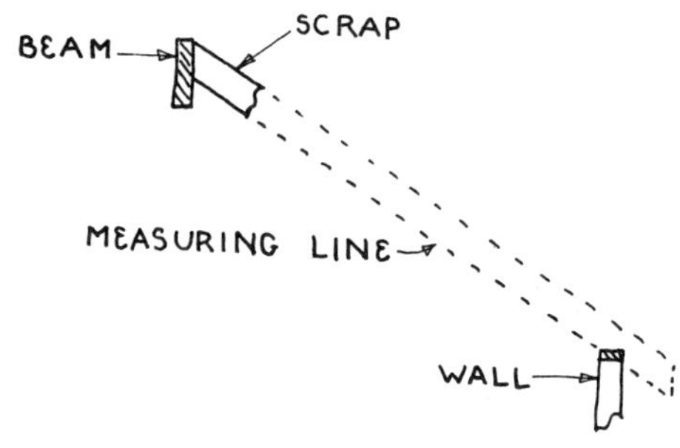

Figure 26-20.

INSTALLING RAFTERS

How you put up your rafters depends on what type of roof structure you have. The process ranges from very simple to tricky and awkward.

How to Put Up Shed Rafters

Shed rafters can be hefted into place, lined up on the right spacing, toe-nailed in with one 8d and then secured with three 16d nails. If the roof overhang is more than 2 feet, the wind could conceivably get underneath in a hurricane and rip the roof off. To prevent this, reinforce the joint with metal straps called rafter tie-downs, which you can buy through lumberyards (figure 26-21).

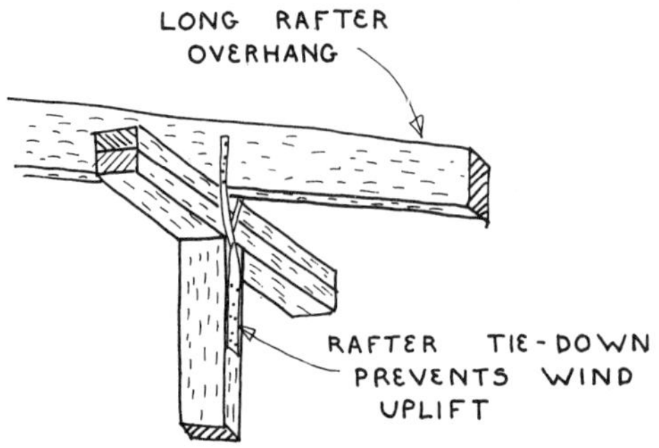

Figure 26-21.

Your local building code may require rafter tie-downs, particularly in high wind or earthquake areas.

How to Put Up Rafter Units

First, assemble your rafter units right on your full-scale drawing. Lay the pieces on the full-scale drawing. Nail some 2 x 4 blocks to the outside of the 3D line shown in figure 26-19. These will help position the rafters while you nail them. Place two rafters on the drawing on the 3D line, against the blocks. Nail a gusset of plywood or 1-inch board (figure 26-22) across the rafters at the ridge. Put a dollop of construction adhesive on each end of the collar tie, and then lay it in place over the right line. If the rafters are 2 inches thick, nail the collar tie in place with the 16d nails arranged in a neat pattern, as shown in figure 26-22. Since the nails may be longer than the combined thickness of the two pieces, drive the nails at an angle that hooks the collar tie into the rafter as shown in the drawing. (If you angled them the opposite way the fastening will be weak.)

If your roof is an exposed-frame type with 3- or 4-inch thick rafters every 3 or 4 feet, the collar ties can still be 2 x 6s, but the joints must be stronger, since a smaller number of ties will have to contain the outward thrust at

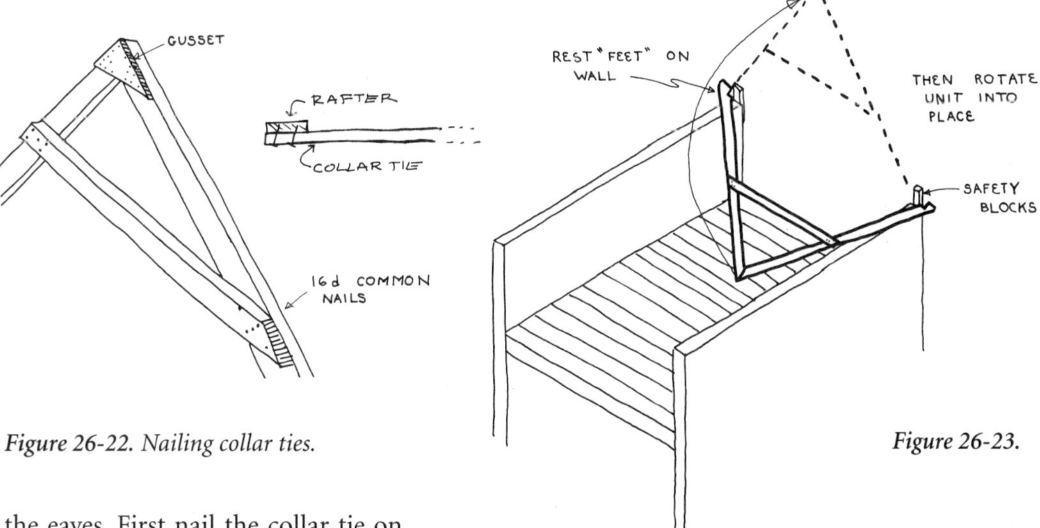

Figure 26-22. Nailing collar ties.

Figure 26-23.

the eaves. First nail the collar tie on just as with 2-inch thick rafters, but using only 4 nails. Then prop the rafter unit up on blocks and drill a ⅜-inch hole through the center of the joint and install a ⅜-inch machine bolt or carriage bolt (and washers) to reinforce the nail joint.

Once assembled, rafter units can be rotated in position much like a studwall. Lay the unit on the floor, and lift up one of the lower corners onto the top of the wall. Then lift the other corner up. The unit is now upside-down, with its feet in the right place, as in figure 26-23. Nail a block at the end of the wall to keep the unit from sliding off. With some friends, rotate the unit up and into place using props, ropes, and any other devices you need. How hard this is and how many people it takes depends on how heavy the rafter units are. If they are heavy, get some experienced help.

When the first unit is up, it can be plumbed with a plumb bob and braced in position.

Take a long piece of strapping or 2 x 4, and mark it with the rafter layout, 24- or 16-inch centers, just as the rafters are laid out along the plate. This will be used to stabilize the remaining rafters as you put them up, and position them correctly at the same time.

As you did the first, tip each rafter unit up into a vertical position. Stabilize it by attaching the marked piece of strapping or 2 x 4 that you just made. First tack the piece to the first rafter unit on the layout lines. Pivot the strapping up and parallel to the ridge, and tack it to the second rafter unit. As you work along, you tack this piece to each rafter unit.

When you get down toward the end of the row you will run out of floor to stand on. When this is about to happen, the rafter units already on the walls can be ganged together and moved back into position later. Once a unit is vertical, it can be moved back and forth along the wall plates fairly easily if one or two people hold the top in position with some kind of brace and someone else walks the bottom ends along with gentle taps of a big hammer, or by walking it along with a flatbar. When the units are all in position, check the layout, and add more bracing to make the rafters secure.

How to Install Gable Rafters Using a Ridge Board

Step 1

Install the ceiling joists (attic floor joists), which are also the collar ties (figure 26-24). When you lay out joist locations on top of the wall, remember that they will rest beside the rafters. Toenail them to the walls with three 16d nails in each joint.

Step 2

Lay a subfloor on top of the joists. This will be your work surface for framing the roof. Leave out the first and last foot or so of the subfloor to avoid obstructing the rafters-joist connection.

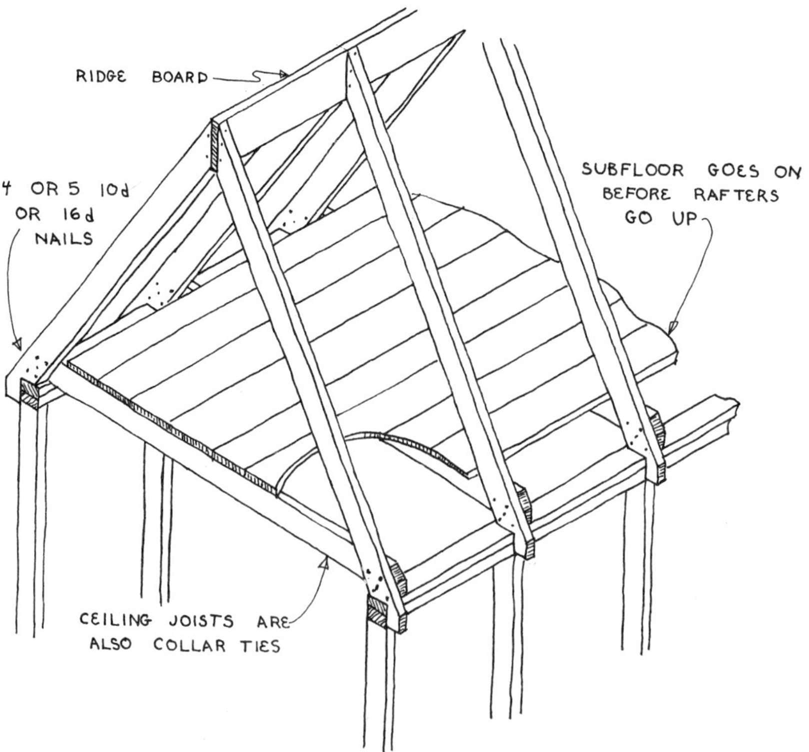

Figure 26-24. *Framing rafters with a ridge board at the peak.*

Step 3
Using scrap lumber, position the ridge board in place temporarily at just the right height, which you can measure on your full-scale drawing.

Step 4
Put up three pairs of rafters, one pair at each end of the ridge beam and one in the middle. These will stabilize the ridge beam. Face-nail the rafters to the floor joists with five 10d or 16d nails for planed lumber. For rough lumber rafters, use 20d nails.

Step 5
Install the rest of the rafters following Step 4. Install metal rafter tie-downs if the overhang exceeds 2 feet, or if codes require it.

COVERING THE ROOF FRAME

Plywood sheathing

Usually the roof will be covered with ⅝-inch plywood, laid horizontally, which is stiffer than running the sheets parallel with the rafters.

Plan a safe way to install the plywood. You may be able to work from the inside on the attic floor. You might build a complete staging along the eaves, using pipe staging, at a convenient height to work on. The important point is to think about how you will handle the large, and heavy, sheets, and keep them under control. The steeper the roof, the trickier this is. This is a day where a professional's help would be well worth the expense.

Normally the plywood edges will be covered by some roof trim, so the plywood shouldn't project beyond the rafter end. Make a chalk line about 48½ inches up from the rafter ends. The first course of plywood can go right below this line. Tack a 10d or 8d nail into the low end of the rafter. These nails will temporarily support the first sheet of plywood, which of course has a tendency to slide off the roof.

The basic procedure is to rest a sheet on the nails, line it up with the chalk line, and tack it down with 8d galvanized box nails every 8 inches. The next row will be easier to handle, because it can rest on the first row. We often tack 2 x 4s to the lower sheets, to create better footing for working on the roof.

The second row of plywood will be easier, because it can rest on the first row. Start the second row with a half-sheet at one end to stagger the joints. You may find it helpful to nail additional 2 x 4 cleats to the plywood already installed as you move up.

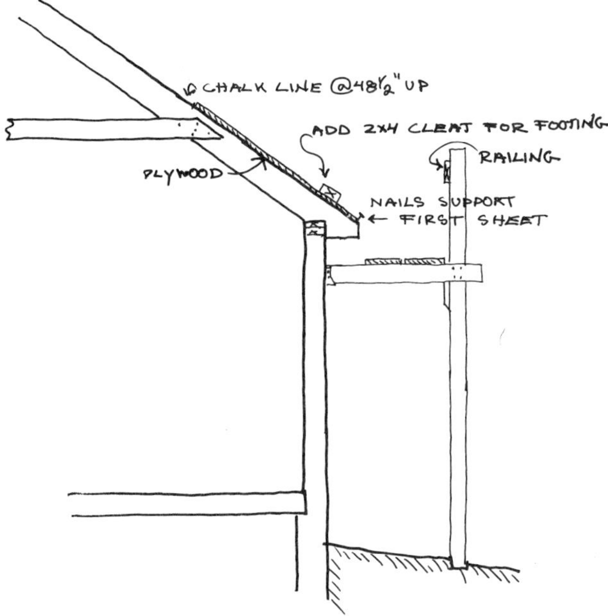

Figure 26-25. Putting up the plywood roof sheathing.

There is increasing emphasis on safety practices in all aspects of building, and this is particularly so when it comes to working high up. Figure 26-26 shows some of the gear now available that is specifically made for roof work. It comes from an excellent article by Howard Stein on this subject, "Working Safely on the Roof," in *Fine Homebuilding* #99, January 1996.

Board Sheathing

You might use a continuous sheathing of boards on your roof, if the boards are significantly cheaper than plywood. Today, this is not usually the case. If it is, follow the same basic approach as with plywood. Chapter 24 discusses some considerations when putting up boards.

Spaced Boards

Galvanized metal roofing is often installed over spaced rough boards called "purlins." Often these would be rough 1 x 4s or 1 x 6s, 16 inches on center. The appeal of this system is that the boards provide excellent fastening for the roofing (better than plywood), and they provide a built-in ladder to work on. If you use spaced boards, add extras wherever heavy snow or ice might fall down from a roof above. Sheath these areas entirely.

Roofing Felt

It usually makes sense to make the roof waterproof right away. Usually this is done using 15-pound "felt" paper, tar paper. First cut pieces of the felt to the length of the roof, plus a little. They'll be easier to handle than the full roll. Unroll the first piece on the roof, and staple it in place along the eaves. Get it straight at the bottom, and nice and flat. Nail it down along the bottom edge every foot or so with 1¼-inch galvanized roofing nails. If the felt will remain un-roofed for more than a few days, use roofing tins, also. These are little tin discs that you nail through, which hold the felt better and protect it from being torn off by the wind.

It is hard to walk on felt. It tends to slip under your feet, particularly if the roof is at all steep. You may want to nail 2 x 4 cleats over the lower row of

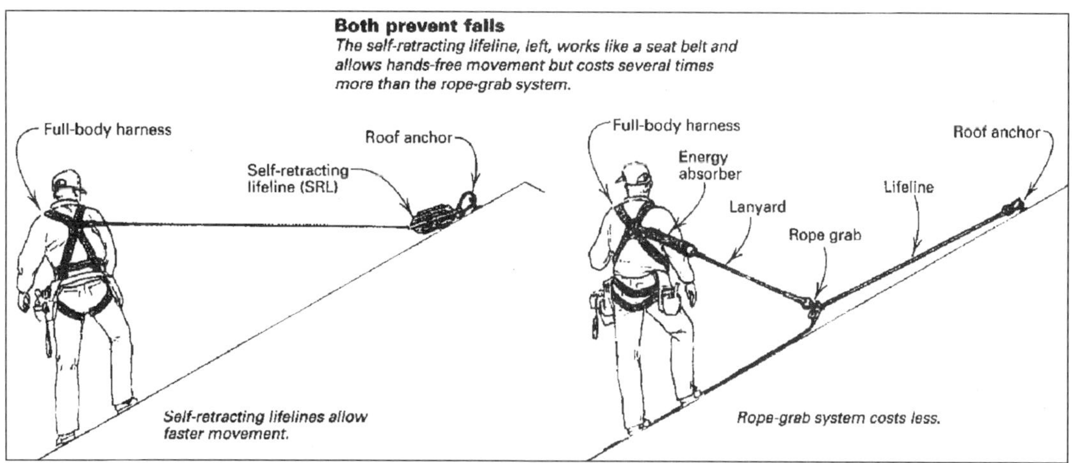

felt as you work your way up the roof. You could also use "shingle brackets" and planks (see chapter 27). Lay out the second sheet of felt, overlapping the first one about 3 inches. The felt has white lines printed on it. Line up the second row of felt with the upper line carefully. Then nail through both layers where they lap. In the same way, work your way up the roof, entirely covering it with felt.

ROOF TRIM

You have already figured out your roof trim details. Since the roof trim is usually hard to reach, it makes sense to set up to do one area, and install all the pieces. As I mentioned earlier, this is easiest if the trim elements are standard sizes of lumber. It is probably best to have a helper who can hold the other end of the tape for measuring, and help support a trim board as it is tacked in place.

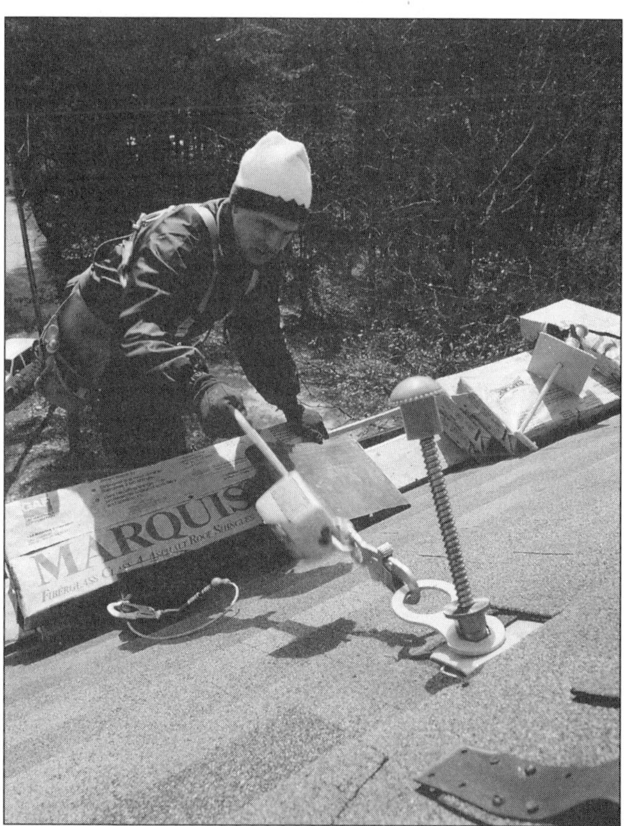

Figure 26-26. *Safety gear for roof work. From* Fine Homebuilding, *January, 1996. Reproduced by permission. (Photo: Steve Culpepper.)*

Create a safe and comfortable work area. Low to the ground, a couple of planks on sawhorses may be enough, but usually real staging is needed. This might be pump staging, complete with railings. It might be pipe staging, positioned perhaps 3 feet below the rafter ends, so the rafters are at a comfortable height.

We usually use #2 pine for our roof trim. (It needs a finish, which can be prepainted on the ground before installation.)

Use 8d galvanized box nails. You will find that 45-degree miter joints look better and are easier to fit. At the corner where the eaves meet the rake, a miter joint is tricky to make because it is a compound angle, but the result will look better and actually be more durable, because exposed end-grain will deteriorate faster than face grain.

27

ROOFING & FLASHING

DESIGNING DURABLE ROOFS

I have read about a church roof in England, consisting of large, thick slates, hung on oak pegs, drilled into the timber frame below. About every 400 years, they have to remove all the slates and replace the rotted oak pegs. Then they put the same slates back up. That's a durable roof!

Though I doubt you'll get 400 years out of your roof, there are a few basic rules for designing a roof that will last a long time.

First, make the roof steep enough; flatter roofs usually leak sooner.

Secondly, keep the roof simple. Most roofs begin to leak at the flashings, the places where one section of roof meets another, or where a roof meets a wall. These are the places where water or snow builds up, and where water weathers the roof material. A simpler roof has fewer of these vulnerable places.

These two rules go together. If the roof is steeper, the complexity may pose less of a problem. If it is simpler, the flatness may pose less of a problem.

Third, respect local conditions. In Vermont, we have a lot of snow, and very variable weather. The snow turns to ice, which works against all the joints in the roof: miniature frost heaves. Sometimes we have ice dams. This is when the snow and ice remain frozen at the eaves, but the warmth from the building melts it above (figure 27-1). When this happens, the water can run between elements of the roof, and enter the structure. To minimize this kind of problem, we try to keep flat areas of roof few and small, and make it as easy as possible for the snow and water to run off. Figure 27-2 shows some of the extremes to which we go to avoid ice dams.

The point is to pay attention to local roofing practices. In your area, extreme rains, hurricanes, or excessive heat may be the problem.

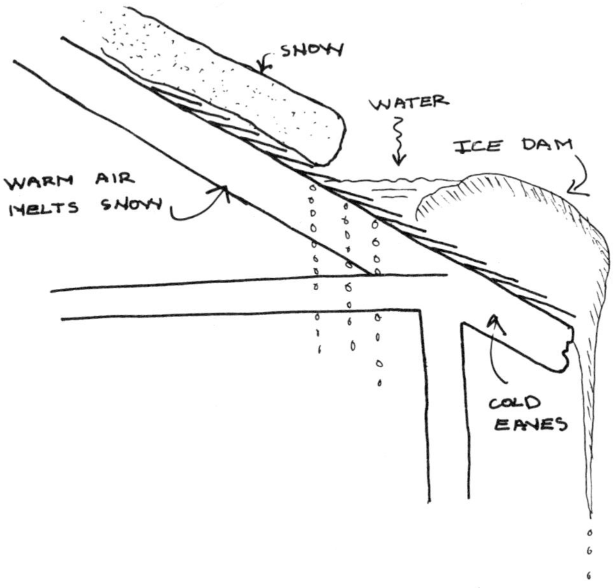

Figure 27-1. Ice dams.

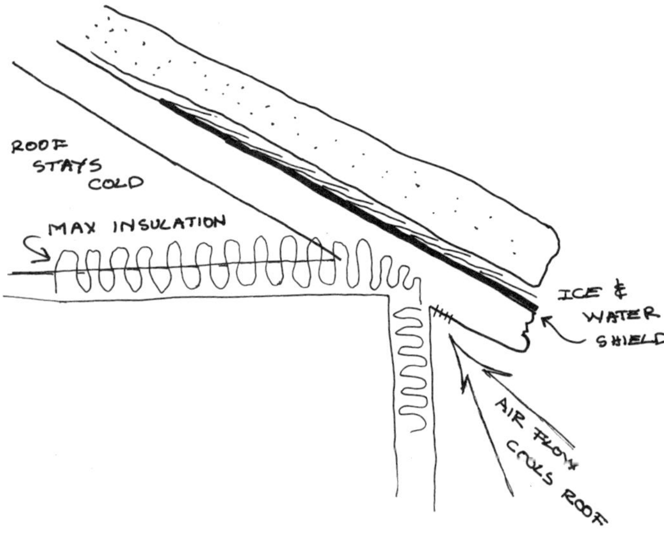

Figure 27-2. Preventing ice dams.

Fourth, when in doubt use an ice and water shield. Ice and Water Shield is a trade name of the Grace Company. There are now several companies that make similar materials. The shield is a thick, heavy, rubberized material that is extremely sticky on one side. It is applied sticky side down under vulnerable places such as eaves or valleys. The sticky material seals around the roofing nails or screws. It provides an excellent second line of defense when the roof is subject to extreme conditions.

Finally, take care with the flashings. Often a roof is done well, but the flashings are not. Since most leaks are really flashing leaks, it pays to study the correct way to put up flashings, and do them carefully, with good materials.

SAFETY

The higher and steeper the roof, the more dangerous it is. The more I roof, the more time I take to set my project up safely with stagings and other gear. I like a solid staging at the eaves, good footing, good shoes, and proper gear for working on the roof, such as roof harnesses, shingle brackets, or hook ladders.

I don't hurry, or work when the roof is icy or wet. I don't go up if I feel apprehensive, which I sometimes do. I try to make sure that I'm safe, and that those below me are safe from falling objects. I'll go the extra mile to create a safe work area, even when there is a less safe, quick way I could use more easily.

Roofing is one area where it makes sense to sub out the work. The more dangerous or complex the roof, the more reasons to have a pro do it. Tricky flashing details are an additional reason to hire a professional.

TYPES OF ROOFING

Roll Roofing

The least expensive roofing is ordinary roll roofing, which consists of 3-foot-wide rolls of asphalt-impregnated felt coated with shiny sand-sized mineral particles (figure 27-3). The mineral surface protects the roofing by reflecting sunlight away, particularly if the mineral is a light color. Roll roofing comes in 45-pound, 60-pound, and 90-pound weights. Heavier roofing lasts longer. It always make sense to use the heaviest weight, unless the roof is temporary.

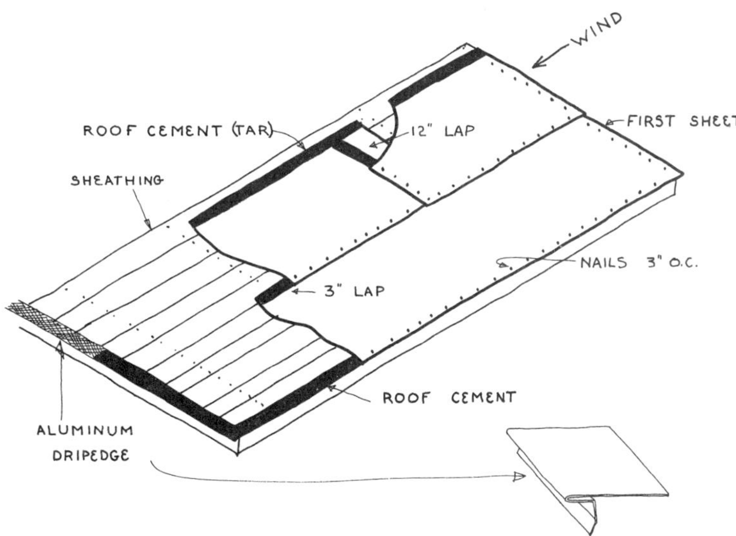

Figure 27-3.
Roll roofing.

Since it comes in such big sheets, roll roofing is quick to install. It can be laid horizontally on shallow roofs with a pitch of 2:12 or greater (figure 27-4). However, for pitches above 3:12 or 4:12, the material slides around so much that it is probably not worth using. It can be used vertically on steep roofs with a pitch of about 10:12 or steeper. Roll roofing costs about $15 for 100 square feet of coverage (one "square").

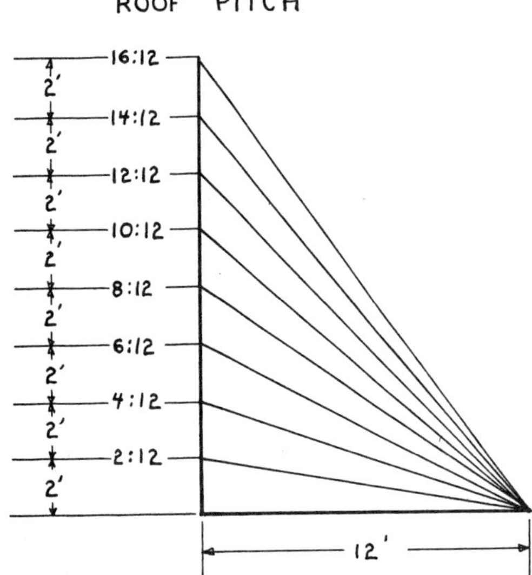

Figure 27-4.
Roof pitches.

Double-Coverage Roll Roofing

Double-coverage roll roofing is similar, but much more durable than single-ply roll roofing. In double coverage, also known as "half-lap," each ply laps the next one halfway (see figure 27-5). The upper half of each sheet is not mineralized, and is covered completely with the lower half of the next sheet up. As the drawing shows, most of the nails are covered. Double-coverage is used on pitches as flat as 1:12, though any roofing will last longer with more pitch than that.

Like ordinary roll roofing, double coverage roll roofing will be awkward to install on steeper roofs. It costs about $30 per square.

Asphalt Shingles

The most common roofing material for houses is asphalt shingles. Rex Roberts, who wrote a great book for owner-builders called *Your Engineered House,* argued that roll roofing was better than asphalt shingles because it had fewer joints. My experience is that good quality shingles are more durable if installed correctly. The sun is hard on any asphalt roof. But where a roll roof will eventually split and crack, shingles can move

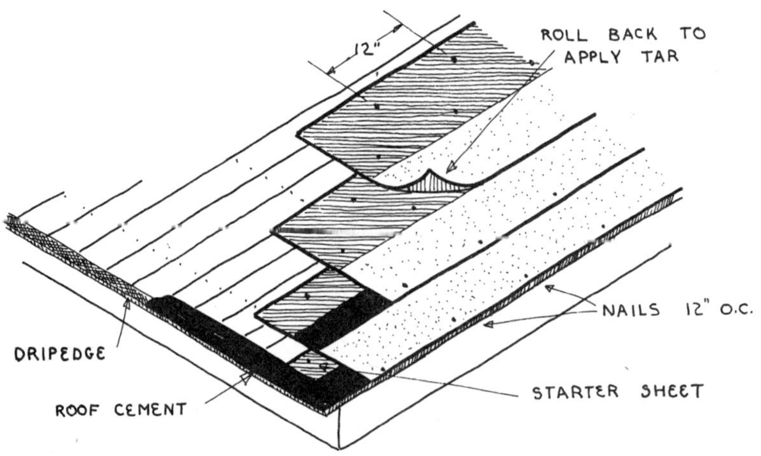

Figure 27-5.
Double-coverage roll roofing.

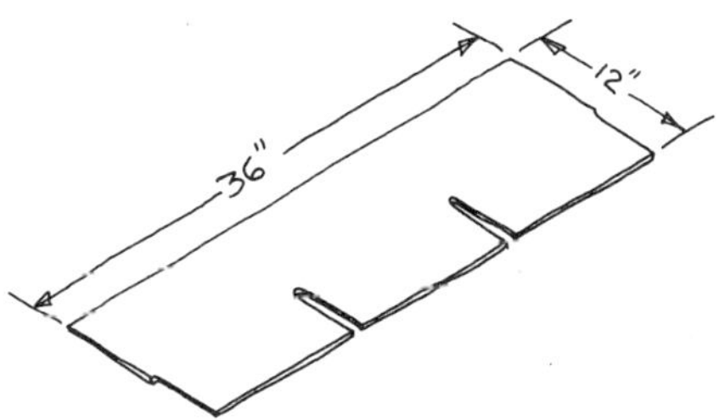

Figure 27-6. Asphalt shingle.

around a little as they deteriorate. They will still shed water even when in poor condition. They are also are easier to flash or step-flash successfully around chimneys, skylights, or other penetrations, and shingles are generally more pleasant to work with. Figure 27-8 illustrates how they go onto the roof deck.

There are many types of asphalt shingles. There are two categories: organic shingles (with a felt and asphalt base), and fiberglass shingles. There have been lots of problems with fiberglass shingles, so I tend to order organic ones, in the heaviest grade available.

Metric-sized shingles imported from Canada are a little larger, making them quite a bit quicker to install.

This chapter pictures "tab shingles," which have slots in them to mimic the look of tiles or wood shingles. There are also shingles with no slots. I prefer the appearance of tab shingles, but it's true that in the long run, tab shingles wear out first at the slots; that's where holes first appear twenty or more years down the road. On the other hand, the tabs may be better able to expand and contract without splitting. Basically, my view is to buy the best quality shingle, whichever style appeals to you.

Asphalt shingles can be used when the pitch is 4:12 or greater, and they cost about the same as "half-lap" roll roofing.

Metal Roofing

Metal roofing is made of various metals, but galvanized steel is the most common (figure 27-11). It is more expensive than asphalt roofings (about $45 per square), but with good planning it can be installed very fast. Since it is quite stiff, it does not need continuous support, so savings on sheathing are sometimes possible. It can be used when the slope is 3:12 or steeper. It is now available in colors.

Figure 26-8 shows a house with an "Onduline" roof. Though it is not used very often, it is a favorite of mine. It's a stiff, corrugated, coated asphalt product, that comes in a variety of colors. It goes up like sheet metal, but I find it much easier to work with. It costs about the same as steel roofing.

There are many other roofing materials, such as wood shingles, standing seam metal roofs, tile, and others. I've chosen to include here the most common and most economical types.

HOW TO PUT UP ROOFING

A first step for all the roofings below is to determine where you want the protection of an ice and water shield, and to apply it according to the instructions that come with it. I would apply a course along the eaves and valleys of any roof, or anywhere there may be snow buildup. The only exception would be roofs installed over spaced boards, which don't really provide anything for the shield to stick to.

Most roofing comes with excellent instructions printed on the packaging. What follows is introductory.

Regular Roll Roofing

If roll roofing is to have a long life, it should be laid over plywood sheathing or dry T&G boards. A smooth, flat surface is essential. Wear your oldest clothes, because you will be working with tar. Have some mechanic's hand cleaner and rags on hand for cleanup.

Figure 27-3 shows the procedures for roll roofing.

Step 1

After the roof has been covered with felt, nail galvanized or aluminum drip edge to the edges of the roof sheathing with roofing nails. Nail the drip edge at the bottom of the roof first, and lap the side pieces over it. With a shed roof, there will also be a drip edge along the top.

Step 2

Precut pieces of roofing slightly longer than the roof is wide, and lay these sheets out on the roof deck so that the sun can warm and flatten them. Each sheet should sit in the sun for about a half-hour. If the roofing is nailed down too soon it may bubble as it expands. As you roof, always keep a few pieces stretched out in the sun. You can cut the roofing with a utility knife. Your blades will last longer if you cut from the back of the sheet. You can also purchase a special roofer's

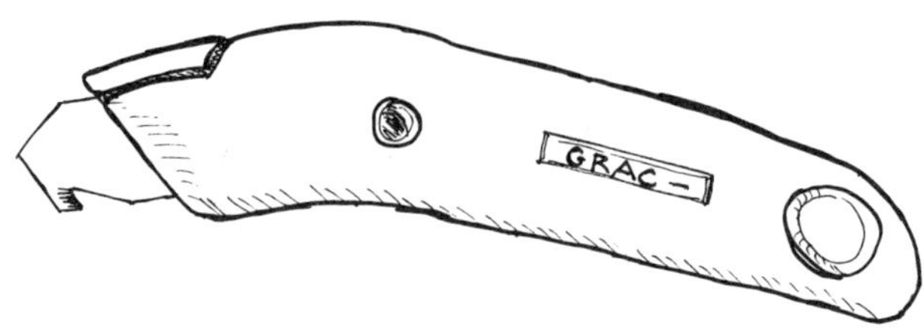

Figure 27-7. Utility knife with a roofer's hook blade.

hook blade for your utility knife, which allows you to cut from the back without flipping the sheet over (figure 27-7).

Step 3

The tar-based roof cement that goes with roll roofing goes by various names, such as "plastic roof cement" or "cold cement." Tell your supplier what you are doing, and he or she will give you the right product. In cold weather applications, store the cans in a warm room to make the cement easier to use.

Spread a bed of the roof cement 3 or 4 inches wide, along the bottom edge of the roof, on top of the drip edge. Do the same along each side, starting at the bottom and going up 3 feet. Spread the tar with a disposable trowel or piece of shingle.

Step 4

Place the first sheet of roofing along the bottom edge of the roof. Let it overhang the drip edge around ⅛ inch at the bottom. Smooth it carefully, and nail it down along the bottom edge with 1¼- or 1½-inch galvanized roofing nails every 3 inches. These nails have large heads to hold the roofing firmly. The row of nails should be close to the bottom edge of the roofing without missing the sheathing. Nail down the ends of the sheet in the same way. Every exposed nail in roll roofing should go through a bed of roof cement after it goes through the roofing, because the roof cement seals the nail hole.

Step 5

With a chalk line, make a line three inches down from the top edge of the first sheet and parallel to it. This line represents where the bottom edge of the second sheet will be, and shows where to put the roof cement that holds down the bottom of the second sheet. Spread a ribbon of roof cement three inches wide above this chalk line. Also put a ribbon of roof cement on the sheathing under where the ends of the next sheet will be, along the rake. Place the second sheet with its bottom edge on the chalk line, smooth it, and nail it as before with one nail every 3 inches. Locate the nails ¾ to 1 inch in from the edge of the sheet.

Step 6

Continue placing, cementing, and nailing sheets until you get to the top of the roof. The ridge of a gable can be covered with a half sheet folded over the top. Vertical joints should be lapped 12 inches and placed so that the prevailing wind will not lift the upper sheet.

Double-Coverage Roll Roofing

Like regular roll roofing, double-coverage is laid in sheets working from the bottom up to the ridge, but many of the procedures are different. Figure 27-5 shows one common procedure. Only use the cement specified for this type of roofing. Apply it with the special long-handled brush specified by the manufacturer.

Step 1

Install a drip edge as described in step 1 for roll roofing.

Step 2

Also as before, unroll a roll or two, cut pieces to length, and stretch the sheets out in the sun, or in a heated room, to warm and flatten them.

Step 3

With the special brush, spread a bed of roof cement on the drip edge at the eave and along the edge of the roof.

Cut one sheet in half lengthwise. Separate the mineralized from the unmineralized part. Set aside the mineralized part for later use. The unmineralized part will be a starter strip. Nail it down along the eave using the procedures given in step 4 for roll roofing.

Step 4

Here is the main difference in method between double-coverage and regular roll roofing: Nail all the sheets in place, through the top half only, before you apply the roof cement to the joints. The only places that get roof cement are where the roofing covers the metal drip edges. After all the sheets are nailed down, you fold them back one at a time to apply the roof cement.

Take a full-width sheet and place it along the eave overlapping the drip edge ⅛ inch and covering the starter sheet. Nail the sheet through the upper half only with 1½-inch galvanized roofing nails every six inches in a staggered pattern.

Step 5

Continue on up the roof, nailing down the sheets through their upper half only and lapping each sheet halfway. Lap vertical joints 6 inches, with the upper sheet to windward.

Step 6
When all the sheets are laid, take a chalk line and mark the bottom edge of each sheet.

Step 7
Starting from the top, fold each sheet back, and spread a band of roof cement just above the chalk line, covering as much of the non-mineralized surface as possible. The more neatly you do this, the better-looking the roof will be. Make sure you use roof cement at all vertical joints and along the drip edge.

Step 8
After you cement the joint, fold the top sheet down, smooth it out, and nail it at the vertical seams where the sheets meet. You can also add a few nails along the bottom edge, but these are often omitted.

An alternative method is to work up the roof one sheet at a time. Coat the top half of one sheet, install the next, and so on. This gives a more complete coverage of cement.

Asphalt Shingles
Figure 27-6 shows a standard asphalt shingle and figure 27-8 shows typical installation details. As before, start with an ice and water shield where it makes sense. Here is a step-by-step procedure for installing asphalt shingles.

Step 1
Install ice and water shield.

Step 2
Install 15-pound felt and drip edge as described above. It is best to lap the felt over the drip edge at the bottom, and lap the drip edge over the felt at the sides.

Step 3
The first row of shingles must be doubled. This is often done by nailing a special starter strip along the eave, over the 15-pound felt and the ice and water shield. The strip is the same as the shingle material, only without slots. It is also common practice to make a starter strip by installing a row of shingles upside down (slots up).

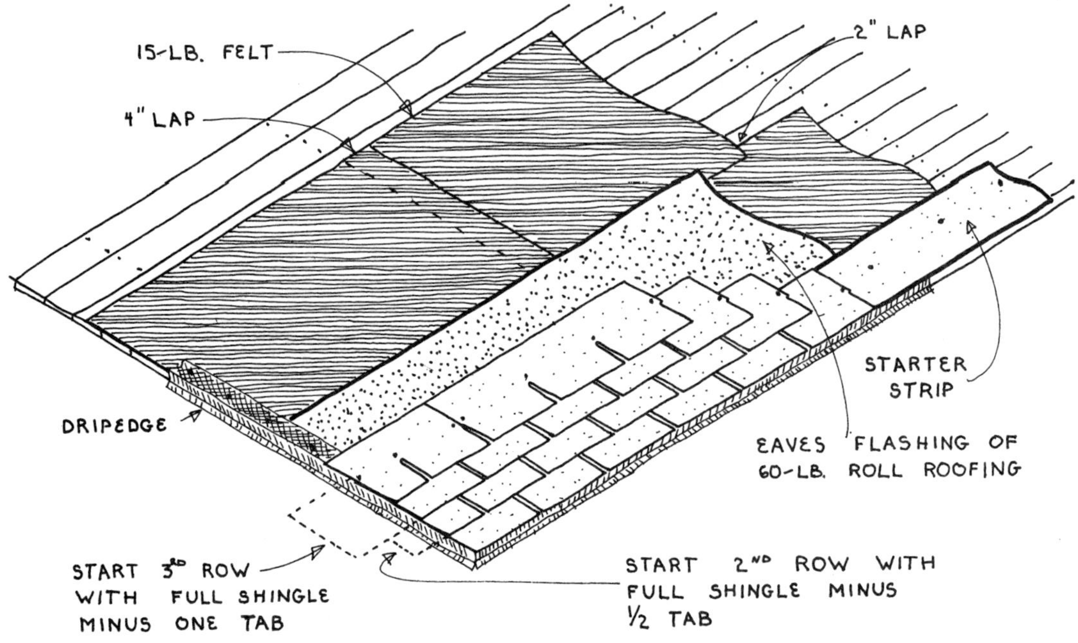

Figure 27-8. Installing asphalt shingles.

Step 4

Nail the first row of shingles over the starter strip with four 1½-inch galvanized roofing nails per shingle, as shown in figure 27-8. Locate the nails about 1 inch above the top of the slots. The shingles should overlap the drip edge side and bottom by about ¼ inch.

Step 5

Establish a few chalk lines, starting from the shingle slots and running up to the top of the roof parallel to the sides of the roof. These will help you keep the slots aligned as you move up the roof.

Step 6

Asphalt shingles are laid with 5 inches of their width exposed to the weather. Five inches comes just below the top of the slots, so the rows can be aligned using the slots as a guide. Periodically run a horizontal chalk line to make sure the rows are straight and parallel to the roof deck. For metric shingles, the exposure will be a little more.

Start the second row with a shingle from which you have removed one half of a tab. Start the third row with a shingle from which you have removed a full tab. This procedure keeps the slots staggered.

Cut the shingles from the back, or get a roofer's hook blade (figure 27-7). I like to have two knives with me, one with each sort of blade.

If the roof is too steep to walk on, you can support planks to work on with pairs of special shingle brackets (figure 27-9). These brackets are nailed through the roof sheathing to the rafters with three 8d nails each. Locate the metal plate of the bracket where a tab from the next course of shingles you install will cover the nails to prevent leaks. Remove the brackets when all the roofing is done. The teardrop-shaped holes in the metal plate allow you to remove the bracket without disturbing the shingles too much. Wiggle or tap the bracket toward the ridge ¼ inch or so to release it. Lift the tab above the nail and drive the nails flush.

Figure 27-9. One type of shingle bracket.

Step 7

If you are venting through the ridge, the ridge vent will complete the job at the top. If not, you can cover the peak of the roof with a Boston ridge as shown in figure 27-10. Cut shingles into thirds, and blindnail them as shown in the drawing. Blind-nailing means the nails you use to hold down one piece are covered by the subsequent piece. To protect the ridge from the wind, start laying the shingles on the leeward end and work toward the prevailing wind. Lay the pieces with 5- to 6-inch exposure to the weather, meaning all but 5 or 6 inches of each piece should be covered by the subsequent piece.

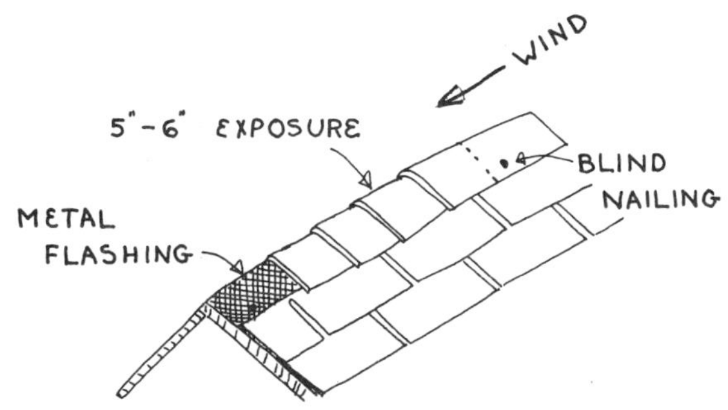

Figure 27-10. Boston ridge.

For extra protection you can put a folded metal or felt flashing under the Boston ridge, as shown.

Galvanized Metal

On roofs pitched below 4:12, the sheathing under galvanized roofing should be plywood or continuous dry boards. Above 4:12, the sheathing can be dry 1 x 4s or 1 x 6s spaced 16 inches on center. However, if there is another roof above, use continuous sheathing wherever ice or snow from above may fall.

This type of roofing is much easier if you figure out everything in advance. You will need a brochure giving all the sizes and accessories of the brand of roofing you buy.

Perhaps the first goal is to figure out how to minimize cutting. Sometimes you can make small adjustments in the roof overhangs or trim details to eliminate almost all cutting of sheets.

If the roof is more than one sheet in length you can vary the overlap to eliminate cutting.

Often the sheets are 38 inches wide, with a 2-inch lap. If the total roof width is dimensioned to be a multiple of three feet, plus 2 inches, including a 1- or 2-inch overhang at either side, you could avoid cutting the sheets, although lengthwise cuts are less of a trial than crosscuts or diagonal cuts.

If the roof is complex, make a careful drawing of exactly how the sheets will lay out, to minimize both waste and cutting.

There are three other interrelated strategic decisions: how to get around on the roof, how to vent it, and whether to paper it. I'm not sure which solution to recommend.

The roof should be vented. If it vents at the ridge, you should probably use a flat ridge, rather than the fitted ridge roll (figure 27-11). It allows air to escape.

Spaced boards make a great ladder for moving about the roof comfortably. If you have plywood sheathing, you will need some way to move about the roof, unless the pitch is flat enough to provide good footing on the sheathing itself. Usually a roof over 5:12 or 6:12 will be a little steep to walk on, even with your hiking boots. The standard solution in this situation is a roof ladder, which is simply a section of aluminum ladder with a special ridge hook attached at one end. It hooks over the ridge of the roof. The awkwardness of this is the best argument for spaced boards as sheathing. They make a much handier walking surface, and provide better fastening for the roofing screws.

On the theory that moisture condenses on the inside of a metal roof when the temperature drops, we sometimes put sheets of felt paper under the roofing. Moisture that collects will flow down the sheet and out at the eaves. These sheets

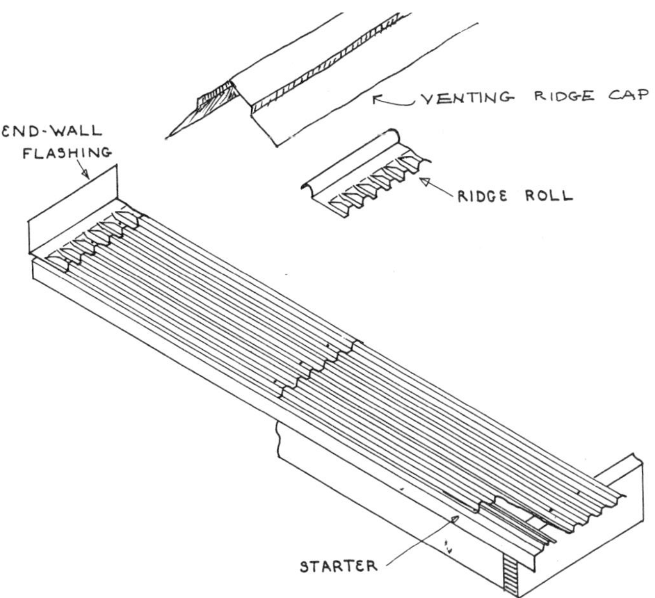

Figure 27-11.
Metal roofing.

can be applied vertically as the roofing goes down, preserving the ladder created by the spaced boards.

However, on a recent project, I chose to omit the felt to save time. I justified this because I had recently removed part of an existing roof built that way, and had found that the ten-year-old 8-inch insulation was in perfect condition. My conclusion from this was that eliminating the paper improved the airflow, so any moisture that did condense could be vented away. The instructions below include the felt, however.

Strategize exactly how the sheets will be handled. You might have one person on the ground (with hard hat), who hands up the sheets. Another person on staging (figure 27-12) would stabilize the sheet at the eaves, and put in the lower screws. The third person would be on the roof aligning and fastening the rest of the sheet. Since things slide so easily on the metal, no one else should work below when this is happening.

Devise ahead of time a safe way for securing the piece of ridge flashing or endwall flashing if there is one. It's quite awkward to straddle the ridge to attach it. My preference is to mount a plank below and parallel to the ridge to work from. To do this, I'd bolt a few standard roofers' shingle brackets with a ⅜-inch bolt to squares of ½-inch plywood about two feet square. Back the bolt up with washers, inside and out. These brackets could then be secured to the installed roofing with four roofing screws. You would attach these in a straight line every

6 or 8 feet, about three to four feet below the ridge, as you installed the roofing, and then rest a plank on them to provide a comfortable walking surface. After the ridge is all done, remove these from the ridge ladder, and fill the holes with roofing screws. Other circumstances might suggest a more convenient or safer way.

You also need an approach to the actual fastening of the roof. Though nails are still available, screws make strong fasteners for metal roofing. Some have built-in gaskets, and are driven with a drill, preferably a cordless drill with a holster, fitted with a ¼-inch hex bit. Use manufacturers' instructions for the location of these screws. These screws are self-tapping: you can put one in the drill, press firmly against the roofing, and the drill will make a hole through the roofing.

However, you may find it is easier to make a small hole ahead of time. One method is to predrill the holes before the sheets go up on the roof. To do this lay out the locations of the holes, and drill through several sheets at a time on a pair of sawhorses. This allows you to make a nice precise row of holes so the fastening will look neat from the ground later.

You can also punch holes when you are up on the roof using a center punch, a big nail, or one of the roofing screws. If you use this method, mark the locations ahead of time carefully with a builder's marking crayon.

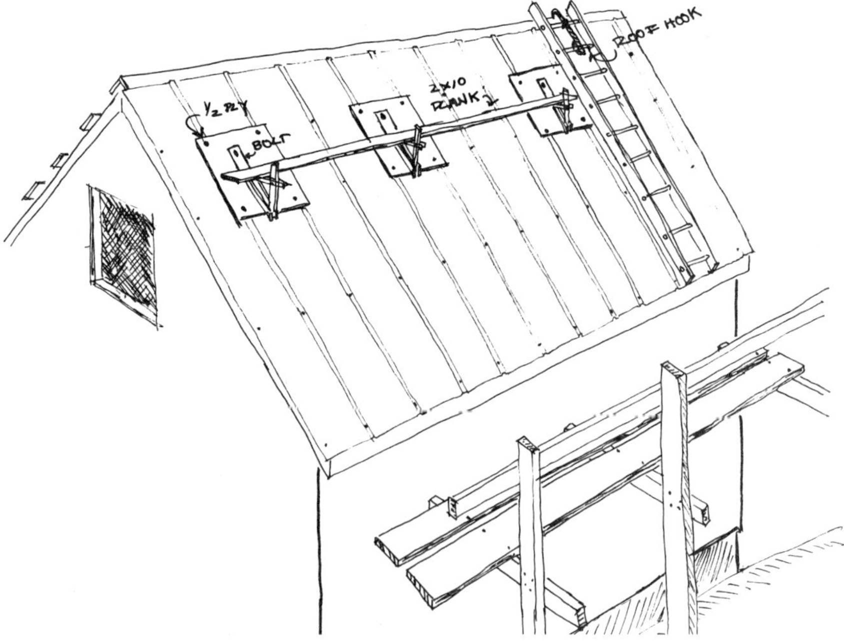

Figure 27-12. Staging for installing ridge.

Here are the procedures for installing metal roofing.

Step 1

Cut the metal sheets to size. Your roofing supplier may cut the sheets to length for you. When that isn't possible, invest in two pairs of tin snips with different cutting actions. Get a large, good-quality straight-cutting snip, and a set of pipe and duct snips. You can make most cuts with one or the other of these. You could also use a sabersaw fitted with a metal-cutting blade.

You can also get a "composite blade" for a circular saw, made for cutting metal. Cutting metal with a circular saw is very tricky and it takes strength. With either power saw, use full goggles, ear protection, and gloves. Since any cutting leaves the edge of the sheets ragged, you may want to use a file to smooth them. Since none of this is much fun, I always go to some lengths to lay out the roof to minimize cutting.

Step 2

If there are any valleys, they go down first. A valley will usually consist of a 36- to 40-inch-wide sheet of metal matching the roofing. Make sure the valley is long enough at the bottom, and nail it down with roofing nails.

Step 3

Run one sheet of felt paper from the bottom to the top, under where the first row of roofing goes. Roll up a precut sheet, then unfurl it from the ridge. Attach it with staples or roofing nails.

Step 4

Starting in a corner, position the first sheet at the bottom, with the correct overhangs, taking pains to make it exactly parallel to the rake edge of the roof. Make sure you've got the edge which goes underneath on the right side (figure 27-13).

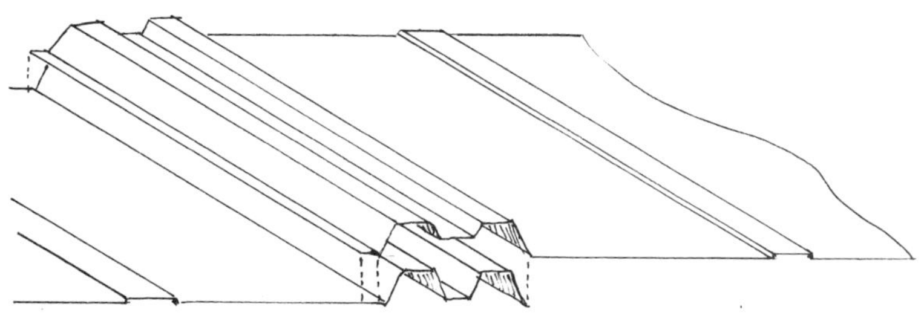

Figure 27-13. Shallow groove over deep groove.

Step 5

Fasten the sheet with gasketed screws except where subsequent sheets will overlap it, side and top. Make sure the sheet sits flat on the deck as you fasten it; you could create a bulge if the center of the sheet isn't completely flat when the edges are fastened.

Step 6

Finish attaching one row of sheets all the way up to the ridge.

Step 7

Install the next row of sheets the same way: felt, bottom sheet, top sheet. Make sure the sheet overlap is tight and square, or the roof can become a little crooked.

Step 8

Install the ridge or endwall flashings.

Step 9

Remove your staging.

FLASHING

Whenever a roof meets a wall or another roof, or is penetrated by a chimney or a vent, there must be some sort of flashing. A flashing is a piece of sheet metal designed to prevent leaks at the joint. Flashing is also sometimes needed on walls where the type of material changes, such as where the siding meets a window or door casing. Figure 27-14 shows several typical flashing details. Each follows the basic principle of roofing: the upper piece laps the lower. Figure 27-14A shows the kind of flashing used above the casings of windows and doors. This flashing, which you can buy ready-made, goes under the siding and over the casing. Figure 27-14B shows the flashing that seals where a roof meets a wall at the top. The flashing goes under the siding and over the shingles. This is a flashing you can make yourself by bending 12-inch sheet aluminum to the right angle. You can buy sheet aluminum in rolls from the lumberyard. When any kind of shingle or tile butts into a wall, use step flashing (figure 27-14C), shown with the siding removed for clarity in the illustration. You can make step flashing squares by cutting up rolls of 8-inch aluminum flashing with tin snips, but it is also possible to buy the squares ready-made.

Most flashing jobs can be done using one of the techniques in figure 27-14. The proper width of lap in any roof flashing depends on the roof pitch, the kind of roofing material, and weather conditions. Your best guide will be local practice.

You will probably not go wrong as long as you keep a few points in mind: (1) Make sure the upper flashing pieces lap the ones below them. (2) To prevent corrosion from electrical action, use nails of the same metal as the flashing used. (3) Avoid exposed nails where possible, and seal exposed nail heads with silicon caulk.

When I come to tricky spots, I often rely on lead flashing, which comes in 8-, 10-, and 12-inch widths and is bought by the pound. It is very malleable, and can be pounded into shapes that would fracture other sheet metals. It is very handy at corners of chimneys or skylights where a simple fold won't work. It is chemically very stable, and won't react with other metals.

Flashing a chimney or a skylight is difficult, and I would suggest seeking the help of an experienced roofer. Although some skylights come in kit form with good instructions and precut flashing, if you are retrofitting the skylight on an existing roof, getting a complete seal can be tricky.

Flashings for metal chimneys and plumbing vents are usually purchased ready-made, and installed as a unit (27-16). The bottom piece of flashing laps the roofing below, and the top half tucks under the roofing above.

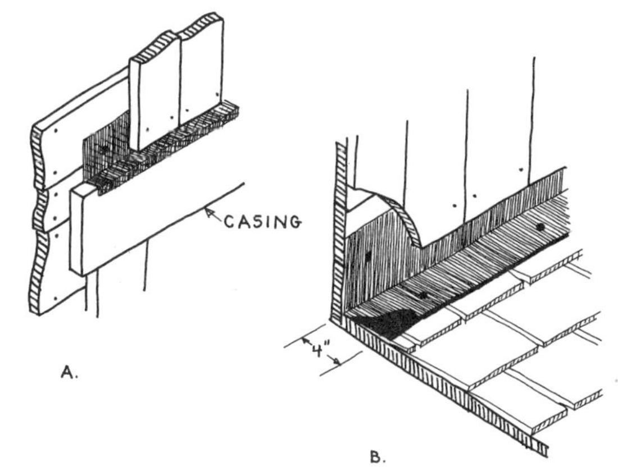

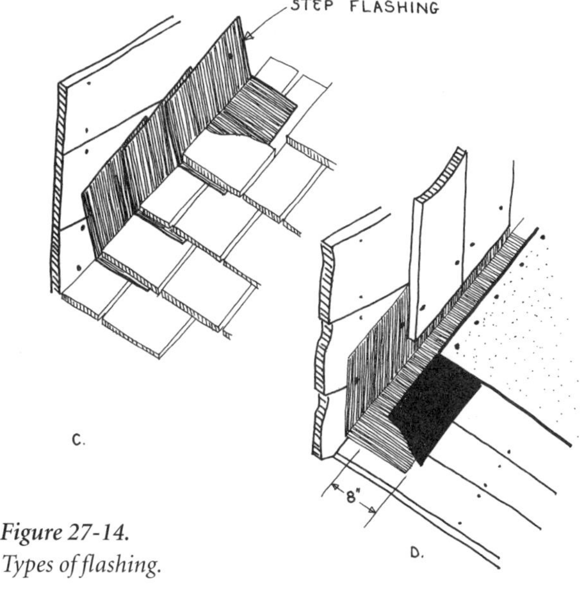

Figure 27-14.
Types of flashing.

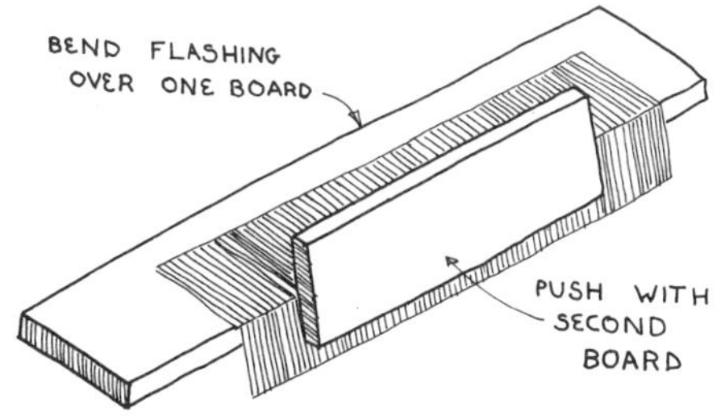

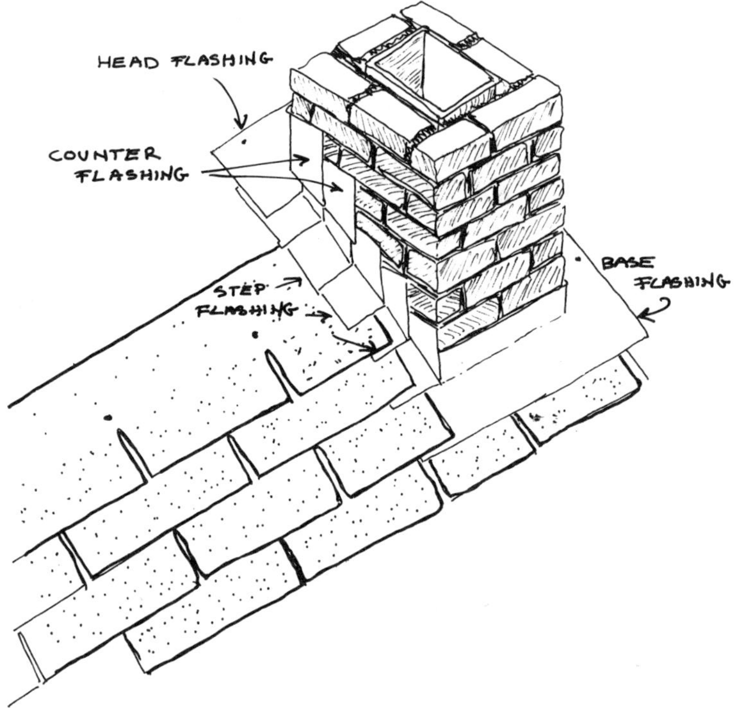

Figure 27-15. A chimney flashing could be made from sheet lead, or soldered up with copper or galvanized sheetmetal. The head and base flashings have to wrap around the corners. Lead is so maleable that it can be stretched into the proper shape with a hammer. The counter flashings (only two are shown) are tucked into the mortar joints. Step flashings are interleafed with the courses of roofing.

Figure 27-16. Flashing for metal chimney.

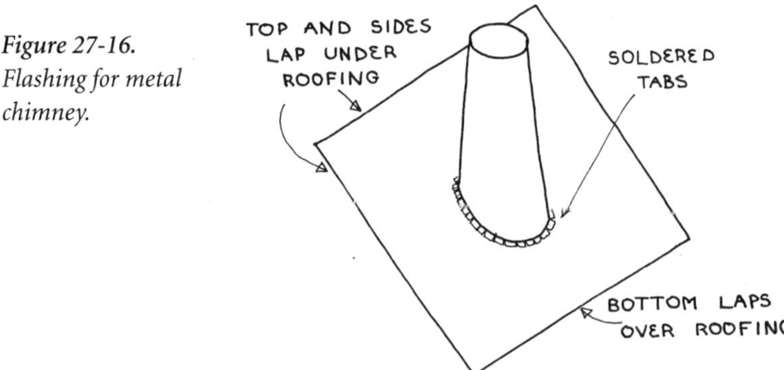

28

WINDOWS & SHUTTERS

WINDOWS ARE THE MOST COMPLICATED AND PRECISELY built part of a house, because they perform many contradictory jobs. They let in the summer breeze, yet keep out the winter wind. They let in the sun's heat, but keep out the winter cold. They let in air but keep out bugs. Because of this complexity, they are expensive to buy, and hard to make well for yourself.

WAYS OF OBTAINING WINDOWS

Because glass is a very poor insulator, windows are a primary avenue for heat to escape. Various methods are used to make windows retain heat better. The most common is double glazing, in which two layers of glass are separated by an airspace. The components vary and include a removable storm sash, a permanently installed second pane of glass, or thermal (insulated) glass, which consists of two or even three panes with a factory-sealed airspace.

In the last few years, Low E glass has been introduced. Low Emissivity or Low E coatings allow light to enter, but regulate heat flow and ultraviolet penetration. Different types and positionings of coatings are used depending on climate and window orientation. Low E windows have an R value of around 2.8, compared to about 2 or less for ordinary double glass.

Even with two layers of glass, windows can account for half of the heat loss in a house. The heating efficiency of windows can be further improved by adding heavy curtains or insulated shutters to insulate the window at night or when not in use.

Windows can be a major part of your building cost. New windows for a 2,000-square-foot house could easily run to $8,000 or more. Lower-quality windows for a smaller house would probably cost $2,500 to $4,000. So it is worth thinking about the alternatives.

The fastest thing to do is buy ready-made units from

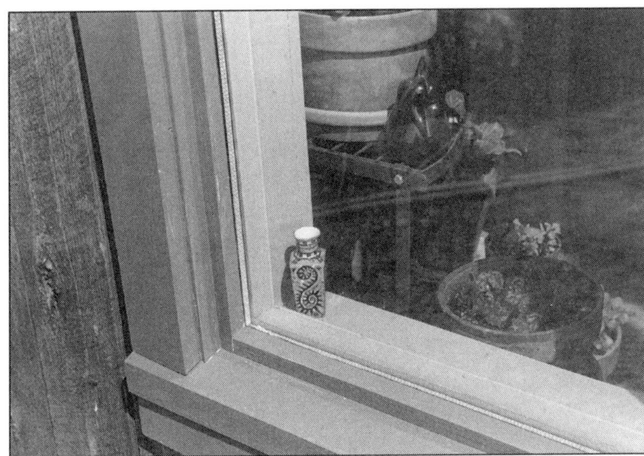

Figure 28-1. Home-made fixed window with insulating glass.

lumberyards or millwork companies. These units cost perhaps $150 to $500 each. The cost depends on size, quality, and type. Horizontal sliding windows tend to be the cheapest, because they are so simple. Double-hung windows cost somewhat more, and casements (hinged at the sides) yet more. The great advantage of new ready-made units is that they can be obtained in the sizes you want, and installed in an hour or two each.

Used window units can be bought through classified ads or from salvage companies for half or less the cost of new units. Sometimes you can scrounge them for almost nothing where people are demolishing or remodeling a building. They will take longer to put in than new windows, because repairs will usually be needed. You can sometimes get new windows at bargain prices, if you check the bargain area of local lumberyards on a regular basis.

Many owner-builders make their own windows to save money. I feel it is relatively practical to build a window unit with a fixed insulating glass pane. Such a window will be cost effective compared to commercial windows. It is also practical to make a fixed window unit that uses recycled window sash, with some sort of storm panel to provide double glazing. This is particularly worth doing if the old sash is beautiful, and doesn't need extensive repairs.

For example, a 30 x 60-inch double-hung window might cost $200 to $350 to buy, depending on quality. The materials to make a fixed window with insulating glass would cost perhaps $80 to $100, but there would be three to six hours of work to manufacture the unit.

I have also built opening windows of all types. This chapter shows the detailing and some of the methods I have learned about. But it takes a long time to build an opening window, and it is unlikely you will be able to make one that functions as well as a commercially manufactured one. It is very difficult to get the weatherstripping right.

There are a couple of situations where I will go to the trouble to make an opening window, even though the savings may be marginal and the window might not move as well as a new one. I might find some particularly beautiful old sash that seems especially worth recycling. More often, I can't always buy a new window that has the look I want. Figure 28-2 shows a a new barn we built in a historic district. The idea was to make it fit right in, as if it might have been there for a hundred years. New windows would have had thicker "muntin bars" separating the panes of glass; the sill would have been thin. To me, this would have been obvious from way down the road. So I bought some old sash from a neighbor (we had taken them out of a wall the year before) and built traditional frames for them.

Usually, though, it's wise to make your own fixed windows to save money, but to buy your opening windows ready-made. Though some of the windows I have made over the years are still in service, a few have been replaced, but most of the fixed windows are still okay.

Figure 28-2. *Here making opening window units for recycled sash was worthwhile, since modern windows would have looked wrong. Design, Paul Hannan, Cornelia Carey; builder, Iron Bridge Woodworkers.*

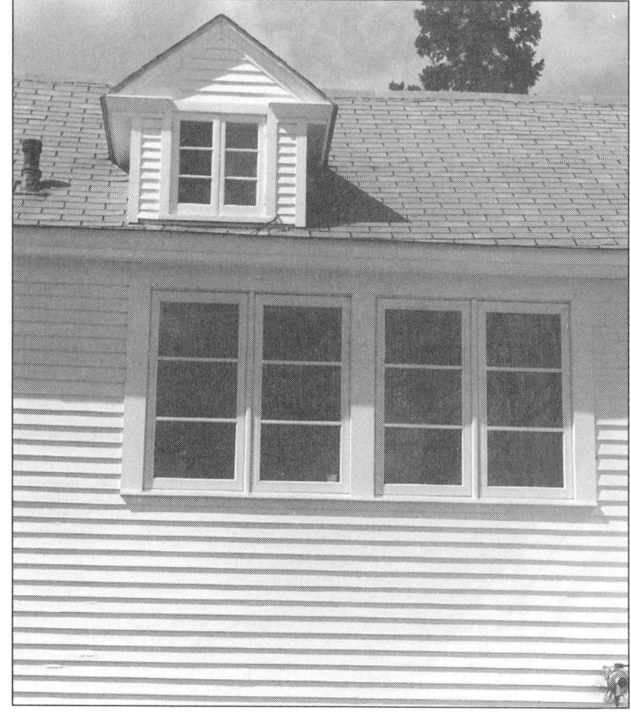

Figure 28-3. *New custom windows (Marvin) match old dormer casement detailing. The Long House.*

Choosing the Shapes and Sizes of Windows for Your House

The earlier design chapters discussed ways to think about the sizes of windows. It also makes sense to choose the best type of window for each application. Double-hung windows are the common windows that slide up and down. Windows that slide horizontally are usually called sliders. Casements are hinged at the side, and usually open outward. Awning windows are hinged at the top, and hoppers, which are uncommon today, are hinged at the bottom.

Choose a style of window that fits your design. I always think modern casements look funny on a cape, and double-hung windows with small panes might look odd on a very modern building.

More important, each type of window works best in a certain shape and in certain applications. Double-hungs work best when they are tall but not too wide. They don't operate as well if they are wide and squat. Similarly, sliders work well for openings that are wide, but not too high. A 30 x 60-inch window (the horizontal dimension is always given first) might make a good double hung, but a 60 x 30-inch would be a better slider. Casements should be relatively tall and narrow. If they are too wide or short, there is too much strain on the hinges. Similarly, a very tall awning window would strain both its hinges and the crank hardware that controls it.

In general, better-quality windows will operate more smoothly than cheap ones. I've put in brand new cheap double-hungs that were weather-tight, but awfully stiff to open. Casements tend to operate more easily than double-hungs. Over a counter or desk, a casement will be more manageable than a double-hung or slider.

Bedrooms should have at least one egress window unless there is a door to the outside. My code defines an egress window as having a sill height no higher than 44 inches, and a net clear opening of 5.7 square feet (820 square inches). The opening must be at least 20 inches wide and 24 inches high. A 20-inch-wide window would have to be 41 inches clear in height; a 24-inch-high window would have to be 34 inches wide. The code also requires that no tools be needed to open the window. Check with your local code officials. But even if there is no local requirement, make sure people can escape in case of fire without injuring themselves in the process.

Window Features

An order form for a new window might look like this:

> D/H 24 x 24, ½" Low E, with screens, primed, 5/4 x 3½ flat casing, 6" sill horns, 6 9/16" jambs.

At one time, ordering windows was simple. If you bought a double-hung, the glass size and number of panes told it all. Andersen, once the primary supplier of casement windows, made all their windows to one design. Now there are many companies with many features to choose from, and different manufacturers offer different combinations of features. One company might sell the perfect window for one application, but not for another. Also, some companies now will make windows to order to any size, for a premium price. This is more likely to be important in remodeling situations than in new construction. Here are a few key variables.

Quality vs. price: In general, you get what you pay for. Better windows cost more. Sometimes one company will be a better deal for a few years, as they introduce their product. On the other hand, it is also possible to get discounts of various kinds on windows. It makes sense to show your window list to different lumberyards, and see what prices they offer.

Clad vs. wood exterior: Though some people buy all-aluminum or all-vinyl windows, most house windows will be made basically of wood. But most manufacturers now offer clad options on their wood windows. This is a plastic or sometimes aluminum covering the outside parts of the window. The cladding provides a more durable surface than paint. Often the windows operate better because they don't swell and warp as all-wood windows can. Cladding usually comes in a few colors, and can't easily be painted, so it may not fit your design scheme.

Figure 28-4. *Clad, trimless window, with trim elements added to create a traditional look. See also figure 9-6.*

Trim styles: Many modern windows, particularly the clad types, come with a ½ inch or ¾ inch flange or projection, and no sill or casings. The siding is intended to be applied right up to the flange. Others come with a narrow casing called "brickmold." Wider flat casings are also available. Some companies will make up windows with any size flat casing you want. You can also order some windows without casings so you can trim them as you wish.

I always belabor this decision, because the mass, shape, and color of the window trim is a major feature of how the building will look from the outside. If I'm designing a house that's supposed to have a Craftsman or Adirondack flavor with dark red trim, I'm not sure I want white plastic coated casings that I can't paint.

Glazing options: Most windows today come with insulating, Low E double glazing. Sometimes there are different Low E configurations available. Argon gas is sometimes used inside the double glass to improve the insulation value. Certain windows are also sold with a removable storm panel instead of double glazing. Others come with triple glass. There are many kinds of mounting grills, some designed to imitate old style small panes. In general, I pick the simplest option that looks right on the building, and stay away from fake small panes.

Ordering Windows

Your window order is complicated. It pays to spend some time reading your window catalogs carefully. Many of the choices you make will affect the window cost, and perhaps how long it takes to get them.

First you specify the type of window, such as casement. The size can be expressed as a unit dimension, glass size, or rough opening. I try to go by the rough opening size—width written first, then height. You will also specify whether it is clad or not, whether it has screens, and how it is trimmed. Usually there will be stock choices that are inexpensive, and deviations will be expensive. I often order windows uncased, so I can do it the way I want to. You also have to identify the type of glazing you want, and the thickness of the jambs. The jamb goes from inside the wall finish to outside the sheathing. Standard jambs are often 6 9/16 inches, which provide for a 2 x 6 wall, ½-inch interior finish, and ½-inch sheathing, plus 1/16 inch extra for "buildup." Many companies use jamb extenders, which they may apply or provide for you to install. These are strips which extend the jamb to match your wall thickness. They can be sawed down to non-standard sizes.

Your builder, designer, or window salesperson can help you put your window order together using the correct lingo.

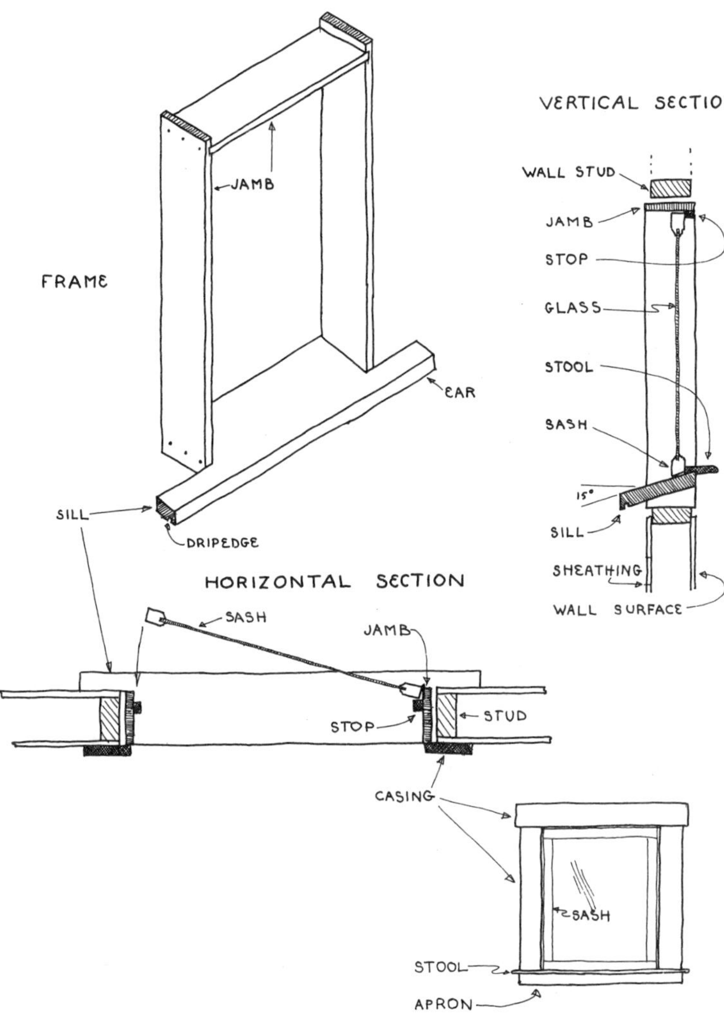

Figure 28-5. Basic window parts.

MAKING WINDOWS

Building Window Frames

The procedure for making a window frame is the same for fixed glass, fixed sash, and some kinds of opening windows. Here are procedures for making and installing the basic window frame. Variations will be discussed later.

Step 1

Take some dry, planed, fairly knot-free boards, and rip them on a table saw to the right width for your jambs. Jambs should come flush with the sheathing on the outside, and with the inside wall surface on the inside (figure 28-6). Add 1/16 inch or even 1/8 inch to account for shrinkage or buildup.

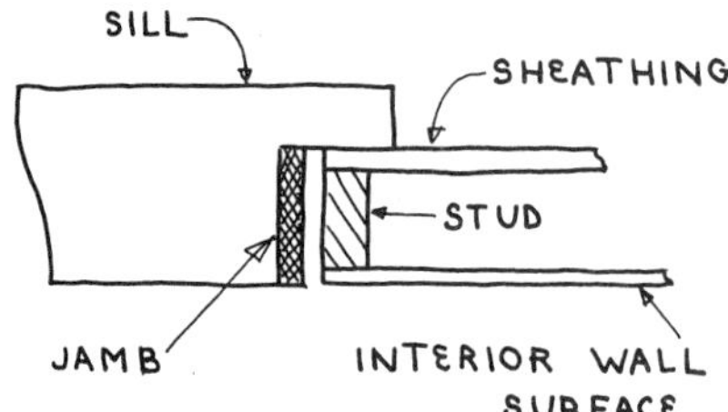

Figure 28-6. *Jamb width.*

STEP 2

For each window, take a piece of the jamb material and draw a picture of the window parts in cross section right on the wood. This picture should look like the cross section in figure 28-5. This is the easiest way to figure out all the dimensions.

STEP 3

Make the sills. The sills should be made of planed, dry, knot-free 2-inch lumber. Two-by-eights or 2 x 10s from the lumberyard are okay if you pick ones with a minumum of knots. You could also use selected pressure-treated stock. Design your sill to stick out about 1 inch past the siding, and to reach at least ½-inch inside the sash. The sill is notched around the studs and sheathing at the ends; this will leave "ears" or "sill-horns" sticking out at each end, which will later support the side casings. Usually double-hung windows have sills pitched at fifteen degrees, and casements pitched at twelve degrees. It shouldn't be flatter than twelve degrees. Make a saw cut on the bottom outside edge for a drip edge.

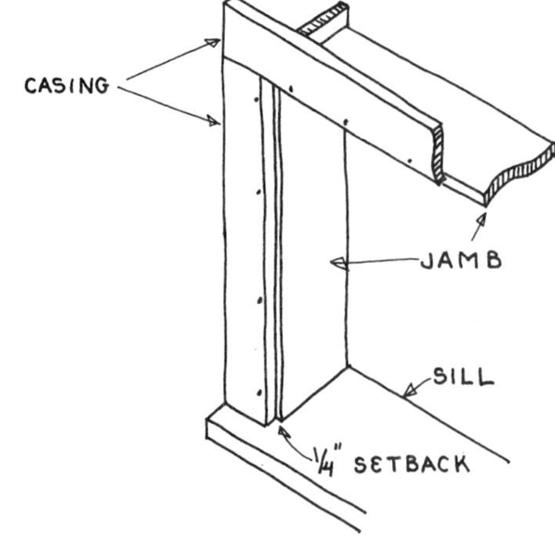

Figure 28-7.

STEP 4

Notch the top jamb and the sill into the side jambs as shown in figure 28-8. These notches can be made with a table saw or router. They can be made somewhat less easily with a circular saw or hand saw and chisel.

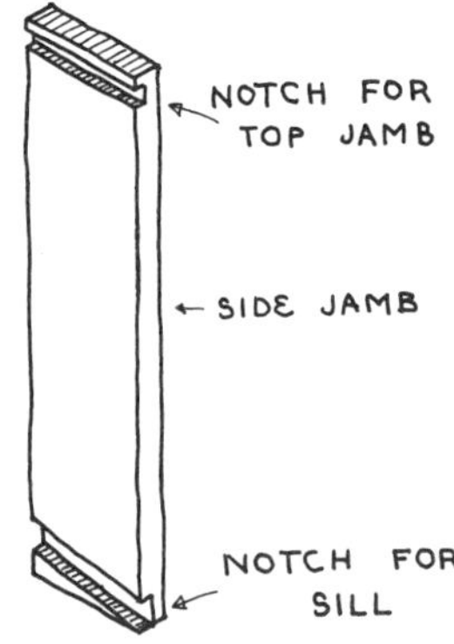

Figure 28-8. Side jambs.

STEP 5

Put some high-quality caulking in the notches and assemble the frame using 1⅝ or 2-inch drywall screws in pre-drilled holes.

Step 6

If you are using vertical boards for siding there will be no outside casing because the boards themselves will case the window. But with most kinds of horizontal siding there will be a casing attached to the frame before it is installed (figure 28-7).

Casing will be easier if you lay the window unit down flat on a table or a pair of sawhorses. Cut the casings from dry, straight, planed boards ¾-inch or preferably 1-inch thick. Shingles and clapboard siding fits better against the thicker casings. Often casings will be 1 x 4 or 1 x 5 boards, but the width is really a question of appearance. Notice that the casings are usually set back ¼ inch on the jamb (¼ inch "reveal"). Cut the side casings first. Make a fifteen-degree cut on the bottom end, then cut the top square to overlap the top jamb ¼ inch. Nail each side casing to the jamb with four or five 6d galvanized box nails. Then cut the top casing flush at the ends with the outside edges of the side casings, and nail it to the jambs in the same way. I make the top casing flush to avoid having to notch the siding later. As you nail, make sure the frame is square. Turn the completed unit over and caulk the joints on the inside with silicone, butyl, or some other high-quality caulking.

Step 7

The next step is usually to attach the sash. We'll discuss how to attach sash for different types of windows in a moment. For now, assume the sash is attached and proceed to the installation of the assembled unit, since installation is the same with any type of sash.

Step 8

Installation. Place the unit in its rough opening from the outside. The casing will automatically align the jambs flush with the sheathing. Use a level to make sure the sill is level. If not, slide a shingle under the low end. If the window has been built square, the sides will be vertical when the bottom is level. Tack the casing tentatively to the sheathing with one 8d galvanized box nail at the bottom end of each side casing. Leave the nail heads sticking up in case adjustments are necessary. Check sides and bottom again with a level. When it's good, tack the top ends of the side casings, leaving the nail heads up. If the window can be opened, see how it opens and closes. Look at the cracks where the sash meets the jamb. If these look uniform, the window is usually sitting correctly. Make adjustments if needed. Then nail the window permanently with more nails, perhaps four on each side and three across the top.

This procedure can be used with any window or door frame. If you are installing a hinged door or window, shim and attach the hinged side first.

How to Mount Sash

Window unit with a fixed sash: You usually put a fixed sash into a window unit when you want a large-size fixed window, or have some especially beautiful odd-sized sash you want to mount fixed, such as a salvaged stained-glass window. If so, the sash can be mounted simply between a pair of stops as shown in figure 28-9.

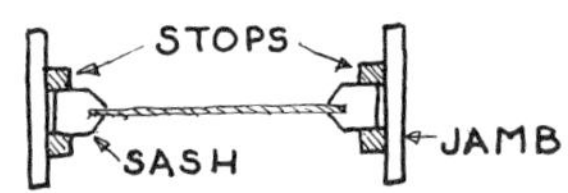

Figure 28-9. Mounting wood sash as fixed window.

These stops can be 1 x ¾ inches or any convenient size. It will be somewhat easier if you put the side stops in before the top. If the sash fits the sill tightly, omit the bottom outer stop and caulk the joint.

Outward opening casement: Casement windows are hinged at the side. For several reasons, an outward opening casement is one of the most widely used type of window. It opens 100 percent, the sash is not in the way when open, and it can act as a scoop to catch the breeze.

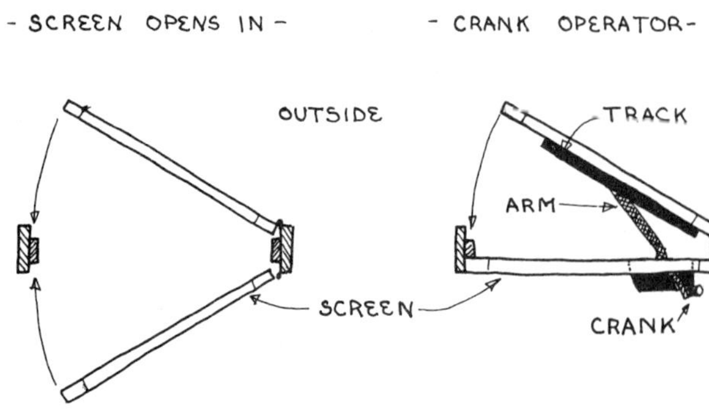

Figure 28-10. Two ways to open casements that have screens.

The problem with an outward opening window is that the screen, which must be on the inside, is always in the way of opening and closing the window. To get around this, the screen must be hinged or removable, or you must use special crank hardware to open the window without disturbing the screen. These choices are pictured in figure 28-10. In either case, all the dimensions must be figured out in advance for the window to work right. Figure 28-11 shows other typical details.

If you use a casement crank operator, you will find it cannot pull the sash tight enough for winter. In winter, the screen is removed (perhaps replaced with a storm sash) and the sash is pulled tight with some type of latch.

All this may seem like a lot of trouble to go through to open a window, but a window that does everything is always complicated. The trick is to find a fairly simple type of window that will do most of what you need well enough.

Inward opening casement: An inward opening casement is simpler than an outward opener because the screen can be mounted fixed on the outside and the sash can open inward

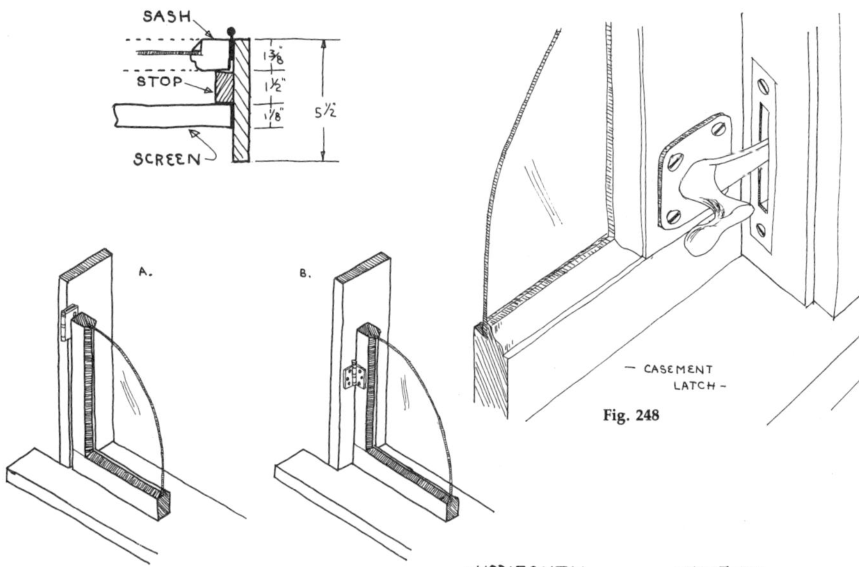

Figure 28-11. *Casement details.*

with no complex hardware. There are two ways to set this up. First, the screen can be installed permanently between stops, as shown in figure 28-12. Second, the screen can be mounted in a frame that screws to the outside of the outside casing. You can remove the screen every year to substitute a matching storm window.

The drawback of the inward opening window is that the sash may be in the way when open. Recess the bottom outer stop into the bottom edge of the sash as shown in figure 28-13 to avoid trapping rainwater behind the stop.

Outward opening hopper: The main advantage of an outward opening hopper window is that it needs no crank. It will open by the force of gravity and can be closed with a nylon rope threaded through a small hole in

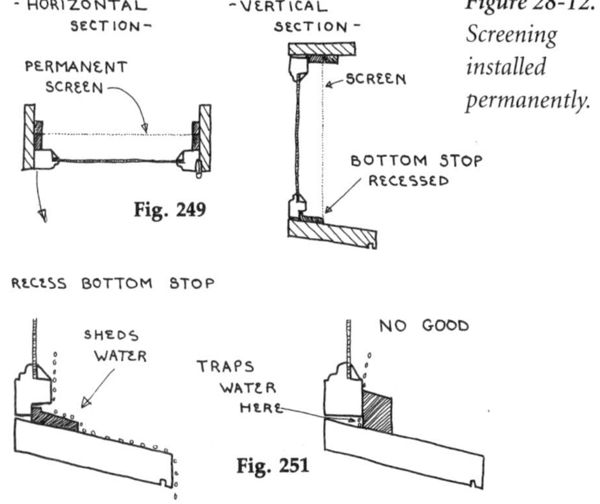

Figure 28-12. *Screening installed permanently.*

Figure 28-13. *Recess bottom stop for drainage.*

the frame of a screen. Its main disadvantage is that it will catch some water if left open in the rain. But it is the simplest type of homemade opening window.

Horizontal sliders: This type of window is tight and does not catch water. The screen can be on the outside without obstructing the sash.

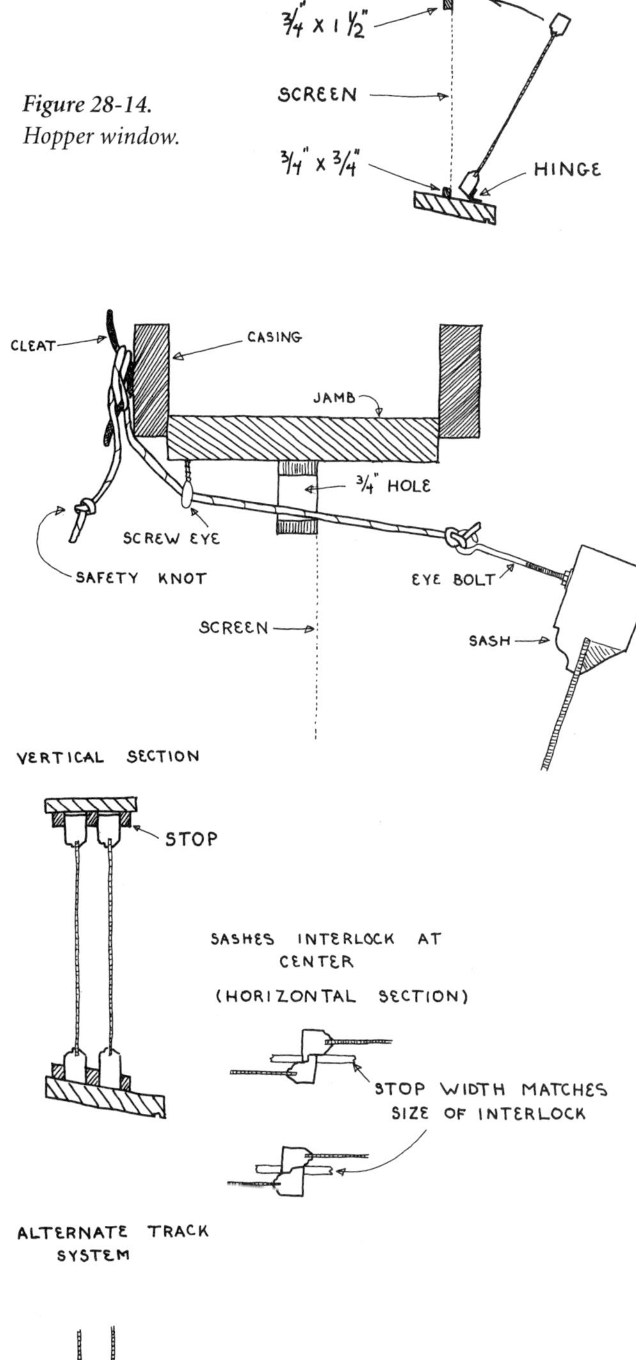

Figure 28-14. Hopper window.

Figure 28-15. Sliding window details.

The sash will slide in two adjacent tracks consisting of three rows of stops, as shown in figure 28-15. It will be easiest to understand exactly how this works if you examine an old wooden double-hung window. There must be some sort of interlock detail to close off the space between the two sash when the window is closed. The middle stop must be the same width as this interlock.

The outer bottom stops, the ones that might get rained on, should be shimmed up on neoprene washers for drainage. Nail the stops down through the hole in the washer.

The best sash for sliders are pairs of old double-hung windows turned sideways, which have the interlock mechanism built in. Some newer double-hung sash are made with a groove all the way down the edge. If you have these, use the alternate track system shown in figure 28-13. Design the screen to be removable so the windows can be washed.

Double glazing. Figure 28-16 shows three kinds of double glazing: first, a storm window or storm panel; secondly, a second piece of glass mounted about ¾ or 1 inch away from the main piece of glass (this solution works best with fixed windows); thirdly, insulated or thermal glass (glass companies can make thermal glass panels for you in any size).

My favorite type of storm panel is a "deadlight," made to size at our local

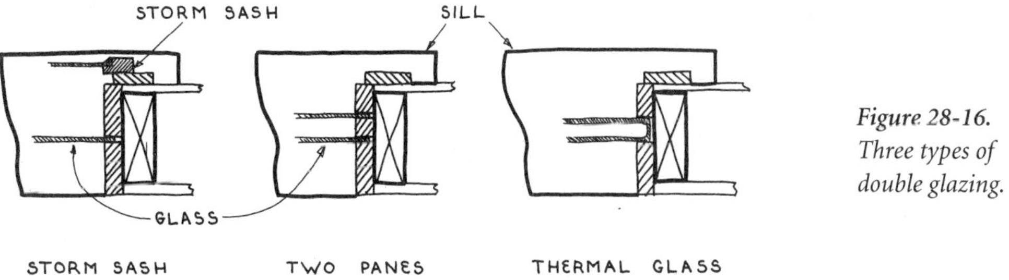

Figure 28-16. Three types of double glazing.

glass company. The panel is simply a piece of glass, rimmed with a thin aluminum edging with an integral weatherstrip. It can be clipped to any sort of window. Usually I fasten it to the outside casing as in figure 28-17.

Figure 28-17. Aluminum dead light makes a simple storm panel, easy to install and remove.

If you are building fixed windows or openers with new sash, I suggest you find out what thermal glass will cost. Often it will not be much more than two panes of ordinary glass, particularly if you have several panels of the same size. Insulated glass also saves work because only one pane need be installed.

Figure 28-18 shows typical details for fixed thermal glass. The bottom sill should have a horizontal section where the glass itself sits. The glass is propped up on "setting blocks," little rubber shims. Special self-adhesive glazing tape goes on either side of the glass to cushion and weatherstrip it. It is important to get installation instructions from the glass vendor, so your warranty on the glass will apply if the seal between the panes fails, which is not uncommon.

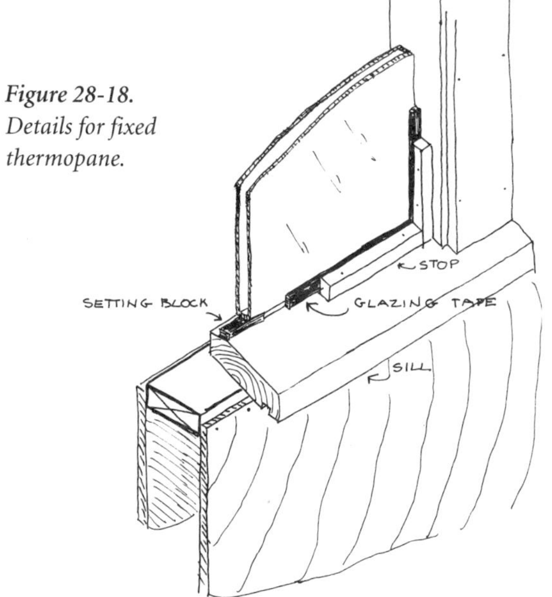

Figure 28-18. Details for fixed thermopane.

Fixed Glass Between the Studs

Though fixed windows can be built with a separate frame, fixed glass can also be mounted right between the regularly spaced wall studs. This simplifies your framing because no rough openings and no headers are needed. You can create a large bank of windows quickly and economically using this method. You provide accurate horizontal blocking to define the openings, then install outer side stops, making sure all the stops line up in a row. Then little wedges are used to force the notched sill tight up to the stop. The glass is installed with setting blocks and glazing tape. Then the inner stops are installed with screws or small

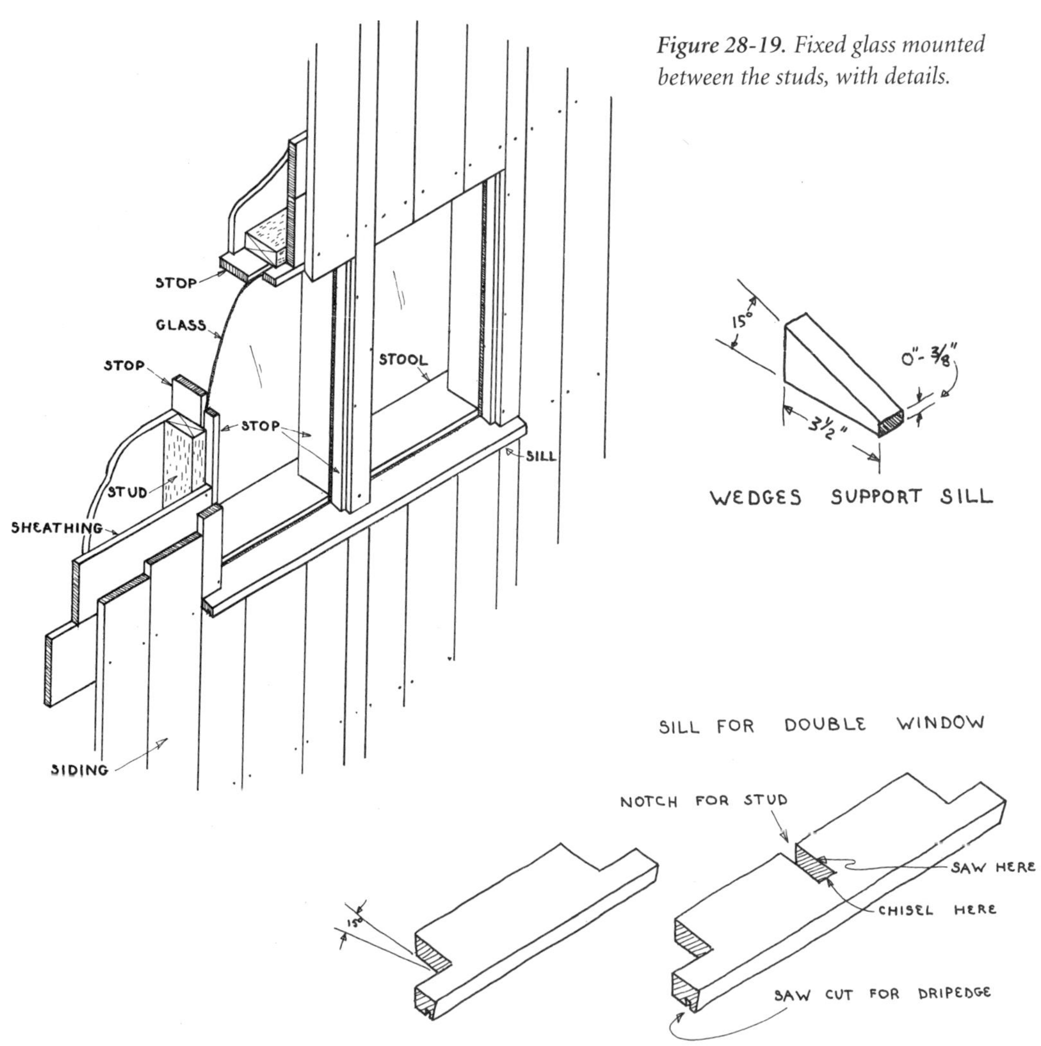

Figure 28-19. *Fixed glass mounted between the studs, with details.*

nails so you can repair the window if needed. Inside and out, it's trimmed in the normal way, except that inner trim boards should be nailed lightly.

Figure 28-19 shows the basic detailing.

How to Build Screens

Good screens can be made out of some knot-free 1 x 2 boards. Figure 28-20A shows a screen design that can be used when the screen will be screwed or fastened in place rather than hinged. It consists of a 1 x 2 frame with 45-degree joints at the corners, fastened with Scotch fasteners, which are large staples you pound in with a hammer. They don't look like much, but are very strong. Put one on each side of each joint.

A hinged screen, or one that you want to look more finished, can be joined at the corners with a halflap joint and glued with waterproof glue. The notch for the halflap can be made with a table saw.

Attach the screening with a staple gun. Cover the edge of the screen by turning the screened side toward the building or by tacking thin wood strips over it.

You can also purchase screens with aluminum frames made up by glazing companies.

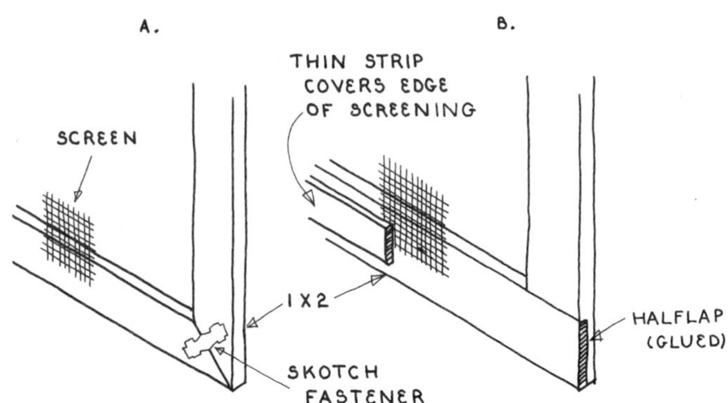

Figure 28-20. *Screens.*

How to Build Insulated Shutters

Even with double glazing, a window will let out about ten times as much heat per square foot as an insulated wall. Heavy curtains will improve this situation if they are closed when the room is not in use and at night. As figure 28-21 shows, they will be more effective if they are boxed at the top to prevent circulation of air between the room and the space behind the curtain.

An actual shutter is even more effective than curtains. Figure 28-22 shows two constructions. The first has ½-inch Homasote nailed to a mitered frame of

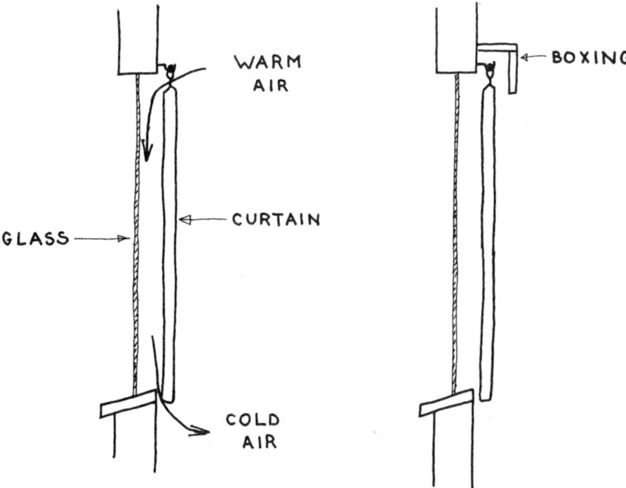

Figure 28-21. Boxing makes curtains work better by reducing drafts.

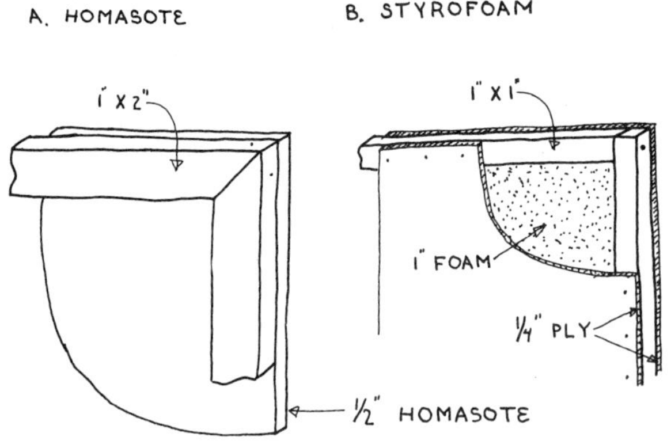

Figure 28-22. Shutters.

1 x 2s. The second, which is more expensive but more effective, consists of 1 x 1 frame with 1-inch foam in the middle, covered by a layer of ¼-inch plywood on either side. You can make a thicker panel for 1½-inch or 2-inch foam the same way. Shutters go on the inside since they will be opened and closed frequently. They can be hinged to swing sideways or upward. They can slide sideways if there is wall space. Or they can clip into place and be stored on a rack or in a closet nearby when not in use (figure 28-23).

Weatherstripping

Weatherstripping is critical to the function of any window. Figure 28-24 shows a good kind of weatherstripping for most situations. It consists of a felt strip protected by a wooden strip. Weatherstripping systems also come with a foam strip (which I don't recommend) or a compressible bulb, which may be the most eVective version. Nail it up after the window or door is installed. Place the weatherstrip on the stop with a little pressure against the closed sash or door. Then tack it to the stop with the nails provided with it.

Another handy type of weatherstrip is folded plastic flange. It comes in rolls, with a self-sticking back. It takes some experimenting to figure out which works best for a particular application. Home centers sell many kinds, and you can usually find one

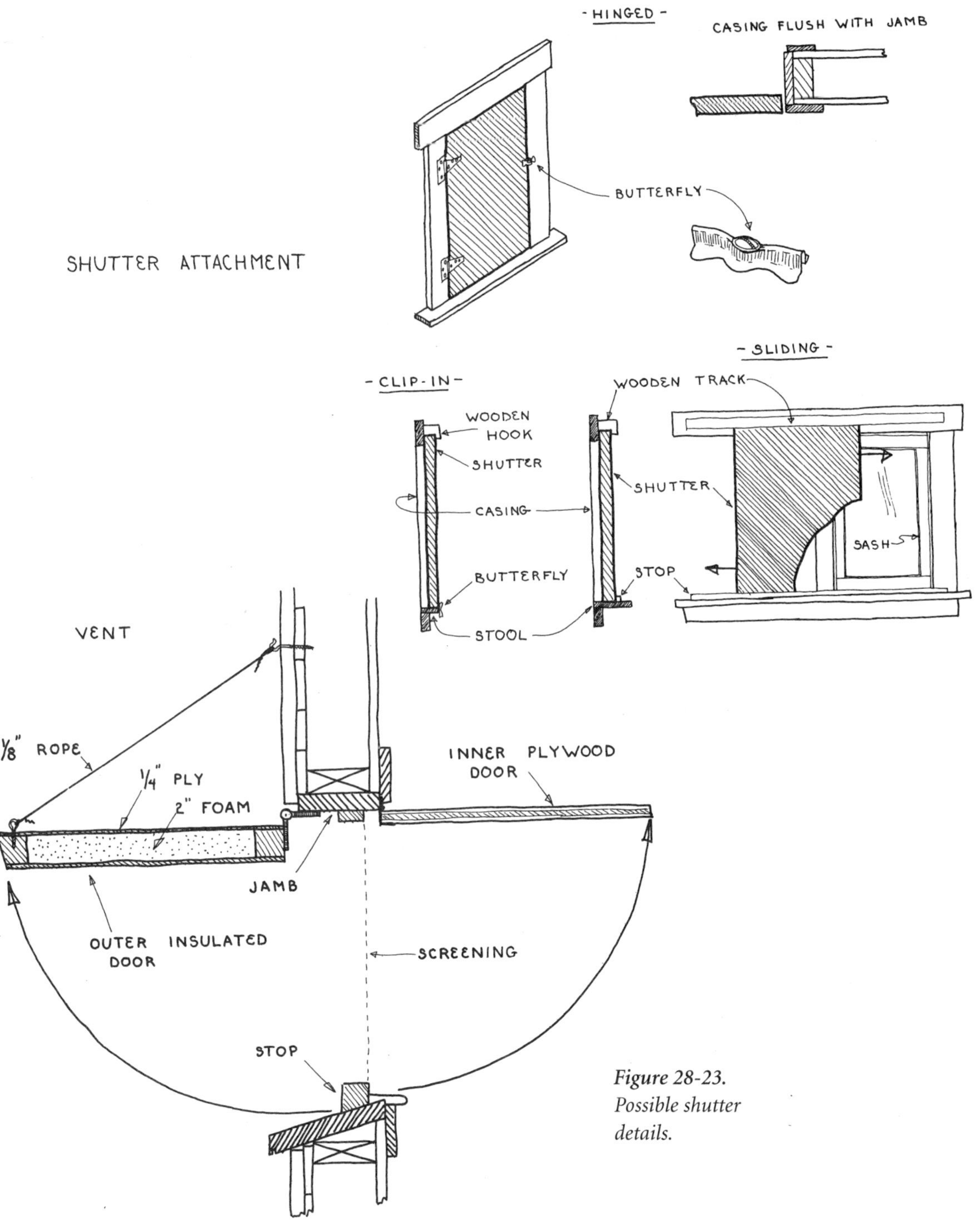

Figure 28-23.
Possible shutter details.

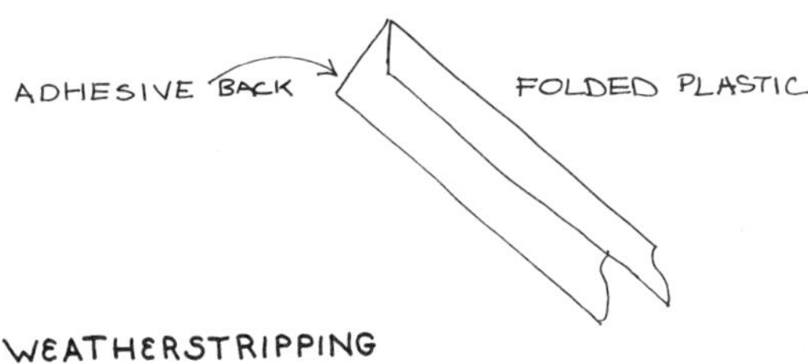

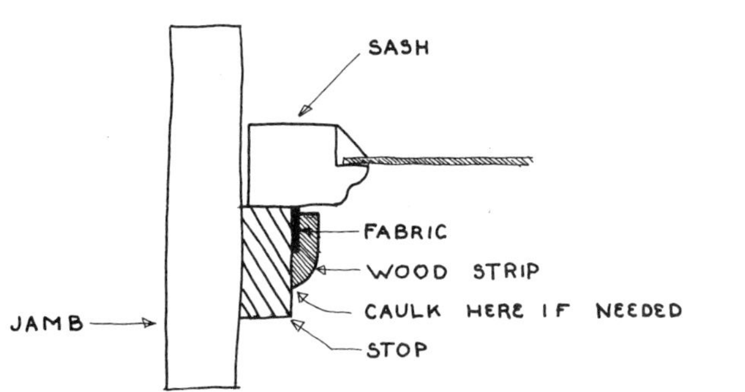

Figure 28-24. *Two of the many good kinds of weatherstripping.*

specially made for your purpose. Special strips are made for door bottoms, for example. One useful type consists simply of a roll of felt about ½-inch wide and about ⅛-inch thick.

SCROUNGING WINDOWS

Used windows can be bought from salvage companies and through classified ads. Often they can be obtained free from people who are remodeling their houses. Old wooden storm windows are plentiful and useful, particularly for fixed windows. Get as many of the same size as you can. You can either create a bank of uniform windows or use them in pairs as double glass. When possible, get used windows with their frames. Check them carefully. Make sure the corner joints are still tight, and there is no evidence of rot. If there is much damage, it's not worth reusing.

A casement sash is under a lot of strain at the joints compared to a sliding, fixed, or hopper window. Many of the sash you scrounge will be storm windows or old double hung sash, which were not designed to take much strain. Old storm windows should never be used as casements, though they can be used as sliders or hoppers. Old double-hung sash can be used as fixed, sliding, or hopper windows, but should not be used as casements if they are more than 20 or 24 inches wide.

29

DOORS

The two most common types of wood doors are panel doors and flush doors (figure 29-1). Panel doors are the old-fashioned kind with a perimeter frame and molded panels in the middle. Flush doors are the flat modern type. Hollow-core flush doors have a cardboard lattice inside and are for light duty or interior uses. Solid-core doors have particle board inside and are for heavy duty or exterior uses. Exterior doors are usually 1¾ inches thick; interior doors are 1⅜ inches thick.

Figure 29-1.

In recent years, metal insulated exterior doors have become popular. I am not particularly fond of the idea of metal doors on my cozy house, and they aren't that beautiful to look at. But they function so well, and cost so little compared to quality wood doors, that we use them often. If carefully selected, nicely painted, and supplied with decent hardware, they are a good choice.

Hollow interior doors are very inexpensive, and not worth purchasing used. Solid-core flush doors cost more, and panel doors even more.

New doors are available prehung in a frame, with hinges, latches, and other hardware all mounted. This saves a huge amount of work.

OBTAINING DOORS

Ordering New Doors

If you are considering new doors, take your plan to local suppliers, or take a list that specifies door sizes, jamb widths, and whether each door is right-hand or left-hand swing. There will usually be a "window and door person" who can show you a variety of choices.

Always buy doors "set up," which means hung in a frame, with the door and jamb drilled out for hardware. Interior doors don't need a sill or threshold, and usually should be ordered without casings.

Exterior doors might be ordered with exterior casing

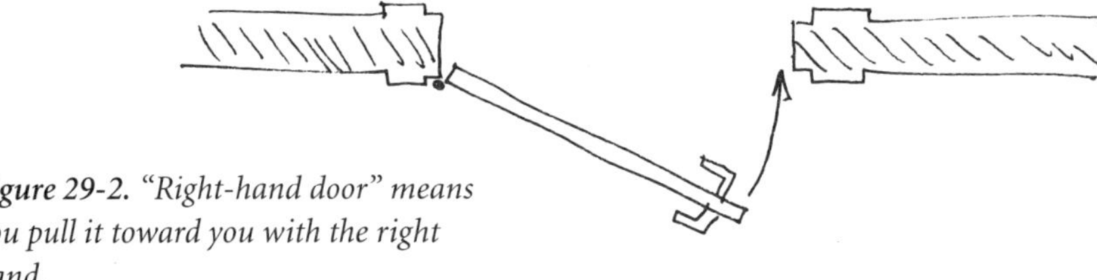

Figure 29-2. "Right-hand door" means you pull it toward you with the right hand.

installed. They also require a threshold. Order exterior doors with a low-profile or accessible aluminum sill if possible.

I also strongly recommend getting lever-style door hardware instead of knobs. It is more convenient, and easier to grasp, and looks great if carefully selected.

Obtaining Used doors

You can often find good, used panel doors for low prices. If you buy a used panel door, make sure there is no rot and that the joints are tight. Sight along the door to make sure it is not warped, because a severely warped door is worthless.

MAKING DOORS ON SITE

Figure 29-3 shows three types of doors to make with materials you may well have on your building site. If you build doors, use straight dry wood and assemble the door on a flat surface. Any crookedness in your work surface will become a permanent feature of the shape of the door.

An exterior door frame is like a window frame, with a sill, jambs, stops, and casings. An interior door is the same, except that there is no need for a sill. Doors are hung flush with the jamb (figure 29-4). If they are set back, they will not open all the way. Interior doors can be hinged to open in the direction that is least obstructive. Exterior doors usually open in so a storm or screen door can open out, though they can swing out too.

Cut the jambs from straight, dry, relatively knot-free 1-inch or 2-inch lumber, or buy special jamb stock from the lumberyard. Ready-made jamb stock has the stops formed by notches taken out of the corners of the jamb (figure 29-4). If the jambs are made from 1-inch lumber, the stop portion will be applied as a separate piece.

You can make sills yourself out of dry, knot-free 2 x 8s or 2 x 10s, or you can buy special sill stock, which has the shape shown in figure 29-5. The little hump at the top makes water that runs under the door run out again. The sill and jamb

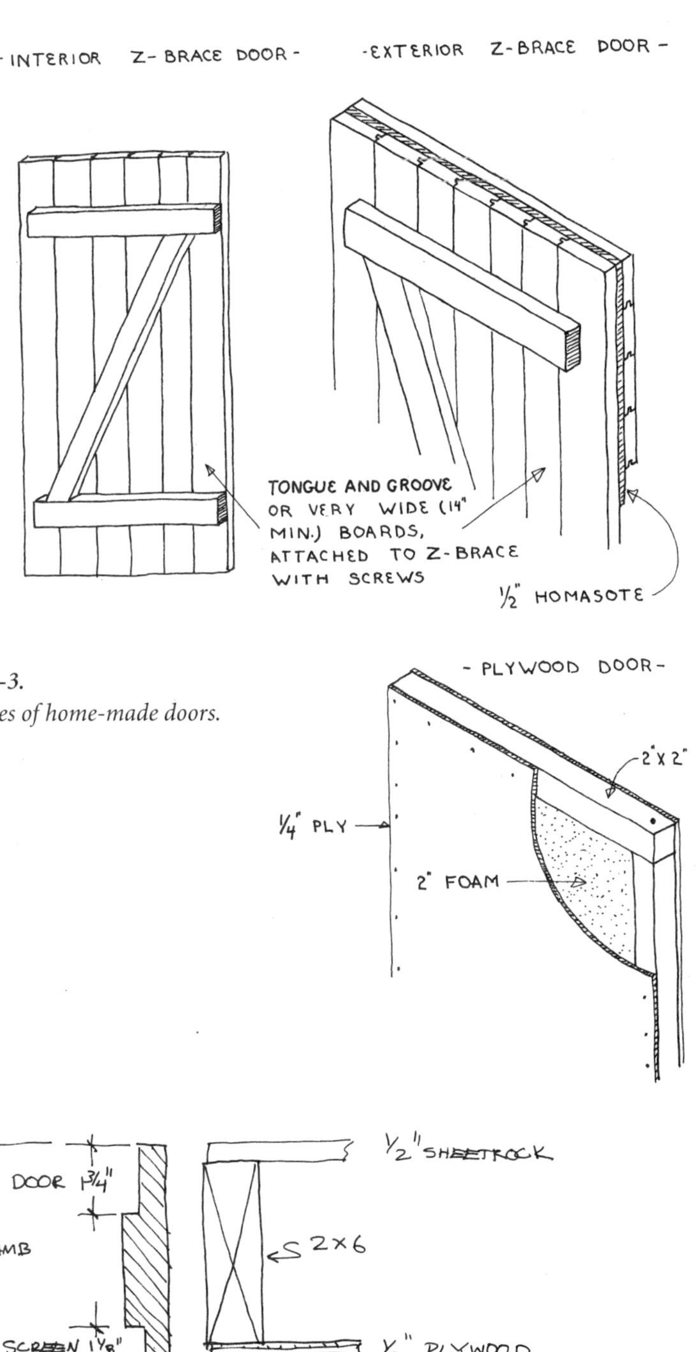

Figure 29-3.
Three types of home-made doors.

Figure 29-4. Typical jamb dimensions.

stock you use must match the size of the door, the thickness of the door (or doors), and the total thickness of the wall.

Figure 29-4 shows a jamb made to fit a wall nominally 6½ inches thick, with a 1¾-inch exterior door, and a 1⅛-inch screen door or storm door. Jamb stock from lumber yards is made to match these and other typical dimensions.

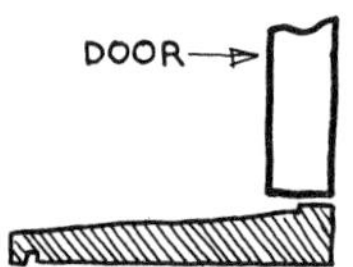

Figure 29-5. Shape of wood sill for exterior doors.

How to Hang Doors on Butt Hinges

Most doors are hung on butt hinges notched or "mortised" into the door and the jamb (figure 29-6).

Step 1

Build the door frame, consisting of jambs, stops, and (for exterior doors) a sill. Follow steps 1 through 5 for windows, pages 381–84. The notch in the jamb for the sill is shown in figure 29-7. Make the frame 3⁄16 inch wider and ¼ inch higher than the door for clearance (figure 29-6). If you are using an old door, square up the door with a skillsaw or a sharp plane before making the frame.

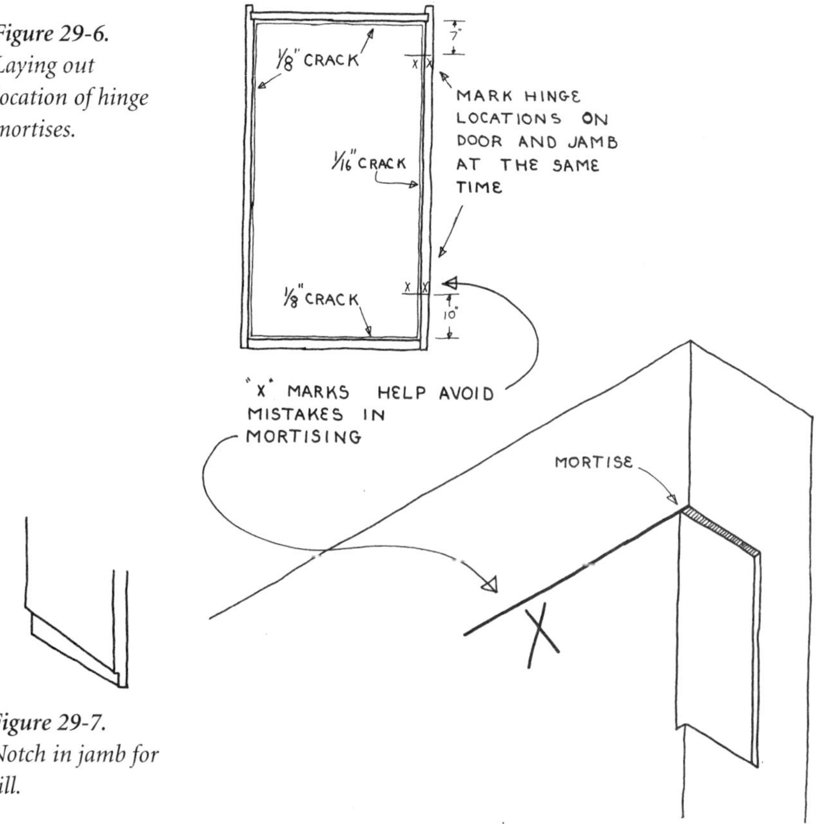

Figure 29-6. Laying out location of hinge mortises.

Figure 29-7. Notch in jamb for sill.

Step 2
When the frame is assembled and the stops in, lay the frame on the floor, hinge side up. Place the door in the frame, with a ⅛-inch crack at the top, and a 1/16 inch crack on the hinge side.

Step 3
Mark the door and the jamb simultaneously for the hinge location, using a sharp pencil or knife and holding a square against the jamb. The line represents the top of each hinge on the door and the jamb. For an exterior door, or any quite heavy door, add another line for a third hinge, locating the hinge somewhat above the middle of the door. Mark X's on the door and the jamb to indicate which side of the line each hinge goes on. This will prevent confusion later.

Step 4
Lay out mortise. Use 3 x 3 loose-pin butt hinges for interior doors. For exterior doors use 3½ x 3½-inch loose-pin butt hinges. Remove the door from the frame and draw each mortise on both the jamb and the door. Each mortise will be as wide as the hinge is wide and as deep as the hinge leaf is thick. The hinge itself can be used as a template for marking the door and jamb. Plan the mortise so that the hinge will stick out beyond the jamb by about ½ inch. This overhang makes room for the pin and provides clearance so that the door can open without hitting the casing.

Step 5
With a sharp knife or a hammer and chisel, cut around the perimeter of the mortise, as shown in figure 29-8. After cutting down on all three sides, remove the scrap as shown in figure 29-8. A sharp chisel is the key to this procedure.

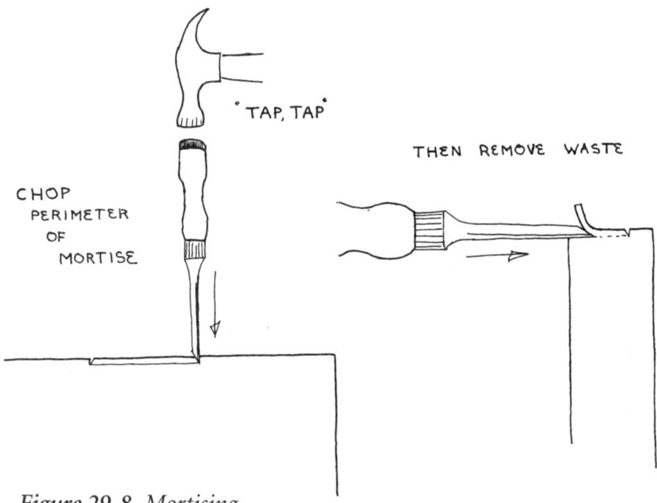

Figure 29-8. Mortising.

Step 6
Remove the hinge pins and screw each leaf into its mortise with only one screw in each hinge.

Step 7
Test-fit the door in the jamb by replacing the pins in the hinges and making any needed adjustments. Then put in the other screws.

Step 8
Hanging the door. Take a long 4- or 6-foot level and check the wall stud on

the side where the hinges will go. It helps a lot if this stud is exactly plumb, but if it isn't, determine which end of the stud projects the farthest into the door opening, and mark it with an X.

Then set the door frame in the opening. I usually leave the door on its hinges for this process, so I can see what's happening, but you can also remove the door at first if it helps. Making sure the jamb is flush on both sides with the wall surface, tack it to the stud on the hinge side where your X is, using an 8d or 10d finish nail, or alternatively, a sheetrock screw through a pre-drilled hole.

Next, slide shingles behind the other parts of the jamb as needed to shim the entire length plumb and straight. You want to keep the jamb pieces square with the wall. If the wall stud is square to the wall, use a pair of shingles together, sliding them back and forth to get the desired thickness. If the stud is crooked, it might work better to use one shingle. The ends of these shingles get trimmed off later. Nail or screw the jamb through the shingles. If you are nailing, leave the heads up in case adjustments are needed.

Use the same approach to shim and secure the other jambs, letting the crack between the jamb and the door guide your adjustments. When it looks right, and the door closes and opens nicely, fasten the door frame permanently, making sure there is good fastening at the hinges, and near the latch.

Figure 28-9. Shimming the jamb plumb.

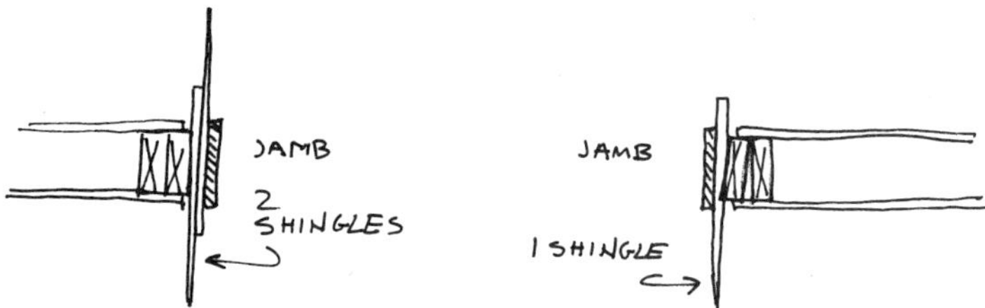

If they are separate pieces, install stops now. An old door will often be a little warped. Installing stops last allows you to make adjustments for the door's imperfections. Cut stops on a table saw to about 1½ x ½ inches. Cut and install the one on the latch side first. Fit it gently up against the closed door, and tack it to the jamb tentatively with a few 4d finish nails. Close and open the door a few times. When it closes nicely, cut and fit the other stops in the same way.

Alternative Method of Hanging an Interior Door

The above section describes the common approach. But as I reviewed it, I realized I don't do this very often, particularly if I am using a recycled door. Since I am the only one I know who uses this approach, choose my other method at your own risk. The basic idea is to build the frame around the door, in place, so that the idiosyncracies of the door can be taken into account.

Make up jamb stock from flat ¾-inch boards, with no rabbet, as described above. Before cutting anything to length, cut the notches that will join the side jambs to the top. Then take the jamb that will carry the hinges, and tack it lightly to the edge of the door, with the notch ⅛ inch above the top of the door. This allows you to mark the hinge locations on door and jamb simultaneously and accurately, as described above. Cut the mortises, install the hinges, and mount the door on this single jamb. Think about where the finish floor or carpet will go, and cut off the bottom end of the jamb accordingly.

Remove the hinge pins. Install the first jamb in the rough opening. This is easiest if the framing on that side has been made perfectly plumb and straight. If not, use shingles to get the jamb perfect. Now hang the door on this one jamb. Take the other vertical jamb, and cut it to length so that the notch will be ⅛ inch above the door. Then install it, using the already hung door to tell you when it lines up right. Shingle behind it until it is basically square to the wall, and ⅛ inch from the door, making sure the relationship of the notch at the top stays correct. Then cut the top jamb to length, and slide it in the notches. When it is in, nail through the vertical jambs into the end of the top jamb. Better yet, use 2-inch drywall screws in pre-drilled holes.

Next, install whatever latch or door hardware you have for the door. Then install stops as described above.

T and Strap Hinges

Doors are also occasionally hung on strap or T-hinges (figure 29-10). They are not mortised in, but simply screwed to the face of the door and the door casing. They can be used with homemade Z-brace doors or panel doors, but flush doors are not designed to be used with them.

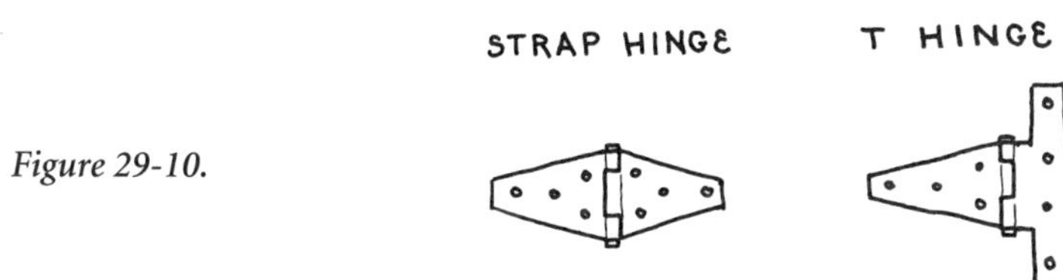

Figure 29-10.

30

SIDING

IF YOUR WORK SCHEDULE PERMITS, SIDE YOUR HOUSE after you roof it and put the windows in, but before you do inside work. If you need to move into your house soon, siding can be delayed and you can proceed to the inside work. Plywood sheathing may be left uncovered indefinitely if you caulk the joints, particularly those above the windows. Board sheathing may be left unsided if you put flashings over the top windows and door casings and then cover the sheathing with 15-pound felt paper (tarpaper) housewrap. Batten the felt down with scraps to keep the wind from tearing it away.

Figure 30-1 shows several common sidings. The top row shows boards used in various ways. These options, which are discussed in more detail later in this chapter, are economical and quick.

Clapboards and shingles cost more and take longer to install, but they are more durable. An option pictured in figure 30-13 is textured plywood, such as T-111. This plywood is grooved on the outside to resemble vertical boards. Textured plywood is the fastest type of siding to put up because it goes up in large sheets. It is the cheapest siding because a single layer serves as siding and sheathing combined.

VERTICAL BOARDS

Vertical boarding is one of the simplest sidings. If you live in an area where there are sawmills, board siding will probably be one of the most economical options. Siding boards should be air-dried or kiln-dried, unless the cracks will be covered with battens. Siding boards should also be planed if possible, because rough boards are not uniform in width and will take forever to put up well. If you want your siding to have a rough texture, a local sawmill may be able to provide you with boards that have been planed on

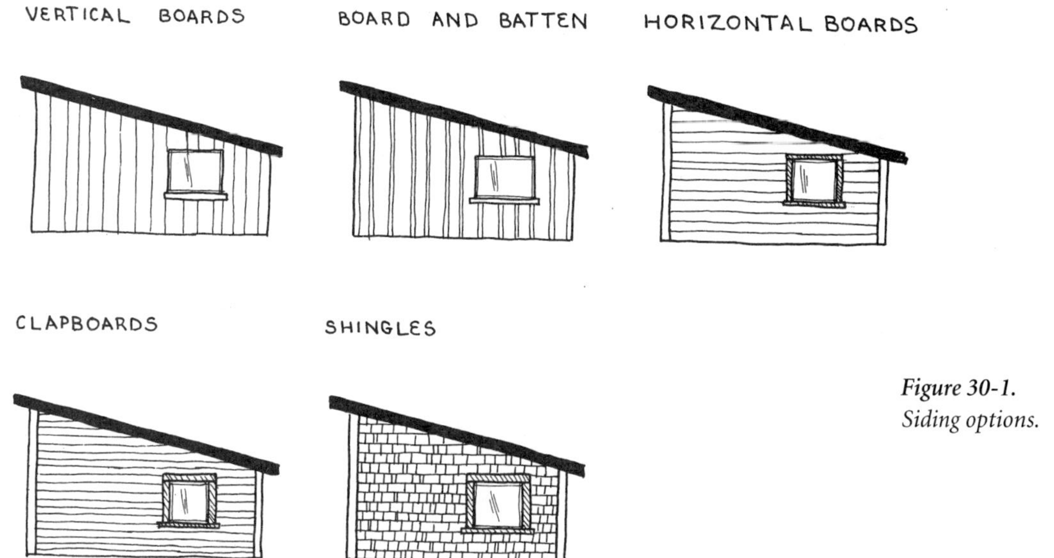

Figure 30-1. *Siding options.*

three sides. Such boards are uniform in thickness and width, but one face is rough for appearance.

Shiplap boards are probably the best type of vertical boarding overall, providing good coverage and quick installation.

Sheathing under vertical boards can be 1-inch boards or exterior-grade plywood at least ½ inch thick. Figure 30-2 shows typical vertical siding details. Notice that there are no casings: The siding itself will cover the cracks around windows neatly.

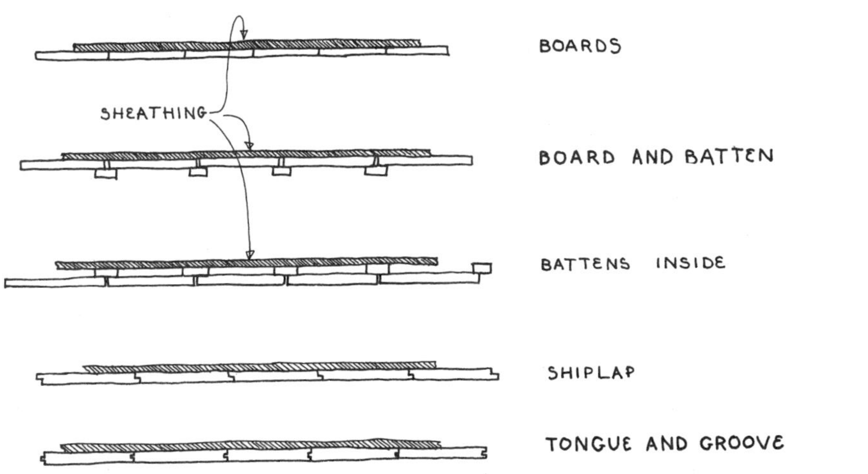

Figure 30-2. *Types of board siding.*

How to Apply Vertical Siding

Step 1

Scaffolding. Make some kind of staging or scaffolding to work from. If your house is low you can work easily from a stepladder or a plank across two sawhorses. But on a high wall, build a scaffolding (see chapter 34).

Step 2

Paper. Unlike shingles, some water may find its way through board siding. Some sort of water barrier below makes sense. This could be ordinary felt paper, red building paper, or housewrap.

Step 3

Siding. Start from one corner and work across the face of the wall. Make sure the first row is very straight and plumb, and nail it with two 8d galvanized common nails every 2 or 3 feet. Board siding has a tendency to work loose over time, especially on the sunny side of the house. This is not too severe a problem over board sheathing, and you can secure the boards by driving the nails at various angles. Angling the nails gives better holding power than driving all nails perpendicular to the wall surface. But with ½-inch plywood sheathing, simply driving the nails at angles will not be enough. One solution is to drive over-long nails and bend them over from the inside with your hammer. Another approach is to screw the plywood to the siding from the inside with 1¼-inch drywall screws.

A third option is to add blocking to the wall framing until you have solid nailing for the siding about every three or four feet.

As you nail your siding, remember that the nails are exposed, and become part of the "finish." Pound them in a neat and consistent way.

If the house is so tall that one board will not make it from bottom to top, join the boards with "scarf joints" at a 45-degree angle to shed water (see figure 30-3).

Every few rows, use a level to make sure the boards are plumb. If they are not, you can correct this by leaving slight cracks between the next few boards at one end only, making the other end as tight as possible. No casing is needed around window and door openings. Above windows and doors, cut the boards at a 45-degree angle to create a drip edge, as in figure

Figure 30-3. Vertical siding details.

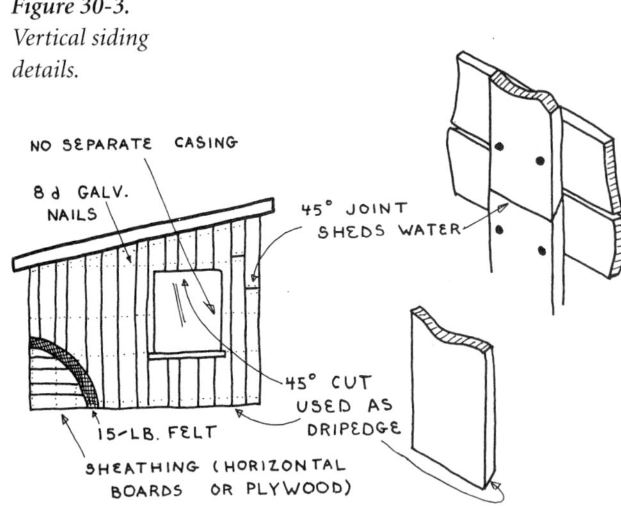

30-3. A 45-degree cut at the bottom of each row is also a good idea. If you have two power saws available, set one square and one at 45 degrees to avoid constant resetting.

HORIZONTAL BOARD SIDING

Figure 30-4 shows various types of horizontal board sidings. Any horizontal board siding must be tongue and groove or shiplapped to shed water, and the boards should be well air-dried or kiln-dried. Since horizontal boards can be nailed to the studs through the sheathing, the sheathing itself does not have to hold nails. Figure 30-5 shows horizontal siding methods.

How to Apply Horizontal Board Siding

STEP 1

Paper. Put up 15-pound felt or housewrap. If the paper will be exposed to the weather for an extended period, fasten it with little discs called "roofing tins" to protect it from being torn off by the wind.

STEP 2

Window and door casings. Plain shiplap or tongue and groove boards may be put up with no casings if you want. But novelty siding, drop siding, and similar types of siding will not look or fit right without casings. Casings go up before the siding. Often the casings are part of the windows and doors. If not, put them up next. As figure 30-6 shows, the outside edge of side casing should be flush at the top with the end of top casing, and at the bottom with the end of the sill. This

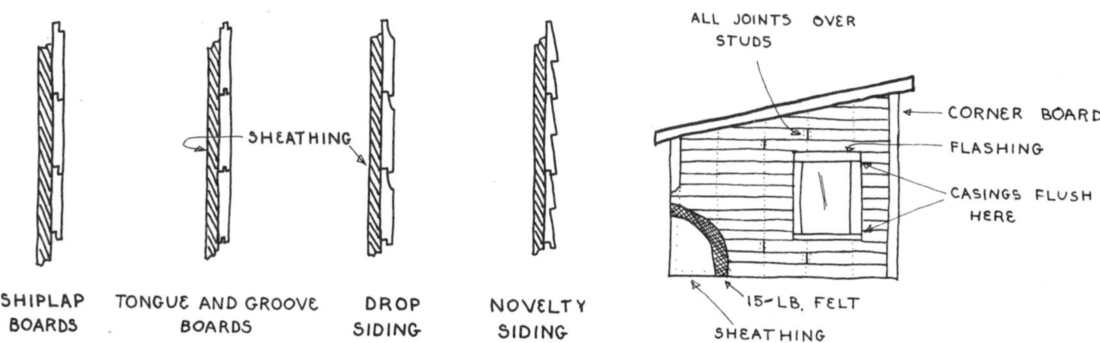

Figure 30-4. Horizontal siding variations.

Figure 30-5. Horizontal siding details.

will avoid notching the siding. Make the casings out of wood somewhat thicker than the siding. If the siding is ¾ inch, the casing can be 1 inch or more. If the thicknesses are the same, you may have trouble getting a neat fit.

Leave a ¼-inch setback, or reveal, all the way around the jamb (figure 30-6). This will look better than if you try to make the casings flush with the jambs.

Put up the side casings first. Cut a 15-degree angle at the bottom to fit the sill. Then cut the side casings to length so that they run from the sill up to the top jamb, lapping over the top jamb by ¼ inch to create the ¼-inch setback. Set the side casings back ¼ inch on the side jamb and nail them on with 8d galvanized common nails. Cut the top casing to come flush with the outside edges of the side casings, and nail it on as you did the sides.

Figure 30-6. Casing details.

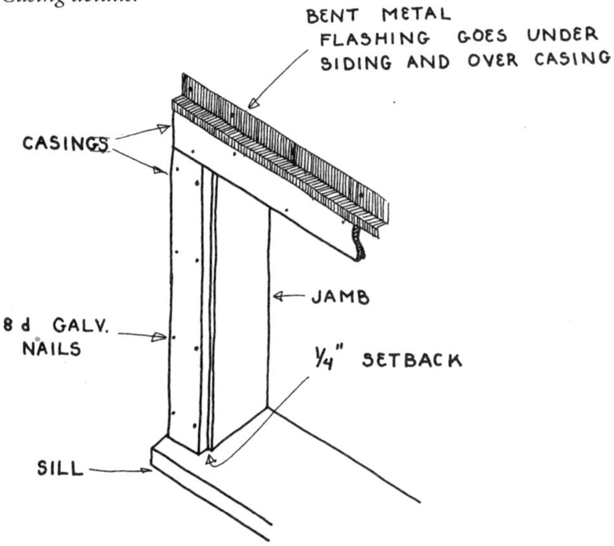

Step 3

Flashing. Install a bent metal drip cap above every top casing that might get rained on. You can make these yourself, by bending strips of aluminum, but they are sold by lumberyards ready-made to fit common thicknesses of casing.

Step 4

Corner boards. Corner boards are shown in figure 30-5. You can put up horizontal sidings with miter joints at the corners, but using corner boards will save time. As with casings, the width of the corner board will affect how the house looks, so choose it carefully. If you use drop siding or some other siding that is not flat on the outside, nail the corner boards up before the siding, just as you did the casings. Like the casings, the corner boards should be ¼ inch thicker than the siding if possible, because the siding will butt to them. Before you install the corner boards, wrap the corners tightly with 15-pound felt. With sidings that are basically flat on the outside, such as plain shiplap boards, or possibly drop siding, the corner boards can be put on after and on top of the siding. In that case, the corner boards need not be thicker wood than the siding.

Step 5

Siding. Horizontal siding can go up much like vertical siding. Start at the bottom of the wall, get the first row level, and work up row by row. Put two 8d galvanized common nails into each stud, driving the nails at varying angles. Fit the boards as tightly as possible, and check periodically to make sure they are parallel with the house frame. If not, make adjustments as you would with vertical boards.

CLAPBOARDS

Clapboards are one of the nicest kinds of siding. The following instructions are for the standard 6-inch clapboards, but you can use the same methods to put up narrow clapboards or the wider sizes that are sometimes called "bevel siding."

There is enormous variation in the cost of clapboards. Clear redwood or cedar clapboards are very expensive. Native clapboards from a local source may be very cheap. Cheaper clapboards will have more waste, so over-order. In general, I seek out a middling grade, mostly "clear to the weather," meaning that the part of the clapboard that will be exposed is knot-free. It definitely pays to spend a morning checking out all the possibilities available.

People today often encourage builders to prefinish the back of clapboards before putting them up. People sometimes dip clapboards or shingles in a bath of stain. I have taken many ancient clapboards off of old buildings which had never received this deluxe treatment. I do believe that any clapboards should have some sort of finish on them, whether it's a clear sealer, a heavy stain, a transparent stain, or a real paint.

Figure 30-7 shows typical clapboard details.

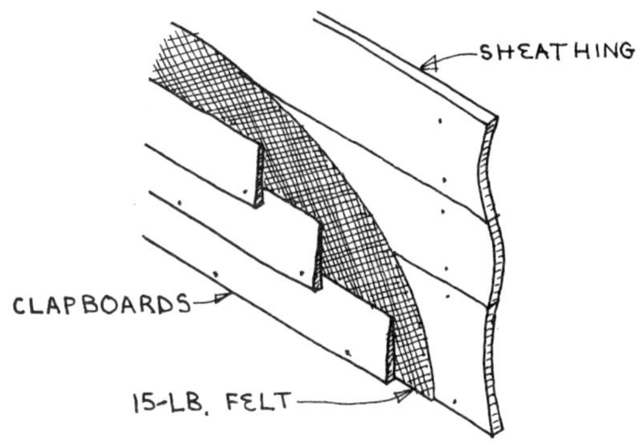

Figure 30-7. Clapboards.

How to Apply Clapboards

Step 1

Paper. Paper or housewrap is essential with clapboards.

Step 2

Casing and corner boards. Case the windows and put up corner boards as described above.

Step 3

Layout. Clapboards come in a variety of sizes, but the common size is ½-inch thick at the butt by 5½ inches wide. This is nominally a 6-inch clapboard. Clapboards must overlap at least ½ inch. So with 6-inch clapboards there will be an exposure of from 3 inches (very uncommon) to 5 inches. I like a 4-inch exposure.

After you decide what exposure you like, figure out where the rows will be and mark the location of each row on the corner boards and casings. Try to make the bottom edge of a row come out even with, or slightly above, the top casing of windows and doors, and even with the bottom of the window sills. This will make the work quicker, easier, and neater. Your chosen weather exposure will usually not fit in perfectly with this objective, but often a small, unnoticeable adjustment in the interval (perhaps ⅛ to ⅝ inches over a few courses) will enable you to align the clapboard neatly with the window, top and bottom.

Step 4

Nailing strip. Nail a narrow strip as thick as the thin end of the clapboards to the bottom edge of the wall. The first row will be shimmed out by this strip.

Step 5

Nailing. Different ways of nailing up clapboards are used in different places. This is the way I know. Use a chalk line to locate the bottom of each row. Nail the clapboards with 5d galvanized box nails every 10 inches (more or less), nailing about ½ inch above the bottom of the clapboard (figure 30-8). Stagger the joints at least 12 inches. The clapboards should fit tight where they butt into one another. Use a small, fine-toothed hand saw, a "chop-saw," or a table saw to cut them. Nail neatly, trying to establish an attractive nailing pattern. As in most

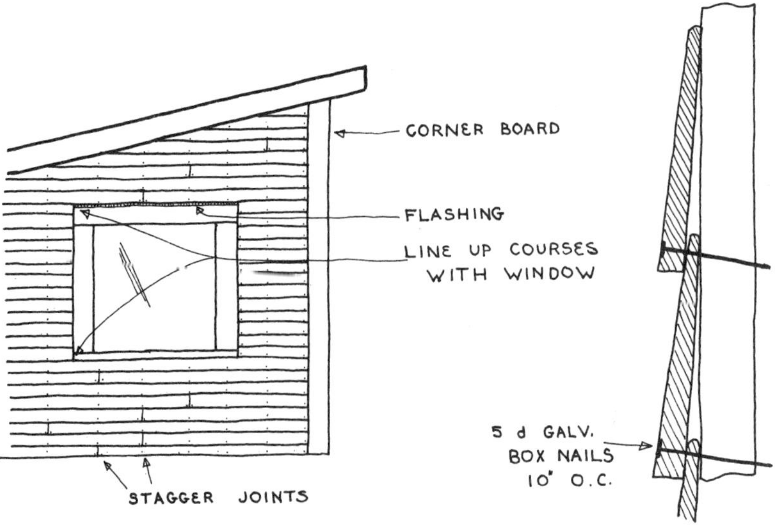

Figure 30-8. Clapboard details.

finish work, the difference between a good job and a sloppy job is the care taken in nailing. Apply caulk behind the clapboard joints at corner boards or casings.

If you are using bevel siding, which is ¾ inch thick at the butt and wider than regular clapboards, nail 8d galvanized nails into each stud, and make all butt joints occur over the studs.

Step 6

Molding. Often there will be a molding to cover the joint between the top row of clapboards and the overhang of the roof. It is usually an intricate shape, but a simple 1 x 1 will do as well as any other milled piece. If the clapboard fits perfectly no molding is needed.

There are several procedures you might choose for a more deluxe siding job. Many people paint or stain the back of clapboards before putting them up. I'm not sure this is needed. Sometimes small strips of felt paper, about 2 x 6 inches, called "babbits," are placed behind all joints between clapboards. I have seen this detail on many old houses; it provides an extra line of defense against water. Such extra precautions might make sense where the height of the wall or weather exposure will result in greater weathering than usual.

SHINGLES

Cedar shingles are one of the must effective and durable siding materials. It is easy to work shingles neatly around odd-shaped moldings or trim elements, or interweave them with flashings. Anyone can learn to shingle. Shingles are fun to put up, and can be used in decorative ways. They are often stained, which will add somewhat to their longevity. But even the cheapest shingles, left completely unfinished, will last many decades before any damage occurs to the house beneath them. This is because with a five-inch or even 5½-inch exposure, which is typical, you get in effect three layers of coverage. For all these reasons, shingles have become our siding of choice.

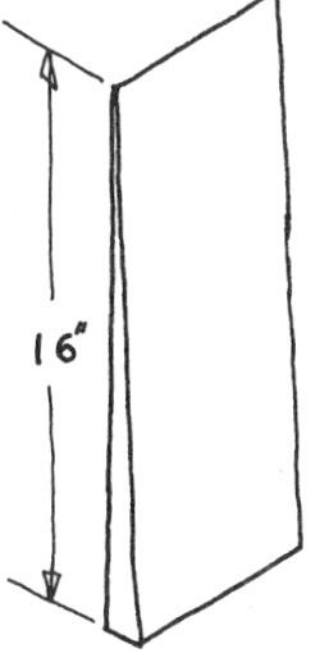

Figure 30-9. Shingle.

Top-quality, clear western red cedar shingles are expensive, but less expensive grades, such as white cedar, only cost between $9 and $18 per bundle, which translates into about $50 to $100 per square (100 square feet). Investigate what your local ven-

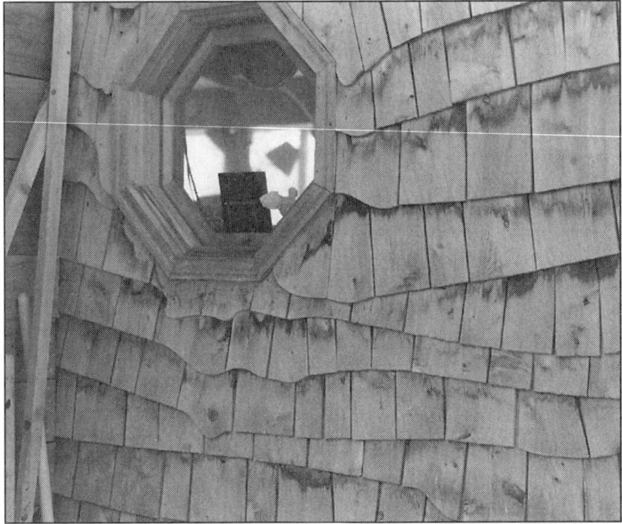

Figure 30-10. "Little Hawaii." *All sorts of decorative patterns are possible with shingles.*

dors sell, and look for shingles which will be basically "clear to the weather." If you choose a cheaper grade, the poorest shingles can be selected out for shims, kindling, or to side an outbuilding.

Shingles can be put up over a sheathing of 1-inch boards or ½-inch exterior-grade plywood.

How to Apply Shingle Siding

For board sheathing, use a housewrap under the shingles. With plywood sheathing, we often only seal the joints, particularly the corners, using felt paper or housewrap, which sometimes can be purchased in a self-adhesive 6-inch-wide tape. The joints could also be caulked. Most builders, it's true, use the housewrap over plywood, or a complete coverage of felt or building paper. That's okay, too.

Step 1

Casing and corner boards. Case windows and doors, and install corner boards as for horizontal board siding. Shingles will build up to a total thickness of about 1 inch, so thicker "five-quarter" (5⁄4) stock for corner boards will do a better job, though standard 1-inch boards (actually ¾-inch thick) will work. To save painting time, and make a crisper job, paint the trim before shingling—at least the first coat—unless the shingles will have the same finish as the trim.

Step 2

Layout. Shingles are laid with as little as 4-inch exposure or as much as 6 or 7 inches. I would suggest something in the 4½- to 5½-inch range; they will go up fast, but provide good protection.

Shingle courses should line up with the top and bottom of windows and door trim (figure 30-11). You accomplish this by adjusting the layout of the courses. For example, if a window is 42 inches high, and you are aiming for a 5-inch exposure, space the courses about 5⅜ inches apart to get the top course to fall where you want it to. Even a ½- or ¾-inch change won't be noticed, while a notched course will be an eyesore (at least to professional carpenters who come visiting).

You can mark the courses on the corner boards or window trim, but we often lay them out on long pieces of strapping called story poles. These can then be carried around the building to mark where courses go. Be sure to position the story pole at the same height each time you use it. Rig it to hang from the top wall plate, or butt up to the rafter overhang or soffit.

The bottom course of shingles is

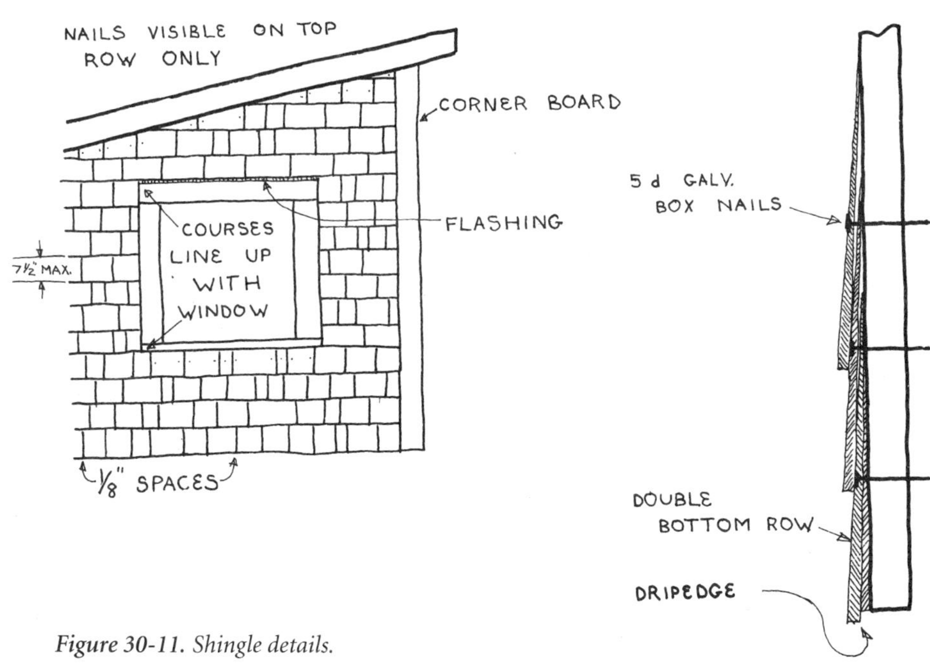

Figure 30-11. Shingle details.

doubled. The inner course can be nailed flush with the wall sheathing, and the outer one overhangs about ¼ inch to create a drip edge.

The first course of shingles is easy. Just nail them up flush with the bottom of the sheathing.

Once this course is up, shingles are easiest to position by resting them on a temporary shelf called a ledger strip, made from 1 x 3 strapping or any straight scrap lumber that is tacked below the layout line.

To provide a place to attach this ledger for the bottom, outer course, extend a few of the inner shingles down about three inches. The ledger can be screwed to these extended shingles, which can be cut off later.

Usually the shingles are nailed with shingle nails, which are 5d galvanized box nails. Use two nails per shingle, located an inch or two above where the next course will fall, and roughly ¾ or 1 inch from the edge of the shingle. Leave a space of ⅛ to ¼ inch between shingles, so they can move around when they get wet.

When the base course is down, screw your ledger to the extended shingles with 1¼-inch drywall screws. Set it ¼ inch below the sheathing to create a drip edge. Now you are ready to begin shingling in earnest.

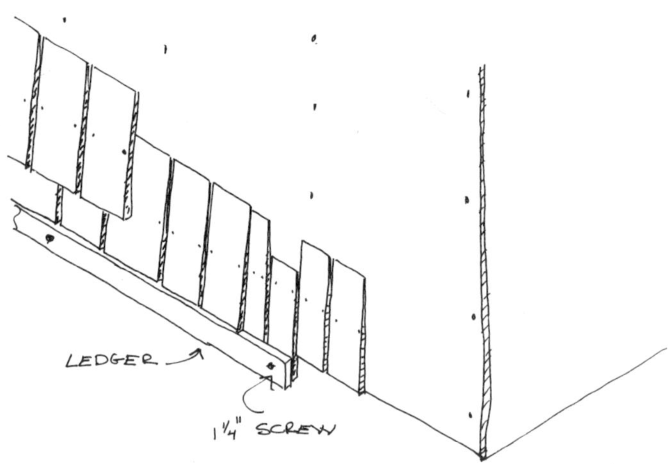

Figure 30-12. Extending a few of the shingles in the starter course gives you a place to secure a ledger strip.

It might seem that shingling is repetitive and uninteresting. But as I contemplate putting up tens of thousands of them to cover a whole house, I quickly find the prospect a most interesting ergonomic study, and I put quite a bit of thought into exactly how I'll do it.

First I try to find the best place to put my bundle of shingles, so I don't have to reach very far to grab the one I want, or lean over awkwardly.

I think about which shingles to do first in a row. You want a tight fit at the ends, where the shingles meet trim boards, so I put them up first. As you place each shingle, you are staggering the joints (figure 30-12). So I make a mental note of where the joints in the course below are located, so that when I reach for the next shingle, it will be of a width that will result in staggered joints most every time. You can adjust the crack between adjacent shingles to some extent to get proper overlap. I have found that if I observe carefully how I maneuver the shingle from pile to wall, and note exactly how I get the nail from my pouch to the wall, I can usually make a lot of little adjustments that result in increased speed or more comfort.

I also think about the best way to cut the shingles, depending on where I am. The most portable way is a simple utility knife, carried in a little scabbard made for the purpose. A sabersaw with a long extention cord is handy. A small, lightweight portable table saw is a big timesaver when a whole bunch of shingles have to be cut to the same length, as happens under windows or at the top of the wall. Different tools will work best in different areas.

At first, you'll be shingling from the ground. Using sawhorses with planks on them, or perhaps a stepladder, you can get up a few feet higher. But eventually, you will usually need a staging. Probably the most convenient staging is pump jack staging, which allows you to create a staging the whole width of the wall which you move up gradually. If you will be using pumps, I would advise getting some experienced guidance at first. Make sure you set up a railing also.

The last few courses of shingles will be cut to length, and the last course will have to be face-nailed, unless a molding of some sort is used to cover the top joint.

Shingles are sold by the bundle. For a 4-inch exposure, order five bundles per 100 square feet, which is known as a "square." For a 5-inch exposure, order four bundles per square, and for a 6-inch exposure, order three and a third bundles per square.

TEXTURED PLYWOOD

Textured plywood comes in a variety of designs. If you want to use it, see what local suppliers have to offer. It can be nailed directly to the studs with no felt paper underneath. Perhaps the main difficulty is shedding water at horizontal joints, such as over a window or door, or where two sheets meet on a tall wall. Figure 30-13 shows common ways of flashing plywood. In the case of the drip cap over windows and doors, you would install the plywood first, leaving it unnailed above each window or door opening. Then slide the flashing into place, also without nailing. Then install the window, slide the flashing tight to it, and complete the nailing.

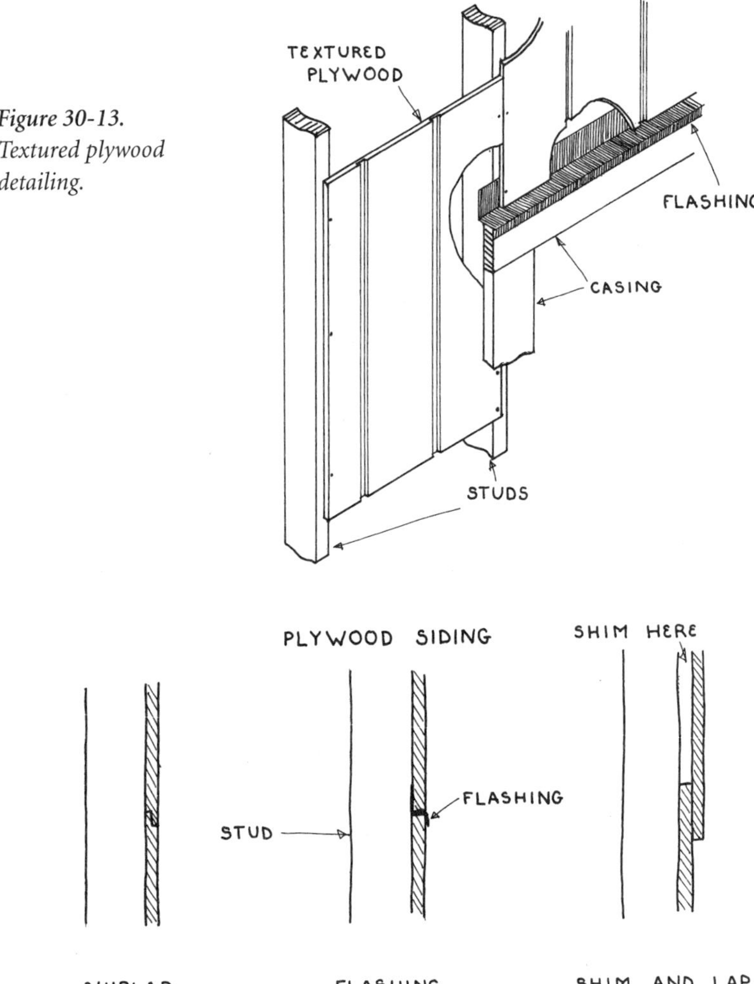

Figure 30-13. Textured plywood detailing.

FINISHING SIDING

Rough boards are often left unfinished. They weather to gray where they get wet, or to brown where exposed only to sun. Boards can also be stained. If I were going to stain them, I would be tempted to set up some sort of plastic trough so I could dip the boards before I nailed them up.

Clapboards are sometimes simply sealed with a penetrating sealer, but I think they will stand up better with a permanent finish. Heavy-bodied, opaque stains are used most often on clapboards and other smooth boards. On my house, I use latex paint, which I prefer working with. Knots have to be treated with "Bin" or "Kilz"—shellac-based sealers that prevent the knots from bleeding through the paint. Paint is more subject to peeling than stains, although heavy stains can also peel.

Shingles can be finished with a penetrating stain, or left to weather, which has the advantage of eliminating expensive maintenance in the future.

Trim boards of all kinds can be painted or stained like clapboards. It is often easiest to paint exterior trim before putting it up. If you are very conscientious, you can then paint the backs as well as the exposed surfaces. If you will be shingling, be sure to paint the trim first to avoid having to cut in the paint where raw shingles meet the trim.

The outside of wood windows and doors, unless clad, have to be very carefully painted. Keeping up this paint surface is perhaps one of the most important long-term maintenance chores.

31

INSULATION & VAPOR BARRIER

In winter, insulation slows down the rate at which the heat of the house is lost to the cooler air outside. In summer, insulation keeps the hotter outside air from heating up the inside of the house. Every material insulates to some extent, but some are far better than others. Good insulators have a maximum of dead-air spaces: small pores or cavities within the structure of the material that provide the obstacle to heat transfer. Good insulators, like warm clothes, are lightweight and porous. Wood and carpeting are effective insulators. Dense materials like glass, metal, and masonry are poor. Materials sold as commercial insulation (foam, fiberglass) are even better insulators than wood because they have many times more dead-air spaces per square inch.

Where to Insulate

Insulation should surround the heated part of the house. Usually all exterior walls will be insulated. If a cellar is unheated, the first floor will be insulated. If the cellar is heated, the first floor will be uninsulated, but the basement walls will be insulated. If an attic is unheated, the attic floor will be insulated but not the roof.

What Kind of Insulation and How Much

When you think about what kind of insulation to use, and how much, you must weigh various factors. First, the insulation value of any insulator depends on the thickness. Two inches are twice as good as one inch of the same insulation. Thus, you want to have as much thickness as is practical. This will usually be limited by how the house is being built. If your wall is four inches thick, for example, you can only fit four inches of insulation into it.

Second, some insulators are more effective than others, per inch of thickness. Two inches of one kind might be as good as four inches of another. Third, sometimes one type of insulation will be much more simple and convenient to install than another. Choose a kind of insulation that fits the way you are building your house.

Finally, think about what a particular insulation will cost for the thickness you need.

Chapter 19 gives the relative insulating value of insulations and building materials and explains how to use these figures to compute your heat loss with different designs. These things might become clearer if we look at the three types of insulation commonly used in building: fiberglass, rigid foam, and cellulose.

Fiberglass: In most cases the easiest, fastest, and most economical insulation is fiberglass blanket insulation placed between framing members. It comes in many thicknesses, and in widths to match the common 16- and 24-inch spacing of house frames. Fiberglass sometimes comes with foil or paper backing, designed to face toward the inside of the house to reflect radiant heat back into the house and serve as a vapor barrier (figure 31-1). But I recommend unfaced insulation, because it is much faster to put up. The usual rule is to put in as much as possible. Minimums might be 6 inches in the walls, and 8 inches in the roof.

Rigid foam insulation: Foam insulation is sometimes sprayed in place, but for the owner-builder the practical choice will usually be foam board in large 2 x 8 or 4 x 8 sheets. Common thicknesses are 1, 1½, and 2 inches.

Since rigid foam insulation is stiff and has some body to it, it can be used under a concrete slab, or on the outside of a foundation. Sometimes foams are used on top of a framed wall or roof which has also been stuffed with fiberglass, to achieve particularly high insulation standards. Continuous insulation bridges over the gaps in the fiberglass that occur at framing members. Foam is primarily used where its rigidity and compactness provide special design economies and advantages.

Though foams are more effective than fiberglass per inch of thickness, they are rarely used as the primary wall or roof insulation because their high cost makes them less effective per dollar. The most common exception is in timber framed houses, where foam wall panels with integral sheetrock and sheathing span over the timber frame to constitute the entire wall sandwich.

Your lumberyard or home center can outline the different types of foams available, their R values, and how they are used. Foam is also used with special building systems, particularly post-and-beam systems, where the savings in labor and lumber can outweigh the added cost of foam. If you want to use foam in your design, discuss your plans with an experienced person who can help you make sure your system will work well and can be built easily.

Cellulose: Cellulose insulation is reprocessed shredded newspaper with fire-retardent chemicals and binders added. It is usually blown in by a truck-mounted machine. It is possible to rent this equipment, but usually very economical to have it installed by a specialist.

The most common use of cellulose, which has a similar R value to fiberglass, is in renovations, where it is blown into walls through holes drilled either from inside or outside. It is also used in attics between and on top of the joists.

More recently, "wet process" cellulose has been introduced. This product has adhesives added so that it can be sprayed into open walls. This should make a tighter job than you would achieve with fiberglass. If this service is available locally, it would be worth investigating.

Vapor Barriers and Venting

Any insulated wall, floor, or roof should have a vapor barrier to prevent condensation of water inside the wall, floor, or roof. Condensation occurs when moisture-laden warm air cools. (Warm air can hold more moisture than cold air.) Your own breath on a cold morning is a good example. The warm air in your lungs is very humid. When you exhale, the air cools, and some of this moisture immediately condenses, forming the cloud of mist you see. The temperature at which this condensation occurs is called the dew point. The dew point will vary depending on the heat and humidity of a body of air.

To understand how condensation works inside a wall, we have to look at how moisture migrates through a wall. In winter, the air in your house is warm. Because of cooking, perspiration, respiration, showers, and such, it is also moist. You might wonder how moisture gets into the wall cavities and condenses, since air does not seem to pass through the wall surfaces. But in fact, moisture vapor flows through your wall. A wall surface can be permeable to moisture vapor, even though it is not permeable to either air or water as such. Moisture will migrate through most surfaces as long as the humidity on one side of the surface is higher than on the other side. This is called vapor pressure.

In winter, the moisture inside a warm house is always flowing through the wall surface into the cooler cavities inside the wall. At some point in the migration, it will reach the dew point and condensation will take place. If the condensation is great, the insulation can become soaked and the framing can start to rot. How bad the condensation will be depends on exactly how the wall is constructed. Some materials are highly permeable to moisture migration and some are almost impermeable. Relatively impermeable materials are called vapor barriers.

When you design a wall or other surface, you have two goals. The first is to make the inner surface (the warm, moist side) as impermeable as possible. This minimizes the moisture that migrates into a wall. The second goal is to make the outer surface (the sheathing and siding) fairly permeable, so that the moisture that does find its way into the wall can eventually flow to the outside. This does not mean that the sheathing and siding are drafty. They can be relatively airtight but be made of materials permeable to moisture vapor.

To keep the inside surface impermeable, staple a plastic vapor barrier behind the finish surface. In a floor, the vapor barrier will go under the flooring. In a wall or ceiling, it will go just under the wallboard or paneling material.

The next problem is to provide an escape route for the moisture that does find its way into wall, floor, or roof. Walls and floors are usually covered on the outside by some combination of wood, plywood, or housewrap. These materials are sufficiently permeable to let excess moisture out, as long as there is a vapor barrier on the inside.

Roofs are often subject to more moisture penetration since the heat in the house tends to make them hotter. Also, many roofing materials, particularly metal and asphalt roofs, are effective vapor barriers that may not let excess moisture out. Therefore, it is common practice in building to ventilate the roof. This means providing a flow of outside air between the roofing and the insulation to carry off the excess moisture. Figure 31-1 shows the basic idea. Usually there will be a screened inlet vent at the eave and an outlet at the peak or ridge of the roof. The air flows in the lower vent, rises in the cavity between the insulation and the sheathing, and exits at the top. This will not work if the entire

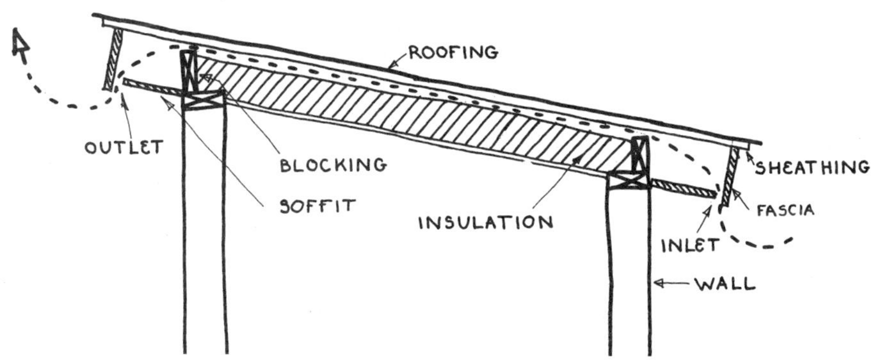

Figure 31-1. Venting a shed roof.

cavity is stuffed with insulation, because there will be no path for the air to flow along. Chapter 26 has further discussion of venting.

The Air Barrier

In recent years there has been a shift in emphasis in this area of design, as more research is done on moisture problems in building and on highly insulated building systems. I consider this a fairly controversial subject. However, the key idea seems to be that most moisture problems come not so much from moisture migration through materials, but through actual holes into the wall or roof cavity, because much more air and moisture can flow through a hole in a piece of drywall or other surface than can migrate through vapor pressure. This makes intuitive sense to me. To the extent this is true, the vapor barrier is more importantly an "air barrier," with few or no leaks.

So, proper detailing of the plastic sheeting for vapor barriers is crucial. The sheets should be large, the joints as few as possible. There should be a 6-inch minimum overlap at any unavoidable joints. We take great pains to minimize air gaps around outlets, windows, or any other unavoidable penetrations of the plastic sheeting.

But there are further measures some experts recommend, such as caulking joints between plastic sheets, or using special electrical boxes or shields for conventional boxes which permit an airtight seal.

The box on pages 322–23 shows a superinsulated wall system designed to have a continuous air barrier all the way around. Details have been worked out to make the surface continuous between floors, at the basement level, and at the transition from wall to roof. A particularly interesting feature is the 1-inch air space inside the plastic film. All wiring can be run in this space, and thus holes in the plastic are avoided.

An important point to mention is that often moisture problems in buildings aren't caused so much by errors in construction, but by an excess of moisture in the first place, coming from uncovered earth in the basement or crawlspace, or from unventilated showers, dryers, or other equipment in the house.

INSTALLING FIBERGLASS INSULATION

Fiberglass is an incredibly itchy and unpleasant material to work with. Wear gloves, a long-sleeved shirt, goggles, and some sort of respirator over your mouth. Open the windows. Order unfaced insulation; it's quicker to work with.

The walls will be insulated with 6-inch insulation made in widths to fit the space between the studs. Fiberglass insulation comes in rolls or short 4-foot batts. It is easiest to cut if you compress it by using a 6-inch board as a straightedge to cut along.

Cut notches to fit around the electric boxes. Stuff all cracks with scrap, particularly the cracks between window and door jambs and the adjacent studs. Your objective should be a tight, draft-free job.

Use the same method to insulate the roof, but use as much insulation as possible. If you have 2 x 10 rafters, you could install 8 or 9 inches of insulation. If the attic is open, you could then add another 3 or 4 inches of insulation on top of the joists, running perpendicular to the joists.

Make sure, though, that any venting path is not obstructed.

INSTALLING A PLASTIC VAPOR BARRIER

Install the vapor barrier right away, because the exposed fiberglass is very irritating.

Plastic film comes in rolls of various sizes. It is also available in 6-mil and 4-mil thicknesses. I would choose the 6-mil if available. Pick a size that will give you a minimum of joints but that will be convenient to handle. Staple the plastic on the inside of the framing after insulating is completed. Join sheets over framing members, overlapping the pieces at least 6 inches. This is easier with two people.

Lap the plastic down onto the floor at the bottom, so that the wall materials can help seal the joint there. Similarly, leave an extra lap at the ceiling. The vapor barrier at the ceiling should lap down onto the walls.

One method to reduce leakage at outlets is to put plastic over the outlet, then make an x-shaped cut in the middle of the outlet, and carefully peel the plastic back until it just pulls back over the sides of the outlet box.

There are a variety of approaches to getting a tight air barrier at the gap between doors or windows and adjacent wall frames. Stuff wider spaces with fiberglass scrap. Very narrow cracks (⅛ to ¼ inch) can be caulked. Cracks in the 1-inch range can be sealed with spray foam that comes in cans. I install the plastic vapor barrier right over the windows. Then, after the window is cased on the inside, I trim the plastic back to the casing with a sharp utility knife. When I need to open the window before doing the trim, I cut the plastic a couple of inches inside the window jamb.

32

FINISH WORK

FINISH WORK MEANS ALL THE PARTS OF THE HOUSE THAT show from the inside. It includes the floor, wall, and ceiling surfaces, and all the trim such as casings, baseboards, and moldings. Figure 32-1 shows these basic elements. Finish work represents the last and most visible stage in building. Its success depends on doing a neat, careful job. The cheapest materials, used carefully, can look terrific. The best materials put up carelessly will not. The finish-work stage is the time to slow down the pace, take pains, and work carefully.

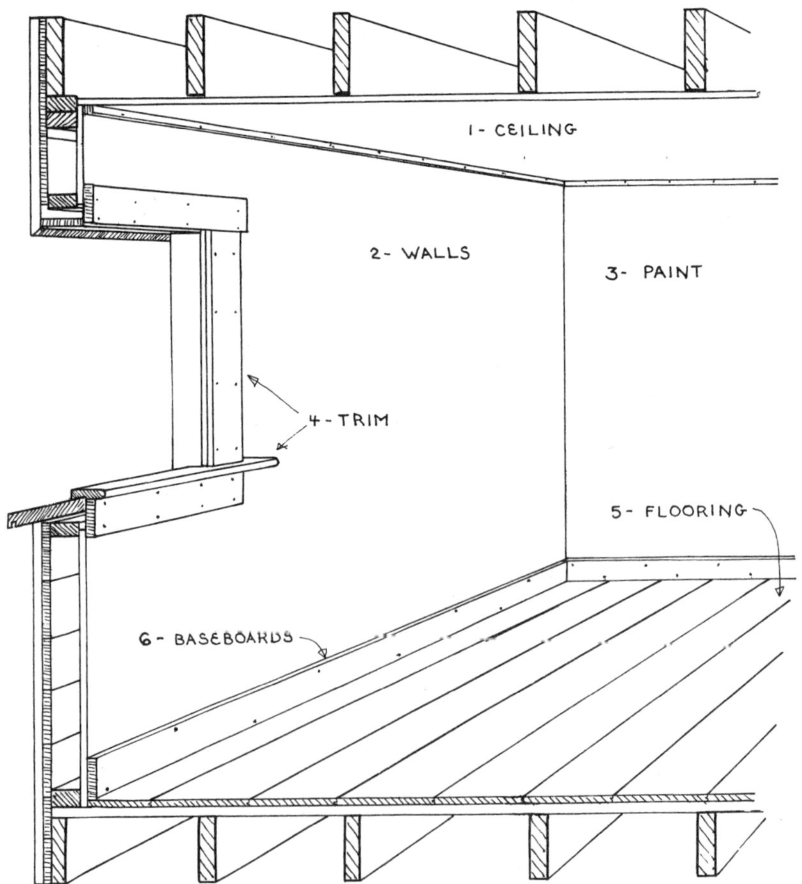

Figure 32-1. *Usual order of finish work.*

Fortunately, finish work is not quite as hard as it seems. If you look at the inside of a well-built house, everything will look as though it fits perfectly. In fact, many of the joints in finish work are covered up by window casings, baseboards, or other trim, so they do not have to fit that well. For example, the baseboard covers the joint between the wall and the ends of the flooring. As long as the flooring comes to within ½ inch or so of the wall, it will look perfect once the baseboard is installed.

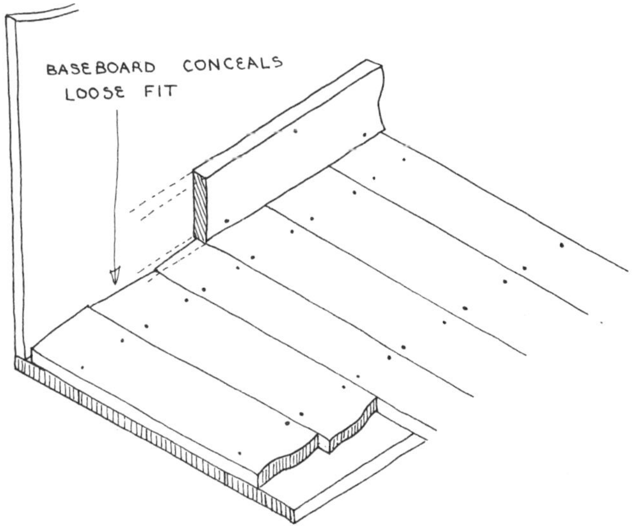

Figure 32-2. In finish work, the last pieces often cover loose fits on the joints they cover.

You could go years without window trim, baseboards, finish floors, or cabinet doors, as many owner-builders have done. But to me the woodwork is one of the fun parts. Once the basic trim is done, I can relax, put up my feet, sip my coffee, and enjoy looking over what I have done.

Figure 32-3. There is always room for more finish work.

Finish work is on-going. There is always room for more shelves or built-in benches, bands of trim, or wainscotting. Often I think a space is done, then get a feeling later that something is missing—if only a different shade of off-white.

This chapter describes the simple way finish work gets done in most houses. In a well-designed house, you might not need to do anything beyond what is described by this simple approach.

There are many trim and finish details that are complex and difficult to do. You can take classes or read entire books on how to pickle your oak finishes, or apply your wall paint with a sponge, or for that matter, make your own wall paint. There are molding and trim details that require special tools and years of study.

But there is a very wide range of effects you can get and moods you can

Figure 32-4. It looks complex, but isn't that hard to do. Iron Bridge Woodworkers. (Photo: Nancy Wasserman.)

create using the same simple methods described here. Many of these options are described and pictured in sidebars.

However much finish work you decide to do, the basic strategy is to concentrate on the joints that will ultimately be exposed. The hidden ones can be a loose or sloppy fit.

To take advantage of this principle, you must do things in the right order. If you put the baseboard down before the flooring, you would have to laboriously measure every piece of flooring and the fit would never be as good. Or if you tried to install drywall around the window trim instead of letting the trim cover the edges of the drywall, you would have a hopeless time fitting it properly.

The ideal sequence for the finish work is shown in figure 32-1.

1. Cover the ceiling.
2. Cover the walls. The top edge of the wall masks minor inaccuracies at the edges of the ceiling. The finish material, which will often be drywall, is usually fitted roughly around windows and doors, and at the floor, since these joints will be covered later.

3. Paint the ceiling and walls if they are to be painted. Since the flooring, window trim, and baseboards aren't up yet, you don't need to mask anything, and you don't need dropcloths.
4. Put on the window and door trim, hiding the rough edges of the drywall around these openings.
5. Lay the floor.
6. Finally, the baseboards are put down, covering the joints between wall and floor.

Of course, it is not always possible to follow this order, and differences in design may make your work sequence vary.

WALLS AND CEILINGS

The same materials can be used on walls and ceilings, so both are discussed together. Wall and ceiling finish of course go up after plumbing, wiring, and insulating are completed.

Drywall

"Drywall" is gypsum wallboard, often referred to as Sheetrock, which is actually a common brand of the material. It consists of a core of gypsum, a powdery white mineral, with a coating of paper on either side. It comes in ⅜-, ½-, and ⅝-inch thicknesses, and in sheets of 4 x 8, 4 x 10, 4 x 12, and 4 x 16 feet.

Usually ½-inch is used on both walls and ceilings. Joints between sheets are invisible when properly covered with paper tape and a special plaster called joint compound.

This is the least expensive surface you can use, with the possible exception of the cheapest grades of prefinished plywood wall paneling. It is a relatively quick way to put up an economical ceiling or wall.

Even though drywall is heavy and clumsy, use the largest available sheet wherever possible. This minimizes the number of joints you must finish. Covering the joints is the most time-consuming part of drywalling. Never piece together small scraps if you can help it. It may save you the $7 another sheet costs, but the added labor of taping will offset the savings many times.

The long edges of the sheets are tapered, so that the paper tape and joint compound that conceal the joints will end up about even with the surrounding surface. Orient all the sheets on a given ceiling or wall the same way, so that tapered edge meets tapered edge for best results.

Step 1

Strapping and returns. Your wall framing is on 24-inch centers. The sheets can run horizontally or vertically, whichever arrangment results in the fewest joints to tape. On ceilings, using 24-inch centers could result in some noticeable sagging. A common practice is to convert to 16-inch centers by running strapping, which is a cheap grade of 1 x 3-inch boards, on 16-inch centers, perpendicular to the framing. The strapping provides an opportunity to make the ceiling surface flatter. If there are uneven spots in the ceiling plane, you can shim beneath the strapping to make it level or flat. Secure the strapping with 8d nails or 1⅝-inch drywall screws. Run the drywall sheets perpendicular to the strapping, or the joists if you decide to omit the strapping.

Before you start to hang the drywall, make sure all insulation, vapor barriers, plumbing, and wiring are in place. Also make sure you have provided sufficient support for the edges of the sheets. Joints that are perpendicular to the framing can be unsupported because drywall will span the 16 inches or 24 inches between framing members. But joints parallel to the framing or at the ends of the sheets must be supported. Usually the sheets will be nailed or screwed right to the framing, but occasionally there will be no framing where you need it. This often occurs at gable ends or near partitions. Nail in stud scraps to provide support. These added supports are called "returns" or "nailers" (figure 32-5).

Figure 32-5. A: Drywall return (nailer) at inside corner. The clamp holds the piece firmly as it's nailed in. B: Strapping converts 24-inches-on-center to 16-inches-on-center, and allows for leveling.

A

B

Step 2

Measuring and marking. Though some sheets will go up full size, many will have to be measured and cut. If the space is rectangular, you can simply measure the space and transfer these measurements to the sheet. Longer lines for cutting can most easily be made by snapping a chalk line. If one edge of the sheet must be cut at an angle, you can find the angle by measuring all four sides of the area empirically and transferring the measurements to the sheet. As long as two corners of the area are square, you will have an accurate shape. You can also measure angles with a rafter square. Place the rafter square in the corner that isn't square. See how much the line deviates from 90 degrees at the end of the square. Then reproduce this deviation on the piece of drywall (figure 32-6).

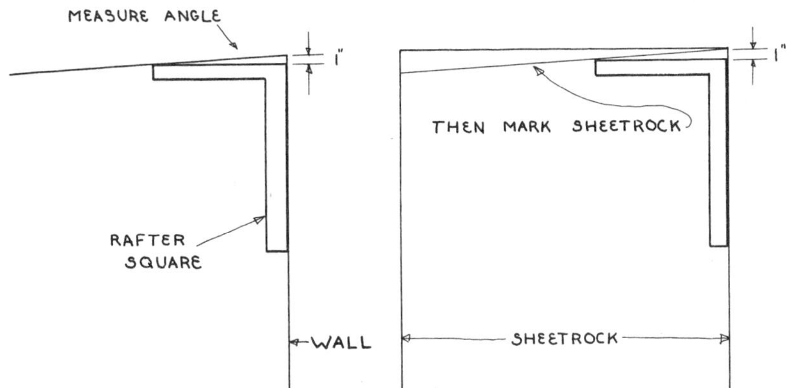

Figure 32-6.
Measuring angles.

Mark the sheets to fit about ¼ inch smaller than the space to avoid having to recut or trim the drywall. You don't want to handle the heavy sheets more times than necessary.

Locate the holes for electrical outlets by measuring up and over to the box from established lines. This can also be done empirically. Rub some carpenter's chalk along the edges of the box. When the drywall has been cut to size, hold it up in place and push it firmly against the chalked edges. The outline of the box will be marked on the back of the drywall. Whichever method you use, fit the hole fairly tight so that the box's cover plate, which goes on later, will conceal the rough edges of the hole.

Step 3

Cutting. First run along the the guide lines with a utility knife to cut the paper surface. Use a straightedge or follow the line freehand if you have a steady hand. Go clear across the sheet, even if the line you want stops short of the edge. Now snap the joint back as in figure 32-7 to break the gypsum core. Keeping the drywall folded back, cut the paper backing at the crease with the utility knife. Notches, holes for electrical boxes, and curves can be cut with a special little saw

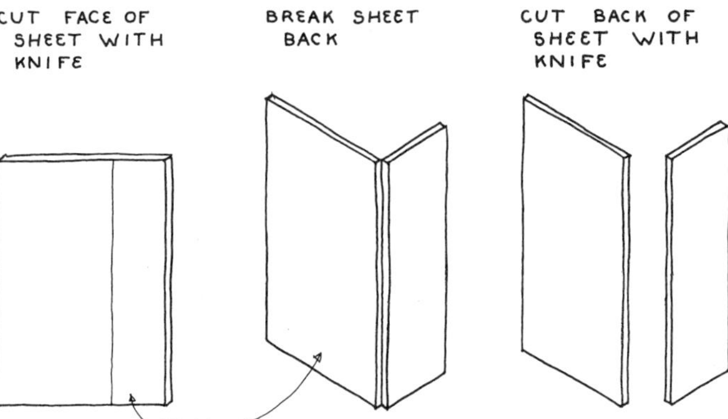

Figure 32-7. Cutting drywall.

called a "keyhole saw" made for this purpose. A keyhole saw is also handy for minor trimming. An electric sabersaw is even handier.

Since you will be doing these operations many times, tool up properly. Though they make utility knives with retractable blades, I suggest getting a fixed-blade model and a little holster to carry it in safely. Buy yourself a 4-foot-long drywall T-square. It's invaluable for measuring and cutting sheets.

Sometimes when you try to fit a piece you have cut, it will be slightly too large in spots. If a piece needs to be trimmed along an edge, you can shave it down with a rasp-type plane called a Surform, or with the utility knife.

Step 4

Hanging. As you begin hanging the drywall you will need a supply of 1¼-inch drywall screws and a couple of pounds of 1¼-inch drywall nails. The nails are for installing corner bead. Some people like to use them for initially securing the drywall in place. But for most of the fastening, the drywall screws are a tremendous convenience. A ⅜-inch variable-speed, reversing, cordless drill is a great tool for driving the screws. It's much handier than dragging a cord drill around. Get a 9-volt or larger model. Whichever drill you use, fit it with a magnetic bit-holder, and buy plenty of bits for it. They get used up quickly when putting up drywall.

You will also need stepladders, extension cords, and good lighting. In general, two people can do the walls—one in a pinch—but three people are needed for ceilings.

Space nails and screws about 8 inches apart, sinking them a little below the surface of the drywall, just enough to create a slight dimple, but not enough to break the paper surface (figure 32-8). The hole gets plastered over later.

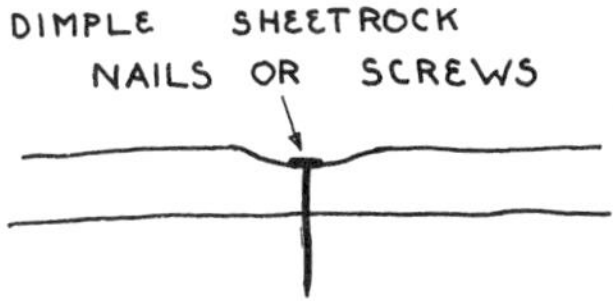

Figure 32-8.

Ceilings come first. Usually two people hold the sheet in place while a third nails or screws it into the strapping.

Spend some time imagining the easiest way to "hang the board." Depending on ceiling height, you might use just a couple of step ladders, or several strong planks supported on sawhorses. It often helps to make a "dead man," a T-shaped prop to help support the sheet while it is being secured (figure 32-9). This should be a little longer than the distance from floor to ceiling, so you can jam it between the floor and ceiling. Also find a good place to put your drill, so you can reach it when you have the heavy sheet in place. I like to load it with a screw, and set it somewhere I can reach it while I'm supporting the sheet with one hand or even my head.

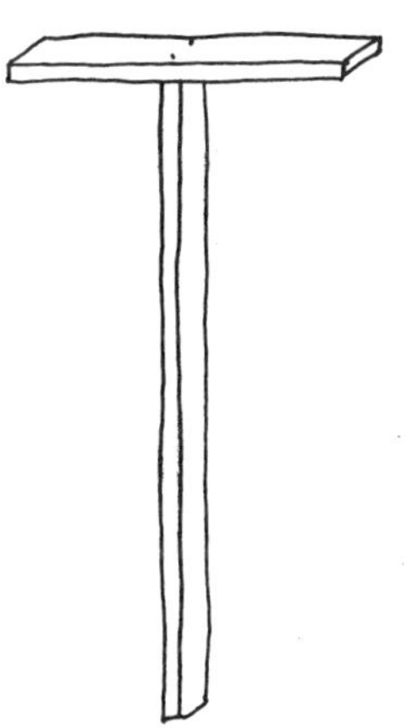

Figure 32-9. *2 x 4 prop or 'dead man' for hanging ceilings.*

Often the best strategy for hanging drywall is to put up each sheet with six or eight screws, and come back later to add the rest.

When hanging drywall on walls, a tight fit is more important at the top than at the bottom, where the drywall will be covered by baseboards. When a single piece runs from floor to ceiling, cut it a little short. Using a flat bar or other levering "see-saw" tool between the sheet and the floor as a lever, lift the piece up tight against the ceiling. This is easier as a two-person job, one lifting, one nailing. Align the edges of the sheet along the centers of studs. Overlap the stud enough so that it can be screwed in all along, but leave enough of it exposed so that the next piece can also be attached. Tack each piece with one or two screws, then examine the fit before finishing. The studs that don't come at joints will be hidden behind the sheets, so mark the stud location on the floor and ceiling in advance to find where the nails for these studs will go.

It may seem impossible to install drywall over a pipe coming through a wall. Cut the whole sheet to size, and cut the hole for the pipe. Cut a corner of the sheet off as shown in figure 32-

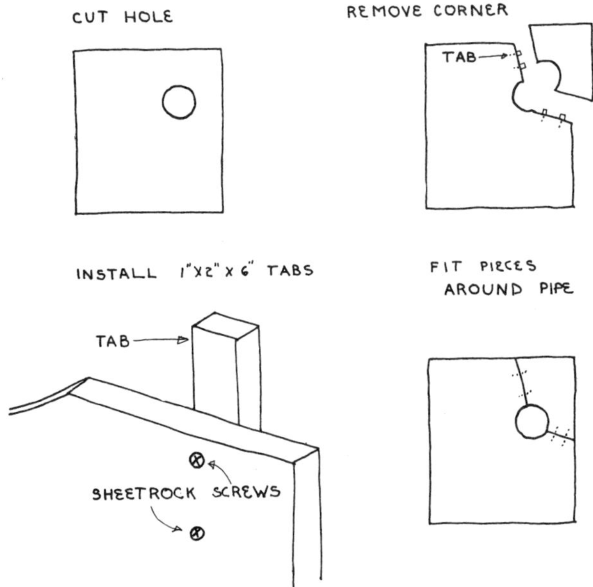

Figure 32-10. Fitting drywall around a pipe.

10. Install the larger part of the sheet in the usual way. The problem may now be that there is no framing to which to attach the second smaller piece. If so, attach some 1 x 2 scraps to the back of the large part of the sheet with drywall screws. The small part of the sheet can then be screwed to these wooden tabs.

Step 5

Installing beads. Drywall edges will usually be concealed by joint compound or trim. But in some cases the edges need to be covered by special metal moldings, or "beads," which are in turn covered with joint compound. One such situation is when two planes of drywall form an outside corner. Figure 32-11 shows an outside corner bead, sometimes known by the trade name Durabead. It is nailed over the outside corners with drywall nails approximately every 10 inches.

Where drywall butts into an exposed wooden piece and the joint will not be covered by molding or trim, a "stop bead" is used. There are two kinds of stop beads, J-bead and L-bead, both shown in figure 32-11. L-bead is attached after the drywall is up, while J-bead has to be put up first and the drywall slid into it. The joint is later covered with joint compound. L bead is much handier if it is available.

Step 6

Taping drywall joints. Taping joints, like tying shoes, is easy to show but hard to explain in words. A little expert instruction is essential. Don't be surprised if different people have different methods, because everyone has his or her own way. Here I will summarize my method to give you the general idea of how it is done.

You will need a variety of tools for taping drywall. A 6-inch knife is indispensable, and my experience is that the Hyde brand is worth the premium cost; it has just the right flex. There are several sizes of wide trowels. You may have to try out more than one to find the one you like. A corner trowel is designed to butter the joint compound onto both sides of an inside corner at once. Sometimes I find

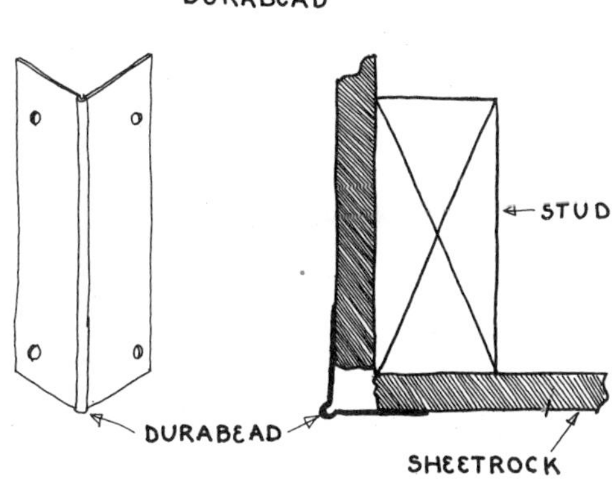

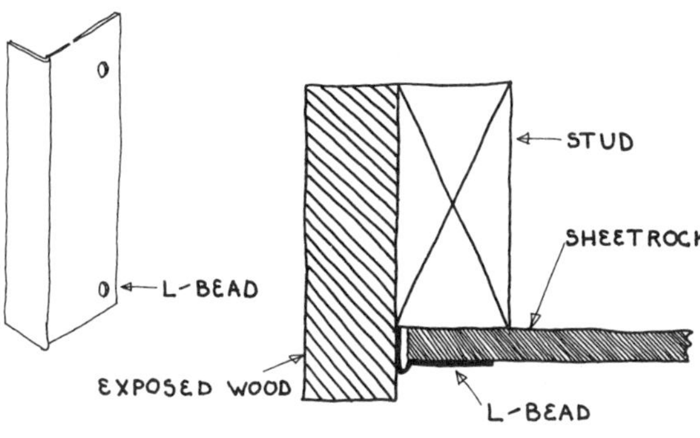

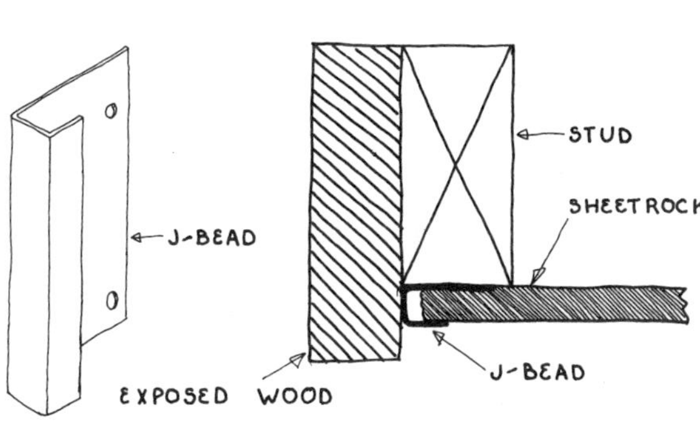

Figure 32-11. Metal beads are used at outside corners, or where drywall butts into wood.

a corner trowel more trouble than it's worth.

Amateurs often use one trowel like a pallet to hold the compound while they tool the joint compound with the other. But most professionals use a hawk, which holds a nice big glob of compound. You might want to invest in the smaller-size hawk and see if you like using it.

I use two kinds of drywall compound. The standard product is called ready-mixed joint compound. It's a smooth, soupy plaster, which dries on the wall through exposure to air. As it dries, it shrinks somewhat; several coats are necessary to get a flat wall. The whole process can be done with this one mix, and often is. With regular compound you would imbed paper joint tape in the joints.

The other type of compound is Durabond 90 (there may be other brands with similar properties) which is mixed from a powder, and dries in 90 minutes. It is much stronger and harder than regular compound, can be mixed to any consistency, and doesn't shrink as it dries. It is great for filling cracks or large holes, but it is also used for the first coat taping, the coat where the tape is bedded. It makes a stronger wall, and can speed the process because of the quick drying time. Because it is so hard, it's not a good finish coat. It's too hard to sand. Usually Durabond 90 is used in combination with a fi-

Figure 32-12. Drywall tools: hawk, wide knife, basic 6-inch knife, drywall saw for cutting holes, and two other styles of wide knife. (Keep your tools cleaner than I do mine).

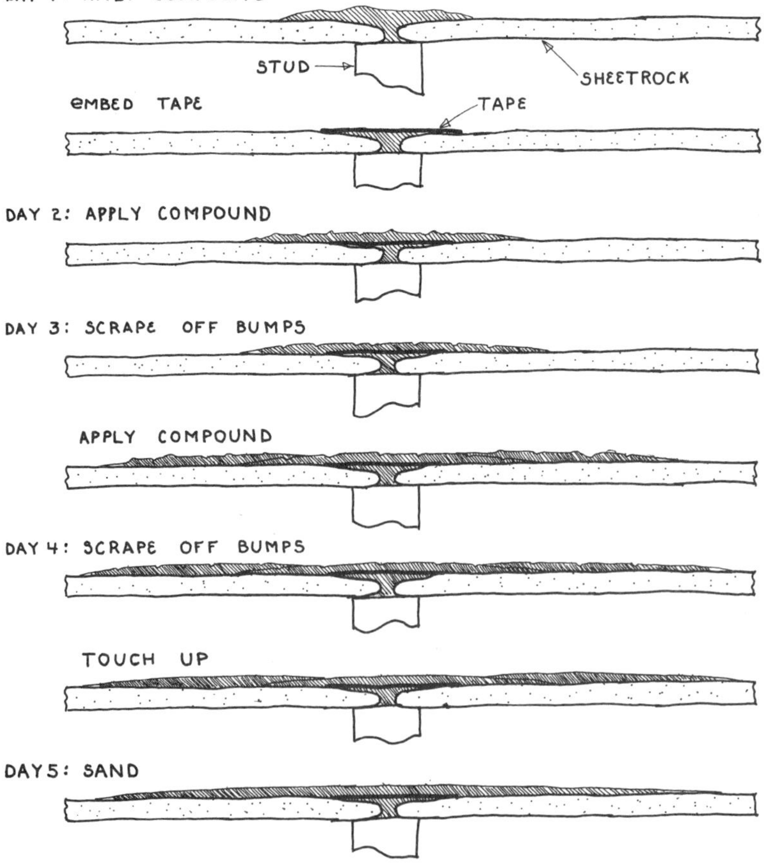

Figure 32-13. Taping drywall joints.

berglass mesh tape, but it can also be used with paper tape. In the instructions below, I describe both approaches for the first coat.

Figure 32-13 shows, in somewhat exaggerated form, how I build up a drywall joint. The basic idea is to bed tape in compound, and smooth over it with successive coats of compound. Then use sanding, scraping, or sponging to eliminate the imperfections.

Day 1: Lay a bed of drywall compound on the joint with the 6-inch trowel. Make sure this coat has no gaps, which will result in bubbles under the tape. Rip off a piece of tape the right length, and lay it along the joint over the wet compound. Use the 6-inch trowel to embed the tape firmly into the compound with one long, firm, steady, stroke. This will clean off the excess compound, which you can reuse. Let this dry overnight. Clean your tools throughly in water. A heavy bristle brush is handy for this. Don't pour the water from this cleanup down the drain.

It is important to prepare the partially used bucket of joint compound for the night. Scrape down the sides, smooth the top of the compound, and close the lid tight.

Day 1 with Durabond 90: With Durabond, the first step might be to apply fiberglass mesh tape to all the joints except the corners. This tape is self-adhesive; you just cut to length and press it in place. Then mix perhap half a bag of durabond with water. You might want to begin with a smaller batch for practice. It can be mixed in a bucket with a stick, but it's worth it to either buy or make a mixer. I take a piece of ¼-inch aluminum rod and bend it to an S curve. Mounted in a heavy-duty drill (not the cordless one), it makes an adequate mixer. You have to mix quickly (since you have only 90 minutes before the material is rock-hard), but mix it to a nice smooth consistency, smooth but not drippy.

Then coat all the joints over the tape with a relatively thin coat. The coat should be thin, because any bumps will be difficult to sand or scrape away. Clean the tools and bucket scrupulously, because the Durabond is hard to remove later. Here too, don't pour the cleaning water down the drain.

The remaining coats will be done with regular compound.

Day 2: Use the 6-inch trowel to spread a bed of compound over the tape, perhaps 6 or 8 inches wide and ⅛ to $\frac{3}{16}$ inch thick. Then use the 10-inch trowel to smooth it. Hold the trowel at a 30-degree angle, use moderate pressure, and go over the joint in one steady stroke. Do not dab at the joint, because each time you do, your trowel will leave a mark. Do your best with one or two long strokes, and then leave the joint alone. It can be scraped down later. Where you have intersecting joints, it may work best to do the verticals one day and the horizontals the next. Let this coat dry.

Day 3: Use the 6-inch trowel as a scraper to remove any bumps sticking up. Then apply another 6- to 8-inch bed of compound on either side of the centered coat using the 6-inch trowel. Smooth these beds individually with the 10-inch trowel as you did the central coat, using a little more pressure. One end of the trowel should be riding right on the drywall, and the other end should be riding down the center of the joint, which you have scraped smooth. Since this track theoretically will be quite straight, the joint should be looking quite good overall, but pockmarked with small imperfections. Let this dry overnight.

Day 4: Use the 6-inch trowel to scrape the bumps off again. Then use either the 6- or 10-inch trowel to smooth in all the holes and dips and uneven places. Put a dab of compound over the hole, then wipe over the hole with a strong squeegee stroke, holding the trowel at about 45 degrees. This should clean off the compound except where the hole is. You can touch up several times in one day, since each coat will be thin and will dry quickly.

Day 5: Keep touching up until the joint looks very flat. There are several approaches to final smoothing. Most professionals sand the wall, using a special drywall sander—basically a sanding block on a stick. This is quick and effective. Many amateurs use a damp sponge. It is also possible to simply repeat the scraping and touching-up until you are satisfied. Whatever method you use, a bright light, held at an angle, will help you see where the imperfections are.

Nail holes: The nail holes can be filled using the same method as for touchups. Put a dab of compound over the hole, and then swipe over the hole with the 6-inch trowel. Use the trowel like a squeegee, cleaning off all the compound except what is in the nail hole. Since the compound shrinks as it dries, each nail hole should be gone over three times, with drying time in between.

Corners and beaded joints: Inside corners are taped like regular joints, except that the tape is folded lengthwise to fit the corner. There is a special corner trowel that will make corners a lot easier.

If you don't like the corner trowel, you could use your 6-inch trowel, and coat one side of the corners one day, and the other side another day.

Corner bead and stop bead are covered with compound only. No tape is needed. First use the 6-inch trowel to cover the bead with a bed of compound about 4 inches wide. Then use the 6-inch trowel to clean off the excess in one long stroke, with firm pressure, holding the trowel at about 45 degrees. One side of the trowel should ride right on the drywall, and the other side should ride on the ridge of the bead. This ridge is there to give you a smooth, straight guide to slide your trowel along. These beads will need at least three coats. Scrape the bumps off between coats.

When the drywall is hung, beaded, taped, and sanded, put a primer coat of latex paint on it. You can't really tell how smooth and invisible the joints are until you look at them painted. Set up lights that approximate the lighting you will have when your house is done. If the joints are invisible under these circumstances, the drywalling is done. If some joints seem too noticeable, you can touch them up on top of the primer coat.

Molding and battens: It is not always easy to get good drywall joints when you are inexperienced. Some people seem to pick it up very quickly with a little instruction. Others find the joints taking forever to complete. One way to get around the problem is to reduce the number of joints you will tape. The first step in this direction is to put molding strips in the inside corners instead of taping them. You can do this between ceiling and wall or between two walls (figure 32-1). The molding strip can be a 1 x 1-inch strip nailed to the studs with occasional 6d finish nails.

Sometimes non-corner joints are molded too. This is called battening. For example, in a ceiling the joints perpendicular to the sheets can be taped, but the parallel joints are covered with 1 x 3-inch battens applied after the ceiling is painted. Battening can look good if it is carefully planned.

Wood Paneling

Paneling made with real boards can never be as cheap or as fast as drywall. But if you live in an area where sawmills sell local wood, paneling isn't too expensive. If you plan your work carefully, wood paneling need not be much more time-consuming than drywall, especially if you have never hung drywall before.

Figure 32-14 shows cross sections of different kinds of boards used for

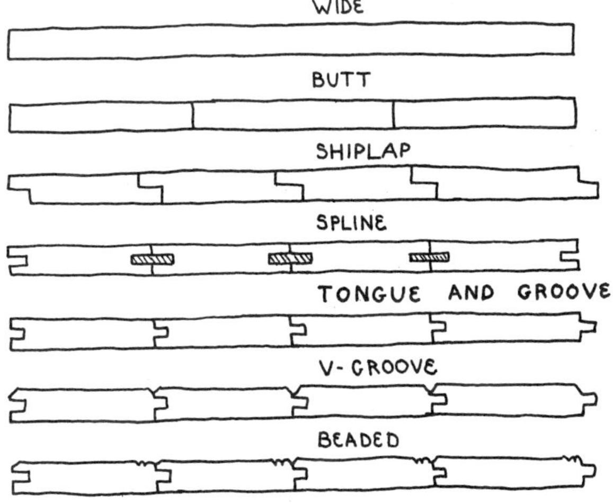

Figure 32-14. *Types of boards.*

paneling. Theoretically, wood paneling can be used on either walls or ceilings, but in practice nailing boards up to a ceiling is time-consuming and awkward. If you want a wooden ceiling, it is better to build the exposed-frame type of ceiling or roof, in which the ceiling boards are nailed on top of exposed second-floor joists or roof rafters.

Paneling can cover all the walls in a room, or just one or two; it can be vertical, horizontal, or diagonal; it can be rough or planed, sanded or unsanded, varnished, oiled, or completely unfinished. One of my favorite variations is the wainscotting shown in figure 32-15, where the lower part of a wall is covered with boards and the upper part drywalled and painted.

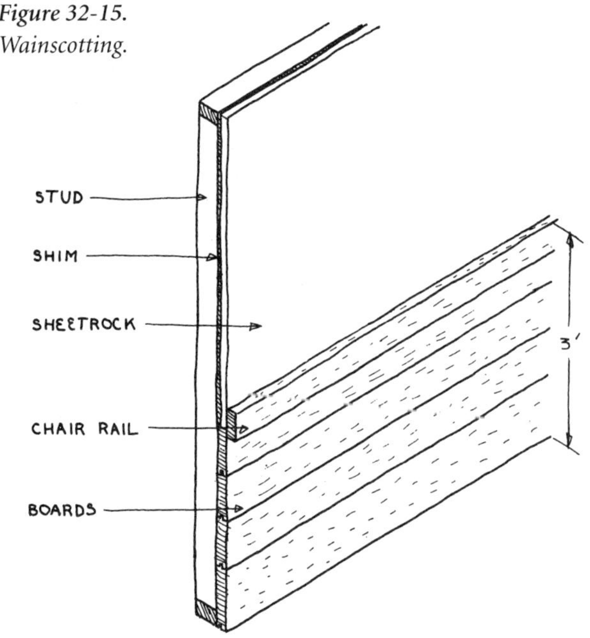

Figure 32-15. Wainscotting.

Paneling boards should be well-planed, clean, very dry, and good-looking. They will probably come from a local sawmill rather than from a lumberyard, because paneling from a lumberyard is expensive. To avoid big cracks during the winter, the boards must be very dry. This means finding a source of air-dried boards or air-drying the boards yourself. But in this case, you must dry the boards even more by stacking them in a heated place for about six weeks before use. Even supposedly kiln-dried paneling from the lumberyard may need this final treatment. Chapter 14 discusses these drying procedures in detail.

Boards can be milled for paneling in many ways. The patterns illustrated are among the most common. Shiplapping or tongue and grooving makes the wall tighter and conceals the insulation or plastic behind the boards if some shrinkage does occur in winter. Not all of the patterns shown are available everywhere. Most vendors sell only one or two types.

Step 1

Blocking. If you are putting up vertical boards, you need horizontal blocking between the studs as a nailing surface. Tongue and groove boards can be supported every 4 feet, but square-edged or shiplapped boards need support at least every 3 feet. Any blocking you do should of course be done before you do the insulation and vapor barrier. If you are putting up boards horizontally, no blocking is necessary.

STEP 2

Measuring and cutting. You will be putting the boards up one row at a time. Select the boards that will make up a row, making sure all the boards in the same row are of the same width. Lay these out, or measure them roughly, to determine which studs the joints between boards will fall on. Stagger the joints from row to row. If possible, position the boards with the heart side facing out, as in figure 32-16, for a smoother wall. Also position boards to hide splits, bad knots, or other defects.

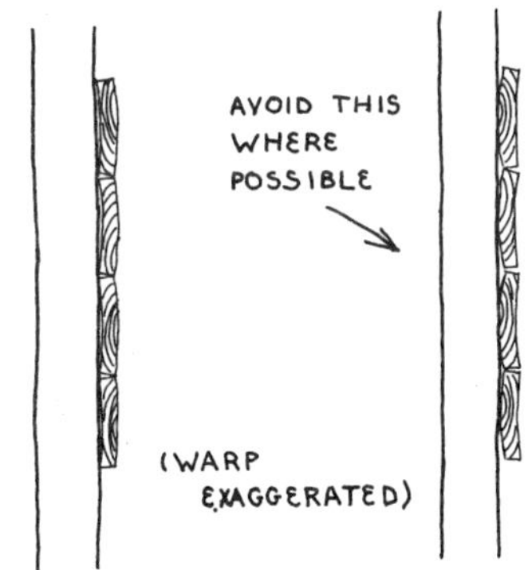

Figure 32-16. "Heart-side out."

When you have planned the row of boards and measured the pieces for cutting, you can cut them with a table saw, a radial arm saw, a chop saw, a skill saw, or a fine-toothed hand saw. See chapter 35 for cutting techniques.

STEP 3

Sanding. Usually you can avoid sanding almost completely if the boards are kept clean and have been smoothly planed by the sawmill. Use 100-grit paper on an orbital sander for any touching-up necessary. If heavier sanding is necessary, a 3 x 21-inch size belt sander with 60-grit or 80-grit would be the tool to use.

STEP 4

Nailing. If your boards are horizontal, start from the bottom and work up. Nail square-edge boards to the studs with 8d finish nails. Sink the nail heads about ⅛ inch below the surface using a nail set (figure 32-17). Install shiplap boards the same way, making sure the row above laps over the row below. Tongue and groove boards have a special nailing procedure. Face-nail the first row with the tongue up. Fit the groove on the second row over the tongue of the first row. The bottom or grooved edge of the second row will be held in place by

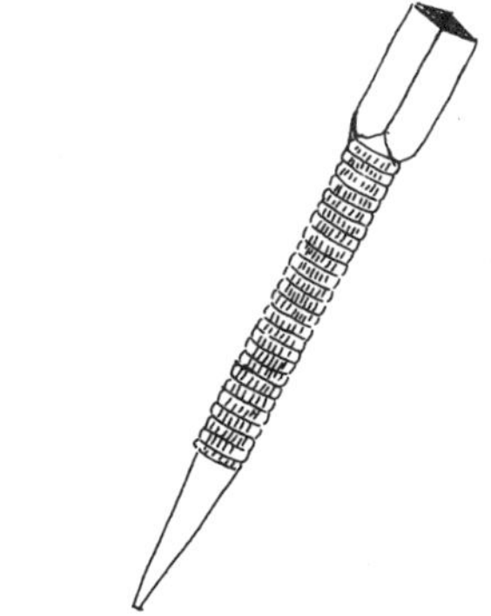

Figure 32-17. Nail set.

the tongue of the first row. Then blind-nail the top edge of the second row diagonally through the tongue, using a 6d finish nail (figure 32-19). If the nails tend to split the wood, predrill with a bit a little smaller than the nail. If a tongue and groove or shiplap board must be pounded into place, protect the edge with a scrap of the same type of board, fitted over the tongue you are pounding against. The top row of boards will have to be ripped to width and face-nailed.

Figure 32-18. Nailing tongue and groove boards.

Step 5

Finishing. Finishes are evolving quickly now as companies try to make more environmentally friendly products. So far, my experience is that a latex semigloss paint will work well if the knots are sealed with a naphtha-based stain killer. For a clear finish, an oil-based polyurethane will provide a durable finish with two coats, and a third for horizontal surfaces, which should be nice and washable. Sand between the first and second coats with 120-grit sandpaper.

The newer water-based polyurethanes are less noxious to work with, dry more quickly, and have water cleanup. They provide adequate protection for trim and paneling, though in my experience they are not rugged enough for kitchen cabinets, counters, or other surfaces likely to receive rough treatment. Three coats of water-based poly is roughly equivalent to two of solvent-based products. Some water-based polyurethanes resist ultra-violet penetration. This means that the wood below doesn't darken with time the way it usually does, or it does so more slowly.

With either type, dust or vacuum thoroughly, sand between coats, and use good light to get a decent job.

There are new non-toxic finishes, usually based on linseed oil using citrus solvents. On one recent project, all trim was finished first with Bioshield Penetrating Oil Sealer, then Livos' Meldos Hard Oil. This combination seemed to provide some real protection for trim, paneling, or for that matter, counters. The Hard Oil by itself looked good, but did not protect as well.

These products come from the *Nature's Choice Catalog* (800/621-2591). Real Goods Trading Company (800/762-7325) specializes in environmentally friendly products, including finishes. Another good source of information is Terre Verde, in New York City (212/925-4533).

WINDOW AND DOOR TRIM

Figure 32-20 shows basic trim pieces. Casings go on the sides and top of a window or door. They nail to the jamb on one side and to the stud on the other. They seal the crack around the window and conceal the edge of the finish-wall material. Casings are the same for both doors and windows, except windows also have a stool and an apron on the bottom. The stool is the protruding piece that hooks over the outside sill, and the apron is the piece underneath that supports it. The stool serves as a window stop and as a support for the casings. The apron reinforces the stool and covers the crack under the window. An alternative plan is to eliminate the stool and put casings on all four sides.

The width of trim is arbitrary, as long as the pieces cover the cracks and can be nailed through into the studs. You could use scraps to mock up the trim in different widths to find the size that looks good to you. One-by-four is often used, but I often like a wider piece, and though the top often apes the sides,

Figure 32-19. *Trim options: Simple techniques can be used to create many nice trim effects. In these examples, different thicknesses of wood, moldings, edge details such as quarter-rounding, and fiddling with the size and location of the elements creates varied looks.*

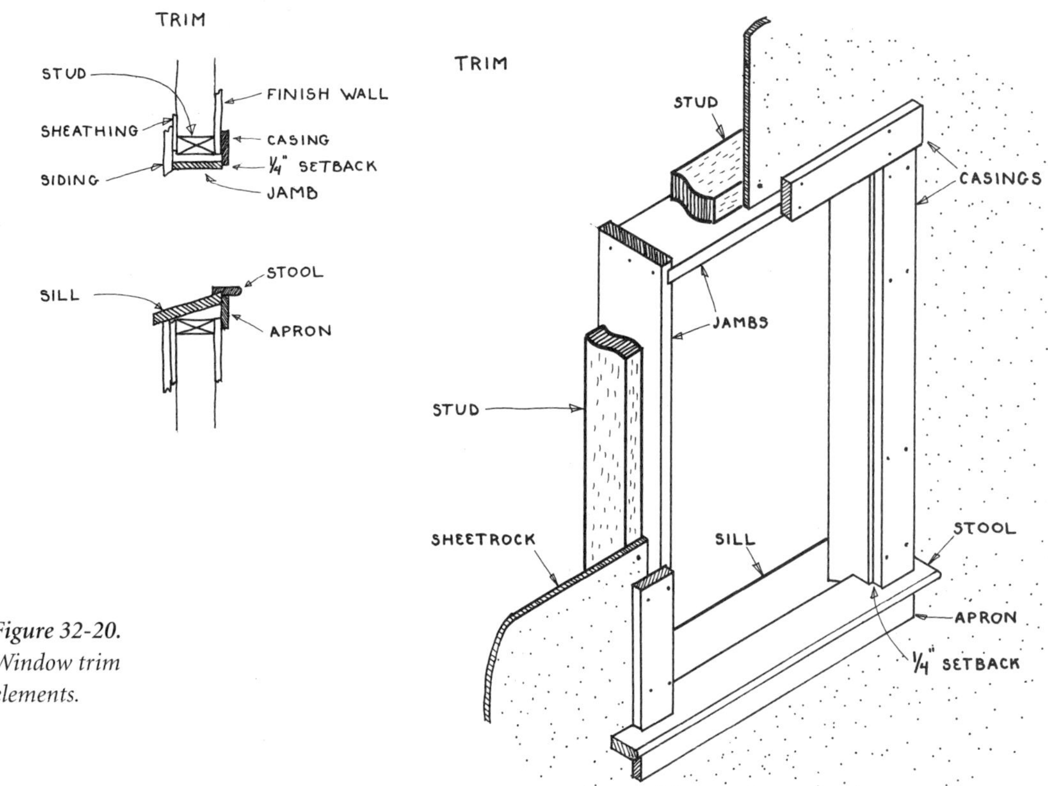

Figure 32-20. Window trim elements.

there is no practical reason it must be the same size. In short, you could just put up 1 x 4s for simplicity, but it is also an opportunity to do something different.

Before beginning, make sure the cracks around the window or door are stuffed with fiberglass, caulked, or sealed with spray foam (which comes in cans). Sometimes I let the plastic vapor barrier close off this space, and trim it with a knife after the window or door trim is up. That creates a nice tight seal.

Step 1

Jamb extenders. Theoretically, the jambs come exactly flush with the inside finish-wall surface, or perhaps stick into the room 1⁄16 inch. The casing will span the gap between these two flush surfaces. But it sometimes happens that they will not be flush. If the jambs stick out too far into the room, they can be planed down to size with a sharp hand plane. However, it is more likely that the jambs will not stick out far enough, because the window was originally made for a thinner wall, or because of errors in building. If so, jamb extenders should be nailed on with finish nails. For example, if the jambs are ¼ inch shy, nail on

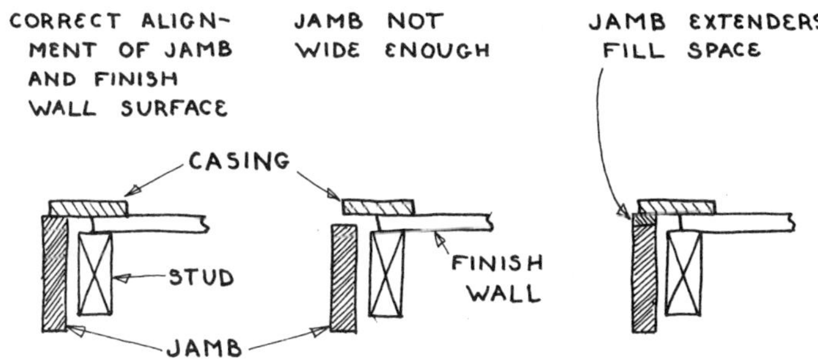

Figure 32-21. Jamb extenders.

¼-inch strips to make up the difference. If the discrepancy varies, set the jamb extenders to the widest dimension.

Step 2

Stool. Figure 32-20 shows a typical stool, which is the first piece of window trim to be put up. Carefully cut and notch the stool to fit the inside of the sill. It should stick past the casings about ½-inch on either side, or perhaps ¾-inch. The inmost edge can be rounded off with a hand plane or a router, and then the whole piece can be sanded smooth. Nail the stool to the windowsill with 4d or 6d finish nails.

Step 3

Casings. The casings can be ¾ or ⅞ inches thick, and 3 to 6 inches wide. Usually there is a ¼-inch "reveal" on the jambs, which means that both top and side casings are set back ¼ inch from the inner edge of the jamb. This hides any irregularity and provides a space for hinge pins if there is sash or screen opening in. Therefore, the length of the side casings will be the distance from the stool to the bottom of the top jamb, plus ¼ inch. Cut the two side casings to this length, as square as possible. Using one 4d finish nail, temporarily tack the side casings to the jambs to hold them in position ¼ inch back from the edge of the jamb.

Often the top casing will overhang the side casings by ¼ to ¾ inch. Measure the length you want your top casing to be, and cut it. Set the piece in place to see how it fits. Perhaps it will fit perfectly, but if it doesn't, trim it with a very sharp block plane. You can plane either the tops of the side casings, or the bottom edge of the top casing.

When the fit is perfect, nail the side casings permanently. Nail to the jamb with 6d finish nails and to the stud with 8d finish nails. Make the nailing pattern neat, and try not to mash the wood with the head of the hammer.

For now, just nail the head casing in with one 4d finish nail in the middle. If you nail it off now, it will shrink just a hair, even if it has been carefully dried. You will lose your good fit between side and top casings. Let it sit there until the room has been heated for a week or so, then nail it down as you did the others.

Step 4

Apron. Line the apron up with the casings. Push it up as you nail it so it will give good support to the stool. Recess all nails about ⅛ inch with a 2⁄32 inch nail set.

Step 5

Sand. Sand roughness and dirt from the casing with 100-grit sandpaper. Round off the sharp edges. These little touchups make a lot of difference in the appearance of the work.

Step 6

Remove accidental hammer marks. If you have left an occasional hammer mark on the wood, take a sewing needle and puncture the spot about twenty times and wet it. The water will seep in, swell the wood, and presto! With a little light sanding after it dries, the mark will (usually) disappear completely.

Step 7

Finish. Get at least one coat of finish on the trim soon. If you leave it for weeks, the wood gets dirty and has to be resanded in places.

WOOD FLOORING

Many kinds of boards are used for flooring. Narrow 1 x 3-inch T&G hardwood flooring comes in oak, birch, and other woods, and is durable, beautiful, and expensive. Softwood boards can also be used, either T&G or square-edge. They will be less expensive and more subject to wear but beautiful in their own way. Pine boards 14 to 20 inches wide are available from some sawmills and make a very nice floor. You will have to find out what is available in your area.

Whatever boards you use should be well-planed, straight, and very dry. If you are using native boards from a sawmill, chapter 7 explains how to dry the boards well enough to avoid or at least minimize cracks.

Commercial T&G flooring is usually "blind-nailed" through the tongue with a special nailing machine. But if you are using square-edged boards of any kind, they will be nailed or screwed from the top. While you could use unobtrusive finish nails, it is better to use nails with some holding power, which means nails with some sort of head. It would make sense to visit your lumberyard to see

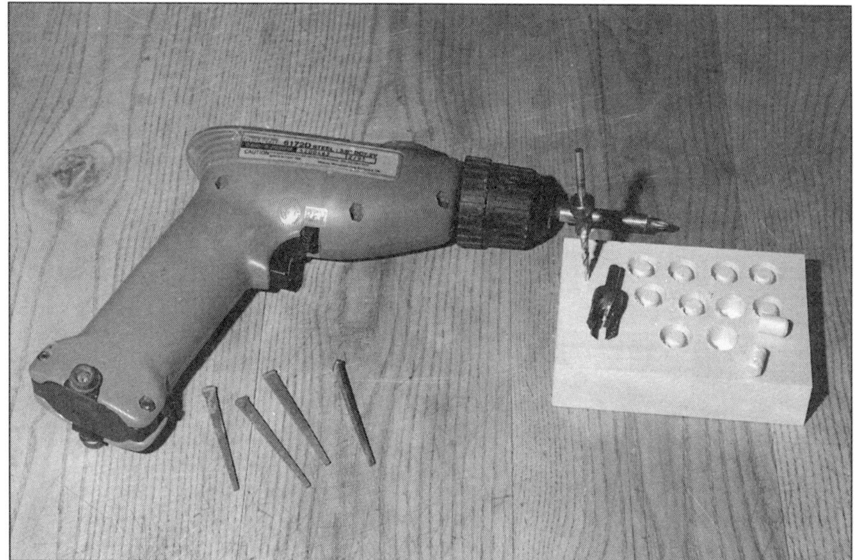

Figure 32-22. Cut nail, drywall screws, pilot bit, plug cutter, and wood plugs.

what's available. It is also possible to use drywall screws covered with wood plugs. My preference is to use cut nails (figure 32-22), supplemented by a few plugged screws where the floor doesn't seem to sit down firmly. Those are the instructions I will provide here.

Begin by stapling a layer of 15-pound felt paper down on the subfloor to reduce drafts and prevent squeaks.

Pick a nice straight board for your first row. Cut it to length, but if possible cut it short by perhaps ¼ or ½ inch if a baseboard will cover the joint. With this and subsequent rows, put the heart side of the board up if it's in good shape. With many inexpensive woods, you will be trying to put the prettiest pieces where they will show the most, and the poorer pieces where they won't be noticed, like under the table or fridge. Buy extra wood so you can cull the sections of boards you don't want to be looking at for the next few decades.

Where possible, align the nailing with the framing below. Devise a neat nailing pattern. As you lay the floor, you will get a sense of how well your system is holding, and if more fastening is needed.

The cut nails go perpendicular to the grain to minimize splitting. Place them in from the boards' end, not right at the end, for the same reason. Sometimes a particular part of a board may not sit down, or a row of nails won't hold it down. Take a #8 pilot bit with a ⅜-inch countersink, and drill a pilot hole where a screw is needed, leaving about a ¼-inch countersink. Install a 2-inch drywall screw.

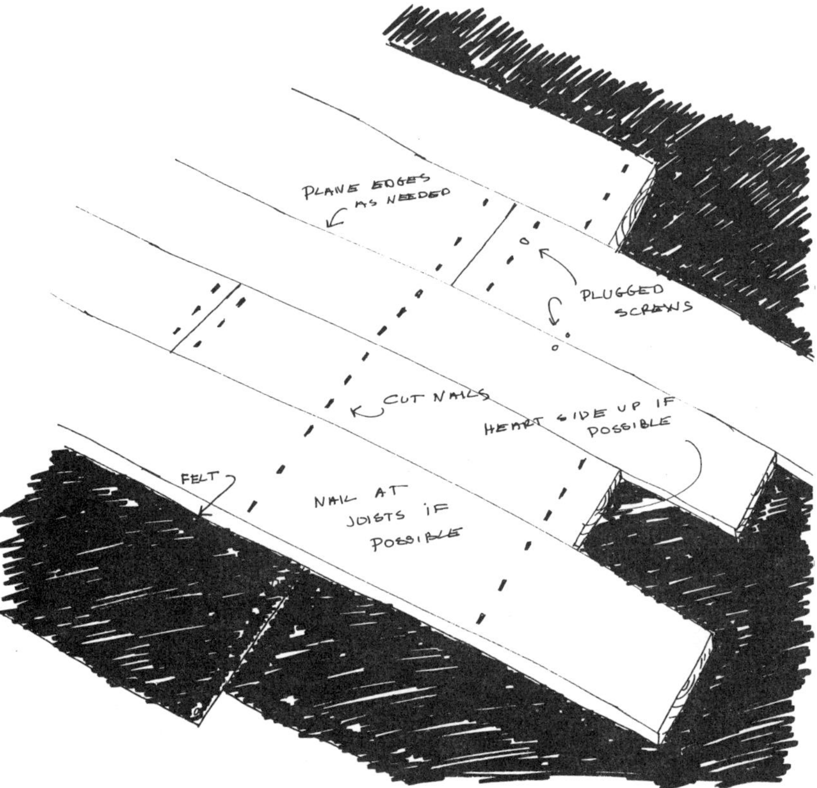

Figure 32-23. Laying wide boards.

Later you can come back and plug these holes with plugs made with the plug cutter. You will need access to a drill press to make up a batch of these plugs. The basic idea is to glue in the plug with yellow carpenter's glue, then sand it flush.

Get as tight a fit as you can. With boards up to 1 x 10, you can usually force them into compliance with a chisel, as shown in figure 24-14. Add a scrap between the chisel and the board to avoid denting it if you are prying with a chisel. With wider boards, you may have to use a hand plane to trim a board to fit.

With clean pine boards, sometimes you can simply touch-up with hand sanding, or with a small portable belt sander. Often, however, it is worthwhile to rent a power floor sander and sand the whole floor with medium and fine paper. These machines are big and unwieldy. Go with the grain, and move the sander very smoothly. Never let the sander stop in one place for even a second; it will gouge the floor. A special edger is used around the edges of the room. If possible, get an experienced person to show you how to run these machines.

It is hard to make a clear recommendation here about floor finishing. As I mention above, I lean toward a good brand of solvent-based satin polyurethane finish, with at least three coats. There are also commercial-grade water-based finishes. I haven't found a regular water-based product that is tough enough for a floor, although there are very rugged commercial water-based products available.

Floor finishes can be applied with a large brush, or a special applicator, which is more like a mop. In either case, the essential point is to vacuum the floor scrupulously before beginning. Between coats, sand lightly by hand with fine sandpaper, and vacuum again.

Any hard varnish-like finish wears. My experience is that in low-traffic rooms such as bedrooms, a polyurethane finish will last many years. In a kitchen or entry area, they wear through quickly. With softwood flooring, some spots will wear through in a year or two. Any dirt or sand that tracks in becomes an abrasive to speed this process. One solution is to lightly sand and recoat these much-abused floors every two to four years.

Another approach is to use one of the oil finishes sold for floors. Some of these are marketed as environmentally safe. An oil finish will usually be quick to apply, but less resistant to staining than a non-porous finish. The floor can be re-oiled every year or two with no sanding.

With either style of finish, some people simply have everyone take their shoes off at the door. This is probably the only measure that will really keep those floors like new. My approach, though, is to use the oil-based polyurethane, and then let the wear occur as it will.

Other Kinds of Flooring

Carpeting can be applied right over a ¾-inch T&G plywood subfloor, or 2 x 6 decking. You could probably carpet over ¾-inch T&G boards over 16-inches-on-center framing. Linoleum, sheet vinyl, vinyl tile, or ceramic tile will last longer over a very stable, stiff, and smooth subfloor, normally in two layers. We usually use ¾-inch T&G plywood as the bottom layer, although ¾-inch boards will also work. The top layer would usually be a ¼ or ⅜-inch plywood underlayment designed to go under tile or vinyl. With 24-inch-on-center framing, I would suggest ⅜-inch plywood for added stiffness.

As I write this, I realize that such floors raise many environmental questions. Most if not all of the underlayment that works best is sold as lauan. It is made from a variety of hardwoods imported from rainforests in Asia, Central America, or South America. Vinyl flooring or carpeting will be at least partly petrochemical in origin. Ceramic tile may come from the earth, but its manufacture certainly embodies a lot of energy. And all these materials are likely to have been transported great distances.

If you want these types of floors, you could end up doing quite a bit of research. One approach would be to revert to technologies that were common fifty years ago. Instead of vinyl tile, use linoleum, which is a linseed-oil-based product. Instead of putting it over plywood, use narrow T&G boards, perhaps with a layer of felt paper under the linoleum. In the course of my remodeling jobs, I've removed many old floors built this way, which lasted many years. A ceramic floor might be a "mud job"—bedded in mortar, using locally produced, low-fire tiles. Carpet might be second-hand, or a natural fiber carpet of some sort.

Your Natural Home, by Janet Marinelli and Paul Bierman-Lytle (Little, Brown, Boston, 1995), has a good discussion of toxics in the home and gives sources for non-toxic materials such as traditional linoleum.

BASEBOARDS

Baseboards are often made from ¾-inch boards 3½ inches or wider. Install them after the trim is done, the walls are painted, and the floor is finished. Start at one corner and work around the room.

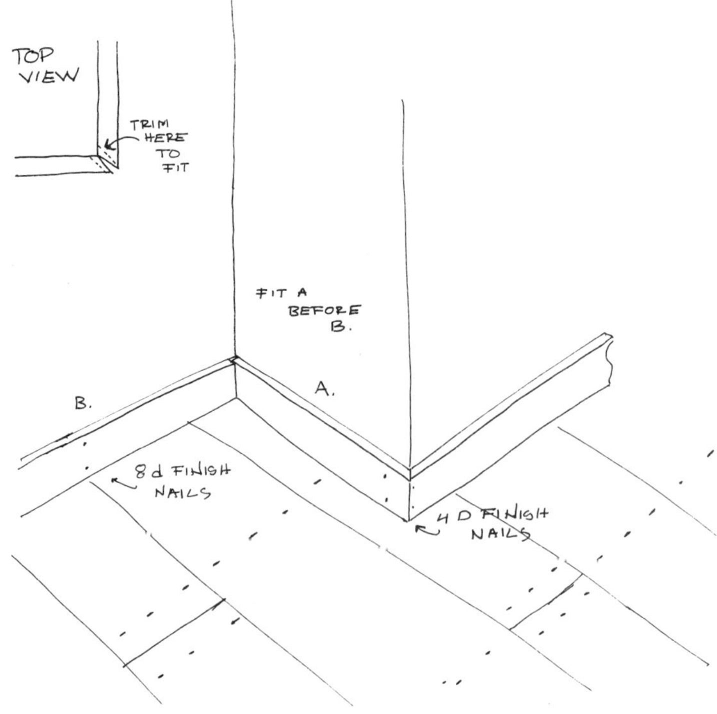

Figure 32-24. Fitting outside corners of baseboards. Putting piece A in before B allows you trim the outside corner as needed for a good fit.

Nail them to the wall studs with 8d finish nails, and countersink the nails with your nail set.

The baseboards can just be butted at outside corners, but a nice miter joint looks better. A table saw or a "chop saw" can make a nice 45-degree cut. However, you will find that these joints will not fit perfectly the first time. To get really good fits, you will have to have a very sharp block plane for trimming the ends.

FINISH CARPENTRY TECHNIQUES

The techniques of doing window and door trim, baseboards, paneling, and other visible woodwork belong to what is known as finish carpentry. Finish carpentry techniques are, in a way, techniques of deception. When you look at a room that has been well finished, the woodwork will look as if every cut is square, every board straight, and every cut carefully made. But in reality, few houses are really quite square or level, and few boards are quite straight, square, or flat. The art of finish carpentry consists first of knowing how to fit things tight so that minor variations are not noticeable; second, of planning work to minimize the number of cuts that need to be perfect; and third, of concealing building faults through visual deception. Finish carpentry is a whole art or trade in itself, and it is not possible to provide a complete description here. I will just list some practical hints that I have found useful:

Plan your sequence to minimize tight joints. This is the idea behind a molding. A single piece put on later covers several loose joints made earlier. The crack between the flooring and the wall is left sloppy; then a baseboard covers this joint to make it invisible.

Avoid flush edges where possible. You can keep from having to make edges line up perfectly by using a setback, or "reveal." In figure 32-20, the casing is set back on the jamb ¼ inch to leave a ¼-inch reveal. This looks good and leaves a place for hinges; it also saves a lot of work. If the casing and jamb were flush, the slightest discrepancy between them would be noticeable. You would have to fit them very laboriously to make the joint look right. With the reveal, even a 1⁄16-inch error will go unnoticed.

Figure 32-25 shows another example. Top window casings can be made flush with the side casings. But if they are, the slightest error in cutting will mean a sloppy job or a wasted piece, because even 1⁄16-inch errors will show. If the top casings are allowed to overhang the side ones ¼ inch or so, such cutting errors become completely unnoticeable.

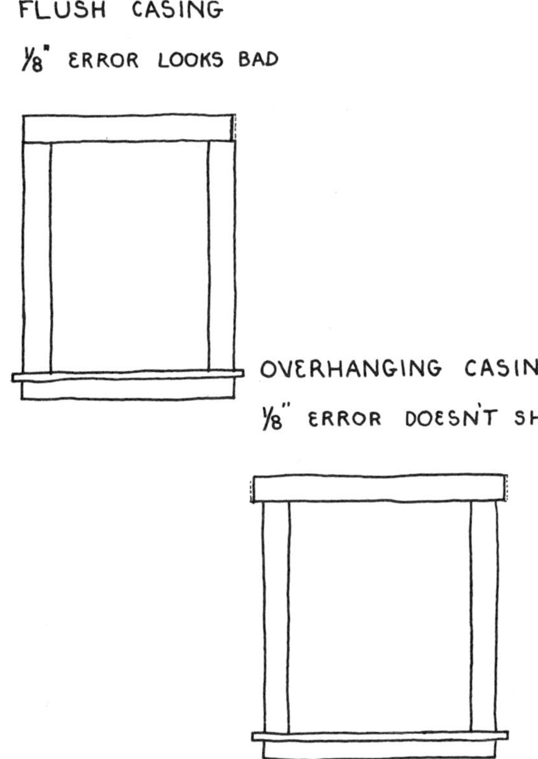

Figure 32-25. Design your trim elements to minimize the number of pieces which have to fit exactly.

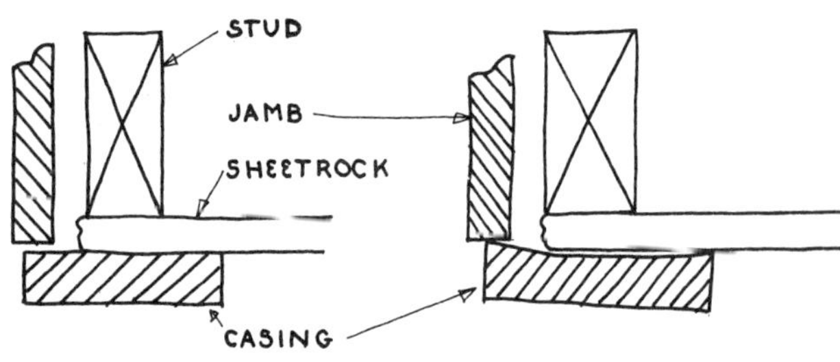

Figure 32-26. Hogging out the back of casing conceals misalignment of jamb.

Minimize double-ended cuts. It is much easier to cut a piece to fit perfectly at one end only than to cut it to fit perfectly at both ends. If you think ahead two or three steps, you can minimize these more difficult tasks. In figure 32-24, I would do baseboard A before baseboard B, because if I had to do some trimming to get the miter joint to fit, baseboard B would cover my tracks.

Don't be fussy where you don't have to be. Only the exposed parts of a piece need to look finished. Those that will be hidden can be done roughly and quickly. Often in an old house, a piece of trim that looks immaculate on its exposed side will, when removed, prove to be a mass of hatchet work or rough chiseling on the back where it had to fit around an obstruction, such as a nail head or a lump of plaster. Casings or baseboards can be scooped out roughly on the back side so that their edges will fit tightly against the walls or jambs. This is especially useful when a jamb is not quite flush with a wall surface (figure 32-26). This "hogging out" used to be done with a hatchet, but is now more easily done with a table saw, skill saw, or router.

Keep discrepancies far apart. If two lines are askew, the closer they are to each other the more crooked they will look. If you have a sloping floor but a level chair rail above it, the farther up the chair rail is on the wall, the harder it will be to tell that the two are out of line. If you run the boards parallel to the bottom, they will look crooked in relation to the top. But if as you work up you taper each board slightly, the discrepancy will be averaged out over many boards. No two adjacent lines will be noticeably out of parallel, and the overall crookedness will not be seen. This is called splitting the difference.

NOTES ON PURCHASING MATERIALS.

The rule of thumb for ordering T&G boards is: The area to be covered plus one-third equals the number of board feet needed.

For T&G wall paneling, order two pounds of nails per 100 feet of surface.

Order one pound of drywall screws for every six sheets of 4 x 8 drywall.

33

STAIRS

Building a staircase is difficult; the geometry is hard to visualize, and the carpentry must be precise. But the staircase has a large impact on how a house looks. A well-thought-out and well-made staircase makes the whole house more interesting and beautiful. A main reason many older houses are so appealing is that the stairs are more than just a way to get to the second floor.

PLANNING

Locating the Stairs

Plan the geometry and location of the stairs when making floor plans and elevation drawings. Find a general location on your first- and second-floor plans that fits the overall plan. That means the stairs will start and end in locations that fit with the traffic pattern you want. People should be able to get up and down stairs without disrupting parts of the house that are supposed to be private or quiet. Make the staircase an integral part of the house plan. Often in a big, old house an elaborate staircase will be in a central hall, right inside the front door (figure 33-1). When you enter you can go directly to any area of the house. Such a staircase ties the house together visually, while preserving the privacy a family needs. The same layout with the stairs moved would be less practical and more disruptive.

Figure 33-1.
Stair locations.

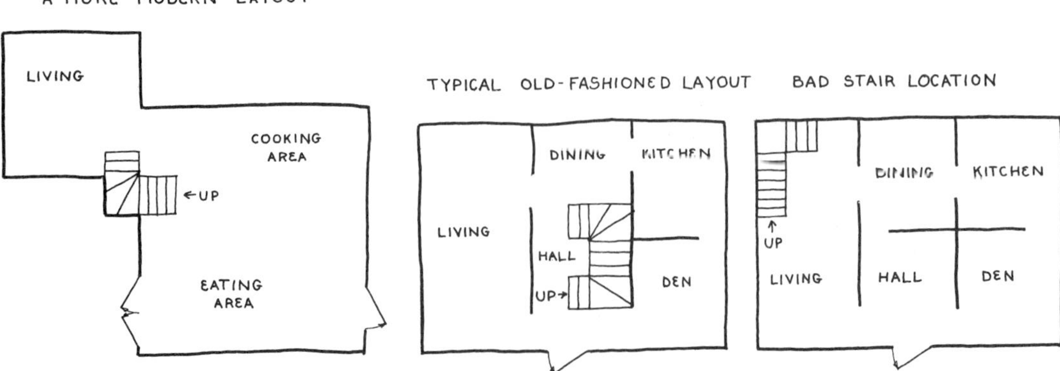

Figure 33-2. *Dudley house. Built by Breslaw, Clark, and Duberstein.*

The staircase in a house you build can be in harmony with the design you have. In the "more modern layout" in figure 33-1, the large kitchen is the center of activity for the family. There is no hall, but traffic flows through the large kitchen along paths that divide the eating and cooking areas.

Chapters 2 and 4 will help you find the best place for your stairs.

Layout

When you know roughly where the stairs will go, figure out a tentative layout. Figure 33-3 shows the most common stair layouts. A straight staircase is the most economical and easiest to build. Build a straight staircase if possible if you have never built a staircase before. But often, particularly in a small house, the floor plan will not accommodate a straight layout. You may have to build L-shaped or U-shaped stairs with a landing. This is like building two staircases that meet at a platform. Sometimes even a landing staircase will not fit. The landing will have to be replaced with two or more triangular winding stairs, called winders, to compress the staircase. A winding staircase is harder to build and more dangerous to use.

The layout you choose has to fit in with the floor framing as well as the floor plan, because you have to make a very large hole in the second floor for the staircase. If the hole is parallel with the floor joists, it is fairly easy to leave one

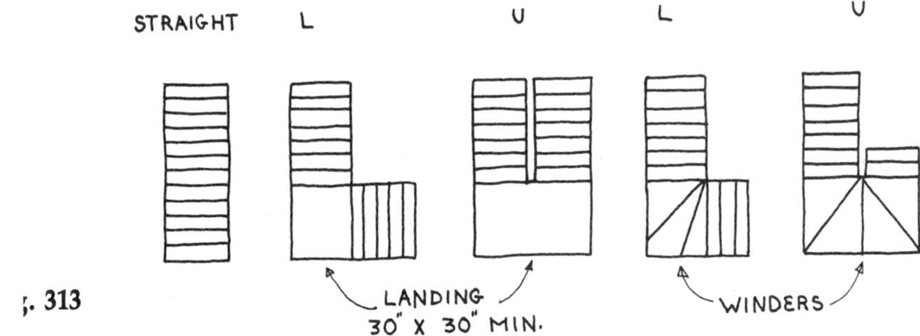

Figure 33-3. *Most common stair layouts.*

or two joists out to make room for the stairs. But if the staircase is perpendicular to the floor joists, you might have to cut six or eight joists off in midspan to make the hole. Try to minimize the disruption of the framing. (Chapter 24 shows how stairwells are framed.)

Scale drawings: Once you have a tentative idea of the general shape and location of the stairs, make scale floor plans for elevations or cross sections. Show each step in scale so that you can make sure everything will work.

Staircase angles, like rafter angles, are not expressed in degrees, but in rise and run. In figure 33-4, the distance from the finish floor downstairs to the finish floor upstairs is 8 feet. This is the total rise. The floor area used up by the stairs

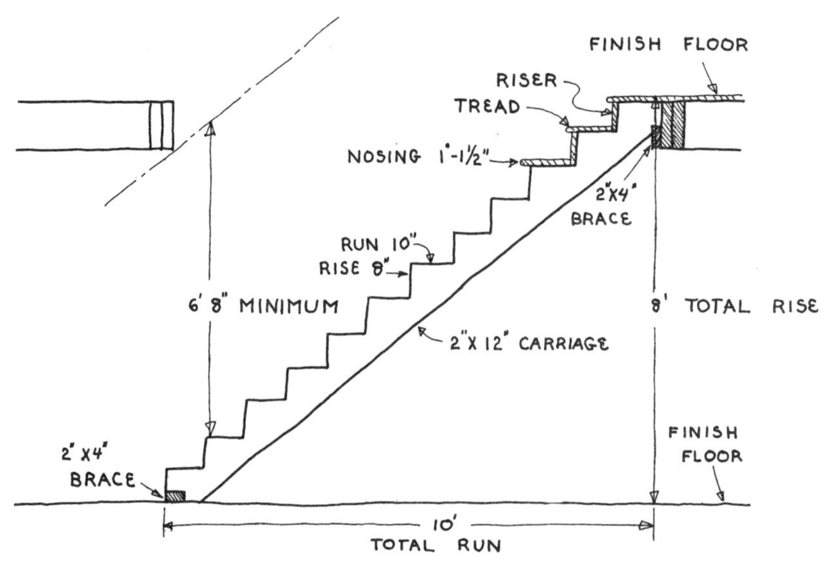

Figure 33-4. *Elements of a staircase.*

is 10 feet. This is the total run. So this is a staircase with a pitch of 8:10. Each step is a triangle that is a small version of this large triangle. The rise of each step is 8 inches, and the run is 10 inches. The rise of each step, times the number of steps, gives the total rise. Similarly, the run of each step times the number of steps (including the top step) gives the total run.

If the distance between floors in figure 33-4 were greater, the rise and run of the individual steps might stay the same but the number of steps would be greater. The numbers don't usually come out so even.

Certain step sizes have been found to be safe, comfortable, and gradual enough. One rule of thumb is that the rise plus the run should be a number between 17 and 18. Aim for a rise of 7 or 8 inches and a run of 9 to 11 inches. This will yield a fairly gentle pitch, a good wide tread, and a reasonable step height.

Start with the total rise, which is a given. Find a figure between 7 and 8 inches that divides evenly into the total rise. In figure 33-5, 8 inches goes into 8 feet exactly 12 times. This would give 12 steps, each with a rise of 8 inches. Suppose your total rise was 126 inches:

126 divided by 15 equals 8.4 inches
126 divided by 16 equals 7.8 inches
126 divided by 17 equals 7.4 inches

You have a choice between 15, 16, or 17 steps. If you had 15 steps 8.4 inches high, the steps would definitely feel a little tall. Either 16 or 17 will give a more comfortable rise.

To find a good run to match the rise, subtract the rise from 17½ inches. In the example above, subtracting 7.4 from 17.5 inches would give 10.1 inches. This could be fixed in practice at 10 or maybe 10¼ inches. Now that you have the run (say it is 10 inches), multiply it times the number of steps (17) to get the total run, 170 inches. You will need 170 inches of floor space for the stairs.

Draw the staircase on your floor plan in scale, with the right number of steps. Make a main staircase 30 inches or 3 feet wide if possible, after consulting local building codes.

If you are building L-shaped or U-shaped stairs, this process will be a little complex. First try to find a design that will fit without triangular winding stairs. A landing should be thought of as an extra wide step. Its height off the floor should be an even multiple of the stair rise, so that all steps will be the same, including the step up and down from the landing. Any step can be widened into a landing. Try several positions for the landing to find the one that gives the best floor plan without sacrificing headroom.

Draw in winders if a cramped layout necessitates them. For safety, the width

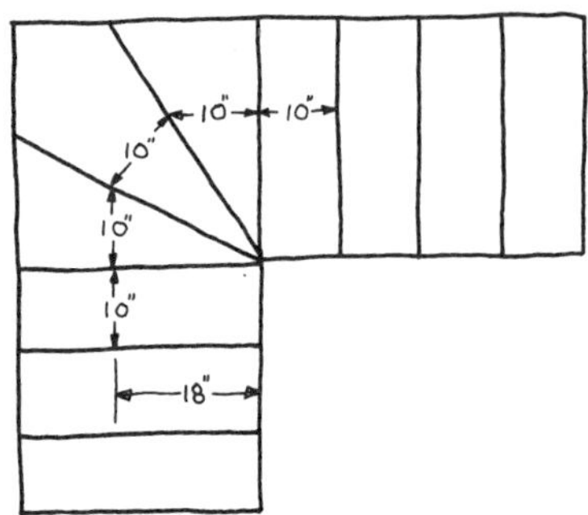

Figure 33-5.
Winder layout.

of the winders should equal the run of the regular stairs at a point 18 inches out from the corner (figure 33-5). It is particularly important that the winders be consistent in size.

When you have a tentative floor plan, use it to add your staircase to the elevation or cross section drawings you have made of your design. For turning staircases, you may need a drawing from two different angles. Probably the easiest way to make these drawings is to lay a tracing of the appropriate elevation over the stair plan. These drawings will help you visualize how the stairs will be built, and enable you to make sure there is sufficient headroom below and above the stairs.

The minimum headroom in building codes is 6 feet, 8 inches. Owner-builders often miscalculate, here. I feel 6 feet, 6 inches is acceptable, and 6 feet 4 inches is marginal: there will be occasional bumped heads coming down the stairs.

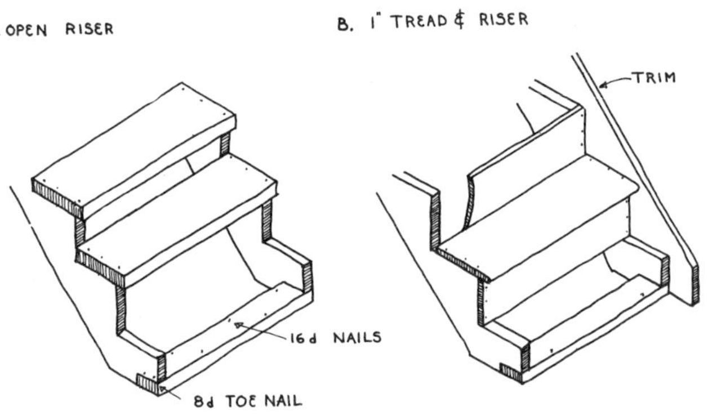

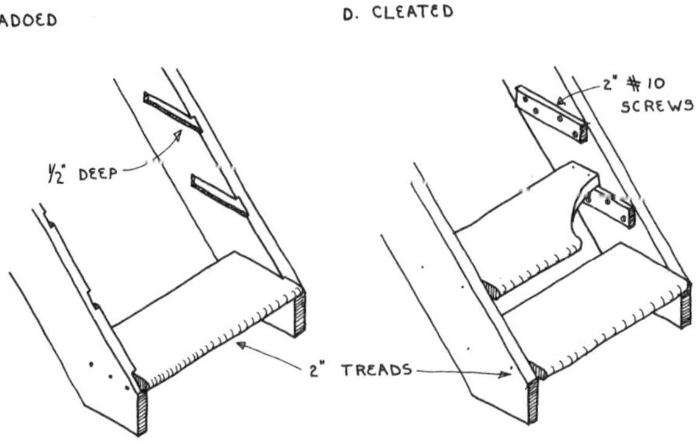

Figure 33-6.
Types of stairs.

BUILDING STAIRS

The basic parts of the staircases are the carriages, treads, risers (if any), and two 2 x 4-inch cleats or braces, notched into the stringers at top and bottom to hold the ends of the carriages in position.

Figure 33-6 shows the most common ways to build stairs. Type A is an open-riser type, which means it has treads but no risers. The timbers that support the treads are called carriages or stringers and are cut from 2 x 12s. The sawtooth-shape shown is the most common type. With 2-inch-thick treads, two of these carriages can support treads 3 feet, 6 inches or 4 feet wide. Wider stairs will need an extra carriage in the middle.

The staircase pictured in Type B is like Type A but dressed up to be more finished. It has risers to close off the toe space and help support the treads front and back. Hard pine a full inch thick, or oak treads ¾-inch thick can span three feet safely, but I would suggest a middle carriage between treads more than 32 inches wide.

Type C is a more finished open-riser staircase in which the carriages are notched or dadoed out for each tread.

The cleated carriage in Type D is perhaps the easiest type to build of all the stairs shown.

This chapter describes how to build the kind of staircase with sawtoothed carriage. Adapt the directions as needed if you are building any of the other types. If you choose Type C, with dadoed treads, make sure there is some method of positively holding the two sides together, beyond screws or nails into the ends of the treads. The stairs in figure 33-1, for example, have two threaded rods going all the way across, concealed in the structure.

For good-looking but economical stairs, use carefully selected framing lumber for the treads. It should be rounded-over at the front edge, called "nosing." Another good option is 5⁄4 ("five-quarter") yellow pine tread stock, which finishes to about 1 inch thick. Risers can be common ¾-inch pine boards.

Step 1

Lay out carriages. Make the carriage from the best 2 x 12s you have. You will mark these using a rafter square and a sharp pencil. Figure 33-8 shows the procedures for laying out a stringer for a staircase with a rise of 8 inches and a run of 10 inches. You can substitute the rise-and-run numbers from your own design. Put a 2 x 12 on a pair of sawhorses. Measure back from the top end 10 inches along the bottom edge of the 2 x 12 and make a mark. Line up the square on the stringer as shown in part A. Here the eight on the long side of the square is on

Accessible Stairs

AN ACCESSIBLE STAIRCASE has several features a standard stair may not. It's gentler—perhaps a 7-inch rise and 11-inch run. It has 1¼- to 1½-inch round railings on both sides, spaced 1½ inches from the wall. These are easy to grasp, and a person's arm won't get caught in the space between railing and wall. To minimize the danger of tripping, it definitely has a riser, and the area under the nosing is slanted.

So far, the building industry has resisted this standard, which uses quite a bit of extra floor space. Codes usually limit stairs to a rise of no more than 8¼ inches, and a run of no more than 9 inches. Minimum width would be 36 inches, wall to wall.

I like the 7:11 pitch where space allows, though this is not always possible in small houses. I believe that graspable railings on both sides are quite important.

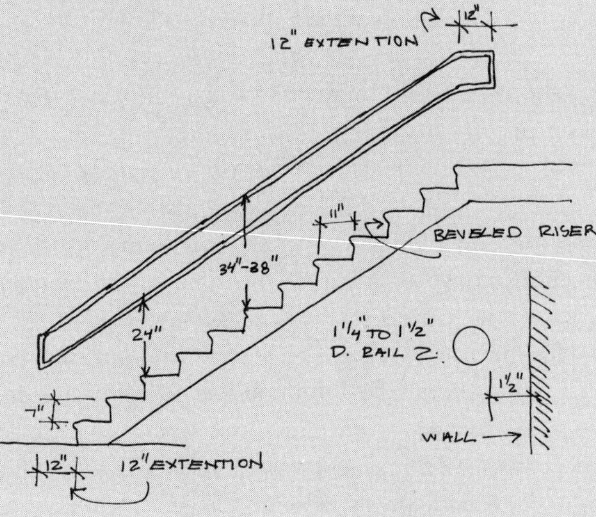

Figure 33-7. "Accessible staircase details: gentle pitch, ergonomic railings, minimal tripping hazard, and non-slip surface."

the mark you just made. The 10 on the short side of the square is lined up with the bottom edge of the 2 x 12. With the square in this position, mark line A, which represents the topmost vertical cut on the stringer.

It may not be clear now why lining up the square this way enables you to mark the stringer at the correct angles, but as you continue with the procedure, the logic of the rafter square will become clearer.

The next step is to mark the topmost horizontal line of the stringer, line B. Flip the rafter square into the position shown in B. The 10 on the long side and

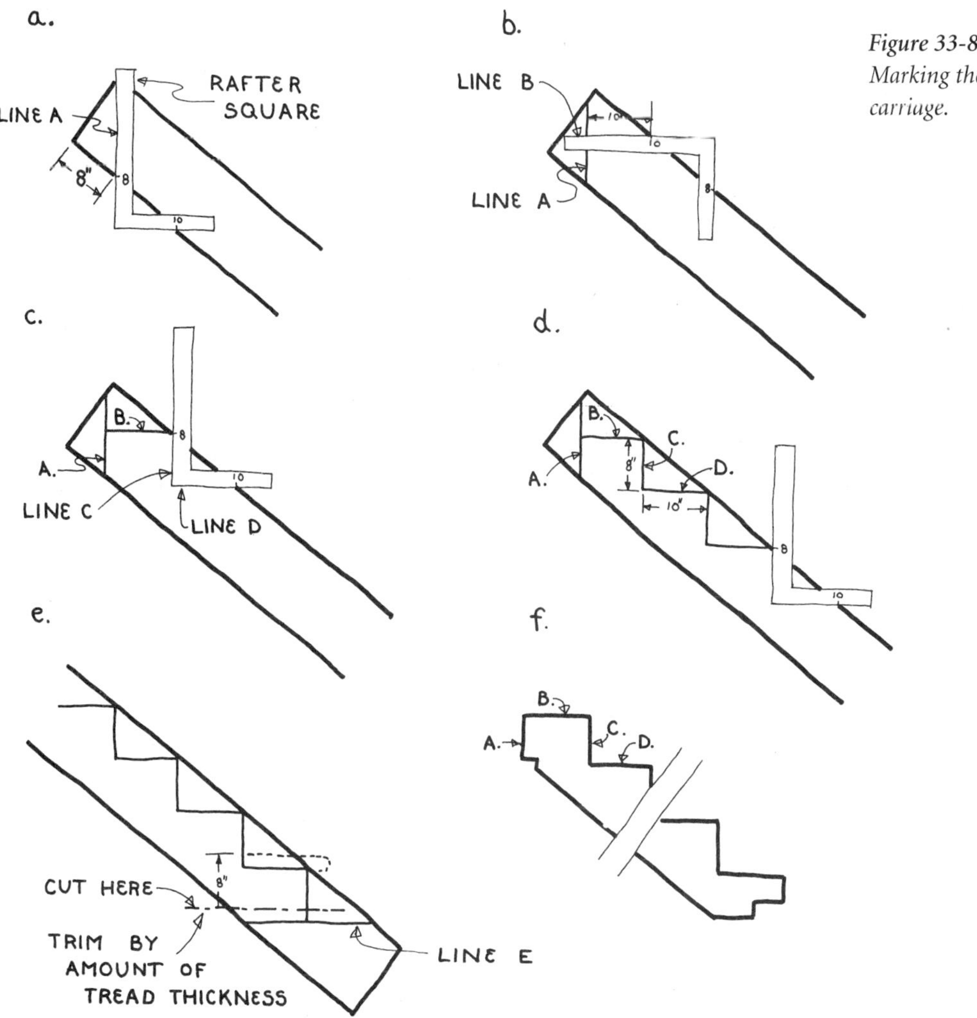

Figure 33-8. Marking the carriage.

the 8 on the short side are lined up with the top edge of the 2 x 12. Now slide the square along the top edge of the 2 x 12 until the distance from line A to the top edge of the 2 x 12, where the long side of the square crosses it, equals the run of the stairs. In the example shown, this is 10 inches. Then mark line B as shown. This is the top step.

You next mark the first notch to be taken out of the stringer, which represents the first step down. Place the square as shown in part C, again aligning the rise and run along the top edge of the 2 x 12. The 8-inch mark on the long side of the square should just intersect the right-hand end of line B. The 10-inch mark on the short side of the square should intersect the top edge of the stringer, farther down. When the square is carefully aligned, mark line C and line D, which represent the rise and run of the first step down. Soon you will notch out by cutting along these lines.

Keeping the square aligned in the same way, slide it down the 2 x 12 one full step, that is, until the long side of the square lines up with the right hand end of line D. You can now mark the second step down. Continue sliding the square down, laying out the steps one at a time, as in part D.

The marking procedure will be easier if you buy a set of stair fixtures or "stair buttons," little brass clamps that tighten down onto the square and act as stops to align it. Finish off at the bottom by drawing line E, the horizontal line at the bottom of the last rise line (part E). Your carriage is laid out. Check your elevation drawing to make sure you have marked the right number of steps.

The next step is called dropping the carriage. If you cut the bottom of the carriage along line E and used it that way, you would have a problem. Once the treads were nailed on, the bottom step would be higher than the others by the thickness of one tread. Therefore, you have to trim off the bottom of the carriage by an amount equal to the tread thickness. Measure up from line E that distance and draw a line parallel to it. This is where your carriage will be cut off. However, if you are installing the staircase on the subfloor, that is, if your finish floor is not yet laid when you do your staircase, you have a further complication. You must draw a new cutoff line below the old one by a distance equal to the anticipated thickness of the finish floor.

Carpenters traditionally are somewhat casual about how they actually attach the stairs to the house. I like to leave a project with no uncertainties about how long the stairs will stay where I put them. In part F of figure 33-8, there is a notch at the bottom for a 2 x 4, nailed to the floor. With that or a similarly strong connection at the bottom, the staircase cannot slide forward. A similar notch at the top end could rest on a 2 x 4 cleat nailed to the header, as shown in figure 33-4.

Sometimes the top step of the stringer isn't flush with the upper

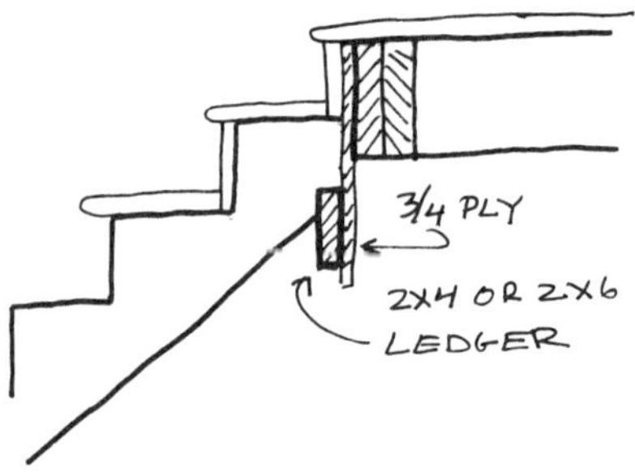

Figure 33-9. Details for top of stairs, when top of carriage is one step down.

floor, but is one step down. In that case, I avoid the traditional and somewhat cavalier solution, which is to hang the staircase from a couple of pieces of strapping. As figure 33-9 shows, I extend the header down with a piece of ¾-inch plywood, which supports either a cleat below the stringer or heavy blocking between the carriages. In either case, I use plenty of nails or screws to join all these pieces.

Step 2

Cut stringer. Cut out the stringer on the lines you have made. The cuts for the steps can be started with a circular saw but will have to be finished with a hand saw or saber saw. Sometimes you will have to use a sharp chisel to neaten the inside corners.

Step 3

Check carriage. Take the stringer over to the opening in the floor for the stairs and put it in place. Check a number of treads for level. If the level reads level here, presumably you have done everything right. If the treads are off level, make what adjustment you can to correct the situation. Check that the carriage is sitting at the level of the finish floor (if it is planned that way) and not on the subfloor. Take a scrap of the material you will use for treads (or any lumber of the same thickness) and stick it on the top and bottom step. At the top, the tread should come flush with the finish flooring. At the bottom, the tread should be exactly one rise distance above the finish flooring level. Sometimes a little trimming or shimming at the top or bottom will make the carriage linc up right.

Step 4

Cut other carriages. When you are sure the first carriage is right, use it as a template to mark the other carriage or carriages. If you have three, use the first as the template for both of the others. Use a very sharp pencil for this operation.

Step 5

Test carriages. Put all the carriages up to make sure they are right. Put a level across from one to the other to make sure the treads will be level. If there are three, use a straightedge to make sure all three are lined up in the same plane. Make any adjustments needed. If your staircase is built up against the wall, and will have a trim board at the side as in figure 33-6B, mark the top of the carriage on the wall now.

Step 6

Cleats. When everything is checked, and before you take the carriage down, mark the floor at the bottom and the framing at the top to indicate exactly where the 2 x 4 braces or other braces should be located. Take the stringers down, cut the 2 x 4s, and nail them with 16d nails on the marks you have just made.

Step 7

Trim boards. Trim boards at the side, if any, can go in before the carriages. This can be a little tricky, particularly with a turning staircase. Starting from the marks you transferred to the wall, draw in the tread locations at the top and bottom of each section of the stairs. Measure up vertically an amount equal to the width of the baseboard. Connect these marks. Using 1 x 12s, fit boards to the wall below these lines, and nail them firmly to the studs. Where the nails will show, use 10d finish nails. Elsewhere use 10d common nails.

Step 8

Install carriages. Nail the carriages in place. Against a wall, nail directly to the framing. At top and bottom, use 16d nails to nail firmly to the framing, but pre-drill with a 3⁄32 nail bit to prevent splitting.

Step 9

Treads and risers. The nosing on the treads should overhang the risers (if any) by 1 to 1½ inches. If there are no risers, the treads should overhang the notches in the carriages by that amount. Thus, the treads will actually be wider than the nominal run by the amount of the nosing. Risers should run past the treads at the bottom and butt up to the underside of the treads at the top. At the base of the riser, pre-drill and screw in place 1⅝-inch drywall screws through each riser into the back of the tread for strength. One-inch treads and risers can be nailed to the carriages with 8d finish nails. Two-inch treads will need 16d finish nails. If the nails split the wood, pre-drill with a bit slightly smaller than the nails. Construction adhesive at all joints will make the stairs stronger and prevent squeaking.

Step 10

Protecting and finishing the stairs. A staircase can be left unfinished, but you may want to oil or varnish it to match the floors. This should be done as soon as possible after the treads go on so dirt is not ground into the wood. If the staircase goes in while heavy building is still going on, tack or tie protective masonite or plywood covers over the freshly finished stairs. You may also install temporary treads made from scrap lumber that can be replaced with good treads after the heavy building is over.

Step 11

Handrails and guardrails. The railing should be graspable and very strong, and not spaced too far from the wall. If you use standard brackets to support a handrail, they should be attached at every other wall stud, and secured firmly to the stud. Figure 33-10 shows handrail configurations that work. Open stairs—those not enclosed by a wall—need guardrails to prevent people from toppling off the edge. The stairway enclosure should be at least 36 inches high, and the vertical rails should be no more than 4 inches apart, to prevent children from getting caught in between. Consult your local building code. Guardrails are also needed for lofts, balconies, and decks.

Landings and Winders

Structurally, think of a landing as a floor, framed like a regular floor, with two staircases attached. Figure 33-11 shows a typical landing. It consists of a box of 2 x 6s (joists), which can be supported by doubled 2 x 4 posts to the floor or nailed to adjacent walls with numerous 20d common nails. This can be topped with a plywood subfloor and then a flooring that matches your treads. Run this flooring parallel with the lower stairs to provide a nosing where you step up on the landing. Notice that the landing is wider than the bottom stairs so that it can fully support the upper staircase.

This same approach can be adapted to building a set of winders, as shown in figure 32-13. First, build a full-size rectangular landing as just described. This will be the first

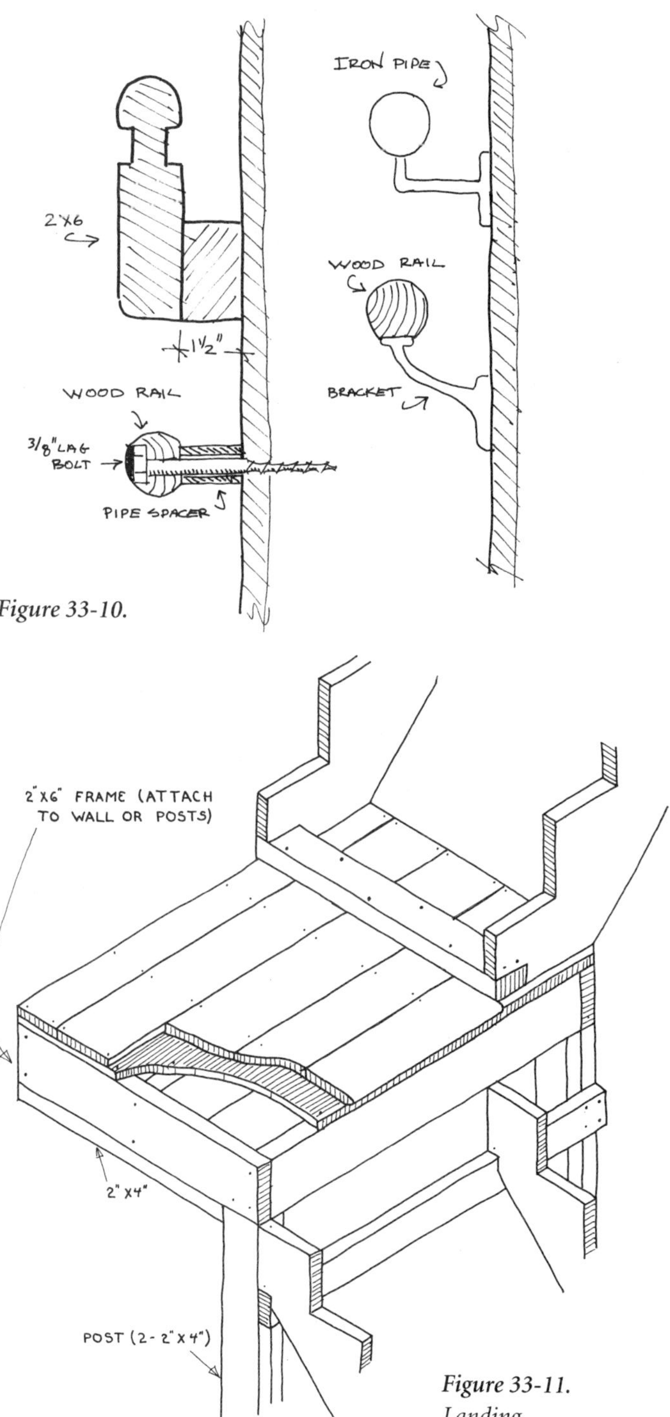

Figure 33-10.

Figure 33-11. *Landing construction.*

A.

B.

C.

Figure 33-12. *Elegant examples. A: Built by David Palmer, Iron Bridge Woodworkers (photo: Nancy Wasserman). B: Built by Geert Burger (photo: Geert Burger). C: Segal method stairs (photo from* The Self-Build Book, *by the authors, reprinted courtesy of Green Books.)*

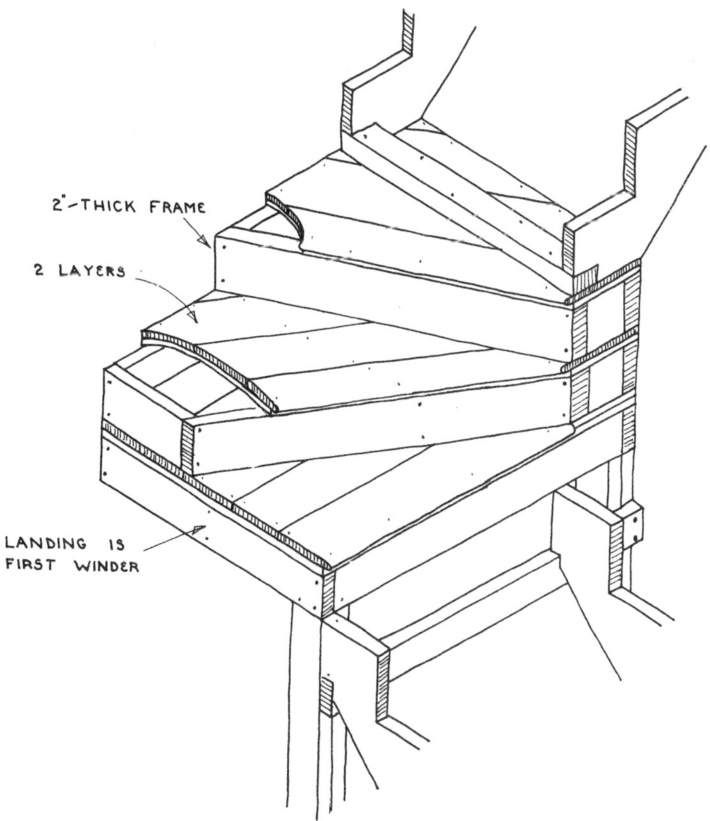

Figure 33-13.
One way to construct winders.

winder. Then build a stout platform with a 2-inch-thick frame on top of the landing, with a plywood subfloor and then a flooring of your tread material. Run the flooring parallel to the leading edge or riser of the winder. If there is a third winder, it can be built the same way on top of the second. Notice that the landing and both winders go under the upper run of stairs to give it good support. Then the upper staircases can be built on the winders just as though on a floor.

Winding staircases that are right up against walls (as most winding stairs are) are often built with a lighter framework of cleats nailed to the adjacent walls. Instead of the steps being stacked like blocks, each winder is supported at the ends by 2 x 4s or 2 x 6s nailed to the wall, and in the middle by the risers that connect the treads together. These frames are usually improvised in a somewhat haphazard way, since they will not be exposed to view in the main part of the house. In this case, the upper staircase cannot be supported by the winders, but must be nailed directly into the walls. If you want to build winders this way, study examples in existing houses and get some experienced help.

34

SCAFFOLDING

THE BEST SCAFFOLDING IS THE EARTH, OR THE PARTS OF the building you have already completed. One of the advantages of a one-story house is that it is relatively easy to reach every part of it. In the same vein, though a steep roof sheds water better than a flatter one, it is hard to work on. We often design for a 6:12 pitch, fairly easy to walk on, but steep enough to shed water well.

The simplest ways to get higher than you can reach from the ground or floor is to use stepladders or sawhorses. A couple of planks across a pair of stout sawhorses will work well for many interior tasks. One often sees old, split, or rotten stepladders on building sites. Make sure yours are relatively new, and follow the safety precautions that are often marked on them. In fact, make sure all ladders or other staging equipment are in good repair.

Most builders use pipe staging (figure 34-1). It can be seen on most any building site. Each section consists of a rigid pipe frame at each end, and scissor-shaped diagonal braces on each side, which make up a platform. Heavy 2-inch planks can be put at various levels depending on the work at hand. You can stack these units up to reach higher levels, as long as the frame is tied to the building periodically. There are a variety of accessories for special situations, including safety equipment needed for pipe staging to be safe. Since properly set up pipe staging is so strong, it might make sense to rent a few sections to use for various purposes about your site.

There are several other types of useful commercial staging to use on almost any site situation. Pump staging (figure 34-2) is used for siding projects. It allows you to move up and down on the wall as work progresses. Pumps also come with special planks, railing supports, work stations, brackets, foot pieces, and other useful accessories.

With either of these types of staging, or any of the other staging gadgetry that is seen on building sites, get a lesson from an experienced and preferably an apprehensive builder. There are techniques that will make your staging both safer and more convenient.

Basic rules include:

- Make sure the staging has a wide, solid base beneath each vertical support. You don't want it sinking into the mud or sliding down the hill.
- Add rugged railings if more than 6 feet above the ground.
- Use strong planks, and don't let them span until they are more than a little springy. Planks should be made of "plank grade" wood.
- Inspect your staging daily.
- Protect those below the staging.

Every type of staging requires caution on the part of users. I think it is important to know your limits. I'm comfortable up to certain heights, but above those, I let someone else take over. If you feel nervous about a particular staging arrangement, it's telling you something: let someone else do the staged work, or improve the staging. That might mean adding railings, more planks, thicker planks, or taking the trouble to build really good staging rather than making do.

Anyone working under or passing by staging is in some danger, too. Often a person working up on the staging is carrying a lot of tools and other gear, and materials may be stored on the planks. Occasionally something heavy may fall off even with careful

Figure 34-1. Pipe staging.

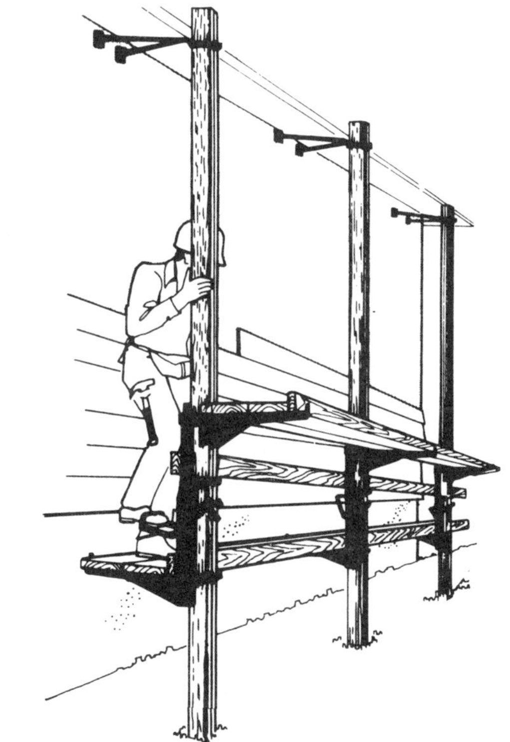

Figure 34-2. Pump staging. (Drawing courtesy of Alan Kline, Lynn Ladder & Scaffolding Company, West Lynn, Massachusetts.)

work. There are several precautions to take against an injury from a falling object. No one should be working directly under staging in use. Anyone walking under or by staging should be looking up. The person above should secure things carefully and watch out below, particularly when moving around or setting up. When in doubt, wear a hard hat, or install netting to catch falling objects. It is also a good idea to restrict access to the area below the scaffold when someone is working above. A professional builder may also be subject to additional OSHA rules.

HOW TO BUILD A SCAFFOLD

There will be situations where it is best to build staging with lumber from the site. Apply the same care to planning your staging as you would to building the house itself; think about the loads and forces that apply. The main difference between a scaffold and a house frame is that you want to be able to take the scaffold apart easily.

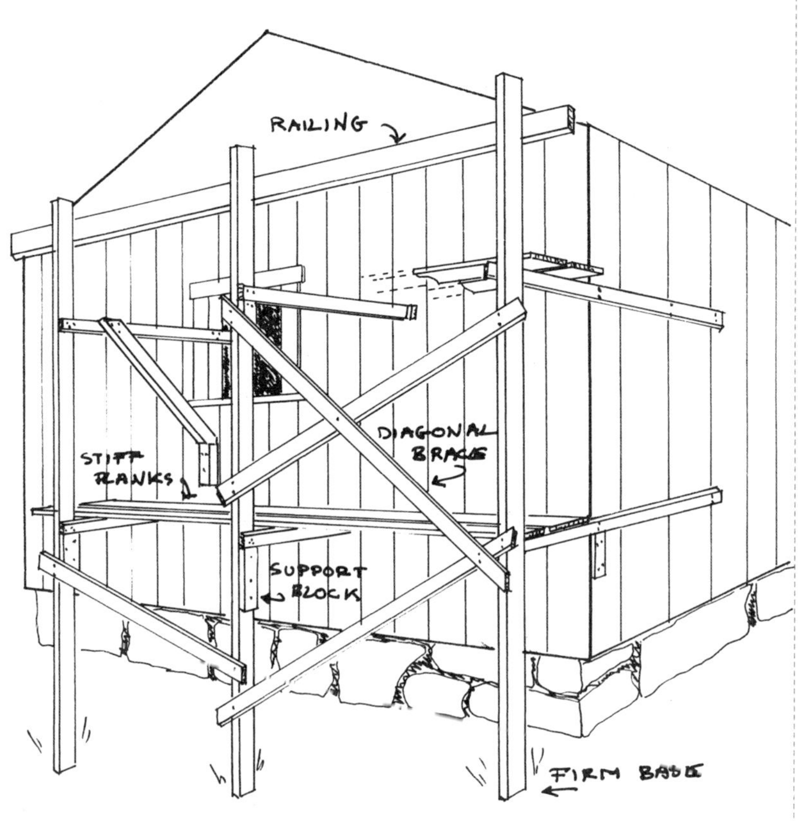

Figure 34-3. *Staging you make yourself.*

Scaffolding is usually built quickly, but it must be strong, because a fall from 20 feet is no joke. I have had scaffolding that looked strong collapse under me more than once, and only luck, in the form of something handy to jump to, prevented me from being hurt. Check scaffolding carefully before you trust your life to it. Inspect it every day or two. Look for loose joints and nails pulling out. Never climb on a scaffolding you didn't build yourself until you have made sure it is safe.

Figure 34-3 shows a common way stagings are built on site. The planks are strong 2 x 10s or 2 x 12s with no major defects. Purchase a few rough-cut planks that are a full 2 inches thick. They can span about 8 to 10 feet safely. If they feel springy, double them up or shorten the span. The 2 x 6 or 2 x 8 horizontals that support them have to be well secured at the house. I like to hang them in joist hangers, well nailed to the building's frame.

The scaffold should be supported by and nailed to 2 x 4 or 2 x 6 posts with at least three 16d nails, driven in a downward direction from the horizontal into the vertical. Special double-headed staging nails are helpful for this. They can be easily removed, yet they hold well. You can add a ledger block below any joint to create more strength. At the bottom, all posts have to be on a firm base. I like to dig them into the ground about 4 to 6 inches, so they will stay in position. This is particularly important on any kind of slope. If the ground is muddy or soft, you may need to create a little footing of some sort also.

Strapping or 2 x 4s serve as diagonal bracing: After the planks are in place, tack them down. You can toenail them through the edge so that the nail heads don't stick up where you will be walking. One or two horizontal 2 x 4s nailed to the inside of the posts make a good railing, and provide a place to hang tools.

35

USING TOOLS

BUYING TOOLS

You can buy good, used hand tools through classified ads or from stores that specialize in secondhand tools. New tools can be purchased at hardware stores and chains, such as Sears. Because of the enormous increase in interest in building and woodworking, there are now several excellent tool catalogs and tool discount store chains. We buy most of our tools through these sources. In general, buy good quality tools from known brands. When I have a favorite brand, I will mention it.

Hand Saws

While the carpenter once packed a whole box full of hand saws, most carpentry today is done with electric tools. Yet there are still many tasks where the hand saw is invaluable. I'd suggest the new Stanley Toolbox saw, which is a short, western-style saw shape with Japanese-style saw teeth. It's a great tool. You might later purchase a 12-point finish saw for trim work. At some point, you will also need a hack saw.

Measuring Tools

Rafter square: Figure 35-1 shows a rafter square, also known as a steel square. It's a handy tool for measuring and marking operations, especially during framing. It is also handy as a straightedge and for innumerable little tasks. Beyond this, the rafter square is the carpenter's computer. The tables and numbers on the square enable you to figure rafter lengths, find angles, and solve countless other carpentry problems. Stanley rafter squares come with a little book called *How to Use the Stanley Steel Square,* by L. Perth, which provides an excellent introduction to the use of this tool. This is not essential knowledge for someone building just one house, but if you enjoy geometry, it may be worth learning about the rafter square.

Fig. 323 OLS

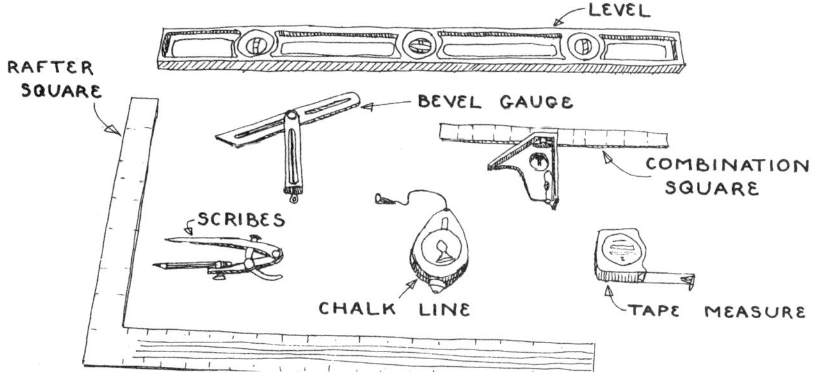

Figure 35-1. Measuring and marking tools.

Levels: The level is used to make sure surfaces are exactly level (horizontal) or plumb (vertical). Often there will be two vials for horizontal measurement and two for vertical measurement. I would suggest investing in a 24- and a 48-inch model. Buy good-quality levels, but test them for accuracy. To test a level, find a very flat, nearly horizontal surface. A smooth kitchen counter works well. Read the level, and mark the spot it sits on exactly with a pencil. Flip it end for end. If the reading is identical (even though not level), that vial is accurate. The vials for reading plumb are tested against a nearly vertical surface in the same way. Often there will be one really good level in a stack at the store, and it is worth searching for.

Combination square: The combination square is a good tool to carry while working, because it is so versatile. It measures 90-degree and 45-degree angles, which you have to do continually. By sliding the 12-inch ruler back and forth in the cast-iron frame, you can use the combination square to make many identical measurements quickly. If you want to make a series of marks 6 inches in from the edge of a board, set the ruler to protrude from the frame by 6 inches. Then slide the square along the board, and make your series of measurements quickly and accurately.

Speed square: The speed square is one of the great new tools of the last twenty years. Today, every carpenter carries one at all times. It takes measurements as a combination square does, but it also can be used to read or mark any angle, expressed either in degrees or pitch (rise and run). It also serves as a cut-off guide with a power saw. Read the instruction booklet to get the most out of it.

Bevel gauge: Also known as an angle bevel, the bevel gauge pivots at the joint to measure and mark any angle.

Tape measure: Always carry a tape measure with you. The best ones (in my opinion) are Powerlock tapes by Stanley. Probably the most versatile sizes are the ¾-inch x 16-foot or 1-inch x 25-foot models. You also will need a 50-foot tape for laying out your foundation, but this is a good tool to borrow since you will only need it for a few days.

Chalk line: This is a reel of string inside a case filled with powdered chalk used to make long, straight lines. Stretch the string tight between two points and snap the line in the middle. This will leave a line of chalk on the work.

Scribes: A set of scribes is like a compass, except that there is a lock to hold the setting. Scribes are used to transfer shapes for curved or irregular cuts.

Protractor guide: A miter gauge, or protractor guide, is an adjustable T-square used as a guide or fence with skill saws. You can also buy a large speed-square for this. Figure 35-1 shows a protractor guide being used.

Planes: You need a large plane for smoothing or trimming the edges of boards and a small one for rounding edges and trimming end-grain. The large plane can be either a smoothing plane (8-10 inches) or a jack plane (13-14 inches). The short plane should be a block plane (about 6 inches).

Chisels: I suggest buying a ½-, a 1-, and a 2-inch chisel. Stanley and Millers Falls are good brands, but also keep your eye out for Buck, Greenley, Marples, and Witherby.

Combination stone: To sharpen your edge tools, you will need a combination sharpening stone with a coarse and also a medium side. Finer stones can be obtained through the tool catalogs, for instance Woodcraft Supply (800/535-4482). I very much prefer water stones over oil stones (see page 476).

Hammers: The regular hammer you see everywhere is called a 16-ounce claw hammer. I think most people will have the best luck with this size hammer, but some people may prefer the larger 20-ounce framing hammer. Buy a top-quality hammer, because cheaper hammers may break and will be more fatiguing to use. Stanley, Plumb, Vaughn, Hart, and Estwing, among others, make good

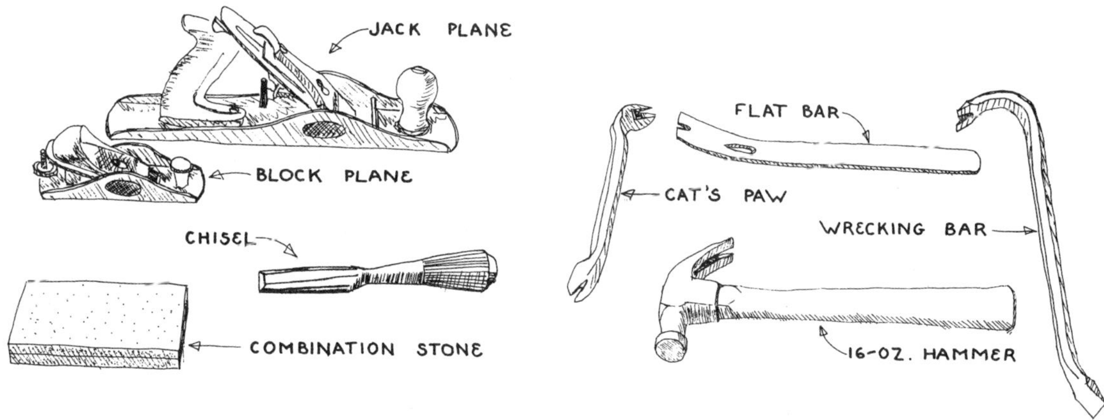

Figure 35-2. Edge tools.

Figure 35-3. Hammers and bars.

hammers. Heft several brands and pick the one that feels most comfortable in your hand. You may also want to invest in a medium-weight sledgehammer.

Bars: You need a hook-shaped wrecking bar for heavy-duty levering and a flat bar for levering objects when you don't want to mar their surfaces. For foundation work you may want to borrow or rent a 5-foot crowbar.

Cat's paw: This tool is for removing nails that have been driven all the way in. It is essential. Pound the notched end into the wood until the notch catches the nail head, then lever the nail out. Get a steel, not cast iron, cat's paw.

Miscellaneous Tools

In addition to these basics, you will need a utility knife, a staple gun, good nail aprons, two 25-foot heavy-duty extension cords, a 2⁄32-inch nail set, screwdrivers (regular and phillips head), tin snips, and at least two large clamps. Figure 35-4 shows a handy style.

Power Tools to Buy

Portable circular saws: There are many excellent portable circular saws, also called skill saws. The Skil brand makes good, economical, lightweight models. I use a Makita, which is one of the less expensive professional brands, and a Milwaukee, which is more heavy-duty.

The standard saw has a 7¼-inch blade, great for all-purpose framing. But my favorite saw is a smaller 5½-inch trim saw. Makita, Skil, and Rockwell make these. Because it's so light, it's great for beginners. As long as the blade is sharp, you can build a whole house with it.

Buy carbide-tipped combination blades for these saws with 24 or more teeth.

Saber saw: A saber saw is a portable reciprocating saw for curved cuts and scrollwork. This is not an essential tool, but it is cheap and very handy for small jobs. Some inexperienced people, particularly those who help you out for short periods, may find it easier to use than a circular saw, though less accurate for straight cuts.

Drills: A ⅜-inch reversing, variable-speed drill (figure 35-7) is essential. In fact, you will need two, one of which is cordless. The plug-in model doesn't have to be an expensive model. The cordless should be at least a 9-volt model, which can recharge in no more than one hour. Buy an extra battery unless it has a quick-charge design.

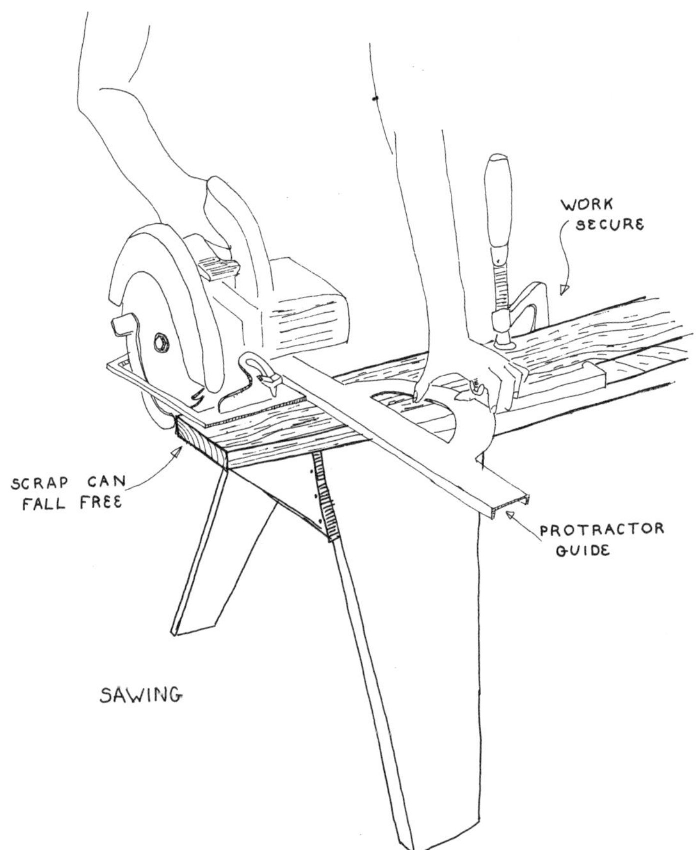

Figure 35-4. Portable circular saw.

Figure 35-5: Skil's 5½-inch trim saw is very light, easier to use than full-sized saws.

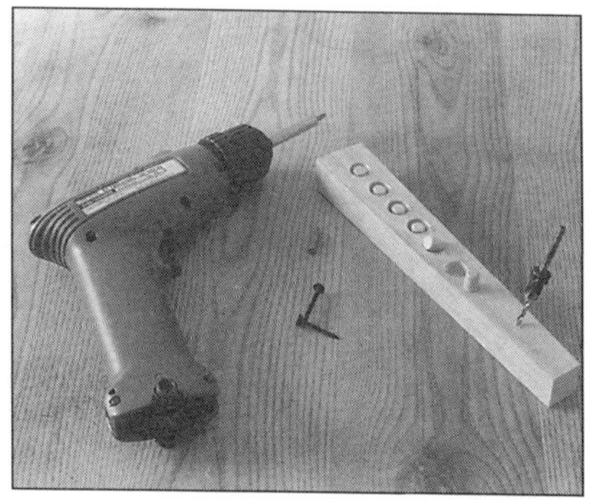

Figure 35-8. Cordless drill, bitholder, replaceable tip, drywall screws, pilot bit, plug cutter, and wood plugs.

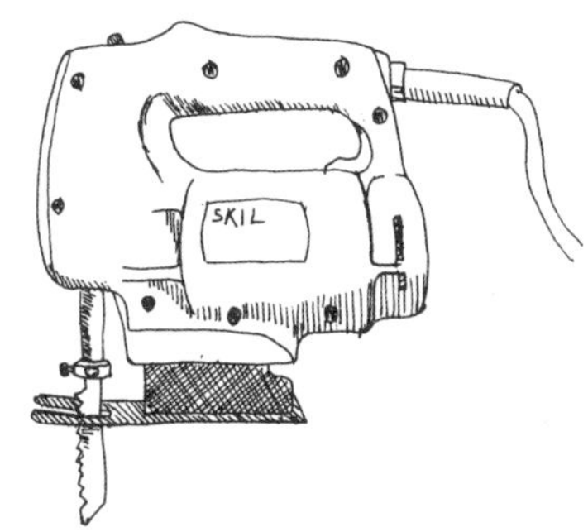

Figure 35-6. Saber saw.

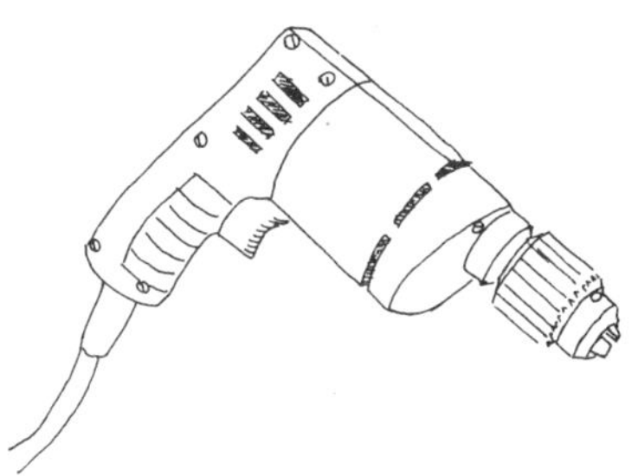

Figure 35-7. Electric drill.

These drills have become even more important, because of the proliferation of the drywall screw. These screws come in many lengths, and are often the best way to fasten things together. Drywall screws are self-tapping. You fit your drill with a magnetic bit holder, insert a nice fresh phillips tip, put the screw on the tip, hang on tight, and just drive the screw. For fancier projects, you can pre-drill with a pilot bit, and conceal the screw with a wood plug. With this simple technology you can build shelves, cabinets, and perform many carpentry tasks which are much more troublesome with hammer and nails.

You will also need a set of flat "spade bits."

Table saw: With a table saw, you move the work across the blade, instead of the other way around. Lengthwise cuts (rips) in boards are made by sliding the board along the rip fence, which can be moved side to side for different desired widths. For crosscuts, the board is held against the crosscut guide (also called the miter

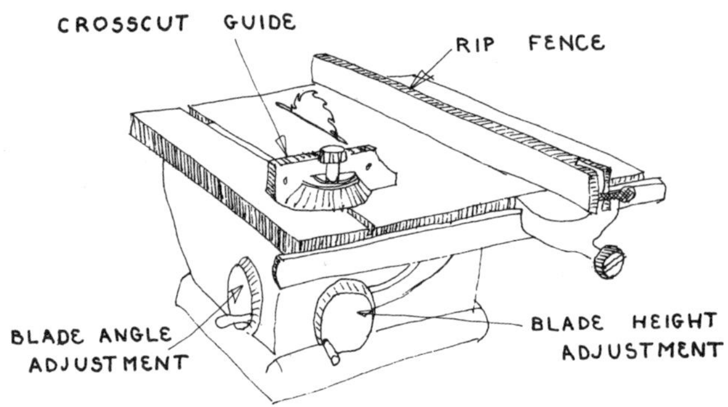

Figure 35-9. *Table saw.*

fence), which slides neatly in the notch of the saw table. Both the blade and the crosscut guide can be tilted to any angle up to 45 degrees. A special gadget called a dado is used for making notches. It consists of several blades that stack together in various combinations for cuts of various widths. If you buy a dado, buy the stacking kind, rather than the adjustable one-piece kind, which in my view is very unsafe.

I strongly recommend buying a table saw. It permits you to cut boards accurately both with and against the grain and to make cuts at almost any angle, and it makes notches and joints of many kinds. All this can be done with hand saws or a portable circular saw, but the table saw will do them more quickly, easily, and accurately. The savings in time will be particularly great for finish work, window making, cabinet making, and other fine work. You can buy a used saw for $100 to $200 and resell it later at the same price. Look for an old Craftsman 10-inch model, such as is shown in figure 35-9, or a Rockwell or Delta 8- or 9-inch saw. When you buy a used saw, have a knowledgeable person look it over to make sure it is all there and in good shape.

You can now purchase small, very light, portable tablesaws. They are handy and powerful, but not as accurate as a cast iron saw.

You will see many professional woodworkers (I admit: including myself) walking around with missing or shortened fingers; most of these people got those injuries on the table saw. A table saw is safe only when used properly. Its hazards are not obvious, and you can risk injury without being aware of it. For each type of cut there are one or two safe techniques, and two or three techniques that appear safe. There are good books on using table saws (available through the Woodcraft Supply Catalog at 800/ 535-4482). However, I think you should have someone who knows how to use one correctly teach you safe cutting methods and watch you use the machine at first.

While you are shopping for tools, someone may recommend that you buy a radial arm saw. This is a type of saw in which the blade and motor move back and forth above the saw table on a track. A radial arm saw does the same work that a table saw does and is strongly promoted by stores and home hobby books. Its advantage is that the work is stationary and the saw moves, so long pieces, which would be unwieldy on a table saw, can be crosscut easily.

Though I haven't checked the statistics, I consider radial saws more dangerous than table saws. If you have one and value your fingers, restrict its use to cross cutting.

Chop-saws are very popular with amateur and professional builders. They make very accurate cross-cuts on narrower boards such as trim or baseboard. But I have never felt they were safe.

Router: A router is a motor mounted on a circular base. A chuck called a collet on the end of the motor shaft holds bits of many shapes to make various cuts. A router is not an essential tool, but it is handy for notches, grooves, and molding shapes. I use mine to make hinge mortises, notches in window and door jambs, and for similar work. It's great for rounding off the edges of boards. Again, the Woodcraft Supply Catalog is a good source of books on router use.

Belt sander: With a belt sander, you can sand your woodwork ten or twenty times as fast as you can by hand. While a belt sander is not essential, it will be a big help if you plan to sand much of your finish work. Get the 3 x 21-inch belt size. A "half-sheet" orbital sander is also great for finish work.

Figure 35-10.
Router.

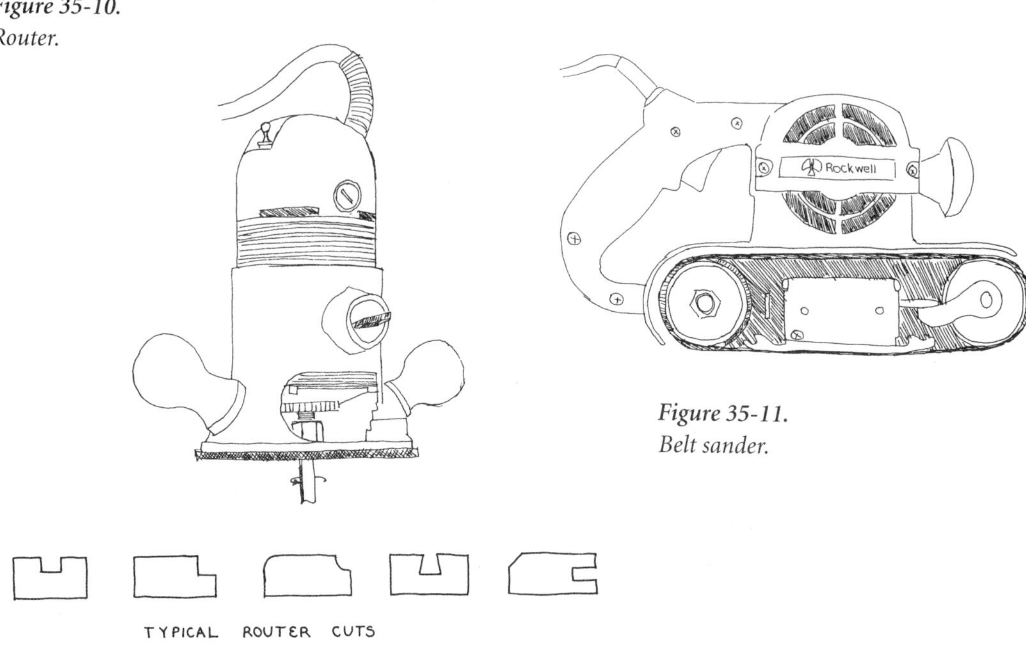

Figure 35-11.
Belt sander.

Tools to Make

Sawhorses: Amateur builders almost never build sawhorses, but they are worth the effort. Sawhorses serve as chair, table, workbench, scaffold, and in general make it possible for you to work at a comfortable height. You can do everything working on the floor, but it is more fatiguing and takes longer.

You will need four, preferably six, sawhorses. Figure 35-12 shows a design that I have found quick to build and fairly strong. It will be relatively easy to figure out the angles if you make a full-scale drawing of the sawhorse on a piece of plywood.

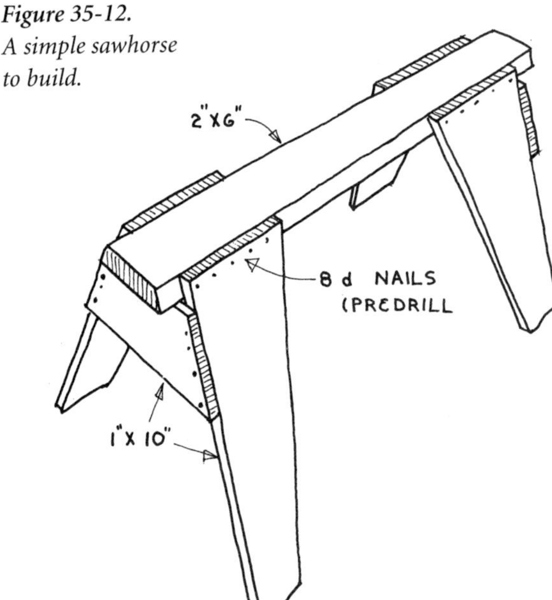

Figure 35-12. A simple sawhorse to build.

Water level: Figure 35-13 shows a homemade water level, a cheap, little-used but essential tool for leveling foundations and other large-scale leveling operations. It consists of a jug and 15 to 30 feet of ¼-inch (inside diameter) clear plastic tubing from a hardware store.

To set up the level, fill the jug with water. Then stick one end of the tube in the jug. It helps to use string or a coathanger to secure the tubing to the jug. Suck on the free end of the tube to fill the tube with water.

The first time you are likely to use your water level is when you put up batter boards for the foundation. These boards have to be the same height, so you have to mark the stakes all at the same level, namely, the foundation height. Here is the procedure for making such a measurement with the water level. You can use the same method with any similar problem.

Set the jug securely somewhere so that the water inside is even with the proposed foundation height. You can make minor adjustments by letting a little water out of the jug through the tube, or by pouring a little more into the jug. Carry the far end of the tube to each stake and hold the tube against the stake. When the water settles down, mark the stake opposite the water level in the tube. Always take measurements within 6 inches of the end of the tube. Keep your finger over the end of the tube while carrying it around to keep water from leaking out and changing the setting.

If the level is not working right, there may be a kink in the tube or air may have gotten into the line back at the jug. Remove the bubbles by letting water run out of the far end of the tube, and start again.

When it is inconvenient to place the jug at the level you want to find, put the reservoir at an arbitrary level and take readings at the various locations. Then measure up or down a uniform amount to the final level you need. Usually a stick or story pole is handier for this than a tape measure.

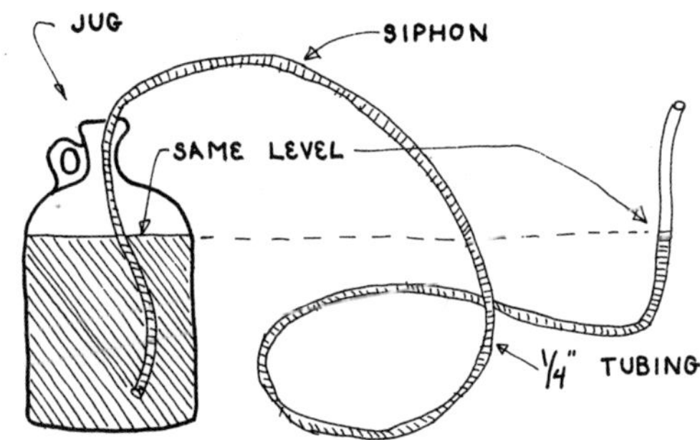

Figure 35-13. *Water level.*

Plywood cutting jig: If you will use a lot of plywood in your design, the jig shown in figure 35-14 is worth making. It is a straight track 8 feet long that your circular saw rides along to make long straight cuts. The jig is made from two strips of straight, unwarped ½-inch plywood, one 6 inches by 8 feet and the other approximately 12 inches by 8 feet. The saw rides on top of the wide piece, with the base sliding against the edge of the narrow piece. This edge, edge A in figure 35-14, must be perfectly

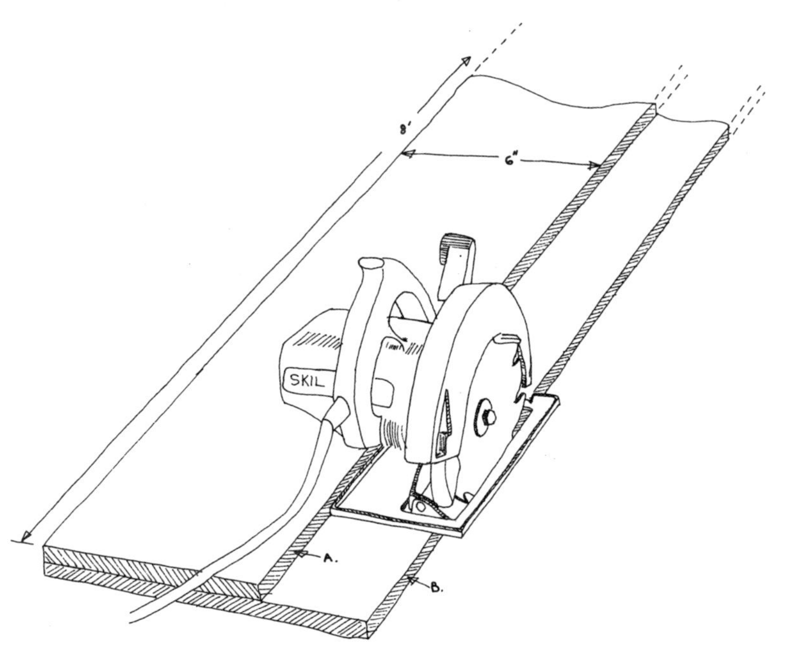

Figure 35-14. *Plywood cutting jig.*

straight, so it should consist of one of the factory-cut plywood edges, which will be perfectly straight. Edge B is precisely as far away from edge A as the saw blade is from the left edge of the saw base. Thus edge B shows exactly where the saw will cut.

The accuracy of the jig depends on the precision of the distance between edge A and edge B. Do not try to align these two edges by measurement. Instead, build the jig with the bottom piece oversize an inch or two on the right. The first time you run the saw down the track, being careful to keep the saw smoothly against edge A, the saw itself will cut edge B perfectly.

To use the jig, place the plywood to be cut on sawhorses. Place edge B on your marks, and clamp or tack the jig to the work. Set the saw blade just deep enough to cut through the plywood, and carefully run the portable circular saw down the track.

USING TOOLS

Sharpening

A sharp saw or chisel will work ten times faster than a dull one. Often the difference between the results achieved by an experienced person and an inexperienced person is the difference in the sharpness of their tools.

Lumberyards, hardware stores, and saw shops will sharpen saw blades for you on short notice, or exchange sharp blades for dull ones.

You can sharpen chisels and planes yourself with a combination stone, which has a medium and a coarse side (figure 35-15). I prefer a water stone to an oil stone.

Start sharpening by using the coarse side of the stone. Wet the stone thoroughly with the proper fluid—water or light oil—and keep it wet. Grind the bevel side only and grind the whole bevel surface at once, holding the blade as in part A. Stroke the blade on the stone in a broad figure-eight pattern. After a while, you can feel a slight burr on the back side of the sharpened edge. When you feel this burr all along the edge, remove it by turning the stone to the medium side and rubbing the chisel a few times, holding the chisel flat as shown in part B. Then grind the beveled side again, this time using the fine side of the stone until another burr forms. Remove the burr as before.

This sharpening method will give a sharp, usable edge. If nicks are avoided, you will not need to use the coarse side again. Just touch up with the medium side. If a blade gets nicked, it can be taken down on a grinder and then sharpened with a stone. A razor-sharp edge can be obtained with finer sharpening stones

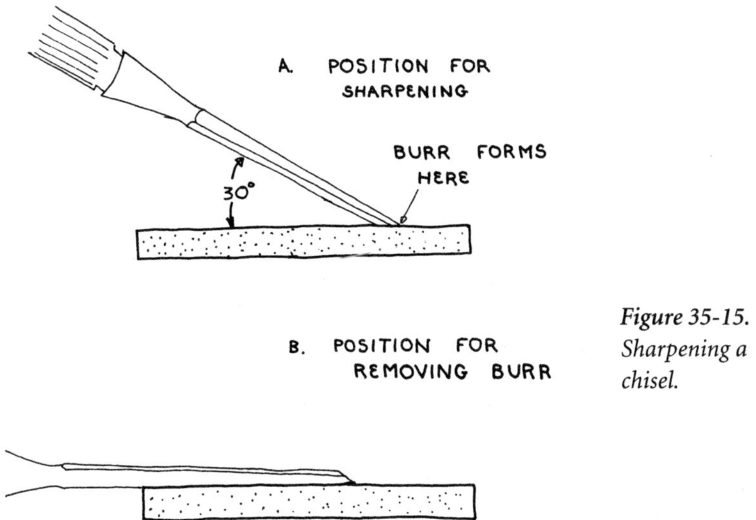

Figure 35-15. Sharpening a chisel.

purchased through mail-order tool dealers. A really sharp plane is a joy to use, and any energy or money spent on sharpening will be amply returned to you.

Safety

Both hand and power tools can seriously injure you. Some safety rules apply equally to both:

- No part of your body should be in the line of action of a tool; cut away from yourself.
- Secure the work with nails, clamps, or a firm knee before you cut (figure 35-4).
- Make sure you have a sure footing.
- Don't work in an awkward position.

Some rules are particularly important with power tools:

- Do not wear loose clothing, which can get tangled in the tool.
- Keep long hair from dangling loose.
- Learn how to use a new power tool from someone experienced.

Always use three-prong, grounded cords with three-prong grounded tools, and make sure the third prong, the ground line, is working. Someone who knows wiring can show how to check the ground with a circuit tester. An additional precaution against shock is to use a ground-fault circuit interrupter. This is a device that automatically disconnects the power supply instantly if the power isn't routing normally back through the cord, as when the power is grounding through a tool body to you, and then to the earth. Kitchen and bath outlets sometimes have GFCI outlets, and GFCI breakers can be installed at the service panel. I have a portable GFCI that I can plug into any outlet. Then I plug my cords into it. I make a particular point of using GFCIs when working on wet ground, in a basement, or near plumbing that grounds to the earth. I would suggest having GFCI breakers installed at your temporary electrical hookup, and make sure there are GFCI-protected sources for your power tools.

How to Saw

Positioning: Position a piece of wood where you can work on it comfortably. A lumber pile or low table works well. Sawhorses often provide the best support. Position the piece so that the part you will be using is supported but the part that will be cut off, the waste, is free to drop to the ground. Otherwise the wood will pinch the blade as it buckles. With a short piece, you can position the wood on a single sawhorse, as shown in figure 35-4. Long pieces can be supported by two horses. In that case do not saw between the horses. Cut at one end, so that the scrap can fall free.

Secure the work: Heavy pieces will stay put by their own weight. Some people can comfortably hold down medium-sized pieces with one knee. The most reliable method is to clamp the piece down with a large clamp, particularly if the piece is small.

Marking for a cut: Cut on the waste side of the line. Suppose you want a piece four feet long. In figure 35-16, if you measured four feet from the

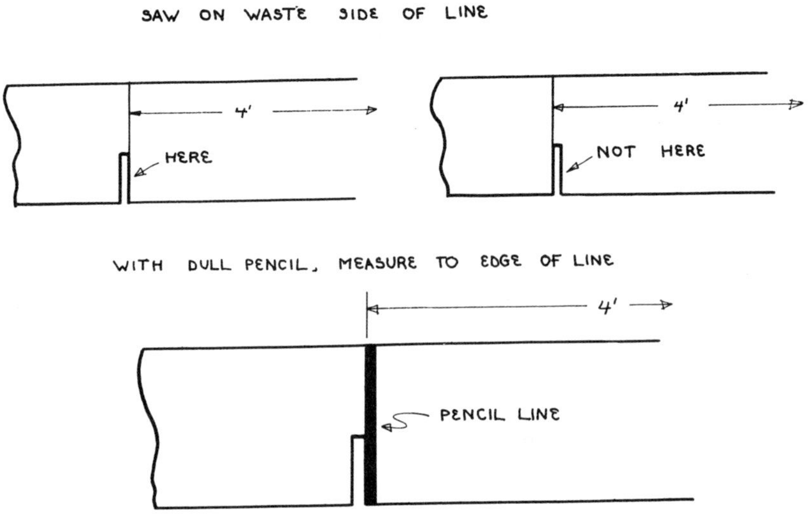

Figure 35-16. *Marking for a cut. Saw on the waste side of the line.*

right-hand end of the piece, made a line, and then sawed to the right of the line, you would be cutting the piece too short by the width of the saw blade. With a dull pencil, even sawing on the correct side of the line may leave you with some inaccuracy, because the line itself has thickness, called a "kerf." The method I use is quite accurate even with a dull pencil. The line is drawn so that its left-hand edge is at four feet. The saw cuts to the left of the mark, leaving the line on the piece.

Cutting: Take a wide, stable, and secure stance, with firm control. With a hand saw, follow the line by eye and use long, strong, angled strokes of the saw. Keep the saw vertical by feel or by eye. If you are using a power saw, you can also follow the line by eye. However, for an automatically straight and smooth cut that does not depend as much on your skill, use a miter gauge. This large T-square can be set at any angle. Hold it against the edge of the wood, then run the metal base of the saw against it. For smaller pieces, up to 6 or 8 inches wide, you can do this with a speed square.

How to make notches: Even big notches can be made quickly with a power saw and chisel. To make a through-notch, one going all the way across a piece, set the circular saw to the same depth as the notch should be and make a series of cuts ¼-inch to ½-inch apart in the area to be notched out. Then chisel out the waste, beginning with the chisel in the position shown in figure 35-17. Position the chisel bevel up. The saw cuts guide the chisel along the right line and make the work easier for the chisel. Chisel out the visible part of the joint first, then knock out the rest more quickly. Usually a firm hand and sharp blows will give better results than a hesitant approach.

The blade of a chisel tends to follow the grain of the wood, and sometimes the chisel will follow the grain in the wrong direction and make a messy job. When this starts to happen, or you anticipate that it will, approach the cut from another direction or angle. Watch the grain as you work, and you will soon learn which angles work best.

If a notch stops partway across the piece, drill a series of holes as shown in figure 35-17. Use a convenient-size spade bit and drill to the right depth. A piece of tape on the bit shaft will help mark the right depth to drill. Then saw and chisel out the notch as before. The drilling removes enough of the wood to make the chiseling easier and guides the chisel in the right direction. Notice that usually the beveled side of the chisel faces the wood to be removed.

Certain notches may call for variations on this method. Sometimes a hand saw will work better, because the power saw will not reach deep enough. Some-

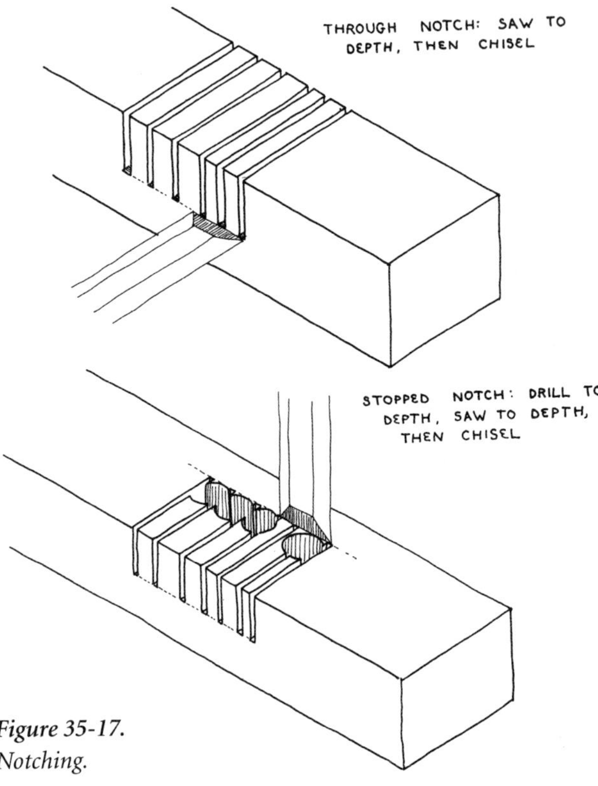
THROUGH NOTCH: SAW TO DEPTH, THEN CHISEL
STOPPED NOTCH: DRILL TO DEPTH, SAW TO DEPTH, THEN CHISEL

Figure 35-17. *Notching.*

times a saber saw will do the whole notch. The basic idea is to use saw cuts and drilled holes to guide the chisel and reduce the amount of work the chisel must do.

Planing: Planes are used for trimming and smoothing. Keep all planes sharp and adjusted properly. Hold a plane upside-down near a light to see if its blade is parallel to the sole of the plane where it sticks through. If not, the lever near the handle is for adjusting the blade side to side. The knob is for adjusting how much the blade sticks out. For a very smooth cut, or for trimming end-grain, you will need a fine cut. Keep adjusting between trial cuts until the plane cuts well.

Secure the board you are working on. Figure 35-18 shows a simple method. Nail a block to a sawhorse and clamp your board to the block. For longitudinal planing you will find that the plane usually cuts much better in one direction relative to the grain pattern. The illustration shows the right direction. When you begin the stroke, put a little more weight on the forward end of the plane to avoid rounding off the near end of the board. At the far end of the cut, put more weight on the rear end of the plane to avoid rounding off the far end of the board. Use a firm stroke.

When planing end-grain, avoid splitting the wood at the edge by planing from both sides toward the middle as shown in figure 35-19.

Scribing

Scribes are for marking irregular or curved cuts (figure 35-20). Position the piece you will mark as close as possible to its final location. Set the scribe to a convenient distance. In the illustration this is the maximum width of the crack. Then transfer the curve to the piece to be cut by running the metal point along the curve while the pencil marks the other piece.

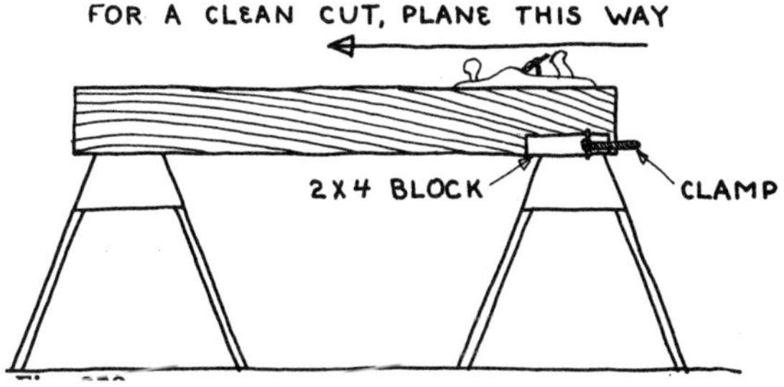

Figure 35-18. *Planing with the grain.*

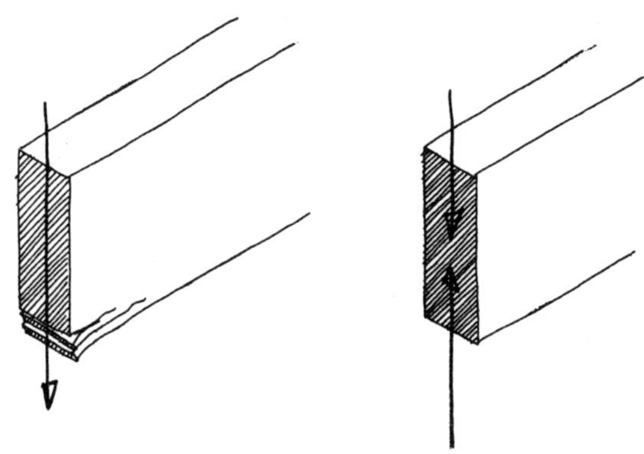

Figure 35-19. *When block-planing end grain, plane from both edges to avoid splitting out.*

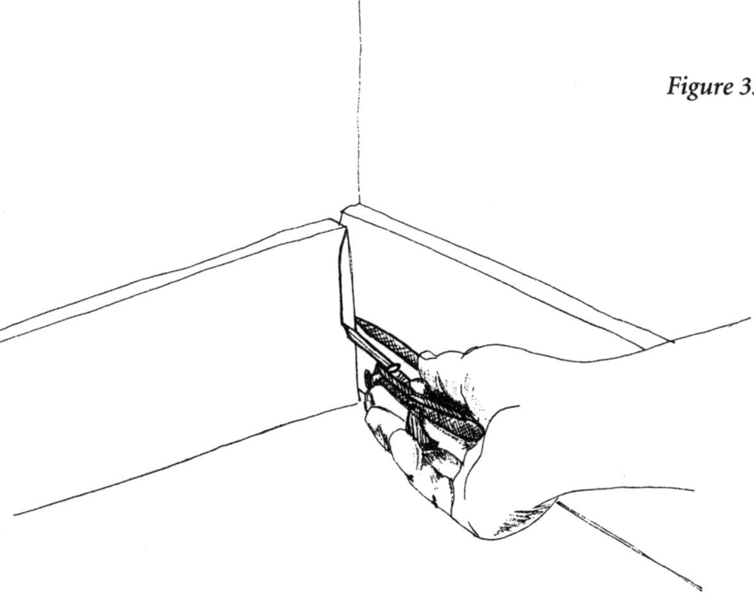

Figure 35-20. *Scribing.*

APPENDIX A

F AND E VALUES FOR TIMBER CALCULATIONS

To use the formulas in chapter 16, you need to know the F value or fiber stress rating of the wood you are using. E (modulus of elasticity) ratings are needed for computing timber sizes for stiffness (deflection) as opposed to strength, and are needed for tables 4 and 5 in appendix B.

E values for ungraded hardwoods are given in table 1. E values for ungraded softwoods are in table 2.

Most lumberyards sell graded lumber from the great lumber regions of the south, the west, or Canada. For grading purposes, three species with similar structural characteristics are grouped together. The most common categories in our area are Hem-Fir and SPF, or Spruce-Pine-Fir. Hem-Fir, for example, includes western hemlock, noble fir, California red fir, white fir, and a couple of other species. SPF includes Engelmann spruce, Sitka spruce, and lodgepole pine.

There isn't one simple F or E number for these woods. First, they depend on the grade. Within each category or species there are eight or more structural grades. At the top is "Select Structural," which is stronger than common grades, more expensive, and may require a special order. The lumber commonly referred to at the lumber counter as "KD" or "construction grade" is often really #1, or #2. Although #1 is usually significantly stronger than #2, they are often sold together in a mixed grade known as "#2 and better." These are the strong but economical grades to use for most framing purposes. Below these in strength are #3, standard, utility, and stud grades. A mid-strength grade called "construction grade" is a lower quality wood which often has relatively high F and E values, but less load-bearing capacity depending on specific applications.

Southern pine and Douglas fir are among the stronger species, SPF and Hem-Fir are in the middle, and cedar and white pine are not nearly as strong.

As examples, here are a few base values for lumber commonly used in housebuilding. Fb is fiber stress in bending, which is what is referred to as F in chapter 16. E is modulus of elasticity, and Fc// is fiber stress in compression parallel to grain, used for calculating post capacities. All these values are for "#2 and better" grade wood for 2-inch, 3-inch, and 4-inch thick lumber, such as 2 x 8s or 4 x 8s.

NAME	Fb	E	Fc//
Douglas fir	875	1,600,000	1,300
Hem-Fir	850	1,300,000	1,250
SPF	750	1,100,000	975
White Pine	575	1,100,000	825

Heavy timbers can have quite different ratings compared to the same 2x grade and species. For example, here are base ratings for the same woods listed above, but for timbers over 5" thick.

NAME	Fb	E	Fc//
Douglas fir	875	1,600,000	600
Hem-Fir	675	1,100,000	475
SPF	575	1,000,000	350
White Pine	575	1,100,000	825

The industry grading rules were updated in 1991, when new tables and manuals came out. These new rules give several additional "adjustment factors" which have to be taken into account when arriving at an F number to use in a particular case. Among the most important are:

Size factor: The Fb (our F) in 2-inch thick to 4-inch thick framing is adjusted for the exact width (d in the formula). In Select Structural, #1, #2, and #3 grades, the Fb for a 2 x 4 would be increased by a factor of 1.5. A 2 x 6 is increased by a factor of 1.3, a 2 x 8 by a factor of 1.2, a 2 x 10 by a factor of 1.1. A 2 x 12 receives no adjustment. Construction, standard, utility, and stud grades are not adjusted.

Repetitive member factor: Timbers used repetitively, such as floor joists or rafters no more than 24-inch on center (and tied together with sheathing or flooring) share the load, and Fb can be increased by 1.2.

Duration of load: Timbers can carry larger loads for short periods. For example, a snow load limited to two months can allow you to increase Fb by 1.15. On the other hand, extra long-term loads (over 10 years) call for decreasing Fb by multiplying by .9.

Thus, while Hem-Fir #2 and better starts with a base value Fb of 850, a series of 2 x 8 floor joists would actually be increased to an Fb of 1,224 because of the effects of size and repetitive factors.

The first edition of this book covered Fb, E, and Fc// and other factors succinctly in two tables reprinted from industry sources. The increased complexity and specificity of the new 1991 rules makes it worthwhile for the reader to send away for industry literature that explains these concepts at length and gives exhaustive ratings for many commercial wood products, including glu-lam timbers and hardwoods. These documents will make it possible to find out what your lumberyard is really selling, how strong it is, and how to get the best value from the products available in your area.

The most accessible document is known as "Western Lumber Product Use Manual: base values for dimension lumber," or Product and Technical manual A, which costs $2.00 from the Western Wood Products Association, Yeon Building, 522 SW Fifth Ave., Portland, OR 97204-2122. Their phone is (503) 224-3930. You may want to purchase the "Ingrade packet," a collection of other very useful related information on such subjects as framing for earthquakes, use of treated lumber, and common framing errors. This packet also includes tables of rafter and joist spans.

A more wide-ranging publication, which also includes other structural design information, is "The National Design Specification," including the "NDS supplement," which actually carries base values for woods not marketed by the WWPA. This currently costs $25, plus shipping, from the American Wood Council, P.O. Box 5364, Madison, WI 53705-5364. Their phone number is (800) 890-7732.

Native lumber you buy from a sawmill will not be officially graded. In my experience, the common softwoods used for framing in the Northeast, such as spruce, hemlock, or fir, can be assumed to have an F of 1,000. Pine would be around 800, although some pieces would be weaker because of large knots, and others stronger if almost free of knots. In other regions, you might be able to determine usable F values by talking to local builders or sawmill operators. Local building codes in some areas may give values you can apply to native wood.

TABLE 1

Modulus of elasticity (E) for ungraded hardwoods

COMMON NAME OF SPECIES	SPECIFIC GRAVITY	MODULUS OF ELASTICITY (X 1,000,000)
Maple		
Black maple	.52	1.33
	.57	1.62
Red maple	.49	1.39
	.54	1.64
Silver maple	.44	.94
	.47	1.14
Sugar maple	.56	1.55
	.63	1.83
Oak, red		
Black	.56	1.18
	.61	1.64
Northern red	.56	1.35
	.63	1.82
Pin	.58	1.32
	.63	7.73
Scarlet	.60	1.48
	.67	1.91
Southern red	.52	1.14
	.59	1.49
Water	.56	1.55
	.63	2.02
Willow	.56	1.29
	.69	1.90
Oak, white		
Bur	.58	.88
	.64	1.03
Chestnut	.57	1.37
	.66	1.59
Live	.80	1.58
	.88	1.98
Overcup	.57	1.15
	.63	1.42

COMMON NAME OF SPECIES	SPECIFIC GRAVITY	MODULUS OF ELASTICITY (X 1,000,000)
Oak, white *cont.*		
Post	.60	1.09
	.67	1.51
Swamp chestnut	.60	1.35
	.67	1.77
Swampy white	.64	1.59
	.72	2.05
White	.60	1.25
	.68	1.75

TABLE 2

Modulus of elasticity (E) of ungraded softwoods

COMMON NAME OF SPECIES	SPECIFIC GRAVITY	MODULUS OF ELASTICITY (X 1,000,000)
Bald cypress	.42	1.18
	.46	1.44
Cedar		
Alaska	.42	1.14
	.44	1.42
Atlantic white	.31	.75
	.32	.93
Eastern red cedar	.44	.65
	.47	.88
Incense	.35	.84
	.37	1.04
Northern white	.29	.64
	.31	.80
Western red cedar	.31	.94
		1.11
Fir		
Balsam	.34	.96
	.36	1.23
California red	.36	1.17
	.38	1.49

COMMON NAME OF SPECIES	SPECIFIC GRAVITY	MODULUS OF ELASTICITY (X 1,000,000)
Fir		
Grand	.35	1.25
	.37	1.57
Noble	.37	1.38
	.39	1.73
Pacific silver	.40	1.42
	.43	1.72
Subalpine	.31	1.05
	.32	1.29
White	.37	1.16
	.39	1.49
Hemlock		
Eastern	.38	1.07
	.40	1.20
Mountain	.42	1.04
	.45	1.33
Western	.42	1.31
	.45	1.64
Larch, western	.48	.96
	.52	1.87
Pine		
Eastern white	.34	.99
	.35	1.24
Jack	.40	1.07
	.43	1.35
Loblolly	.47	1.40
	.51	1.79
Lodgepole	.38	1.08
	.41	1.34
Longleaf	.54	1.59
	.59	1.98
Pitch	.47	1.20
	.52	1.43
Pond	.51	1.28
	.56	1.75
Ponderosa	.38	1.00
	.40	1.29

COMMON NAME OF SPECIES	SPECIFIC GRAVITY	MODULUS OF ELASTICITY (X 1,000,000)
Red	.41	1.28
	.46	1.63
Sand	.46	1.02
	.48	1.41
Shortleaf	.47	1.39
	.51	1.75
Slash	.54	1.53
	.59	1.98
Spruce	.41	1.00
	.44	1.23
Sugar	.34	1.03
	.36	1.19
Virginia	.45	1.22
	.48	1.52
Western white	.35	1.19
	.38	1.46
Spruce		
Black	.38	1.06
	.40	1.53
Engelmann	.33	1.03
	.35	1.30
Red	.38	1.19
	.41	1.52
Sitka	.37	1.23
	.40	1.57
White	.37	1.07
	.40	1.34
Tamarack	.49	1.24
	.53	1.64

Source of tables 1 and 2: *Wood Handbook (Agriculture Handbook No. 72)* by U.S. Forest Products Laboratory, USDA, 1974, pp. 4-7ff.

APPENDIX B

JOIST, RAFTER, HEADER, SILL, & COLUMN SIZE TABLES

List of Tables:

1. Maximum spans for first-floor joists and for rafters supporting over 3 feet of snow
2. Maximum spans for second-floor joists and for rafters supporting 3 feet of snow or less
3. Sizes for heavy timber rafters
4. Maximum spans for joists and rafters when deflection is limited to $1/360$ of span
5. Maximum spans for heavy timbers when deflection is limited to $1/360$ of span
6. Header sizes
7. Sill sizes
8. Safe loads on solid wood columns

TABLE 1

Maximum spans for first-floor joists and for rafters supporting over 3 feet of snow under worst conditions (total load 50 lbs./sq. ft.)

Spans can be used with typical lumberyard #1 and #2 timbers (F approximately 1,200) or rough sawmill timbers (F approximately 1000). Note rafter spans are measured horizontally, not along rafter length (see figure 353).

SIZE	MAXIMUM SPAN 24" O.C.	MAXIMUM SPAN 16" O.C.
2 x 4	4'11"	6'0"
2 x 6	7'9"	9'6"
2 x 8	10'3"	12'6"
2 x 10	13'1"	16'0"
2 x 12	15'11"	19'6"

Source: Table computed by author.

TABLE 2

Maximum spans for second-floor joists and for rafters supporting 3 feet of snow or less under worst conditions (total load 40 lbs./sq. ft.)

Spans can be used with typical lumberyard #1 or #2 timbers (F approximately 1,200) or typical rough-sawn sawmill lumber (F approximately 1,000). Note rafter spans are measured horizontally, not along rafter length (see figure 353).

SIZE	MAXIMUM SPAN WITH 24" CENTERS	MAXIMUM SPAN WITH 16" CENTERS
2 x 4	5'6"	6'9"
2 x 6	8'8"	10'8"
2 x 8	11'5"	14'0"
2 x 10	14'8"	17'11"
2 x 12	17'9"	21'9"

Source: Table computed by author.

TABLE 3

Sizes for heavy timber rafters

Rafter sizes for exposed frame, widely spaced roofs, using rough native lumber, F 1,000. Notice that rafter sizes are based on span measured horizontally rather than rafter length.

	TOTAL LOAD 35 LBS./SQ. FT. MAX. 30" SNOW ACCUMULATION			TOTAL LOAD 50 LBS./SQ. FT. 3-5' SNOW ACCUMULATION		
SPAN	*2' o.c.*	*3' o.c.*	*4' o.c.*	*2' o.c.*	*3' o.c.*	*4' o.c.*
7'	2 x 4	3 x 4	4 x 4	3 x 4	2 x 6	3 x 6
		2 x 6	2 x 6	2 x 6		2 x 8
8'	3 x 4	4 x 4	3 x 6	4 x 4	3 x 6	2 x 8
	2 x 6	2 x 6	2 x 8	2 x 6	2 x 8	
10'	4 x 4	3 x 6	2 x 8	3 x 6	4 x 6	3 x 8
	2 x 6	2 x 8		2 x 8	3 x 8	2 x 10
					2 x 10	
12'	3 x 6	4 x 6	3 x 8	4 x 6	2 x 10	2 x 12
	2 x 8	3 x 8	2 x 10	3 x 8		
		2 x 10		2 x 10		
14'	2 x 8	3 x 8	4 x 8	3 x 8	2 x 12	4 x 10
		2 x 10	2 x 12	2 x 10		3 x 12
16'	3 x 8	4 x 8	4 x 10	4 x 8	4 x 10	4 x 12
	2 x 10	2 x 12	3 x 12	2 x 12	3 x 12	
18'	4 x 8	4 x 10	3 x 12	3 x 10	4 x 12	4 x 14
	2 x 12	3 x 12				

Source: Computed by author and Jim Rader from formulas in chapter 16.

TABLE 4

Maximum spans for joists and rafters when deflection is limited to 1/360th of span

In table 4, E values (modulus of elasticity, see appendix A) are expressed in scientific notation. For example, if E is 1,600,000, it will be shown in the chart as 1.6, meaning 1.6×10^6. This chart was computed by the author for planed timber sizes. The maximum spans given will therefore be conservative for full-size, rough-cut timbers. These figures depend on live load alone, not total load.

TIMBER	LIVE LOAD = 30 lbs./sq. ft. (E x 10^6)	max span	LIVE LOAD = 40 lbs./sq. ft. (E x 10^6)	max span
2 x 4, 24" o.c.	.8	4'8"	.8	4'3"
	1.0	5'1"	1.0	4'7"
	1.2	5'5"	1.2	4'11"
	1.4	5'8"	1.4	5'2"
	1.6	5'11"	1.6	5'5"
	1.8	6'2"	1.8	5'7"
2 x 4, 16" o.c.	.8	5'5"	.8	4'11"
	1.0	5'10"	1.0	5'3"
	1.2	6'2"	1.2	5'7"
	1.4	6'6"	1.4	5'11"
	1.6	6'10"	1.6	6'2"
	1.8	7'1"	1.8	6'5"
2 x 6, 24" o.c.	.8	7'5"	.8	6'9"
	1.0	8'0"	1.0	7'3"
	1.2	8'6"	1.2	7'9"
	1.4	8'11"	1.4	8'2"
	1.6	9'4"	1.6	8'6"
	1.8	9'9"	1.8	8'10"
2 x 6, 16" o.c.	.8	8'6"	.8	7'8"
	1.0	9'2"	1.0	8'4"
	1.2	9'9"	1.2	8'10"
	1.4	10'3"	1.4	9'4"
	1.6	10'8"	1.6	9'9"
	1.8	11'1"	1.8	10'1"
2 x 8, 24" o.c.	.8	9'10"	.8	8'11"
	1.0	10'6"	1.0	9'7"
	1.2	11'3"	1.2	10'3"
	1.4	11'9"	1.4	10'9"
	1.6	12'4"	1.6	11'3"
	1.8	12'10"	1.8	11'8"

TIMBER	LIVE LOAD = 30 lbs./sq. ft. (E x 10^6)	max span	LIVE LOAD = 40 lbs./sq. ft. (E x 10^6)	max span
2 x 8, 16" o.c.	.8	11'2"	.8	10'2"
	1.0	12'0"	1.0	11'0"
	1.2	12'10"	1.2	11'8"
	1.4	13'6"	1.4	12'3"
	1.6	14'1"	1.6	12'10"
	1.8	14'8"	1.8	13'4"
2 x 10, 24" o.c.	.8	12'6"	.8	11'4"
	1.0	13'6"	1.0	12'3"
	1.2	14'3"	1.2	13'0"
	1.4	15'0"	1.4	13'8"
	1.6	15'9"	1.6	14'4"
	1.8	16'4"	1.8	14'10"
2 x 10, 16" o.c.	.8	14'3"	.8	13'0"
	1.0	15'5"	1.0	14'0"
	1.2	16'4"	1.2	14'10"
	1.4	17'3"	1.4	15'7"
	1.6	18'0"	1.6	16'4"
	1.8	18'9"	1.8	17'0"
2 x 12, 24" o.c.	.8	15'2"	.8	13'10"
	1.0	16'4"	1.0	14'10"
	1.2	17'5"	1.2	15'9"
	1.4	18'3"	1.4	16'7"
	1.6	19'1"	1.6	17'5"
	1.8	19'11"	1.8	18'1"
2 x 12, 16" o.c.	.8	17'5"	.8	15'9"
	1.0	18'9"	1.0	17'0"
	1.2	19'11"	1.2	18'1"
	1.4	20'11"	1.4	19'0"
	1.6	21'10"	1.6	19'11"
	1.8	22'9"	1.8	20'8"

TABLE 5

Maximum spans of heavy timbers when deflection is limited to 1⁄360 of span

As in table 4, E values are expressed in scientific notation for convenience. For example, an E of 1,600,000 is expressed as 1.6, meaning 1.6 x 10^6. The figures given were computed by the author using the WWPA span computer for planed timber sizes, but the spans given will work also for rough timbers. Loads are expressed not in lbs./sq. ft., but in lbs./running foot, or more simply lbs./foot of beam. Thus a timber 10 feet long with a load of 200 lbs./ft. will carry a total of 2,000 pounds.

The table gives 3-inch-wide and 4-inch-wide timbers only, but maximum spans for 6-inch-wide or 8-inch-wide timbers can be determined by dividing the load/ft. by 2 and finding the appropriate 3-inch-wide or 4-inch-wide timber. For example, to find the maximum span of a 6 x 8 of E 1.2 x 10^6, when the load is 400 lbs./ft., look up the maximum span of a 3 x 8 of E 1.2 x 10^6 with a load of 200 lbs./ft.

	LIVE LOAD					
	200 lbs./ft.		300 lbs./ft.		400 lbs./ft.	
TIMBER	(E X 10^6)	MAX SPAN	(E X 10^6)	MAX SPAN	(E X 10^6)	MAX SPAN
3 x 6	.8	5'10"	.8	5'2"	.8	4'8"
	1.0	6'4"	1.0	5'6"	1.0	5'0"
	1.2	6'9"	1.2	5'10"	1.2	5'4"
	1.4	7'1"	1.4	6'2"	1.4	5'7"
	1.6	7'5"	1.6	6'6"	1.6	5'10"
	1.8	7'9"	1.8	6'9"	1.8	6'1"
4 x 6	.8	6'7"	.8	5'9"	.8	5'3"
	1.0	7'1"	1.0	6'2"	1.0	5'7"
	1.2	7'6"	1.2	6'7"	1.2	6'0"
	1.4	7'11"	1.4	6'11"	1.4	6'4"
	1.6	8'4"	1.6	7'3"	1.6	6'7"
	1.8	8'8"	1.8	7'6"	1.8	6'9"
3 x 8	.8	7'9"	.8	6'9"	.8	6'2"
	1.0	8'4"	1.0	7'4"	1.0	6'7"
	1.2	8'10"	1.2	7'9"	1.2	7'0"
	1.4	9'4"	1.4	8'2"	1.4	7'5"
	1.6	9'9"	1.6	8'6"	1.6	7'9"
	1.8	10'2"	1.8	8'10"	1.8	8'0"
4 x 8	.8	8'8"	.8	7'7"	.8	6'10"
	1.0	9'4"	1.0	8'2"	1.0	7'5"
	1.2	9'11"	1.2	8'8"	1.2	7'10"
	1.4	10'6"	1.4	9'2"	1.4	8'4"
	1.6	11'0"	1.6	9'6"	1.6	8'8"
	1.8	11'5"	1.8	9'11"	1.8	9'0"

	LIVE LOAD					
	200 lbs./ft.		300 lbs./ft.		400 lbs./ft.	
TIMBER	(E X 10^6)	MAX SPAN	(E X 10^6)	MAX SPAN	(E X 10^6)	MAX SPAN
3 x 10	.8	9'11"	.8	8'8"	.8	7'10"
	1.0	10'8"	1.0	9'4"	1.0	8'5"
	1.2	11'4"	1.2	9'10"	1.2	9'0"
	1.4	11'11"	1.4	10'5"	1.4	9'5"
	1.6	12'6"	1.6	10'11"	1.6	9'10"
	1.8	13'0"	1.8	11'4"	1.8	10'3"
4 x 10	.8	11'0"	.8	9'8"	.8	8'9"
	1.0	11'11"	1.0	10'5"	1.0	9'5"
	1.2	12'8"	1.2	11'1"	1.2	10'0"
	1.4	13'4"	1.4	11'8"	1.4	10'7"
	1.6	14'0"	1.6	12'2"	1.6	11'1"
	1.8	14'6"	1.8	12'8"	1.8	11'6"
2 x 12	.8	10'2"	.8	8'10"	.8	8'0"
	1.0	10'11"	1.0	9'6"	1.0	8'8"
	1.2	11'7"	1.2	10'2"	1.2	9'2"
	1.4	12'3"	1.4	10'8"	1.4	9'8"
	1.6	12'9"	1.6	11'2"	1.6	10'2"
	1.8	13'4"	1.8	11'7"	1.8	10'6"
3 x 12	.8	12'0"	.8	10'6"	.8	9'6"
	1.0	13'0"	1.0	11'4"	1.0	10'3"
	1.2	13'9"	1.2	12'0"	1.2	10'11"
	1.4	14'6"	1.4	12'8"	1.4	11'6"
	1.6	15'2"	1.6	13'3"	1.6	12'0"
	1.8	15'9"	1.8	13'9"	1.8	12'6"
4 x 12	.8	13'6"	.8	11'9"	.8	10'8"
	1.0	14'6"	1.0	12'8"	1.0	11'6"
	1.2	15'5"	1.2	13'6"	1.2	12'3"
	1.4	16'3"	1.4	14'2"	1.4	12'10"
	1.6	17'0"	1.6	14'9"	1.6	13'6"
	1.8	17'8"	1.8	15'5"	1.8	14'0"

TABLE 6

Header sizes

This table assumes #1 or #2 graded lumber or rough-sawn native lumber, F 1,000, and that headers are supporting joists or rafters spanning approximately 10 feet.

SPAN	SIZE IF SUPPORTING ROOF ONLY	SIZE IF SUPPORTING ONE FLOOR PLUS ROOF	SIZE IF SUPPORTING TWO FLOORS PLUS ROOF
4'	2-2 x 4	2-2 x 6	2-2 x 8
6'	2-2 x 6	2-2 x 8	2-2 x 10
8'	2-2 x 8	2-2 x 10	2-2 x 12
10'	2-2 x 10	2-2 x 12	

Source: Massachusetts Building Code

TABLE 7

Sill sizes

Sill sizes can be taken from the table, but a computation for your exact design (chapter 16) might well enable you to use smaller timbers than those shown. The next size smaller sill may usually be used if the wall directly above is covered with ½-inch or thicker plywood sheathing, well-nailed, and the wall is not interrupted by doorways. The table assumes full-size native lumber, F 1,000.

ONE-STORY HOUSES					
	DISTANCE BETWEEN ROWS OF POSTS (JOIST SPAN)				
Sill Span	**8'**	**10'**	**12'**	**14'**	**16'**
8'	4 x 10	6 x 8	6 x 10	6 x 10	6 x 10
	6 x 8		6 x 8	6 x 8	
10'	4 x 12	8 x 8	8 x 10	10 x 10	10 x 10
	8 x 8			6 x 12	6 x 12
12'	6 x 12	8 x 10	10 x 10	10 x 10	10 x 12
	8 x 10	6 x 12	6 x 12		

TWO-STORY HOUSES					
	DISTANCE BETWEEN ROWS OF POSTS (JOIST SPAN)				
Sill Span	**8'**	**10'**	**12'**	**14'**	**16'**
8'	6 x 10	6 x 10	6 x 10	8 x 10	8 x 10
	8 x 8	8 x 8	8 x 10	6 x 12	6 x 12
10'	8 x 10	8 x 10	10 x 10	10 x 10	10 x 12
		6 x 12	6 x 12		
12'	10 x 10	10 x 12	10 x 12	10 x 12	12 x 12

Computed by author from formula in chapter 16.

TABLE 8

Safe loads on solid wood columns (thousands of pounds)

The maximum loads in the table apply to either planed or rough timbers. The values given under A are for relatively strong woods; those under B are for relatively weak woods. A woods are those with an Fc parallel to grain of 1,000 to 1,600, and an E of 1.2 to 1.6 x 10^6. B woods have an Fc of about 800 and an E of about .8 x 10^6.

NOMINAL COLUMN SIZE (INCHES)	6		8		10		12		14	
	A	B	A	B	A	B	A	B	A	B
4 x 4	13.5	9.7	7.5	5.4	4.8	3.4	3.5	2.5	2.5	1.8
4 x 6	21.2	15.2	11.7	8.4	7.5	5.4	5.3	3.8	3.9	2.8
4 x 8	28.0	20.0	15.6	11.1	9.9	7.1	7.0	5.0	5.1	3.6
6 x 6	39.4	30.3	46.0	32.8	29.2	20.9	20.5	14.6	14.9	10.6
6 x 8	51.0	39.7	55.0	39.3	38.1	27.2	26.4	18.9	19.3	13.8
6 x 10	66.0	50.3	71.0	50.7	49.0	35.0	34.2	24.4	25.0	17.8
6 x 12	80.6	60.9	80.5	62.0	60.0	42.7	41.8	29.8	30.6	21.8
8 x 8	68.5	52.5	68.2	52.5	58.2	52.5	61.5	43.8	44.8	32.0
8 x 10	87.0	67.0	87.0	67.0	87.5	67.0	78.5	56.0	65.8	47.0
8 x 12	106.5	83.0	105.5	81.7	105.5	81.7	105.5	81.7	70.0	50.0
10 x 10	111.5	86.6	111.0	86.6	111.0	86.6	111.0	86.6	111.0	84.5
10 x 12	135.5	105.0	135.0	105.0	135.0	105.0	135.0	105.0	135.0	103.0
12 x 12	164.0	127.0	164.0	127.0	164.0	127.0	164.0	127.0	164.0	127.0

Source: *Practical Farm Buildings*, 3d ed., by James S. Boyd, Danville, IL: Interstate Publishers, Inc., 1993, p. 82. Used with permission of the publisher.

APPENDIX C

BEARING CAPACITY OF SOILS & BEDROCK

A SNOWSHOE KEEPS YOU FROM SINKING INTO SOFT SNOW by distributing your weight over a broad area. Similarly, your foundation must be broad enough to prevent serious settling in the earth. Most foundations widen into footings at the bottom to provide the needed area. Usually the footing size is chosen by rule of thumb. For masonry wall foundations the footing is generally twice as wide as the wall is thick. An 8-inch wall will have a footing 16 inches wide. Usually a one-story house on a column foundation can have 20- or 24-inch-diameter footings under each column, and a two-story house can have 30- to 36-inch diameter footings.

If you want to compute approximately how big in area the footings should be, you can divide the presumed bearing capacity of the soil into the presumed total weight of the building to get the total area of the footings in square feet. If your soil will bear 1,000 pounds per square foot, and the house weighs 50,000 pounds, you will need 50 square feet of footing in total. If there are 10 columns supporting the house, each needs a bottom surface area of 5 square feet.

A wooden house with a column foundation weighs about 100 pounds per square foot of living space, including live loads. If you have large masonry masses, such as a fireplace or concrete walls, their weights should be added to the total. Masonry weighs about 150 pounds per cubic foot.

You can find the weight of your house more accurately by adding up its live and dead loads according to the load assumptions in chapter 16. Divide live loads (except heavy storage) by 2, because the maximum live load will never occur throughout an entire house, even during a dance.

Building codes have established standards for the bearing capacity of various bedrocks, clays, sands, and gravels. The chart below provides the values in tons/sq. ft. given in the Massachusetts code. Even the poorest bedrock base will support a lot of weight. Usually on bedrock a footing is needed not to provide surface area, but simply to make a connection between the foundation column and the rock.

Sands, gravels, and clays (or mixtures of them) also provide good support. For example, a 1,000-square-foot house on stiff clay (2 tons capacity per square foot) would need a total footing area of about 25 square feet. If there were 10 columns, each could have footings about 20 inches in diameter.

Building codes do not commit themselves on the bearing capacity of topsoils or soils containing organic material. They leave the question in particular cases up to engineers or building inspectors. However, if a soil is well drained, that is, if it doesn't turn to mud part of the year, you can safely presume its capacity to be ½ ton or 1000 pounds per square foot. With a significant proportion of gravel or rock in the soil, the capacity will be higher. You can tell a lot by looking inside the foundation holes you dig. If the soil is dry — and you think it will stay that way — and seems solid at the bottom of the hole, it will provide a good base for your house. If it looks strong, it probably is.

On a site with more than a gentle slope, the footings rigidify the foundation as well as carry the vertical load. On a sloping site the footings should therefore not be too small, even if the ground has a high bearing capacity. If in doubt, get some experienced advice.

Bearing capacity of soils and bedrock

CLASS OF MATERIAL	TONS PER SQUARE FOOT
1. Massive crystalline bedrock, including granite, diorite, gneiss, trap rock, and dolomite (hard limestone).	60
2. Foliated rock, including limestone, schist, and slate in sound condition.	40
3. Sedimentary rock, including hard shales, sandstones, and thoroughly cemented conglomerates.	20
4. Soft or broken bedrock (excluding shale) and soft limestone.	20
5. Compacted, partially cemented gravels, and sand and hardpan overlaying rock.	10
6. Gravel, well-graded sand, and gravel mixtures.	6
7. Loose gravel, compact course sand.	4
8. Loose coarse sand and sand and gravel mixtures and compact fine sand (confined).	2
9. Loose medium sand (confined).	1
10. Hard clay.	4
11. Medium stiff clay.	2
12. Soft clay, soft broken shale.	1
13. Compacted granular fill.	2-5

Source: Massachusetts Building Code.

APPENDIX D

FROST LINES

THE FROST LINE IS THE DEPTH TO WHICH THE WATER IN the ground freezes. When water freezes, it expands with tremendous power. Water freezing below your foundation can lift it up or sideways, throwing the whole house out of shape. Therefore, the bottom of your foundation should be below the frost line.

The chart below gives the frost lines for different locations. It is taken from *Foundations for Farm Buildings*, Farmers' Bulletin No. 1869, U.S. Dept. of Agriculture, an excellent pamphlet on building foundations correctly. Frost lines will also be specified in local building codes, and there may also be a rule of thumb agreed upon by farmers and builders that may differ from the official one.

These standards are really just rules of thumb or averages. Actual frost penetration will vary greatly. Uncompacted snow is good insulation, so snow-covered ground will often stay completely unfrozen. Where paths have compacted the snow or where it has been plowed away, the frost will penetrate much deeper. In cold climates you can see the evidence of this in summer, because plants that can't take extreme cold — like alfalfa — will be killed off within a few feet of any well-traveled winter path. Wet soil will freeze below the nominal frost line because the water helps conduct the warmth up out of the soil. The drier the soil is, the better it will insulate.

Sometimes a gravelly or sandy soil will drain so effectively that even when freezing temperatures go deep frost heaves will not occur because almost no water remains in the soil to freeze. In general, however, dig below the frost line and choose a well-drained soil.

Notes for frost lines table

1 Where depth is 48 inches and over, basements are generally used.
2 For temporary buildings.
3 Use buttress on outside face of wall or use a footing; less depth is required in gravelly soils.
4 For snow-protected ground.
5 Wooden barns.
6 Masonry barns.
7 Depth to uniform soil.
8 Footings for storage structures reinforced.
9 48 inches if building is unheated.
10 54 inches if building is unheated.
11 Conditions vary (climate, elevation, soil and soil moisture) seek local advice.
12 Frame buildings.
13 Masonry buildings.

Suggested depths for placing bottoms of footings (in inches)

Figures in column A apply to milder areas; those in B to colder areas

	Light Buildings		Farmhouse[1]		Heavy Permanent Barns & Storage		
STATE	A	B	A	B	A	B	LOCAL CONSIDERATIONS
Alabama	12	12	18	18	18	18	Reinforce footings and floor, and use piles in Blackbelt area.
Alaska	48 to 60	60 to 72	48 to 60	60 to 72	48 to 60	60 to 72	In nonpermafrost areas place polystyrene on outside of the foundations walls.
Arizona	12	20	18	36	18	36	Closeness of irrigation a factor.
Arkansas	12	12	16	16	18 to 24	18 to 24	Continuous foundations preferred.
California	6	12 to 18	8 to 12	18 to 24	12	24 to 30	
Colorado	12	18	18	24	18	24	Protect from roof water.
Connecticut	—	24(2)	—	30 to 48	—	30 to 48	
Delaware	18	24	24	30	30	30	Consult county building code.
Florida	surf	surf	surf	6 to 12	surf	6 to 12	Wide footings near surface; sandy soil.
Georgia	6	12	12 to 18	18	12 to 18	18	Conditions variable; seek local advice.
Idaho	12	18	24	36	36	48	Reinforce in wet cold locations.
Illinois	12	18	24	36	36	48	Reinforcement advised.
Indiana	18 to 24	18 to 24	24 to 36	24 to 36	36	36	
Iowa	18	20	36	42	36	42	
Kansas	24	24	60	60	48	48	Reinforce; heavy footings on swelling and shrinking soils.
Kentucky	18 to 24	18 to 24	18 to 24	30	30	30	
Louisiana	2 to 12	2 to 12	2 to 12	2 to 12	2 to 12	2 to 12	Wide footings on alluvial soils.
Maine[3]	48 to 60	60 to 72	48 to 60	60 to 72	48 to 60	60 to 72	
Maryland[4]	—	—	—	—	—	—	Conditions variable; seek local advice.
Massachusetts	24 to 48	24 to 48	24 to 48	24 to 48	24 to 48	24 to 48	Soil conditions fairly uniform.

	Light Buildings		Farmhouse[1]		Heavy Permanent Barns & Storage		
STATE	A	B	A	B	A	B	LOCAL CONSIDERATIONS
Michigan	18(4)	24	36	36	36	36	
Minnesota	12	18	60	60	18(5)	36(6)	
Mississippi	9	9	(7)	(7)	(7)	(7)	
Missouri	12	18	18	24	24	30	
Montana	18	18	44	44	30	40	Seek local advice.
Nebraska	12	18	18	24	18	24	Guard against roof water and rooting animals.
Nevada	0 to 6	18	0 to 6	18	12	24	
New Hampshire	36	48	72 to 96	72 to 96	48	72	Greater depth is for masonry.
New jersey	6 to 8	24 to 30	16	36	16(8)	36(8)	
New Mexico	12	(4)	15 to 18	20 to 24	18 to 20	24 to 30	
North Carolina	12	12	12	18 to 24	12	18 to 24	
North Dakota	18	18	—	—	—	—	Reinforce.
Ohio	18	24	36	42	36	42	Reinforce
Oklahoma	12	18	18	24	24	24	Reinforce masonry for swelling, shrinking and heaving soil.
Oregon[11]	—	—	—	—	—	—	
Pennsylvania	36	—	48 to 72	48 to 72	48	48	
South Carolina	10 to 12	12	14	18	14	18	
South Dakota[12]	18	18	54	54	24	24	For frame buildings use continuous foundations.
	42	42	60	60	48	48	For masonry buildings use continuous foundations.
Tennessee	12	12	24	24	24	24	Guard against termites.
Texas	12	20	20	30	30	34	
Vermont	12	12	60	60	60	60	Conditions vary widely; carry to firm soil.
Virginia	24	24	24	24	24	24	
Washington	—	—	—	—	—	—	Conditions variable; seek local advice.
West Virginia	18 to 24	24 to 30	18 to 24	24 to 30	24	30	
Wisconsin	30	42	36	48	36(9)	42(10)	
Wyoming	24	30	36	42	36	42	

APPENDIX E

SNOW LOADS

Reprinted from *Time-Saver Standards for Architectural Design Data* by John H. Callender. Copyright © 1974 by McGraw-Hill Inc. Used with permission of McGraw-Hill Book Company.

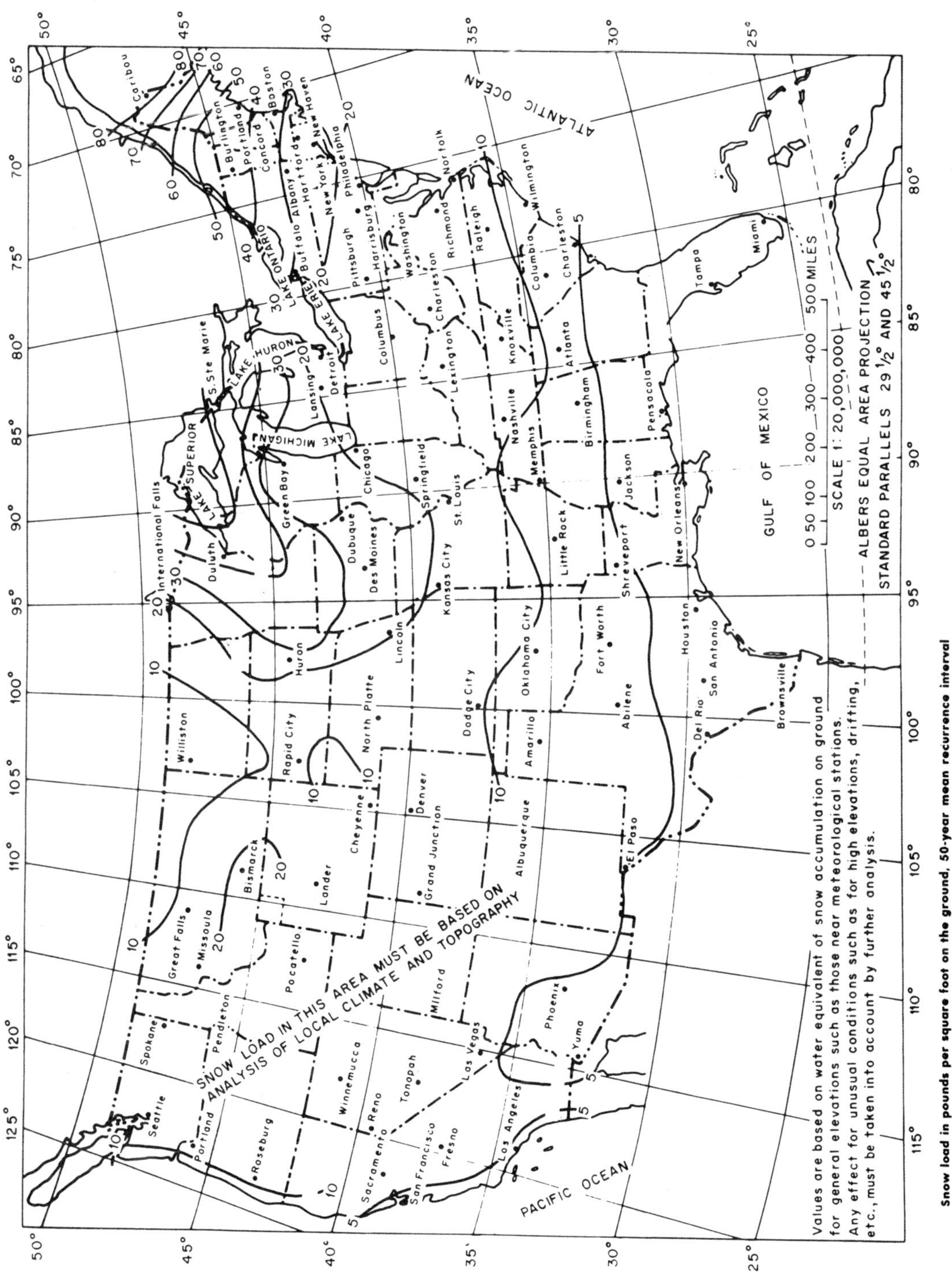

Snow load in pounds per square foot on the ground, 50-year mean recurrence interval

APPENDIX F

WEIGHTS OF VARIOUS MATERIALS

MATERIAL	LBS./CU. FT.
Concrete	144
Glass	160
Seasoned Timbers	
Ash	41
Cedar	22
Douglas fir	34
Eastern fir	26
Hemlock	28
Maple	42
Red oak	44
White oak	46
Norway pine	32
Ponderosa pine	28
White pine	25
Yellow pine (long-leaf)	41
Yellow pine (short-leaf)	35
Poplar	30
Redwood	28
Spruce	28
Walnut	38
Liquids	
Water	62.4
Ice	57.2
Snow (fresh)	8
Earths	
Earth	76-115
Sand/gravel	90-120

BUILDING MATERIALS	LBS./SQ. FT.
Sheetrock, ½"	2
Boards, ¾"	2
Boards, 1"	2½
Plaster	5-8
Plywood, ¾"	2.2
Particle board, ¾"	2-3
Asphalt shingles	2
Wood shingles	3
Slate roofing, 3⁄16"	7
Slate roofing, ¼"	10
Tin roofing	1

Source: *Time-Saver Standards for Architectural Design Data*, by John H. Callender (New York: McGraw-Hill Book Company, 1974).

APPENDIX G

NAILS & SCREWS

NAIL LENGTHS

Nail length is indicated by penny size, summarized in the chart below. A 2½-inch nail is an 8 penny nail, abbreviated 8d.

PENNY SIZE	LENGTH	NUMBER OF COMMON NAILS PER POUND
4d	1½"	316
5d	1¾"	271
6d	2"	181
7d	2¼"	161
8d	2½"	106
10d	3"	69
12d	3¼"	63
16d	3½"	49
20d	4"	31
30d	4½"	24
40d	5"	18

KINDS OF NAILS

The ordinary nail (figure G-1) used for framing and many other tasks is called a *common nail.* Common nails come in all the above sizes. Like most nail types, they are also available galvanized, which means they are coated to prevent rust.

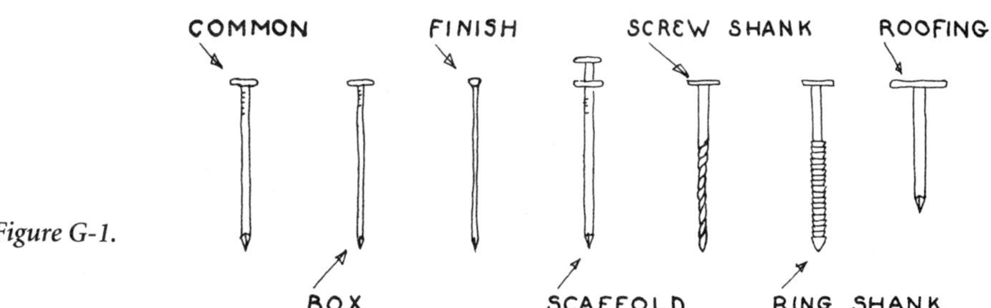

Figure G-1.

Box nails are like common nails, but with a more slender shank, which is less likely to split the wood but more likely to bend during pounding.

Finish nails are even more slender and have a very small head. They are used primarily on interior finish work where you don't want the head of the nail to show. Often the head will be driven about ⅛ of an inch below the surface with a nail set to make the nail unobtrusive. The ⅛-inch hole can be filled with wood putty to completely cover the nail.

Scaffold nails, also known as duplex nails, are used for scaffolding and other structures you want to be able to dismantle. The second head enables you to extract the nail with minimum damage to the wood.

Threaded nails and ring nails have better holding power than common nails, and are often used for flooring, sheathing, and siding. Threaded nails are designed for nailing down boards, and ring nails are designed for nailing down plywood. Sheetrock nails are special ring nails with a blue rust-resistant coating.

Roofing nails are made of copper, aluminum, or galvanized steel, and have extra-large heads.

SCREWS

Figure G-2 shows the most common types of screws. Flathead screws are intended to be countersunk flush with the board they hold down. Roundhead screws usually are used with a washer. Both types are available in many thicknesses and lengths, and with both the regular slotted head or the X-shaped Phillips head.

Lag screws are large screws used to carry heavy loads. They are driven with a wrench instead of a screwdriver. I like the hex-head type, which you can turn with an automotive socket wrench.

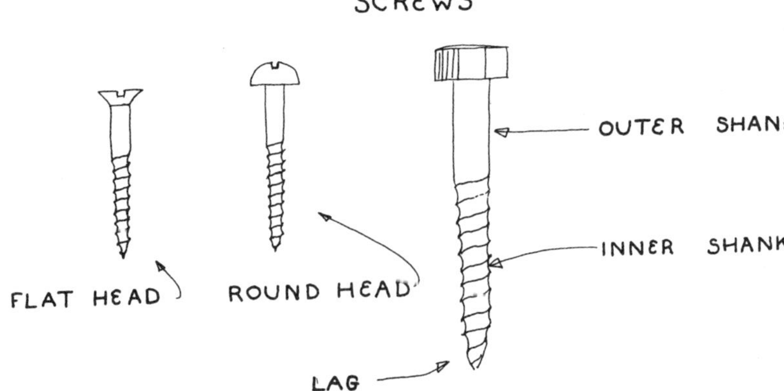

Figure G-2.

It is usually necessary to drill a pilot hole for a wood screw, particularly for the larger sizes or the harder woods. Regardless of the type of screw, the hole in the first piece the screw goes through should be the size of the outer shank of the screw — so the screw slides right through. The hole in the second piece should be approximately the size of the inner shank, the shank inside the threaded portion of the screw. If you plan to use many screws of the same size, you can buy special pilot bits in corresponding sizes that drill both pilot holes at once. Lubricate the screw with soap or wax before screwing it in.

DRYWALL SCREWS

Drywall, especially on ceilings, is now often screwed rather than nailed in place with special thin-shanked, rust-resistant, self-tapping, Phillips head screws. Self-tapping screws need no pilot holes because the thin shank keeps the wood from splitting. They can be driven with an ordinary Phillips head screwdriver, but the best tool is a cordless ⅜-inch variable speed, reversing drill with a magnetic screw-driving bit.

Drywall screws come in sizes up to 4 inches and in standard and galvanized finishes. These screws will be the fastener of choice in many carpentry situations, particularly where nailing is inconvenient or awkward.

APPENDIX H

SOUND ISOLATION

SOUND CAN PASS THROUGH A WALL OR FLOOR IN THREE ways. First, it can leak under doors of through holes or cracks. Second, sound waves in one room can cause the wall itself to vibrate like a drum, which will generate other sound waves in the next room. This is called *airborne noise.* Third, one side of the wall or floor can be struck directly — as by a footfall — again causing the wall to vibrate and create more sound waves on the other side. This is called *impact noise.*

You can control unwanted sound in your house several ways. The really noisy places can be separated from the quiet places by other rooms. The noise level within a room can be reduced by sound-absorbing surfaces in the room. A tile shower is good for singing in because all the surfaces are hard and smooth, which allows the sound waves to keep bouncing around and reverberating. Soft, uneven, textured surfaces limit reverberation by absorbing sound waves. Rugs, pictures on the wall, furniture, hangings, and textured surfaces all help keep noise down.

Sound can be controlled by eliminating air leaks between rooms. The most soundproof wall construction will be ineffective if doors don't fit right, or if sound can find its way through small cracks at the edges of walls. One common leak is back-to-back electrical outlets.

When all these factors have been taken into account, you still may want to make some of the walls or floors in your house more soundproof. The most straightforward way to do this is simply to make the wall or floor heavier so that they are harder to vibrate and therefore less able to transmit sound. You can do this by building masonry walls or floors, but this is expensive. Some of the same results can be achieved by simply locating heavy masses between rooms. In many old houses two bedrooms will often be separated by a space about 2 feet thick. This space will contain a chimney and two closets, one opening into each room. This makes good sense, because it increases the mass of the wall without any extra cost, since the closets and chimney are needed anyway. The closets are particularly good if they have good doors and are full of nice, soft, sound-absorbing clothing. The mass of a wall can also be increased by locating bookcases or other heavy objects on or next to it.

When this approach is inappropriate, you can build a regular-size wall or floor to minimize sound transmission without making it very massive. The idea is to construct the wall or floor so that it makes an ineffective drum. A conventional 2 x 4 wall, consisting of studs with drywall on either side, is a drum that transmits sound from one side to the other very efficiently. The drywall itself is hard and smooth, making a good sounding board, and the studs transfer the sound very well from one side to the other.

A common way to improve this situation is to build a stagger-stud wall, which eliminates the sound transmission through the studs themselves. This does not solve the problem completely, however, because the air inside the wall can itself transmit the sound between the two layers of drywall. One solution is to weave fiberglass insulation between the studs to impede this transfer. This stuffing acts much like clothing in a closet.

With a single row of studs some of the same effect as staggering can be achieved by hanging the drywall on resilient clips or channels. These are springy fasteners that hold the drywall off the stud slightly, preventing the studs from conducting the sound directly through the wall.

The wall surface itself can be designed to vibrate less. It can be done by increasing its mass or by using thicker drywall or more than one layer of drywall. But for a similar cost you can get better results by backing the drywall with ½-inch sound-deadening board, a soft, light material similar in texture to Homasote, which cushions the connection between the drywall and the studs and impedes the sound transmission between the studs. The drywall is glued to the sound-deadening board, and the sound-deadening board is nailed to the studs. Otherwise the nails through the drywall into the studs will transmit the sound to the stud.

Floors over inhabited rooms have to resist impact noises as well as airborne noise. Here again, massiveness, soft materials, and resilient clips will reduce impact noise considerably, but the most effective protection against impact is to have a rug with a pad under it.

Resistance to airborne noise is indicated by the Sound Transmission Class or STC of a wall or floor

STC NUMBER	EFFECTIVENESS
25	Normal speech can be understood quite easily
35	Loud speech audible but not intelligible
45	Must strain to hear loud speech
48	Some loud speech barely audible
50	Loud speech not audible

Impact noise is indicated by the Impact Noise Rating or INR of a floor. The charts below show STCs and INRs for various constructions. Both the charts and the effectiveness ratings above are from the excellent U.S. government book *Woodframe House Construction*, by L.O. Anderson, a government engineer who has written much of the best material for owner-builders.

Sound Transmission Class (STC) and Impact Noise Rating (INR) for floor and ceiling materials

		ESTIMATED VALUES	
DETAIL	DESCRIPTION	STC RATING	APPROX. INR
A (16"; 2 x 8)	FLOOR 7/8" T. & G. FLOORING CEILING 3/8" GYPSUM BOARD	30	-18
B (2 x 8)	FLOOR 3/4" SUBFLOOR 3/4" FINISH FLOOR CEILING 3/4" FIBERBOARD	42	-12
C (2 x 8)	FLOOR 3/4" SUBFLOOR 3/4" FINISH FLOOR CEILING 1/2" FIBERBOARD LATH 1/2" GYPSUM PLASTER 3/4" FIBERBOARD	45	-4

WALL DETAIL	DESCRIPTION	STC RATING
A (16" o.c., 2x4)	½" GYPSUM WALLBOARD	32
	⅝" GYPSUM WALLBOARD	37
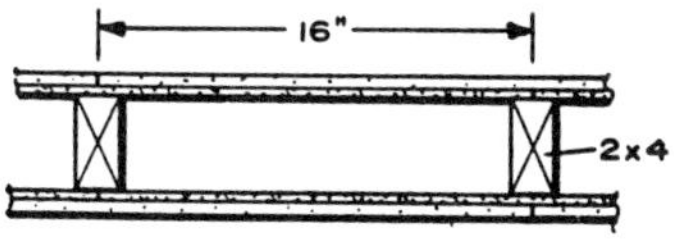 B (16" o.c., 2x4)	⅜" GYPSUM LATH (NAILED) PLUS ½" GYPSUM PLASTER WITH WHITECOAT FINISH (EACH SIDE)	39
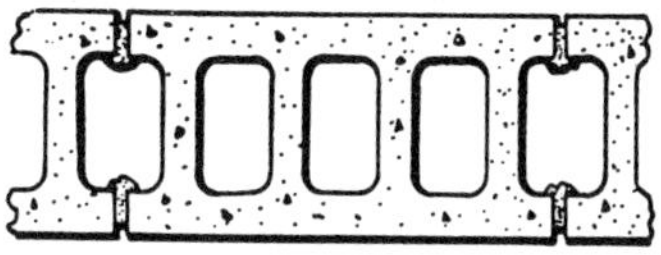 C	8" CONCRETE BLOCK	45
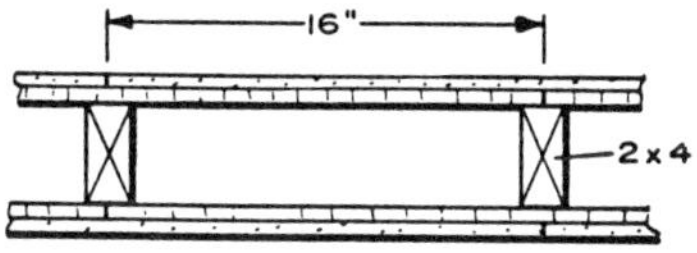 D (16" o.c., 2x4)	½" SOUND DEADENING BOARD (NAILED) ½" GYPSUM WALLBOARD (LAMINATED) (EACH SIDE)	46
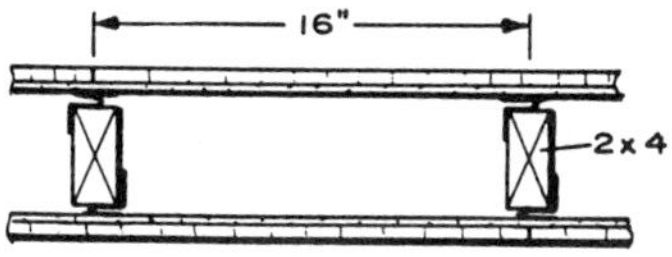 E (16" o.c., 2x4)	RESILIENT CLIPS TO ⅜" GYPSUM BACKER BOARD ½" FIBERBOARD (LAMINATED) (EACH SIDE)	52

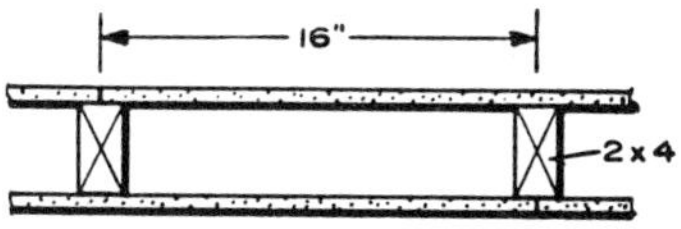

WALL DETAIL	DESCRIPTION	STC RATING
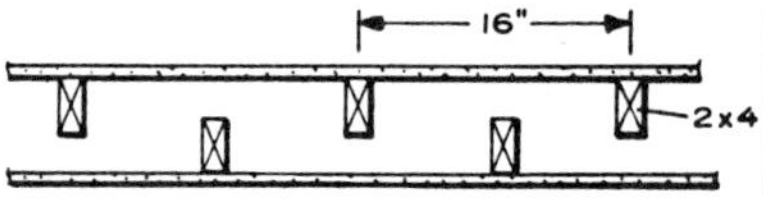 A (16" spacing; 2x4)	1/2" GYPSUM WALLBOARD	45
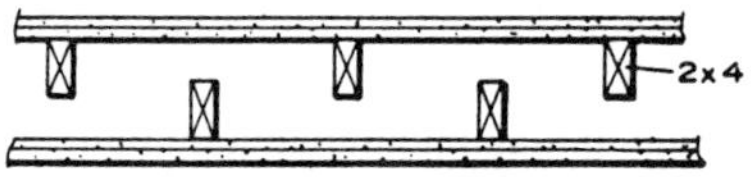 B (2x4)	5/8" GYPSUM WALLBOARD (DOUBLE LAYER EACH SIDE)	45
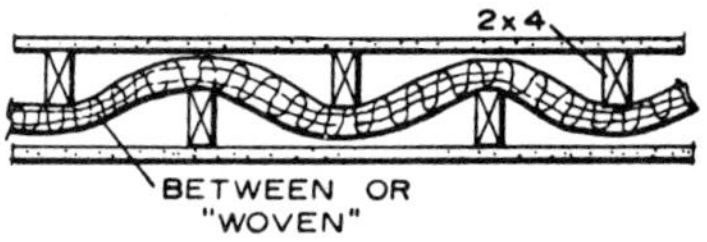 C (2x4; BETWEEN OR "WOVEN")	1/2" GYPSUM WALLBOARD 1 1/2" FIBROUS INSULATION	49
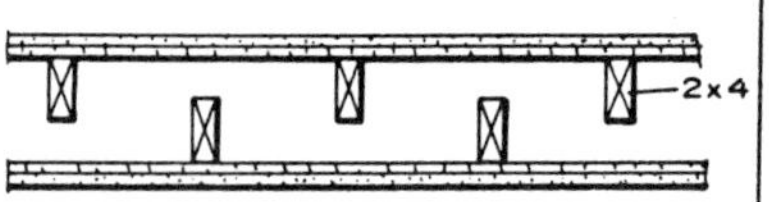 D (2x4)	1/2" SOUND DEADENING BOARD (NAILED) 1/2" GYPSUM WALLBOARD (LAMINATED)	50

INDEX

accessibility
  and bathrooms, 130
  defined, 2, 13
  design applications, 19–24
  design review, 25
  information sources, 24–25
  kitchens, 22, 124, 125–26
  plan (layout) and, 83
  site selection and, 52, 160
  stairs, 454
  sustainable design and, 35
active solar heating systems, 214
Adaptive Environments Center, The (Boston, Mass.), 25
adhesives. *See* glue/adhesives
Adirondack buildings, 48, 108
air barriers, 321, 418
air-dried lumber, 167, 168–69, 170, 175–78
air-locks, 224
air quality. *See* ventilation
air-to-air heat exchangers, 238
alcoves/nooks/bays
  pattern languages and, 3, 27
  plan (layout) and, 93, 94–95
Alexander, Christopher, 2–3, 26–27, 28, 48
aluminum clad windows, 379
aluminum exterior door sills, 394
American Plywood Association, 297
Ames Lettering Guide, 73
anchor bolts, 272, 288
Andersen (company), 378–79
Angel, Shlomo, 26
angle bevel, 468
appearance grade lumber, 170
appliances
  compact, for small houses, 96, 126–27
  plan (layout) and, 95, 114
  recycled, 149–50
apron (window trim), 437, 440
architects
  hiring, 134, 136, 152
  role of, 4, 9, 140
"architect's scale," 73
Architectural Graphic Standards (book), 228
argon gas, 380
ASHRAE Handbook of Fundamentals, 228
asphalt shingles, 360–61, 365–68
attic design gable roofs, 336–38
attics
  roof type and, 104, 336, 338
  vents, 343
Audel's guide, 293
awning windows, 378

"babbits," 407
backdrafts, 236, 237
backhoes. *See* excavation
band joists, 304, 309
banks, contracts and, 143
"barge boards," 108
bars, 469
baseboard radiation, 218
baseboards, finishing, 444–45
basement foundations
  advantages/disadvantages, 274–75
  building techniques for, 290–98
  defined, 274
  floor types and, 303
  types, 290
basements
  flooding, 53, 54
  heat loss and, 222
bathrooms
  accessibility, 21, 22, 23, 130
  plan (layout), 128–30
  ventilation/air quality and, 235
battens/battening, 433
batter boards and strings, 277–80, 474
bays. *See* alcoves/nooks/bays
"beads," 428, 432–33
beams
  cantilevered, strength of, 188–90, 201
  in cellars, 306
  deflection, 207–208
  horizontal, strength of, 186–90
  long span, building second floor, 315
  ridge, 196
  sizing/span calculations, 200–201, 207
  See also post-and-beam construction
bearing surfaces, joints and, 191, 197, 198
bedrock, 54, 284
bedrooms, 89–90, 91, 96, 378
belt sanders, 473
"berms," 21
bevel gauge, 468
"bevel siding," 405
bids, contracts, 139–42
Bierman-Lytle, Paul, 38, 444
"Bin," 412
Bioshield Penetrating Oil Sealer, 436
bird's mouth notch, 197, 344, 345
"blind-nailed," 367, 440
"Block-bond," 294
block foundations, 293–94
blocking
  columns, 198
  floors, 304, 312
  wood paneling, 434–35
  block plane, 468
"blower door tests," 223
"blueboard," 290
blueprints. *See* scale drawings
board and batten siding, 321
"board foot," 169
board sheathing, 321, 329, 330, 354, 400–407
boilers, 217–18
bolts, 191, 193
Boston ridge, 367, 368
bowing, of wood, 167
boxed rafters, 340–42
bracing
  columns/posts, 185
  diagonal, 195–96
  exterior/buttress, 196
  scaffolding/staging, 465
  stairs, 453
  walls, 324, 329–30
Brand, Stewart, 35
breakfast nooks. *See* alcoves/nooks/bays
breezes. *See* wind
"brickmold," 108, 380
bricks. *See* masonry
Brown, Thomas, 322
Btu (British thermal unit), 222
budgeting, financial
  contracts, making good, 139–45
  design costs, 135, 136
  heating costs, 217, 225–28
  resources/goals/priorities, 41
  site selection and, 61
  trim and siding details, 108
  windows, 375, 379
  See also estimates
builders
  collaboration/team work, 133–34, 138

builders, *continued*
contracts, making good, 139–45
hiring, 133–38
plan (layout) review and, 92
relationship with, 144–45
trust, issues of, 137, 143
See also owner-builder
builder's level, 56, 58
building, home
cultural aspects, 3, 26
hiring out, 136
innovation/change and, 4–8
methods, correct order of, 259–61
opportunities/obstacles, new, 3–4
psychological aspects, 145
technology and, 8–9
traditional, 6–8
building codes
egress routes, 90, 378
hiring experts, 136
no-step entry, 111
rafter tie-downs, 350
sheathing, 330
site planning, 97
site selection, 52
timber sizes, 199
wood foundations, 277
building materials
concrete, calculating amounts, 300–301
eliminating superfluous, 36–37
native, sustainable design and, 30
non-toxic/natural, 30, 31, 47, 436–37, 443–44
R values of common, 229
See also lumber; toxic indoor air
building systems
choosing, 70
special, estimating costs, 161
built-ins. *See* storage
Bungalow houses, 84, 108
burst pipes, 250
butt hinges, hanging doors on, 396–98
butt-nailed joints, 192–93
buttress bracing, 196

cabinets, kitchen, 112–15
cantilevered beams
sizing/span calculations, 201
strength of, as structural element, 188–90
Cape houses, 44, 84, 108
carbon monoxide, ventilation and, 127, 235–36
carbon monoxide detectors, 236
carpeting, 443
carriages, 453–57, 458
casement windows, 376, 378, 384–85, 392
casings
doors, 393–94, 394, 403–404, 408, 437, 439–40
wells, drilling, 51
windows, 403–404, 408, 437, 439–40
cat's paw, 469
caulking
plywood sheathing and, 400, 408
trim and, 438
windows and, 383
CDX plywood, 320, 321
cedar clapboards, 405
cedar shingles, 407–408
cedar trees, as swamp vegetation, 53, 61, 243
ceilings
building second floor and, 315–18
deflection, 207
drywall, 427–28
elevation drawings, 102
exposed-frame, 315–18, 434
finish work, 423–37
plan (layout) and, 93
roof design and, 104
cellars. *See* basements
cellulose insulation, 416
cement-coated nails, 193
Center for Environmental Structure, The (Berkeley, Cal.), 26
central heating systems, 214, 215, 217–18
ceramic flooring, 443
chalk line, 468
"change orders," 133, 140
children/teenagers, plan (layout) and, 64, 79–81
chimney flashing, 373
chisels, 468
chop-saws, 473
circuit tester, 477
circulation
in kitchen design, 114, 119
plan (layout) and, 78, 81–83
schematic floor plans/zoning, 68–69
cisterns, 248
cladding options, windows, 379
clapboard siding, 321, 400, 405–407, 412
claw hammers, 13, 468
clay soil, 54
clearances, woodstoves, 215–16
clearstory lighting, 93
"clear to the weather" siding, 405, 408
cleated carriage, 453
cleats, 453, 457–58
clients. *See* owner-builder
Clivus Multrum composting toilet, 255
coal stoves, 219
codes. *See* building codes
co-housing, 37–38
"cold cement," 363
cold water. *See* plumbing
collar ties
as bracing element, 196
ceiling (attic floor) joists as, 352
defined, 336
lap joints, 193–94
sizing, 342
strength of, as structural element, 185–86
structural soundness, evaluating, 197
Colonial houses, 44, 262, 267
column foundations
advantages/disadvantages, 274–75
building techniques, 281–90
defined, 272
floor types and, 303
structural soundness, evaluating, 198
columns and posts
reinforcing blocks, floors, 304, 312
strength of, as structural element, 184–85, 197–98
timber, strength calculations, 209–10
See also post and beam construction
combination squares, 467
combination stones, 468, 476
communities, clustered, 37–38
compost, 124, 125
composting toilets, 255
compression force, 167, 181–84, 191, 210, 308
concentrated loads, 188, 189, 198, 204
concrete
form work, hiring out, 136
mixing/drying, 284–85, 292
tension/compression forces and, 182
volume calculations, 300–301
concrete foundations, 274–75, 281–90, 293
condensation. *See* moisture/humidity
conductive heat loss, 223
connections. *See* joints/connections
constant flow water systems, 248–49
construction. *See* building, home; post-and-beam construction
construction grade lumber, 170
contractors
estimates and, 146, 151, 157
hiring, 136, 138
role of, 4
contracts
author's method, 142–43
bid/cost plus, compared, 139–42
"fixed fee," 142, 143
subcontracts, 150–51
controls, accessibility and, 23–24
conversation groups, 89, 93
cooling. *See* ventilation
copper pipes, 251
corner boards, 108, 404–405, 408
corner posts, 321
corners, drywall taping and, 432–33
cost plus contracts, 140–42
costs. *See* budgeting, financial

counters/counter tops
  accessibility, 21
  ergonomics and, 14, 15
  heights, 21, 122–24
Craftsman cottages, 44, 48, 108
crawl spaces, 274, 275, 304
cross-section drawings
  analyzing, 111
  defined, 102
  foundations, 299–300
  preliminary design, site plan and, 69–70
  technique, 75
  trim and siding details, 108–110
  uses, 107–108
cross ventilation, 234
cupping, of wood, 167

"dado," 472
"deadlight" storm panel, 387
dead loads, 184, 196–97, 202
"dead man," 427
decks
  cantilevers, loads and, 189–90
  square foot costs, estimates, 147
deep-well pumps, 246
deflection, 186, 207–208
"degree days," 225
DeKorne, Clayton, 322
design
  accessibility and, 13, 19–25
  activities, grouping, 63–64
  alternative technologies, 46–47
  building costs and, 161–62
  building system, choosing, 70
  changes, 133, 140
  ergonomics and, 13–18
  floor plans/zoning, 64–69
  goals/activities, listing, 62
  goals/priorities/resources, 41–48, 62–64
  pattern languages and, 26–29
  post and beam construction and, 264–66
  site plans, making, 56–60, 69–70
  site selection, 60–62
  sustainability, 30–38
  See also estimates; plan (layout)
designer/builders
  hiring, 134, 138, 152
  role of, 4, 9
designers
  collaboration/team work, 133–34, 138
  costs, 135–36, 152
  hiring, 133–38
detail drawings, 75–76
dew point, 416
diagonal bracing, 195–96, 324, 329–30, 465
DiBlasio, Michael, 141
dining areas, 88–89, 94
direct vent space heaters, 31, 216–17, 237, 238
disabilities. *See* accessibility
dishwashers, 95, 127
doors
  accessibility, 21, 22, 394
  butt hinges, 396–98
  casings, 403–404, 408
  double, 224
  exterior, 22, 393–94
  flashing, 372
  hanging, 396–99
  heat loss and, 223, 224
  interior, 393, 398–99
  making on site, 394–99
  ordering, 393–94
  recycled, 394
  rough openings, 319–20
  sources, 393–94
  T and strap hinges, 399
  trim, 108, 437–40
  types, 393, 394, 399
doorways, accessibility, 22
dormers, 93, 107
double coverage roll roofing, 360, 364–65
double doors, 224
double glazing, 375, 380, 386–87
double-hung windows, 376, 378, 386
double shear, joints in, 194
dovetails, 191, 192, 267
dowels, 191, 192, 267
down-draft ventilation systems, 128
downspouts, 298
D-pulls, cabinets, 24
drafting techniques/equipment, 72–77
drafts/leaks, 223, 234. *See also* flashing; insulation
drain, waste, and vent (DWV) lines, 251–53
drainage
  foundations, 272, 274, 298–99
  no-step entry and, 111
  site selection and, 52–55, 61
drainage field, septic, 253
drainpipes, 251
drain-waste-vent (DWV) system, 250
drawers, kitchen, 120
drawings. *See* elevation drawings; scale drawings
drills, power, 470
drip cap, 411
driven wells, 243
driveways, 52, 61, 151
"dropping the carriage," 456
drywall screws, 470–71
drywall ("Sheetrock")
  hiring out, 136
  installing/taping, 423–33
  square foot costs, estimating, 150
ductwork, 218
dug wells, 242
"Durabead," 428
"Durabond 90," 429, 431

"earnest money," 50
"ears," 382
earthquakes, 181, 184, 330, 350
eaves
  ice dams, 357, 362
  roof overhang and, 340–42
  spreading of, 193, 196
ecological concerns
  finishing products, 436–37, 443
  floorings, 443–44
  logging practices, 171, 180
  sustainable design and, 3, 30
  See also toxic indoor air
egress routes, 90–91, 378
elasticity (E) rating, wood, 207–208
elderly. *See* accessibility
electric heat tape, 250
electricity
  for building needs, 160
  safety and, 477–78
  site costs, estimating, 151
  site selection and, 51, 61
  See also photovoltaic (PV) electricity
elevation drawings
  analyzing, 111
  defined, 102
  preliminary design, site plan and, 69–70
  roof form, 104–107
  site selection and, 58
  technique, 74–75
  walls, 325
  windows, 103–104
energy efficiency
  design goals and, 47
  site selection and, 61
  sustainable design and, 30, 44
  wall design and, 322–23
  See also heating systems; heat loss
entries. *See* no-step entry
environment. *See* ecological concerns
ergonomic design, 2, 13–18, 83, 122, 125
estimates
  author's method, 154–58
  computers programs for, 158
  detailed, 153–62
  example, step-by-step, 154–58
  global labor estimate, 158–62
  importance of, 143–44, 146–47
  labor time/costs, 41–42, 152, 154, 155–62
  preliminary, 70–71, 147–53, 152–53
  site costs, 151–52
  square foot costs, 147–51
  time budget for owner-builders, 152
  typical house vs. mixed crew measures, 158–62

Euler's formula, 209
European radiators, 218
excavation
  basement foundations, 290–91
  column foundations, 282–84
  foundation drainage, 298–99
  hiring out, 136
  trenches, 243–45, 298–99
  water pipes, 243–45
exits, fire, 90–91, 378
experts, professional
  as design resource, 43–44
  flashing, 373
  hiring, 133–38
  role of, 4
  roof work, 353–54, 359
  site selection, 49
  span calculations, 199
exposed-frame ceilings, 315–18, 434
exposed rafters, 340, 415
exterior doors, 22, 393–94

"facia," 108, 339
fans, 128, 235, 237, 238, 239
fastenings
foundations, 272
  types, 191
  wood flooring, 440–41
faucets, 23, 24
fees. *See* contracts
felt paper
  as air barrier, 321
  roofing, 354–55, 359–60, 368–69
  siding and, 400, 402, 403
  wood flooring, 441
ferns, as swamp vegetation, 53, 61, 243
fiberglass asphalt shingles, 361
fiberglass insulation
  features of, 415
  handling, 238, 312
  installation, 418–19
fiber stress in bending (F or Fb), 200
fiber stress in compression (Fc), 210
Fiksdahl-King, Ingrid, 26
finish flooring, 305
finishing
  backs of clapboards/shingles, 405, 407
  baseboards, 444–45
  carpentry techniques, 445–47
  clapboards, 405
flooring, wood, 440–44
  ordering materials, 447
  paneling, wood, 436–37
  siding, 412–13
  stairs, 458
  trim, 437–40
finish saws, 466
finish work
  ceilings, 423–37
  defined, 420
  estimating costs, 160
  importance of, 420–21
  lumber for, 171
  sequence, correct, 422–23
  walls, 423–37
fire exits, 90–91, 378
fireplaces, 224
"fixed fee contract," 142, 143
fixed windows, 376, 377, 387, 392
flashing
  asphalt shingles and, 361
  door/window casings, 400, 404
  importance of, 358
  installation, 369, 370, 372–73
  siding and, 411
flooding. *See* drainage
floor joists
  about, 303
  attic, 352
  cutting/installing, 308–10
  deflection, 207
  installing, 309–10
  size, determining, 305
  stairs and, 449–50
floor plans. *See* plan (layout)
floors
  blocking, 312
  cutting lengths, figuring, 307
  elevation drawings, 102
  fastening devices, 193
  finishing, 440–44
  framing, 305–307, 311
  headers, 309–10
  insulation, 310, 311, 312
  joist hangers, 310–11
  materials for, ordering, 318
  parts of, 303–305
  radiant heating, 218
  rain cover, 314
  reinforcing blocks, 312
  second, building, 315–18
  soffits, putting in, 311
  subfloors, 312–14
  wiring and, 312
fluorescent light bulbs, 31
flush doors, 393, 399
foam insulation, 269, 415–16
footings, foundation
  basement foundations, 291–92
  column foundations, 284
  as critical design feature, 272
  load distribution of, 198
  pouring, 286
forced hot air heating systems, 217, 218, 219
forced hot water heating systems, 217, 218
forces
  butt-nailed joints and, 192
  structural rigidity and, 195–96
  See also compression force; loads; tension force
foundations
  books and reading, 301
  building, 280–98
  concrete materials, figuring, 300–301
  design features, critical, 272
  drainage, 298–99
  drawings, working, 299–300
  elevation drawings, 102
  hiring out, 136
  holes for waste pipes, 293, 300
  inaccuracies, correcting, 290
  laying out, 277–80
  plan, drawings, 76
  poured concrete walls, 293
  shallow, 274
  square foot costs, estimates, 147
  structural soundness, evaluating, 197–98
  sustainable design and, 32
  types, common, 272–74
  types, selection, 274–77
framing
  doors, 393, 394
  estimating costs, 161
  floors, 305–307, 311
  heavy timber, 44, 261
  post and beam construction, 266–68
  roofs, 349–50
  timber, 271
  walls, 328–29, 332–35
  See also light timber framing; lumber
Franke single sink, 95, 126
"free design," 135
freezing
  pipes, 224, 243, 250–51
  water pumps, 245, 247
frost heaves, foundations, 54, 272, 274
frost line, 222, 243, 272
"frost walls," 274
fuel use, 30. *See also* heating systems
full foundations, 275–76
furnaces, 217–18
furniture, plan (layout) and, 87–90, 101

gable-end vents, 342–43, 344
gable rafters, 347–49, 352–53
gable roofs
  bracing, 195–96
  kneewall design, 332–34
  span distance, 105
  types, 336–38
galley kitchen layout, 117
galvanized metal roofing, 339, 354, 361, 368–72
garages, 147
gas direct vent space heaters, 216, 217
geodesic domes, 6
geothermal heating systems, 219
Gilbreth, Frank, 17–18, 119–20, 142
Gilbreth, Lillian, 119. *See also* ergonomics

"girders," 306
glass/glazing. *See* double glazing
Gleason, Ken and Judy, 129
global labor estimate, 158–62
glue/adhesives, 191, 240, 363
grab bars, 23, 130
Grace Company, 358
grain, of wood, 166, 167
graywater, 254, 255
grid, power. *See* electricity; "off-the-grid"
ground-fault circuit interrupter (GFCI), 477–78
gussets, joints and, 193–94
gutters and downspouts, 298

Habitat for Humanity, 36
hack saws, 466
"half-lap," 360
halls and passages, 21, 68, 81
hammer marks, removing, 440
hammers, 13, 468–69
handles, accessibility and, 23
hand pumps, 246
handrails, 458–59
hand saws, 466
hardwood, 168, 440
"hassle," tolerance for, 42–43, 44
"hawk," 429
headers
  floors, 304, 309–10
  walls, 320, 321
heating systems
  backdrafts and, 236, 237
  backup for, 217
  costs, calculating, 217, 225–28
  number of stories, 67
  recycled equipment, 149–50
  selection, 219, 224–25
  sustainable design and, 31
  types, common, 213–18
  types, complex, 218–19
  ventilation/air quality and, 235, 236–37
heat loss
  computing, 220–25
  design and, 222–25
  by infiltration, 223–25
  minimizing, 211–13
  See also insulation
heat shields, woodstoves, 216
heat tape, electric, 250
hinges, door, 396–98, 399
hip roofs, 107
"hogging out," 446
hollow-core flush doors, 393
"Homasote," 318, 389
home building. *See* building, home; structure, house
home design. *See* design
hoods, range, 127–28, 235
hopper windows, 378, 385, 392
horizontal board siding, 403–405
horizontal sliding windows. *See* sliders
Host-Jablonski, Lou, 4
hot boxes, 250–51, 274
hot water. *See* plumbing
housewrap, 324, 400, 402, 405, 408
How to Use the Stanley Steel Square (book), 466
human factors engineering, 13
humidity. *See* moisture/humidity
hydraulic ram pumps, 246
hydronic heating systems, 217, 218, 219

ice and water shields, 358, 362
ice dams, 357–58
indoor air. *See* ventilation
infiltration, heat loss by, 223–25, 234
inspections, 72, 261
insulation
  cellulose, 416
  design/planning, 47, 414–15
  floors, 304, 310, 311, 312
  foam, 269, 415–16
  foundations, 274–75, 290
  heat loss and, 211, 220, 228
  importance of, 414
  pipes, 250
  roofs, 342, 344
  summer cooling and, 233, 414
  sustainable design and, 31, 32
  types, 415–19
  typical standards, 221, 414
  windows, 375, 389–90
  See also double glazing; fiberglass insulation
insulation supports, 304
interior design. *See* design
interior doors, 393, 398–99
interior partitions, 334–35
inward opening casement windows, 384–85
Iron Bridge Woodworkers, 142
Ishikawa, Sara, 26
island kitchen layout, 117

jack plane, 468
jack studs, 320
Jacobs House, 85
Jacobson, Max, 26
jamb extenders, 438–39
jambs
  doors, 394, 396
  window, 381–82
Japanese houses, 84
J-beads, 428
jet pumps, 247
joint compound, 423, 429, 431–32
joints/connections
  double-shear, 194
  fastening devices/methods, 191–94
  gussets, 193–94
  lap joints, 193–94
  as structural elements, 190–91
  wood-only methods, 191, 267
  See also nails and nailing
joist hangers, 308, 310–11

kerosene direct vent space heaters, 216–17
"key hole saw," 426
kiln-dried lumber, 169, 170, 171
"Kilz," 412
kitchen design
  accessibility, 23, 125–26
  basic principles, 119–20
  cleanup center, 115, 125
  cooking center, 115, 125
  counter heights, 122–24
  drawers, 120
  food flow, 116–17
  L layout, 91
  margin (back of the counter) space, 122
  mix center, 115, 125
  plan (layout), 95, 100, 115–19
  "power kitchen" area, 117–19
  shelves/racks, 120–22
  small/compact, 126–27
  space/cabinets, 112–15
  standard layouts, advantages of, 117
  trash/compost and, 124
  U layout, 91
  ventilation, 127–28, 235–36
  work center, 115, 119–28, 125
kneespace, accessibility and, 21, 22, 125, 126, 130
kneewall design, 332–34, 338
knobs, accessibility and, 23, 24
knots, in wood, 167, 308, 412

labor, estimating. *See* estimates
ladders, 90, 462
lag bolts, 193
land. *See* site, house
landings, stair, 449, 451, 459–61
lap joints, 193–94
Larsson, Carl, 48
latitude. *See* solar orientation
lauan (wood), 443
laundry rooms, 236
layout. *See* plan (layout)
L-beads, 428
lead flashing, 373
leaks. *See* drafts/leaks
ledger strips, 309
let-in braces, 329, 330
"level cut," 344
levels
  buying, 467
  using, 56, 58
  See also water levels; transit
lever door handles, 22, 23–24, 394
light. *See* solar orientation

light timber framing (stick-building), 35, 46, 70, 161, 260
line level, 56, 58
link measurements (ergonomics), 13–14
linoleum flooring, 443
live loads, 184, 196–97, 202, 203
living areas, 89, 93
Livos' Meldos Hard Oil, 436
L kitchen layout, 91, 117, 126
loads
  actual/assumptions, 202–5
  calculating, horizontal elements, 186–90
  compression/tension forces and, 181
  computations for, 203–205
  defined, 184
  inside, supporting on posts and beams, 335
  interior partitions and, 334–35
  joints/connections and, 191
  structural rigidity and, 195–96
  structural soundness, evaluating, 196–98
  timber sizing/span calculations, 202–205
  wall headers and, 320
lofts, 104
log homes, 44, 70
longitude. *See* solar orientation
Low E (Emissivity) window glass, 213, 375, 380
L-shaped staircase, 449, 451
lumber
  drying, 167, 168–69, 175–78
  green, use of, 168, 175
  logging your own, 149, 179–80
  and Optimum Value Engineering (O.V.E.), 37
  planing/milling, 169
  pricing, 169–70
  processing, 168–69
  sources, 170–73, 179–80
  square foot costs, estimating, 149
  See also native lumber; timber
lumberyard wood, 170–71, 172–73

make-up air, 236
Makita saws, 469–70
mapping the site, 56–60
Marinelli, Janet, 38, 444
masonry
  foundations, 276–77, 291, 293, 296–97
  tension/compression forces and, 182–83
measuring tools, 466–69
"miter fence," 471–72
miter gauge, 468, 479
mixed crew, estimates and, 158–62
models. *See* scale models
moisture/humidity
  foundations, 272, 274
  ice and water shields, 358, 362
  wood and, 165–66
  See also vapor barriers; ventilation
molding
  "beads," 428
  clapboard siding, 407
  drywall joints and, 433
money. *See* budgeting, financial
mortise and tenon, 191, 267
mortising door hinges, 396–97, 399
motion study, ergonomics and, 13, 17–18, 119
mound (septic) system, 55, 151, 255
movement limits, ergonomics, 14
"mud" flooring, 444
"muntin bars," 377

"nailers," 424–25
nail holes, filling, 432
nails and nailing
  "blind-nailed," 367, 440
  "butt-nailing," 192–93
  clapboards, 406
  as fastening devices, 191, 192–94
  flashing and, 373
  shingles, 367
  "toe-nailed," 193
  types, 193
  wood paneling, 435–36
native lumber
  buying, 173–74
  drying, 175–78
  planning/ordering, 174–75
  shrinkage, estimating, 177–78
  square foot costs, estimating, 149
  sustainable design and, 30, 31
  warping, preventing, 168
natural gas, 216
Nature's Choice Catalog, 436
Nearing, Helen and Scott, 6
Nearing method, stonework, 296
"net clear opening," 22
"neutral zone," 182
New England, 7, 44, 83, 149
nooks. *See* alcoves/nooks/bays
northern exposure. *See* solar orientation
"nosing," 453
no step entry, 21, 83, 111
notches, 191, 267, 396, 479–80. *See also* bird's mouth notch

"off-the-grid," 33, 51, 70
"Onduline" roof, 361
one- or two-story houses
  building second floor, 315–18
  column foundations, 281, 282
  kneewall design, 332–34
  post-and-beam construction, 266
  scale drawings, 75
  schematic floor plans/zoning, 66–68
  space heating systems, 215
  See also stairs
open-riser stairs, 452–53, 453
Optimum Reach Zone (ORZ), 15, 17
Optimum Value Engineering (O.V.E.), 36–37
organic asphalt shingles, 361
oriented strand board (OSB), 269, 321
orienteering compass, 56, 58
ORZ (Optimum Reach Zone), 15, 17
OSB (oriented-strand board), 269, 321
outboard rafters, 340–41
outdoor spaces, 68, 83, 114–15
outhouses, 256
"out to bid," 139–42
outward opening casement windows, 384
O.V.E (Optimum Value Engineering), 36–37
overestimating. *See* estimates
overhangs. *See* roof overhang
owner-builder
  contracts, making good, 139–45
  detailed estimate, preparing, 154
  hiring builders/designers, 133–38
  resources for, 4, 9, 42–44
  role of, 4, 9, 137
  square foot costs, savings for, 148–49
  time budget for, 152, 158
  tolerance for "hassle," 42–43, 44
Owner-Built Home, The (book), 5

paints and painting, 412–13, 433, 436–37. *See also* finishing
Palmer, David, 1
panel (breaker box), 137
panel doors, 393, 399
paneling, wood, 433–37
panels, wall, 269–71
pantries, 114
partitions, interior, 334–35
passages. *See* halls and passages
passive solar heating systems, 213–14
Pattern Language, A (book), 2–3, 26, 28, 48
pattern languages, 26–29
pegs, 191
pellet stoves, 219
peninsula kitchen layout, 117
percolation (perc) tests, 53, 54–55, 151
permits, 52, 152, 255
Perth, L., 466
phone service, site selection and, 52
photovoltaic (PV) electricity
  site costs, estimating, 151
  site selection and, 51
  sustainable design and, 31
  See also off-the-grid
pipes
  freezing, 224, 243, 250–51
  waste system, 250, 253–55
  water supply, 250–51

pipe staging, 462
pitch
roof, 105, 107, 203, 344, 462
stairs, 450
planes and planing, 169, 468, 480
plan (layout)
activities and, 62, 63–64, 79
alternative schemes, devising, 85
bathrooms, 128–30
column foundations, 281–82
egress and safety, 90–91
first scale, making, 84–87
foundations, 277–80, 299–300
furniture, adding, 87–90
good, features of, 79–84
importance of, 78
kitchen, standard, 117
revision, 91–92
schematic floor plans/zoning, 64–69
site, combining with, 97–99
stairs, 448–49
voids, looking for, 91
walls, 325
See also scale drawings; scale models
"plastic roof cement," 363
plastic vapor barriers, 417, 418, 419, 438
plates, wall, 319, 326–28, 331
platform framing, 319
"plumb cut," 344
plumbing
drain, waste, and vent lines (DWV), 251–53
foundation type and, 274
hiring out, 136
recycled fixtures, 149–50
square foot costs, estimating, 150
supply pipes, 250–51
waste disposal, alternative systems, 255–56
waste pipes, 253–55
plumb levels, 467
plywood
estimating costs, 161
textured, siding, 400, 411–12
types, 304, 312–13, 321, 324
plywood cutting jig, 475–76
plywood sheathing
as joint gussets, 193–94
roof frame, 353–54
selection, 321, 324
siding, 400
vertical board siding and, 401
walls, 320, 329–30, 332
pocket doors, 22
pole houses, 70, 161, 268
pollution, 30, 50. *See also* toxic indoor air
polyurethane finishes, 436, 443
porches, 96–97, 147
portable circular saws, 469–70
portable table saws, 472
post-and-beam construction
advantages/disadvantages, 262–66
framing, 266–68
inside loads, supporting, 335
walls, filling in, 269–71
post foundations. *See* column foundations
power lines. *See* electricity
Powerlock tape measures, 468
power-vent heating systems, 237
"prairie" house, 67
pressure tanks, 247
pressure-treated wood, 32, 276, 297
privacy/proximity
circulation and, 68–69
floor plans/zoning, 64, 66
number of stories, 67
plan (layout) and, 79
professional experts. *See* experts, professional
propane (bottled gas), 216
protractor gauge/guide, 468
P-shaped traps, 251–52
public areas. *See* privacy/proximity
pump houses, 245, 247
pumps, 299. *See also* water pumps
pump staging, 411, 462
"purlins," 354
PVC plastic pipes, 251
PVs. *See* photovoltaic (PV) electricity

Rabin, Helen and Jules, 1, 2
racks, kitchen, 120–22, 127
radial arm saw, 472–73
radiant floor heating, 218, 219
radiators, European, 218
rafters
boxed, 340–42
exposed, 340–42, 415
installing, 350–53
lap joints, 193–94
laying out/cutting, 344–50
notches, 344
outboard, 340–41
sizing, 200, 342–44
sloping, studwalls and, 332
span calculations, 200
structural soundness, evaluating, 197
See also collar ties
rafter squares, 344, 466
rafter tie-downs, 350
railings
scaffolding/staging, 463, 465
stair, 454, 459
rain, as water source, 243
rake, roof, 340–42
ramps. *See* accessibility
Real Goods Trading Company, 436
recycled building materials
design goals and, 47
doors, 394
lumber, 179
square foot costs, estimating, 149–50, 160
sustainable design and, 30, 31–32, 33, 38
windows, 376, 377
red building paper, 321, 402
redwood clapboards, 405
reinforcing bars/mesh, steel
basement foundations, 292
column foundations, 284, 286, 288
masonry, 182, 183
poured concrete foundation walls, 293
reinforcing blocks, floors, 304, 312
renovations, sustainable design and, 37–38
resistivity, insulation, 220
resources, natural
site selection and, 55
sustainable design and, 30
resources, owner-builder's
money, 41, 43, 44
stress, tolerance for, 42–43, 44
support of experts, 43–44
time, 41–42, 44
retrofitting. *See* accessibility
"returns," 424–25
ridge beams, 196
ridge boards, 352–53
ridge vents, 343–44
Rife Co., 246
rigid foam insulation, 415–16
rigidity, as structural element, 194–96
ring nails, 193
rise and run
roof, 344
stairs, 450–51
risers, 453, 458
roads, 52, 151
Roberts, Rex, 7, 360
Rockwell saws, 470
roll roofing, 359–60, 362–65
roofing materials
asphalt shingles, 360–61, 365–68
conventional, 339
as design choice, 105, 339
flashing, 372–73
installation, 362–72
metal, 339, 354, 361, 368–72
roll roofing, 359–60, 362–65
types, 359–61
"roofing tins," 403
roof overhang
as design choice, 105, 107, 108, 339–40
solar orientation and, 79, 234
roofs
covering the frame, 353–55
deflection, 207
design, 339–44, 357–58

roofs, *continued*
durability, 357–58
elevation drawings, 102, 104–107
felt paper, 354–55
framing, 349–50
insulation, 233, 342, 344, 417
pitch, 105, 107, 203, 344, 462
rigidity of, 195–96
rise, load assumptions and, 203
structural soundness, evaluating, 196–97
trim, 355–56
types, 336–38
vents, 342–44, 368, 417
See also rafters; safety
rough openings (R.O.), 319–20, 325, 326
routers, 473
runback water systems, 248–49
running foot, defined, 169
Russian masonry stoves, 215
R values
common building materials, 228
glass, 221
heat loss calculations and, 220–23, 228
polystyrene foam insulation, 269
sustainable design and, 31
walls, 319

saber saws, 470
safety
egress routes, 90–91, 378
electrical work, 477–78
fiberglass insulation, 418–19
fiberglass insulation handling, 238, 312
roof work, 353–54, 354, 356, 358–59, 369–70
sizing stair steps, 451
tools, using, 472, 473, 477–78
See also scaffolding/staging, handrails
salvage. *See* recycled building materials
sanding
power tools for, 473
trim, 440
wood flooring, 442
wood paneling, 435
sash (window), mounting, 384–86
sawhorses, 462, 474, 478
sawmill wood, 168, 171–73. *See also* native lumber
saws
buying, 166
power, 469–70, 471–73
using, 478–79
sawtoothed carriage, 453
sawyers, hiring, 149, 179
scaffolding/staging, 356, 402, 411, 462–65
scale drawings
first plan (layout), 84–87
floor plans/zoning, 64–69
stairs, 450–52
techniques/equipment, 72–77
scale models, 72, 76–77, 103
schematics. *See* scale drawings
"screed," 292
screen doors, 394
screen windows, 389
screws
drywall, 470–71
as fastening devices, 191, 193, 441–42
scribes and scribing, 468, 480
section drawings. *See* cross-section drawings
septic systems
alternative waste disposal systems, 255–56
hiring out, 136
replacement fields, 255
site costs, estimating, 151
site selection and, 52–55
waste pipes and, 250, 253–55
"setting blocks," 387
"set-up" door framing, 393
sewer systems, 250, 253. *See also* septic systems
shade/shading, 233, 234, 339. *See also* roof overhang
shallow foundations, 274
sharpening tools, 476–77
shear forces, 192, 194, 208–9
sheathing. *See* plywood sheathing
shed rafters, 345–50
shed roofs, 336, 342, 462
"Sheetrock." *See* drywall ("Sheetrock")
shelves/racks, kitchen, 120–22, 127
shingle brackets, 355, 367
shingles
corner boards and, 108
finishing, 412
shrinkage/expansion, 166
siding, 321, 400, 407–11
ship-lap board siding, 321, 324, 401
showers, accessibility and, 23
shutters, 375, 389–90
sidewall venting, 237
siding
butt-nailing, 193
clapboard, 400, 405–407
defined, 321
as design choice, 108–10
door/window casings, 403–404
elevation drawings, 108
finishing, 412–13
horizontal board, 403–405
selection, 321, 324
shingles, 400, 407–11
textured plywood, 400, 411–12
types, 400
vertical board, 400–403
"sill-horns," 382
sills
column foundations, 282, 288–90
doors, 393–94, 394, 396
floors and, 306–307, 311
walls, 296–97, 320
Silverstein, Murray, 26
sinks
accessibility, 21, 22
plan (layout) and, 95, 125, 126
small/compact, 126
site, house
building costs, estimating, 151–52, 160
evaluating, 49–55
lot size and house size, 55
plan (layout) and, 79, 97–99
selection, 52–53, 60–62
site plans, making, 56–60, 233
sustainable design and, 31
topography, elevation drawings, 102
topography, site selection and, 52–53
size
of stair steps, calculating, 451
of timbers, calculating, 199–210
See also small houses
Skil saws, 469–70
Skotch fasteners, screens, 389
skylights, 93, 111, 373
slab foundations, 274, 276
sleeping porches/shelters, 96–97
sliders, 376, 378, 385–86, 392
slopes, accessibility and, 21
small houses
books/reading, 99
design, 35, 45
kitchen design, 126–27
plan (layout), key points, 93–97
space heating systems, 215
smoothing plane, 468
snow loads
as live loads, 184, 197, 203
roof design and, 105, 357, 362
tension/compression forces and, 181
soffits
floors, 304, 311
roofs, 342, 343, 344
softwood, 168, 177, 440
soil
foundations and, 272, 274
site selection and, 53–55, 61
solar heating systems, 213–14, 219
solar orientation
design, 7, 44
elevation drawings and, 103, 111
floor plans/zoning, 64
heat loss calculations and, 230–32
plan (layout) and, 79
site selection and, 50, 233–34
slab foundations and, 276
solid core doors, 393
"Sonotubes," 281, 287
sound isolation, 317, 508–12
southern exposure. *See* solar orientation
spaced boards, roofs, 354
space heaters, 150, 214–17

"spade bits," 471
span computer, WWPA, 208
spans, long
  building second floor and, 315
  calculations, timber strength and, 199–210
Sparrow, Bob, 92
special clear grade lumber, 170
speed square, 467
S-P-F lumber, 170
"split-the-difference" contract, 141–42
springs. *See* streams/springs
square foot costs, estimates, 147–51
"square," of shingles, 411
S-shaped traps, 251–52
stacks, waste lines, 251, 253
staging. *See* scaffolding/staging
stains and staining, 412–13, 436–37. *See also* finishing
"stair buttons," 456
stairs
  accessibility, 454
  building techniques, 452–61
  circulation and, 68, 81
  as egress routes, 90
  finishing, 458
  floor construction and, 310–11, 449–50
  layout, common, 449–52
  locating, 448–49
  planning, 448–52
  scale drawings, 450–52
  types, 449, 452–53
Stanley Powerlock tape measures, 468
Stanley rafter squares, 344, 466
Stanley Toolbox saw, 466
static measurements, ergonomics, 13–14
steel. *See* galvanized metal roofing; reinforcing bars/mesh
steel square, 466
stepladders, 462
steps, 21. *See also* no-step entry; stairs
stick-building. *See* light timber framing
stickered pile, air-dried lumber, 175–78
"stiffeners," 268
stone houses, 6, 70
stonework
  estimating costs, 160
  foundations, 295–96
stool (window trim), 437, 439
"stop beads," drywall, 428, 432
stops, door, 394
storage
  built-ins, 94–95
  kitchen design, 119–22, 127
  plan (layout) and, 95, 96
stories, number of. *See* one- or two-story houses
storm doors, 394
storm panels, 380, 387
story-and-a-half. *See* kneewall design
stoves
  heating, types, 219
  small/compact, 126–27
  ventilation, 127–28, 235–36
  See also woodstoves
straight staircase, 449
strap hinges, 399
"strapping," 315, 424–25
straw bale houses, 47, 70, 161
streams/springs, as water supply, 50, 51, 243
stringers, 453, 457
structure, house
  joints/connections, strength of, 190–96
  rigidity, 194–96
  soundness, evaluating, 196–98
  tension/compression forces, 181–84
  timbers, strength of, 184–90
studies and study areas, 90
studs, wall, 319, 320, 321, 328, 388
studwalls, 319–21, 332–34
subcontracts, 150–51
subfloors, 193–94, 304, 312–14
submersible water pumps, 242, 245, 247
summer. *See* solar orientation; wind
sumps and sump pumps, 299
sunlight. *See* solar orientation
surface area, joints, 191
surface-bonded block construction, 294
"Surform," 426
sustainable design
  defined, 3, 30
  fundamental practices, 44
  innovative, 32–33
  mundane, 33–38
  standard, 31–32
swamps, 53, 243. *See also* drainage
Swanson Speed Squares, 344
swing, door, 22

table saws, 471–73
"tab shingles," 361
tape measure, 286, 309, 327, 468
tar-based roof cement, 363
teenagers, plan (layout) and, 64, 79–81
templates, plan (layout), 100, 101
tension force, 167, 181–84, 185–86
Terre Verde (company), 437
testing, water, 50
textured plywood siding, 400, 411–12
T&G (tongue and groove) plywood
  flooring, 440, 443, 444
  ordering, 447
  sheathing, 320, 324
  subflooring, 304, 312–13
therbligs, 18, 120
thermal (insulated) glass, 375, 387
thermal resistance. *See* R value
thermal transmission, 221
T-hinges, 399
Thoreau, Henry David, 8
threaded nails, 193
thresholds, 23, 393–94
timber framed houses, 44, 415
timbers
  as columns, 209–10
  deflection, 186
  load calculations, 202–5
  shear, 208–9
  size calculations, 205–8
  strength, structural soundness and, 184–90, 199–210
  strength calculations, 199–202
  See also beams
Timeless Way of Building, The (book), 26–27
toe-nailed joints, 193
toilets, 255. *See also* waste systems
tongue and groove plywood. *See* T&G (tongue and groove) plywood
tools
  buying, 466–76
  making your own, 474–76
  measuring, 466–69
  miscellaneous, 469
  power, to buy, 469–73
  safety, 472, 473, 477–78
  sharpening, 476–77
  using, 476–81
toxic indoor air
  air changes per hour and, 224, 239
  backdrafts, 236, 237
  causes/prevention, 30, 31, 47, 238–41, 443–44
  central heating systems and, 217
tracing velum, 72
traffic patterns. *See* circulation
transit (builder's level), 56, 58, 277
transitional spaces, circulation and, 83
traps, drain, 251–52, 253
trash/compost, 124, 125
treads, 453, 458
trenches. *See* excavation
triangular winding stairs, 449, 451, 459–61
triangulation, 195, 336
trim
  butt-nailing, 193
  as design choice, 108–10
  doors, 437–40
  elevation drawings, 108
  finishing, 412–13, 437–40
  roof, 105, 355–56
  windows, 380, 437–40
"trimmers," 311
trim saws, 470
truss roof, 338
T-squares, 73, 479
T-111 textured plywood, 400
twisting, of wood, 167–68
Tyvek and Typar, 321

U kitchen layout, 91, 117
underestimating. *See* estimates
U.S. Soil Conservation Service, 55
"upset limit" plan, 141
urea-formaldehyde adhesives, 240
used building materials. *See* recycled building materials
user-needs design, 13, 25
U-shaped staircase, 449, 451
Usonian house, 85
U value, heat loss, 221–22

vapor barriers
  floors, 304, 313, 314
  heat loss and, 224
  insulation and, 416–19
  window/door trim, 438
vapor pressure, defined, 416
velum, 72
ventilation
  air quality and, 235–38
  backdrafts, 236, 237
  insulation and, 416–18
  kitchens, 127–28
  make-up air, 236
  mechanical, 235–38
  roofs, 342–44, 368
  site conditions/orientation and, 233–34
  waste lines, 252–53
  See also toxic indoor air
vertical board siding, 400–403
views
  elevation drawings, 102, 103
  schematic floor plans/zoning, 66
vinyl clad windows, 379
vinyl flooring, 443–44
volatile organic compounds (VOC), 240

wainscoting, 434
walls
  accuracy, checking for, 331–32
  bracing, 329–30
  building, 326–32
  construction in place, 332
  designing, 321–25
  energy-efficient, 322–23
  finish work, 423–37
  framing, 328–29, 332–35
  interior partitions, 334–35
  kneewall design, 332–34
  materials, ordering, 335
  parts of, 319–21
  post-and-beam construction, 269–71
  putting up, 330–31
  sheathing, 329–30, 332
  structural soundness, evaluating, 197–98
  variations, special situations, 332–35
  winding stairs and, 461
warps, in wood, 167–68
waste systems
  alternative, 255
  drain, waste, and vent (DWV) lines, 251–53
  pipes, 253–55
water levels, 278, 279, 286–87, 474–75
water pumps
  location/protection, 245
  types, 242, 246–48
water supply
  constant flow/runback systems, 248–49
  costs, estimating, 151
  moving water to house, 243–45
  site selection and, 50–51, 61
  sources, 242–43
  See also plumbing
water table, defined, 54
wavy grain, of wood, 167
weatherstripping
  heat loss and, 224
  windows, 377, 390–92
weight. *See* loads
Weiler, Peter and Kathleen, 84, 92, 93, 94
wells, drilling, 50, 51, 151, 242–43
Western Lumber Product Use Manual, 207, 210
Western Woods Products Association (WWPA), 207, 208
wetlands. *See* drainage
"wet process" cellulose, 416
wheelchair accessibility, 14, 19–24, 126
Whole Earth Catalog, The, 5
wind
  air barriers, 321
  number of stories and, 67
  rafter tie-downs, 350
  roof design and, 105, 203
  site selection and, 52
  solar orientation and, 233–34
  structural rigidity and, 195–96
  structural soundness and, 197, 330
  tension/compression forces and, 181
  See also ventilation
winders, 449, 451–52, 459–61
windmills, 247
windows
  casement, 384–85
  casings, 403–404, 408
  clad, 379
  cost, 379
  design choices, 377–78
  as egress routes, 90
  elevation drawings, 103–404, 111
  features of, 378–80
  fixed glass between studs, 388–89
  flashing, 372
  frame, making, 381–84
  heat loss and, 211, 213, 223, 224, 375
  kitchen, 125
  making on site, 376–77
  ordering, 380
  plan (layout) and, 93, 95
  quality, 379
  ready-made, 375–76
  rough openings, 319–20
  R ratings, 221
  sash, mounting, 384–86
  screens, building, 389
  shutters, 375, 389–90
  solar orientation, 233
  sources, 375–77, 392
  trim, 108, 380, 437–40
  weatherstripping, 390–92
  wood, 379
  See also double glazing
winter. *See* solar orientation; wind
wiring
  floor construction and, 312
  hiring out, 136
  square foot costs, estimating, 150
  withdrawal, force of, 192
wood
  categories of, 168
  drying-out process, 165–66
  elasticity (E) rating, 207–208
  pressure-treated, 32, 276, 297
  problems with, 165
  shrinkage, 165–66, 168, 177–78
  strength, size and, 200, 207
  structural defects, 166–68
  tension/compression forces and, 183–84
  windows, 379
  See also lumber
Woodcraft Supply Catalog, 472, 473
wood flooring, 440–43
wood foundations
  advantages/disadvantages, 276–77
  building techniques, 297–98
wood paneling, 433–37
woodstoves
  advantages/disadvantages, 224–25
  as backup systems, 215–16, 217
  number of stories and, 67
  sustainable design and, 31
work centers, kitchen design, 119–28, 125
"work triangle," kitchen, 116, 117
Wright, Frank Lloyd, 67, 85
Wylde, Margaret, 15

Your Engineered House (book), 5, 360
Your Natural Home (book), 444

Z-brace doors, 399
zones/zoning. *See* building codes; plan (layout)